Human and Machine Vision

Notes and Reports
in
Computer Science and Applied Mathematics

Editor
Werner Rheinboldt
University of Pittsburgh

Human and Machine Vision

EDITED BY

Jacob Beck
Department of Psychology
University of Oregon
Eugene, Oregon

Barbara Hope
Azriel Rosenfeld
Center for Automation Research
University of Maryland
College Park, Maryland

1983

ACADEMIC PRESS

A Subsidiary of Harcourt Brace Jovanovich, Publishers
NEW YORK   LONDON
PARIS   SAN DIEGO   SAN FRANCISCO   SÃO PAULO   SYDNEY   TOKYO   TORONTO

Proceedings of the Conference on Human and Machine Vision held in Denver, Colorado, in August 1981, sponsored by the National Science Foundation.

ACADEMIC PRESS, INC.
111 Fifth Avenue, New York, New York 10003

United Kingdom Edition published by
ACADEMIC PRESS, INC. (LONDON) LTD.
24/28 Oval Road, London NW1 7DX

Library of Congress Cataloging in Publication Data
Main entry under title:

Human and machine vision.

(Notes and reports in computer science and applied
mathematics ;)
    Includes index.
    1. Visual perception--Congresses. 2. Human engineer-
ing--Congresses. 3. Image processing--Congresses.
I. Beck, Jacob. II. Hope, Barbara. III. Rosenfeld,
Azriel, Date. IV. Series. [DNLM: 1. Visual per-
ception--Congresses. 2. Computers--Congresses. 3. Vi-
sion--Congresses. WW 103 H918 1981]
BF241.H85 1983 001.53   83-9976
ISBN 0-12-084320-x

PRINTED IN THE UNITED STATES OF AMERICA

83 84 85 86   9 8 7 6 5 4 3 2 1

Contents

Contributors

Numbers in parentheses indicate the pages on which the authors' contributions begin.

D. H. Ballard (107), *Department of Computer Science, University of Rochester, Rochester, New York 14627*

Jacob Beck (1), *Department of Psychology, University of Oregon, Eugene, Oregon 97403*

Michael Brady (39), *Artificial Intelligence Laboratory, Massachusetts Institute of Technology, Cambridge, Massachusetts 02139*

Myron L. Braunstein (85), *Department of Psychology, University of California, Irvine, California 92717*

Lynn A. Cooper (97), *Learning Research and Development Center, University of Pittsburgh, Pittsburgh, Pennsylvania 15260*

M. Fahle (365), *Universität Tübingen, Tübingen, Federal Republic of Germany*

J. A. Feldman (107), *Department of Computer Science, University of Rochester, Rochester, New York 14627*

Ralph Norman Haber (157), *Department of Psychology, University of Illinois, Chicago, Illinois 60680*

Avishai Henik (395), *Department of Psychology, University of Oregon, Eugene, Oregon 97403*

Takeo Kanade (237), *Department of Computer Science, Carnegie-Mellon University, Pittsburgh, Pennsylvania 15213*

John R. Kender (237), *Department of Computer Science, Columbia University, New York, New York 10027*

Paul R. Kolers (259), *Department of Psychology, University of Toronto, Toronto, Ontario, Canada M5S1A7*

Stephen E. Palmer (269), *Department of Psychology, University of California, Berkeley, California 94720*

D. N. Perkins (341), *Graduate School of Education, Harvard University, Cambridge, Massachusetts 02138*

T. Poggio (365), *Massachusetts Institute of Technology, Cambridge, Massachusetts 02139*

Michael I. Posner (395), *Department of Psychology, University of Oregon, Eugene, Oregon 97403*

K. Prazdny (1, 413), *Fairchild Laboratory for Artificial Intelligence Research, Palo Alto, California 94304*

Azriel Rosenfeld (1), *Center for Automation Research, University of Maryland, College Park, Maryland 20742*

H. A. Sedgwick (425), *Department of Vision Science, State College of Optometry, State University of New York, New York, New York 10010*

Jay M. Tenenbaum (481), *Fairchild Laboratory for Artificial Intelligence Research, Palo Alto, California 94304*

Shimon Ullman (459), *Artificial Intelligence Laboratory, Massachusetts Institute of Technology, Cambridge, Massachusetts 02139*

Andrew P. Witkin (481), *Fairchild Laboratory for Artificial Intelligence Research, Palo Alto, California 94304*

Steven W. Zucker (545), *Department of Electrical Engineering, McGill University, Montreal, Quebec, Canada H3A2A7*

Preface

This book is based on a conference on human and machine vision held in Denver in August 1981. The conference brought together psychologists and computer scientists interested in visual perception.

The conference was marked by a remarkable convergence of interests and goals. Not only were the two groups concerned with similar problems, but each group was sensitive to the approaches taken by the other. Psychologists have come to appreciate the value of computational models over qualitative theories. Computational models reveal unrecognized problems and issues that fail to become apparent until one considers how visual properties such as motion, surface orientation, and texture are computed. Computer vision scientists appreciate the importance of ultimately being able to create general-purpose vision programs rather than programs tailored to specific tasks. The psychophysical study of the human visual system provides important insights on how to model the flexibility required by a general-purpose visual system.

The chapters survey the visual problems of representation, segmentation, organization, motion, and space. The individual authors do not agree on a single point of view, but do generally agree on the important problems to be addressed: the structure of stimulation provided by the environment, the nature of visual processing, and the nature of the representations of visual information.

An interdisciplinary program of research in visual perception is developing among neurophysiologists, perceptual psychologists, and computer vision scientists. This book provides a good example of the ways that perceptual psychologists and computer vision scientists are thinking about visual problems, and what each have to say to the other. The interaction between computer vision and psychology is still maturing. Even as recently as ten years ago, the commonality of interests did not exist for producing a conference in which the two groups would exhibit a strong interaction. The Denver conference demonstrated that the interaction between perceptual psychologists and computer vision scientists will certainly increase in the years to come.

Jacob Beck
Barbara Hope
Azriel Rosenfeld
February 1983

Acknowledgments

The Workshop on Human and Machine Vision was supported by the National Science Foundation under Grant MCS-81-08445. We are grateful to all those who aided in the preparation of this book, particularly Joseph Pallas, Andrew Pilipchuk, and Janet Salzman.

A Theory of Textural Segmentation

Jacob Beck

University of Oregon
Eugene, Oregon

K. Prazdny

Fairchild Camera and Instrument Corporation
Palo Alto, California

Azriel Rosenfeld

University of Maryland
College Park, Maryland

Abstract

Research on textural segmentation is reviewed and a model is presented. The model assumes that textural segmentation occurs as a result of differences in the first-order statistics of local features of textural elements rather than as a result of differences in the global second-order statistics of image points. Experiments are reported which conflict with the original Julesz conjecture that textural segmentation is a function of differences in global second-order statistics. The results support the hypothesis that textural segmentation depends on differences in the local features of textural elements. The results also support the hypothesis that the grouping of features into textural elements can affect textural segmentation by modifying the salience of existing feature differences and by introducing new feature differences.

1. Introduction

This paper is concerned with how a visual display is segmented into components on the basis of textural differences. The problem of textural segmentation has been studied in two ways. One approach has used randomly generated black and white dot textures. Random dot textures differ in dot density (first-order statistics) and in the joint spatial correlation of the dots (second-order statistics). Julesz (1975) has conjectured that textural segmentation does not occur for textures that have the same global first- and second-order statistics. That is, textures which differ only in third- or higher-order statistics are not spontaneously segmented. Gagalowicz (1980) has argued that textural segmentation is a spatially local process and that local rather than global nth-order statistics are important. Gagalowicz presents demonstrations showing that textures with the same global second-order statistics are spontaneously segmented if they differ in local second-order statistics.

A second approach to the problem of textural segmentation has used textures composed of elements or primitives arranged either regularly or randomly (Beck, 1967; Olson & Attneave, 1970; Julesz et al., 1973). The study of such textures has led to an alternative hypothesis concerning textural segmentation. Beck (1972, 1973, 1982) hypothesized that textural segmentation is based on differences in the distribution of the slopes, sizes, colors, and brightnesses of textural elements and their parts. Beck's hypothesis accepts the principle that textural segmentation depends on local statistics rather than global statistics but it differs from earlier hypotheses in two ways:

- Textural segmentation occurs as a result of differences in the first-order statistics of stimulus features rather than as a result of differences in the second-order statistics of image points.

- Stimulus features are not computed directly on the retinal array but are computed as the result of a two-stage process. First, feature detection occurs in terms of receptive fields like the concentric and elliptical receptive fields that physiologists have found in the visual systems of cats and monkeys. Second, local linking operations form higher-order textural elements. Textural segmentation occurs as a result of feature differences between textural elements.

The following two examples illustrate the role of linking operations. In Figure 1a, the horizontal and vertical lines constitute separate textural components. In Figure 1b, the linking of the lines inhibits them from acting as independent textural elements. The horizontal lines no longer are seen as a separate textural component relative to the vertical lines. Figure 2 shows how linking may facilitate textural segmentation. In Figure 2a, the vertical alignment of the opposing Us in the bottom region generates subjective contours. The lower half of the display is readily segmented from the upper half. In Figure 2b, the alternating orientation of the Us interferes with the generation of subjective contours and textural segmentation is decreased.

The proposed model for textural segmentation contains the following components: (1) feature detectors, (2) local operations that link detected features into higher-order textural elements, (3) differencing operations that

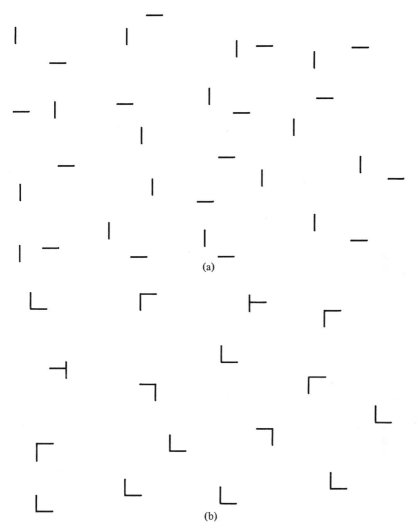

Figure 1. Illustration demonstrating the effect of linking on textural segmentation. In (a) the vertical lines are segmented from the horizontal lines. In (b) the connecting of the vertical and horizontal lines inhibits segmenting the display into vertical and horizontal lines.

encode feature differences in neighboring spatial regions of various sizes, and (4) decision units that segment a pattern into textural components as a function of the magnitude, number, and distribution of difference signals. The model will be described in greater detail in Section 5. At this point, it is useful, however, to indicate that differencing operations are assumed to take place with respect to both features detected by elementary feature detectors (and associated with textural elements) and "emergent" features of the textural elements themselves. In Figure 3, for example, the short vertical lines are detected by ele-

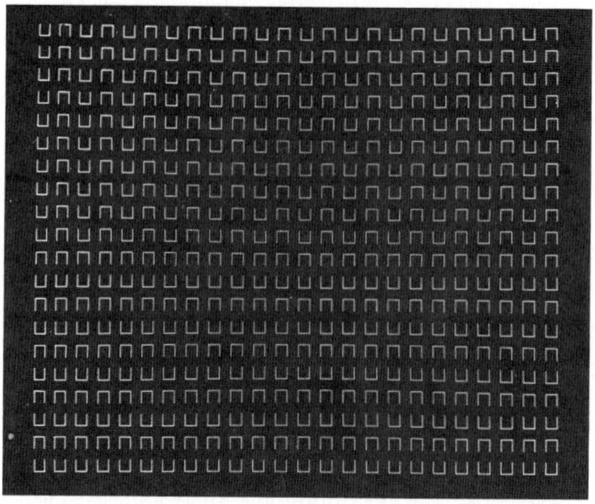

(a)

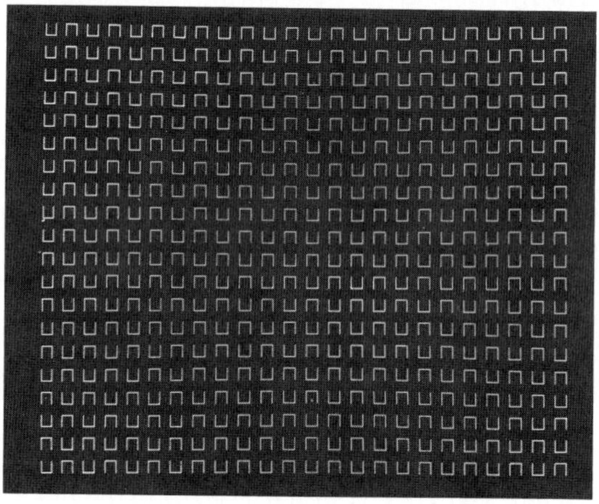

(b)

Figure 2. Illustration demonstrating the facilitation of textural segmentation by subjective contours. In the bottom region of (a), the orientations of the Us in adjacent rows are alternated to produce subjective contours. In the bottom region of (b), the orientations of the Us in adjacent columns are alternated to weaken the subjective contours.

mentary feature detectors. These lines also link to form long lines. The length of the long lines is an "emergent" feature which makes them stand out from the surrounding short lines. Emergent properties may introduce similarities, as well as differences, between textural elements. Figure 4a presents a block diagram of the model.

Julesz's (1975) hypothesis that textural segmentation fails to occur if two textures have identical first- and second-order statistics applies directly to the retinal array. For black and white textural patterns, first-order statistics measure the probability that a monopole, such as a dot, will fall on a black point of the array and second-order statistics measure the probabilities that a dipole, a pair of dots, falls on two black points, two white points, or a black and a white point of the array. Two textures have the same second-order statistics if for each length and orientation the probability that a pair of black points occurs at that separation is the same for both textures, and similarly for a pair of white points or for a combination of a black and a white point. Two textures which have the same second-order statistics have the same first-order statistics. Julesz (1980) has modified his model of textural segmentation to include a process based on feature detection as well as a process based on nth-order statistics.[1] He hypothesized that textures which differ in elongated blobs (e.g., line segments) or in terminators (e.g., line ends) are perceptually segmented even though they have identical second-order statistics. Figure 4b presents a block diagram of the Julesz model. We will further consider that model in Section 6.

2. Properties Producing Textural Segmentation

Not all stimulus differences produce textural segmentation equally strongly. In Section 2, we review evidence that textural segmentation occurs

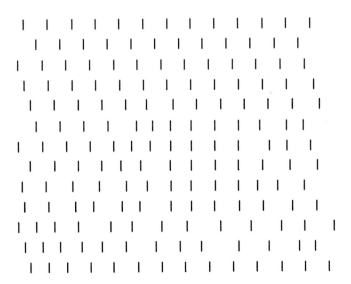

Figure 3. The short vertical lines are linked to form long lines. The length of the long lines is an "emergent feature" which makes them stand out from the surrounding short lines.

---

[1]Recently, Julesz (1981a,b) has reported evidence counter to his conjecture concerning global dipole statistics and has adopted a view similar to that of Beck (1972) and of Marr (1976) that textural segmentation is solely the result of differences in the first-order statistics of local features.

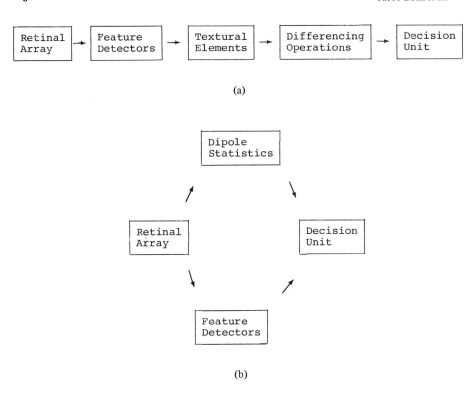

(a)

(b)

Figure 4. Block diagrams of (a) the Beck, Prazdny, and Rosenfeld model, and (b) the Julesz model of textural segmentation.

Figure 5. The tilted Ts (a difference in line slope) are segmented from the upright Ts while the Ls (a difference in line arrangement) are not.

strongly for simple stimulus properties extracted by elementary feature detectors. In Section 3, we will consider how linking operations affect textural segmentation by modifying the salience of existing feature differences or by introducing new feature differences.

Research has shown that textural segmentation occurs strongly in terms of simple properties such as brightness, color, size, and the slopes of lines composing figures (Beck, 1972). These properties differentially excite local feature detectors such as contrast detectors and edge and bar detectors. These properties are also processed automatically and in parallel without focused attention (Beck & Ambler, 1972, 1973). A review of the literature shows that slope is the most important of the variables associated with shape for producing textural segmentation (Beck, 1982). More complex properties, such as differences in the arrangement of the lines of a figure that leave the slopes of the component lines the same, generally do not produce strong textural segmentation. For example, the similarity of individual figures is strongly influenced by line arrangement. A tilted T is judged to be more similar to an upright T than is an L. When these figures are repeated to form textures (Figure 5), the texture made up of Ls is more similar to the texture made up of upright Ts than to the texture made up of tilted Ts. Line slope is the more salient property for textural segmentation while line arrangement is the more salient property for similarity judgments of individual figures (Beck, 1966b).

Figure 6 illustrates the textural segmentation produced by rotating L and U figures. The strongest textural segmentation occurred with a 45° rotation (Displays 6a and 6d), intermediate degrees of textural segmentation with a 90° rotation (Displays 6b and 6e), and little or no textural segmentation with a 180° rotation (Displays 6c and 6f). The rotation of the disparate figures in Displays 6a and 6d altered the slopes of the component lines from vertical and horizontal to 45° and 135°. The rotation of the disparate figures in Displays 6b and 6c and 6e and 6f left the slopes of the component lines the same as in the background figures. Greater textural segmentation, however, occurred for the sideways disparate figures in Displays 6b and 6e than for the inverted disparate figures in Displays 6e and 6f.

Rotating an upright U on its side changes the number of horizontal and vertical lines. Beck (1972) proposed that the greater segmentation of the sideways Us is due to the fact that the sideways Us differ from the upright Us in the number of their horizontal and vertical lines. The upright Us have two vertical lines and the sideways Us have two horizontal lines. A difference in the distribution of line slopes as well as a difference in actual slope is effective in producing textural segmentation. Julesz et al. (1973) report similar results. They explain the segmentation of the sideways Us, but not of the inverted Us, in terms of their dipole differences. Upright and inverted Us have the same dipole statistics while upright and sideways Us differ in their dipole statistics. Upright Us have horizontal dipoles not found in sideways Us while sideways Us

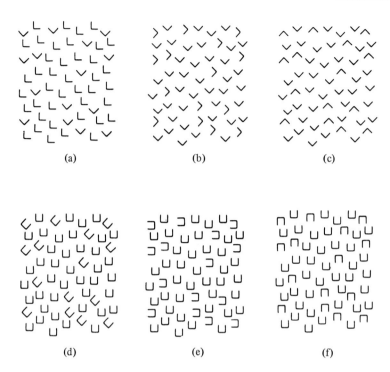

Figure 6. The displays show the textural segmentation produced by rotating L and U figures. Textural segmentation is strongest in displays (a) and (d), intermediate in displays (b) and (e), and is very weak, if it occurs at all, in displays (c) and (f).

have vertical dipoles not found in upright Us. The greater textural segmentation of the sideways Vs than of the inverted Vs appears to be due to the overall orientation of the figures. The upright V has its greatest extent in the horizontal direction and the sideways V has its greatest extent in the vertical direction. One might expect that if the greater textural segmentation of the sideways Vs than of the inverted Vs occurred because the upright and sideways Vs differentially stimulated vertically and horizontally oriented receptive fields, then making the sideways Vs small would improve their segmentation. The segmentation in Display 6b was found, in fact, to be better than in Display 6a when the Vs subtended a visual angle of less than five minutes (Beck, 1972).

Research has also shown that no distinction is to be made between curves and lines. The difference between a curve and line segments that only crudely approximate the curve fails to produce textural segmentation (Beck, 1973). Changes in the slopes of lines made out of dots do not produce as strong textural segmentation as similar changes in the slopes of solid lines (Schatz, 1977), perhaps because solid lines stimulate edge and bar detectors more strongly than dotted lines. It is important to point out that not all changes in slope facilitate textural segmentation equally well. A + rotated 45° to form a × has the same slopes of lines as a T tilted 45°. A × in a background of +s produces much weaker textural segmentation than a tilted T in a background of upright Ts (Beck, 1966a). What is important is not the orientation of lines per se but whether the change in the orientation causes feature detectors to be differentially stimulated. A × and a + because of their central symmetry would strongly stimulate concentric receptive fields and thereby decrease textural segmentation. The stimulation of common feature detectors is assumed, when the model is described in greater detail in Section 5, to decrease the magnitude of the difference signals producing textural segmentation. Another example showing that one must consider the feature detectors stimulated and not simply the slopes of the lines of a figure is due to Schatz (1977). Schatz showed that a change in the slope of a single line from vertical to diagonal produced stronger textural segmentation than a change in the slope of three parallel lines from vertical to diagonal. Multiple lines are blob-like and may be expected again to stimulate common concentric receptive fields.

Differences in arrangement may interfere with textural segmentation even when they themselves do not produce strong textural segmentation. An experiment was conducted using the procedure of Beck (1972) to assess the textural segmentation of displays in which the background and disparate figures were not all the same but varied. The displays were like those shown in Figure 6. In Figure 7, the figures to the left of the dash indicate the background figures in a display; the figures to the right of the dash indicate the disparate figures. Figure 7 shows the mean time it took to count the disparate figures in a display and the mean ratings of segmentation. The mean counting times and the mean ratings that are not significantly different from each other at the .01 level are underscored by the same lines in the figure. The results show that introducing

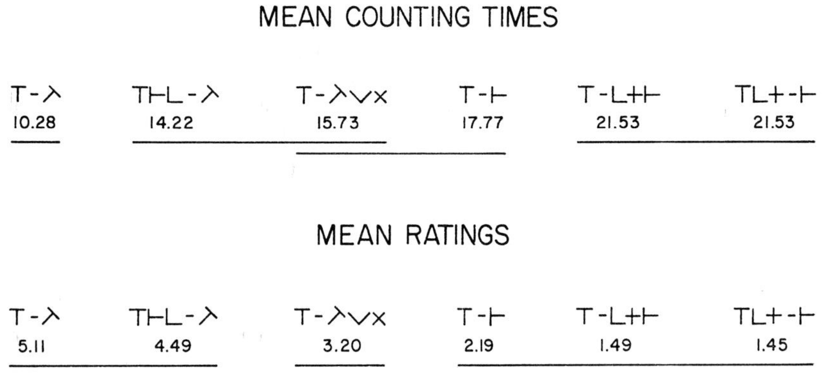

MEAN COUNTING TIMES

T-⅄	T⊢L-⅄	T-⅄∨x	T-⊢	T-L+⊢	TL+-⊢
10.28	14.22	15.73	17.77	21.53	21.53

MEAN RATINGS

T-⅄	T⊢L-⅄	T-⅄∨x	T-⊢	T-L+⊢	TL+-⊢
5.11	4.49	3.20	2.19	1.49	1.45

Figure 7. The mean times to count the disparate figures and the mean ratings for textural segmentation are shown. The figures to the left of the dash indicate the background figures in a display; the figures to the right of the dash indicate the disparate figures. Pairs of background and disparate figures underscored by the same line fail to differ significantly.

arrangement differences into either the background figures or the disparate figures reduced the segmentation of the disparate figures. For example, a tilted T in a background of an upright T, a sideways T, and an L yielded weaker textural segmentation than a tilted T in a background of upright Ts. A tilted T, a ×, and a V in a background of upright Ts yielded weaker textural segmentation than a tilted T in a background of upright Ts.

Though the visual system contains cells sensitive to certain types of differences in line arrangement, textural segmentation occurs less strongly for differences in line arrangement than for differences in line slope. The difference between slope and arrangement differences in producing textural segmentation appears to be quantitative rather than qualitative. One reason, for example, that figures differing only in line arrangement would not give strong textural segmentation is that the lines of the figure stimulate the same slope analyzers. One should note, though, that Foster and Mason (1980) have reported visual adaptation effects to the slopes of the component lines of a figure but not to the arrangement of the component lines. This suggests a possible qualitative difference in the way the visual system processes slope and arrangement differences.

Textural segmentation occurs strongly in terms of difference in line slope and of difference in color. However, textural segmentation fails to occur in terms of a conjunction of the properties of slope and color. A display composed of red and blue diagonal lines and red and blue vertical lines does not give rise to textural segmentation (Beck, 1982). Segmentation in terms of color, i.e., segmenting the display into red and blue lines, is interfered with by slope similarity since there are both vertical and diagonal red lines and blue lines. Similarly, segmentation in terms of line slope, i.e., segmenting the display into vertical and diagonal lines, is interfered with by color similarity since there are both red and blue vertical lines and diagonal lines. Spontaneous textural segmentation fails to occur in terms of a conjunction of properties. One can search out and hold in attention red vertical lines, blue vertical lines, red diagonal lines, or blue diagonal lines. However, the visual processing underlying textural segmentation cannot preattentively gate out the interference due to slope and color similarity and segment a display on the basis of a conjunction of color and slope properties.

3. Linking Operations

The hypothesis proposed is that the retinal intensity array is not operated on directly in textural processing but is transformed into textural elements as a result of local linking operations. Textural elements are subpatterns which occur repeatedly within the overall texture. The formation of textural elements through linking is hierarchical. The elements may be features or aggregates of features. Textural segmentation occurs as a result of feature differences between the textural elements.

Wertheimer (1923) proposed a number of principles or laws of organization to describe how the parts of a visual display tend to be grouped. Wertheimer's laws of organization reflect different levels of visual processing that subserve different functions (Beck, 1982). Only certain of Wertheimer's laws appear to be effective in forming textural elements. For example, the Gestalt law of Prägnanz which reflects the perceptual system's recoding of the information in a visual pattern in a more efficient way does not appear to play a role in textural segmentation. In contrast, there is evidence that proximity, certain kinds of similarity, and good continuation operate at an early stage of visual processing and play a role in the local linking processes that produce textural elements. Fox and Mayhew (1979), for example, have shown that if a dot is placed near upright and sideways Vs (e.g., figures like those in Figure 6b) textural segmentation is interfered with. However, the interference is reduced if the separation between the dot and the V is increased. Our explanation of this finding is that the dot placed near the V becomes part of the elemental unit, i.e., the textural element is the V plus dot. Common features between textural elements, as mentioned earlier, will be assumed to reduce textural segmentation. When the distance between the dot and the V is increased, however, the dot is no longer part of the V textural element and segmentation improves.

Two experiments were undertaken to study the effects of the size and separation of elemental figures on the formation of higher-order textural elements.

EXPERIMENT 1

Experiment 1 investigated the linking of squares on the basis of size and on the basis of lightness. Linking of the elemental figures leads to the formation of vertical columns in the top and bottom, but not in the middle, regions of a stimulus and should facilitate textural segmentation.

Stimuli. A PDP-11 computer was used to generate two sets of eight stimuli each. In both sets the basic figure was a square. In Set 1, a stimulus consisted of 21 columns and 20 rows of squares. Each stimulus was divided into three regions, a central section flanked by sections above and below. The top section consisted of the top 7 rows, the middle section of the next 6 rows and the bottom section of the bottom 7 rows. The squares in Set 1 differed in size. Columns of large and small squares alternated in the top and bottom sections. Large and small squares alternated within a column in the middle section. Figure 8 shows pictures of the stimuli.[2] Stimuli 8a through 8d consisted of white squares on a black background, and stimuli 8e through 8h of black squares on a white background.

---

[2]Unfortunately, the halftone process fails to accurately reproduce the lightness values of the stimuli in Figures 8, 9, and 10.

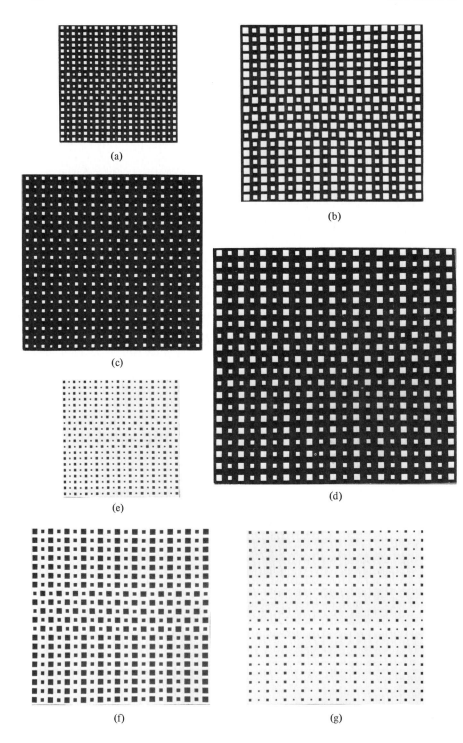

(a)

(b)

(c)

(e)

(d)

(f)

(g)

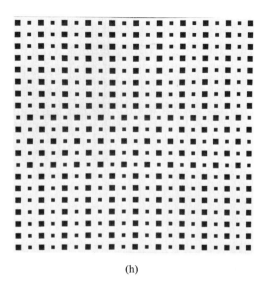

(h)

Figure 8. Size stimuli used in Experiment 1.

The sizes and the separations of the squares on the black and white backgrounds were programmed to be the same. Two sizes and separations were combined to give four stimuli with a black background and four stimuli with a white background. On stimuli 8a and 8e the sizes of the large and small squares were set at 1 mm and .7 mm on a side and the separations between the squares at 1 mm. On stimuli 8b and 8f the sizes of the large and small squares were doubled to 2 mm and 1.4 mm on a side. The separations between the squares were 1 mm. On stimuli 8c and 8g the sizes of the large and small squares were the same as in stimuli 8a and 8e, 1 mm and .7 mm. The separations between the squares were doubled to 2 mm. Stimuli 8d and 8h were 2× magnifications of stimuli 8a and 8e. The sizes of the squares were doubled to 2 mm and 1.4 mm on a side and the separations between the squares to 2 mm.

In Set 2, each stimulus consisted of 18 columns and rows of squares. A stimulus was again divided into three regions of 6 rows each. Figure 9 shows pictures of the stimuli. The squares in Set 2 differed in lightness. Stimuli 9a through 9d consisted of white and gray squares on a black background and stimuli 9e through 9h of black and gray squares on a white background. The black, gray, and white lightnesses approximately matched Munsell Values of 1.5, 5.5, and 9.0. The arrangement of the squares was the same as in Set 1. On a black background, columns of white and gray squares alternated in the top and bottom sections while white and gray squares alternated within a column in the middle section. On a white background, columns of black and gray squares alternated in the top and bottom sections while black and gray squares alternated within a column in the middle section.

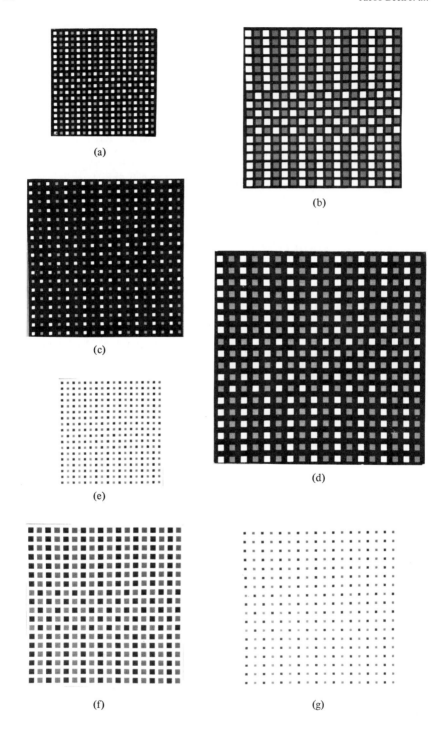

(a)

(b)

(c)

(d)

(e)

(f)

(g)

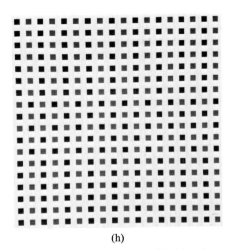

(h)

Figure 9. Lightness stimuli used in Experiment 1.

Two sizes and separations were combined to give four stimuli with a black background and four stimuli with a white background. The sizes of the squares in a stimulus were the same. They were set at 1 mm on a side for stimuli 9a, 9c, 9e, and 9g, and 2 mm on a side for stimuli 9b, 9d, 9f, and 9h. The separations between squares were .9 mm for stimuli. 9a, 9b, 9e, and 9f and 1.8 mm for stimuli 9c, 9d, 9g, and 9h. The stimuli in Sets 1 and 2 were mounted on individual squares of gray (Munsell Value = 6.5) poster board 15 cm on a side.

Procedure. Subjects were instructed to rate the stimuli on a scale of 1 to 5 on how well the center region spontaneously stood out as a whole from the top and bottom regions. They were told to look at a stimulus globally and not to search for the center region. A rating of 1 indicated that the center section did not stand out from the top and bottom sections, i.e., the top, center, and bottom sections appeared to constitute a single pattern. A rating of 5 indicated that the center section stood out strongly from the top and bottom sections. Numbers between 1 and 5 represented intermediate degrees of segmentation. To insure that a subject understood what was to be judged, sample displays were shown in which a center section stood out strongly and in which a center section did not stand out at all. The samples differed from the stimuli in the experiment and consisted of circles drawn on gray cardboard. The circles in the center section differed simply in size from the circles in the top and bottom sections. To familiarize subjects with the stimuli to be scaled, several of the experimental stimuli were shown to subjects before the experiment began. No judgments were obtained.

The stimuli were placed upright on a stand 53 cm from a subject. A cover was removed which exposed a stimulus for approximately 2 seconds. The stimuli in Sets 1 and 2 were combined and the 16 stimuli were presented in different random orders twice to each subject. They were presented once with the center section oriented horizontally and once with the stimulus cards turned 90° and the center section oriented vertically. Half the subjects were presented with the center section oriented horizontally first and half the subjects with the center section oriented vertically first.

Subjects. Twelve subjects served in the experiment. All were naive as to the purposes of the experiment.

Results and Discussion. Table 1 shows the mean segmentation ratings averaged over subjects for the sixteen stimuli in Sets 1 and 2. Examination of Table 1 shows that for the size stimuli (Set 1) the segmentation ratings decreased with increased separation of the squares. To a lesser extent, rated textural segmentation also increased with the size of the squares. The differences in mean ratings were evaluated by a four-way analysis of variance (Size by Separation by Background Lightness by Subjects). The analysis of variance revealed a significant effect of Size [$F(1,11) = 6.8$, $p < .05$], Separation [$F(1,11) = 270.5$, $p < .001$], and the Separation by Background Lightness interaction [$F(1,11) = 15.4$, $p < .01$]. The Separation by Background Lightness interaction reflects a smaller decrease in textural ratings with increased separation on a white background than on a black background. This interaction is most likely an artifact. Though the sizes of the squares on white and black backgrounds were programmed to be the same, the squares on a white background were approximately 10 percent smaller because in making the Polaroid photographs the white of the background encroached on the black areas of the squares.

Table 1 also shows that for the lightness stimuli (Set 2) the mean segmentation ratings again decreased with increased separation of the squares. Background Lightness also had a major effect. Textural segmentation failed to occur for any of the stimuli with a white background. An analysis of variance (Size by Separation by Background Lightness by Subjects) revealed a significant effect of Separation [$F(1,11) = 134.3$, $p < .001$], Background Lightness [$F(1,11) = 662.6$, $p < .001$], and the Separation by Background Lightness interaction [$F(1,11) = 87.3$, $p < .001$]. Figure 10 shows that black squares on a white background do produce textural segmentation if the gray squares are much more similar to the white background than to the black squares. There appears to be a bias to segment a higher intensity from lower intensities unless the middle intensity is much closer to the higher intensity than to the lower intensity. Thus, the gray squares in Set 2 interfere with linking the black squares into vertical columns on a white background but not with linking the white squares into vertical columns on a black background (cf. stimuli 9b and 9f).

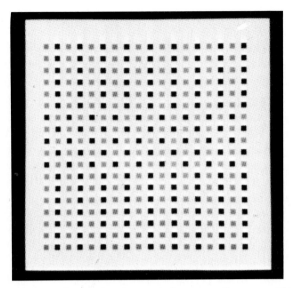

Figure 10. Example showing that textural segmentation occurs if the gray squares are much more similar to the white background than to the black squares.

The finding that interelement distance affects textural segmentation is consistent with the hypothesis that textural elements are formed through proximity linking. The squares in the top and bottom sections of the stimuli more readily link to form vertical columns with the near spacing than with the far spacing. Vertical columns of squares are not present in the center section, and the center section stands out from the flanking top and bottom sections. An alternative hypothesis is that textural segmentation is the result of feature detectors responding directly to the columns of squares in the top and bottom sections. Feature detectors, however, would be expected to scale – that is, main-

Table 1

Mean Segmentation Ratings in Experiment 1

	Near Separation		Far Separation	
	Small Squares	Large Squares	Small Squares	Large Squares
Size Stimuli				
Black Background	3.58	4.04	1.71	2.00
White Background	3.08	3.54	2.06	2.02
Lightness Stimuli				
Black Background	4.58	4.88	2.73	2.71
White Background	1.10	1.04	1.00	1.00

Note: See text for values of sizes and separations. The large figures at a far separation are a 2× magnification of the small figures at a near separation.

tain an approximately constant proportion between their length and width. Consequently, the expectation would be that the large squares at the far separation (i.e., 2× magnification) would yield the same textural segmentation as the small squares at the near separation. Our experimental results showed this not to be true.

Informal observations indicate that the visual angle subtended by a stimulus is the important variable though phenomenal separation may have a secondary effect. When the visual angle is small, the squares tend to link into columns while when the visual angle is large the columns break up into the constituent squares because the distances are too great for the operation of the linking processes. In contrast, Green, Wolf, and White (1959) found that visual angle had no significant effect on the detectability of textural differences based on differences in dot density. The important variable in their study was the ratio of dot size to dot separation. The perception of a density difference does not depend on the formation of higher-order elements. That is, the visual system has evolved so that it can evaluate density differences with equal effectiveness independent of whether the dots merge into gray bars or not. Element separation becomes an important variable when, as in the experiments reported, textural segmentation depends on the formation of higher-order elements.

Another situation in which element size to element separation is the important variable is when similarity grouping is pitted against proximity grouping in an array of figures. When the figures in the columns of the array are all the same while the figures in rows differ but are nearer to one another, alternate groupings are possible. Grouping by proximity leads to perceiving the array as composed of rows while grouping by similarity leads to perceiving the array as composed of columns. The tendency to see rows or columns has been found to remain unaffected by magnification. The perception of rows or of columns depends on the *relative* strengths of the tendencies to link figures into rows and columns. Increasing the separations between the figures when the array is magnified may actually weaken the absolute tendencies to link the figures into either rows or columns but keep the relative tendencies the same.

EXPERIMENT 2

Experiment 2 investigated the linking of lines on the basis of slope. Experiment 2 also investigated whether the Separation by Background Lightness interaction with the size stimuli in Experiment 1 was the result of an artifact.

Stimuli and Procedure. Two sets of eight stimuli each were prepared. The first set consisted of the eight size stimuli (Set 1) used in Experiment 1. The eight size stimuli were output on a printer rather than as Polaroid pictures. The sizes and separations of the squares on a stimulus were approximately 2.7 times their values in Experiment 1. To keep the angles subtended by the stimuli the same as in Experiment 1, the stimuli were presented at a distance of 143 cm.

The stimuli in the second set consisted of 17 columns and rows of lines. Each stimulus was divided into top and bottom sections of 6 rows and a middle section of 5 rows. Two lengths of lines and two interline separations were combined to give four stimuli. The lines making up a stimulus differed in slope and on the first four stimuli were vertical and diagonal. Displays 11a through 11d show pictures of the stimuli. The lengths of the lines were 1.8 and 3.6 mm, respectively. The separations between vertical lines in a column were 3.2 mm and 6.4 mm, respectively. Columns of vertical and diagonal lines alternated in the top and bottom sections. Vertical and diagonal lines alternated within a column in the middle section. The second four stimuli consisted of horizontal and diagonal lines. Displays 11e through 11h show pictures of the stimuli. The lengths of the horizontal lines were approximately equal to those of the vertical lines. The separations between horizontal lines in a column were 4.5 mm (stimulus 11e), 6.1 mm (stimulus 11f), 7.8 mm (stimulus 11g), and 9.5 mm (stimulus 11h), respectively. The stimuli were presented at a distance of 51 cm.

The two sets of stimuli were presented separately. Each set was presented twice. Half of the subjects were presented with the size stimuli first and half the subjects were presented with the slope stimuli first. The stimuli were always presented with the center section oriented horizontally. The instructions and remaining procedure were the same as in Experiment 1.

Subjects. Twelve subjects served in the experiment. Seven subjects had participated in Experiment 1 and five subjects were new. All were naive as to the purposes of the experiment.

Results and Discussion. Table 2 shows the mean segmentation ratings for the size and slope stimuli. The results are similar to Experiment 1. The segmentation ratings decreased with increased separation of the squares, and increased with increased size of the squares. As in Experiment 1, spatial separation has a greater effect than size. The differences in mean ratings were evaluated by a four-way analysis of variance (Size by Separation by Background Lightness by Subjects). The analysis of variance revealed a significant effect of Size [$F(1,11) = 9.8$, $p < .01$] and of Separation [$F(1,11) = 83.5$, $p < .001$]. The Separation by Background Lightness interaction was not significant [$F(1,11) = 2.05$, $p > .1$], supporting the conjecture that in Experiment 1 it was an artifact. The similarity in results of Experiments 1 and 2 supports the informal observation that the angular size subtended by a stimulus rather than phenomenal size is the more significant variable.

A four-way analysis of variance (Size by Separation by Line Direction by Subjects) revealed a significant effect of Separation [$F(1,11) = 24.8$, $p < .001$], Size [$F(1,11) = 25.8$, $p < .001$], Line Direction [$F(1,11) = 23.2$, $p < .001$], and the Separation by Line Direction interaction [$F(1,11) = 22.5$, $p < .001$]. The mean segmentation ratings decreased with line separation and increased with line length. The effects of separation and size are more nearly

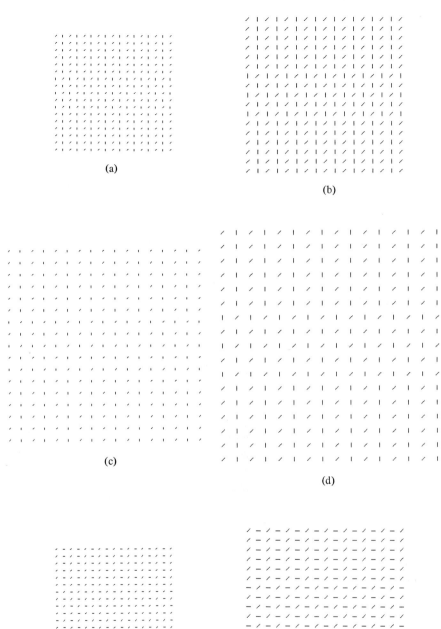

(a)

(b)

(c)

(d)

(e)

(f)

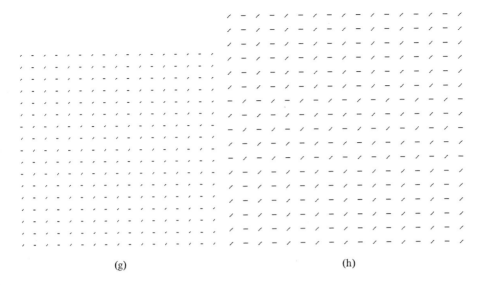

(g)   (h)

Figure 11. Slope stimuli used in Experiment 2.

equal with the line stimuli than with the square stimuli. Unlike the case of the square stimuli, a 2× magnification did not cause the mean segmentation ratings of the line stimuli to decrease. The textural segmentation ratings indicate that a 2× magnification decreased the linking of the squares but left the linking of the lines constant. There appears to be a difference between the linking of nonoriented (square) and oriented (line) elements. The linking of the lines into chains also occurred more strongly when the lines were collinear than when they were parallel, i.e., the linking of horizontal lines to form vertical columns. From Table 2, we see that the Separation by Line Direction interaction is the result of the mean segmentation ratings being uniformly poor with the horizon-

Table 2

Mean Segmentation Ratings in Experiment 2

	Near Separation		Far Separation	
	Small Figures	Large Figures	Small Figures	Large Figures
Size Stimuli				
Black Background	2.81	3.70	1.58	1.71
White Background	3.06	3.46	1.85	2.02
Slope Stimuli				
Vertical Lines	2.67	3.33	1.63	2.35
Horizontal Lines	1.42	1.83	1.27	1.94

Note: See text for values of sizes and separations. The large figures at a far separation are a 2× magnification of the small figures at a near separation.

tal line displays. The horizontal lines in stimulus 11f were actually vertically slightly closer together than the vertical lines in stimulus 11d. The mean segmentation ratings in Table 2 indicate greater linking of the vertical lines than of the horizontal lines. Phenomenally, the horizontal lines tend to link in the direction in which they point. The linking into long horizontal lines competes with the linking of the lines into vertical columns and interferes with textural segmentation. Figure 12 shows a display with Ls and Ts in place of the squares or line segments. Textural segmentation fails to occur. As mentioned earlier, linking is a hierarchical process and may be of features or of textural elements that are aggregates of features. Strong linking, whether of features or of textural elements, appears to be limited to simple properties such as lightness (brightness), size, and slope. Similarity in the arrangement of lines fails to produce strong linking just as a difference in the arrangement of lines fails to produce strong textural segmentation.

U DISPLAYS

Our hypothesis is that textural segmentation occurs in terms of features and properties of textural elements that make the output of feature analyzers not freely available to subsequent processes. Only certain subsets of features and

```
T L T L T L T L T L T L T L T L T L
T L T L T L T L T L T L T L T L T L
T L T L T L T L T L T L T L T L T L
T L T L T L T L T L T L T L T L T L
T L T L T L T L T L T L T L T L T L
T L T L T L T L T L T L T L T L T L
L T L T L T L T L T L T L T L T L T
T L T L T L T L T L T L T L T L T L
L T L T L T L T L T L T L T L T L T
T L T L T L T L T L T L T L T L T L
L T L T L T L T L T L T L T L T L T
T L T L T L T L T L T L T L T L T L
T L T L T L T L T L T L T L T L T L
T L T L T L T L T L T L T L T L T L
T L T L T L T L T L T L T L T L T L
T L T L T L T L T L T L T L T L T L
T L T L T L T L T L T L T L T L T L
T L T L T L T L T L T L T L T L T L
```

Figure 12. Example showing that linking into columns does not occur when the textural elements (Ts and Ls) differ only in the arrangement of their component lines.

properties are effective in producing textural segmentation. Julesz's conjecture that textural segmentation is a function of second-order but not of third- or higher-order statistical differences does not restrict the dipole differences to be considered. (The conjecture originally was formulated with random dot textures in which there are few, if any, figural constraints.) According to his conjecture dipole differences arising from between textural elements – in other words, from pairs of points that belong to different elements – should give textural segmentation as readily as dipole differences arising from within textural elements, i.e., from pairs of points belonging to the same element. For example, in Figure 13a the dipoles in the top 11 rows differ from the dipoles in the bottom 11 rows. In the bottom 11 rows, the Us in adjacent rows are oriented in opposite directions. The bottom 11 rows, thus, have long and short dipoles that arise from the base lines of opposing Us that are not present in the top 11 rows. The presence of these dipoles might be expected to give rise to textural segmentation. In fact, however, the dipoles arising from between Us appear not to be salient, and a spontaneous segmentation of the display into two textural regions does not occur. Julesz might explain the results by arguing that long dipoles are less salient than short dipoles. However, if the texture is viewed from a distance textural segmentation is still weak.

The arrangement and spacing of the Us in Figures 13 through 16 were varied to form higher-order textural elements through proximity, good continuation, and closure. In Figure 13b, the U figures in the top 11 rows are all upright and aligned. The bottom lines of the Us link on the basis of collinearity (a special case of good continuation) into long horizontal lines. The U figures and the long horizontal lines are competing textural elements. The connecting of the lines into U figures reduces but does not completely inhibit the linking of the horizontal lines into long lines. The top half of the display is segmented from the bottom half on the basis of length differences. There is also a tendency for opposing upright and inverted Us in the bottom halves of both displays to link. If the opposing Us link, a new textural element is formed which differs in its number of lines and size from the textural elements in the top region. The tendency for opposing Us to link is weak since it fails to produce textural segmentation in Figure 13a. It is, however, increased if the Us are brought closer together, as in Figures 14a and 14b, and textural segmentation is strengthened. The arrangement of the Us in Figures 14a and 14b are the same as in Figures 13a and 13b except that the Us are spaced closer together to strengthen their tendency to link.

Increasing the vertical and horizontal separations between the Us in Figures 15a and b decreases the tendency for the Us to link. In Figure 15b, the Us have been vertically jittered so that the rows as well as the columns are not aligned. In Figures 16a and 16b, the Us are approximately two and a half times closer vertically than horizontally. This causes strong vertical linking and textural segmentation does not occur. Linking processes can change the sali-

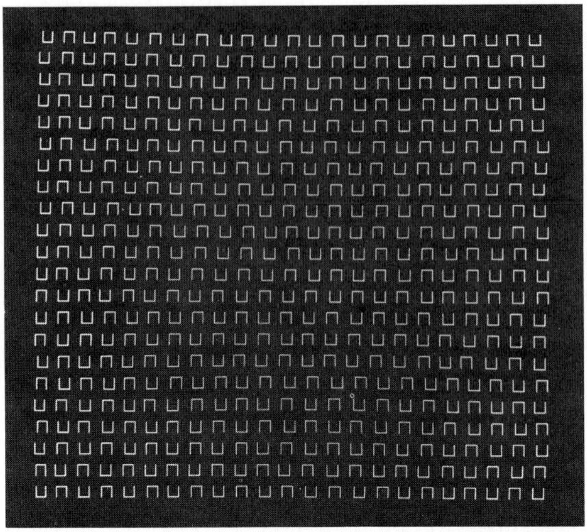

(a)

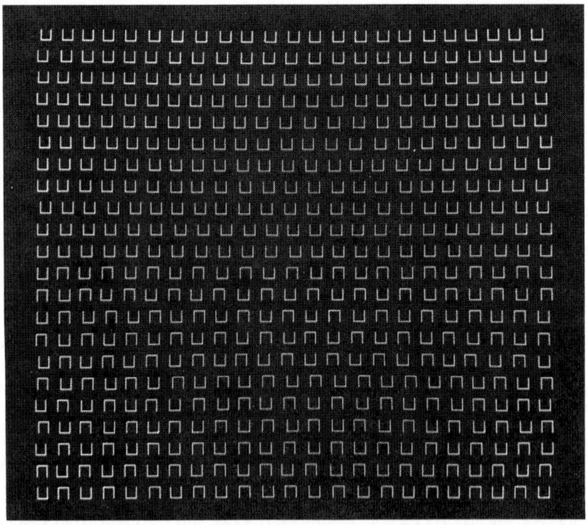

(b)

Figure 13. Examples in which (a) textural segmentation does not occur though interfigural dipole statistics differ, and (b) textural segmentation is facilitated by linking based on the collinearity of the base lines of the Us.

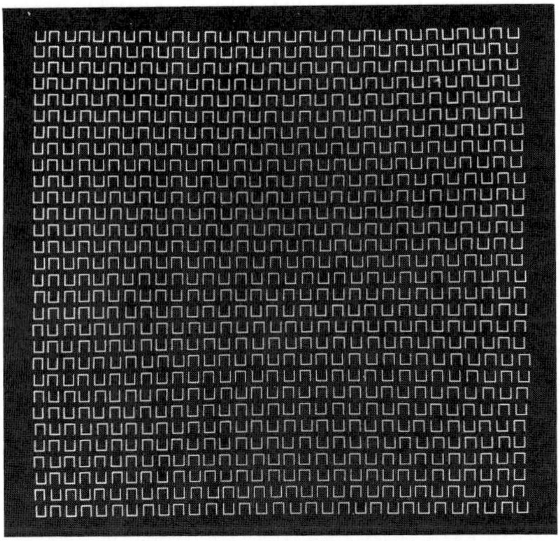

(a)

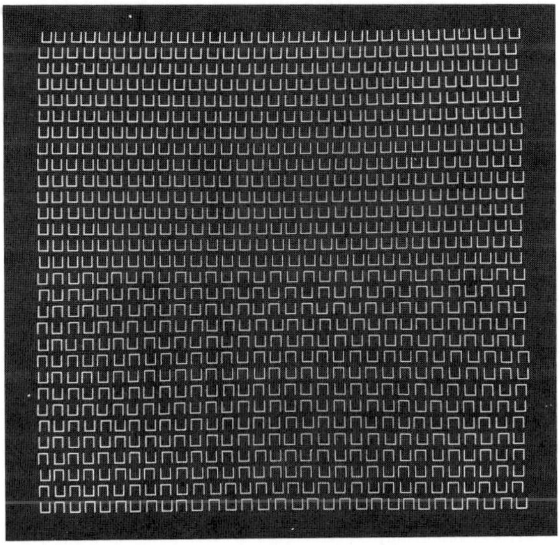

(b)

Figure 14. Bringing the Us in Figure 13 closer together strengthens textural segmentation in (a) through linking based on closure and in (b) through linking based on closure and good continuation.

(a)

(b)

Figure 15. Examples in which (a) increasing the spacing between the Us and (b) increasing the spacing and misaligning the Us weakens linking based on good continuation. Textural segmentation is decreased.

ence of features through the formation of textural elements. Features which break up textural elements do not produce textural segmentation as readily as features which do not do so.

4. Second-order Statistics

If two textures have the same first-order statistics, Julesz's original conjecture implies that textural segmentation is a function of the amount of difference in second-order statistics. One needs to note that the conjecture does not include its converse. That is, the absence of textural segmentation does not imply that two textures have the same second-order statistics. Julesz, however, has not specified either the conditions under which differences in second-order statistics fail to produce textural segmentation or how textural segmentation varies as a function of the amount of difference in second-order statistics. Without such specification, it is difficult to definitively test the implication of the conjecture that textural segmentation is a function of the amount of difference in second-order statistics. We examined whether textural segmentation is a linear, logarithmic, or power function of dipole differences. Dipole differences are calculated for three ranges of dipole lengths. The dipole lengths ranged from approximately 1.1' to 43.2', 21.6', and 10.8' for the viewing distances in Experiments 1 and 2. Caelli (1980) has reported that the visual system is more sensitive to shorter dipoles than to longer ones, particularly dipoles that subtend less than 30' in length.

The proportions of black and white, or first-order statistics, are approximately the same for the different regions on each of the displays. The second-order or dipole statistics of textures composed of two brightness levels are determined by the correlation function. If we regard such a texture as a array of 0s and 1s, the value of its autocorrelation for a given displacement (distance and direction) is proportional to the number of pairs of 1s at that displacement. Similarly, the autocorrelation of its complement is proportional to the number of pairs of 0s, and the cross correlation of the texture with its complement is proportional to the number of (1,0) and (0,1) pairs. These correlations may therefore be used to assess the differences in dipole statistics. For the top and middle regions of the size and slope stimuli used in Experiments 1 and 2, and the top and bottom regions of the U displays, a window containing a 7×7 array of textural elements was displaced relative to itself in a larger area of the same texture. Dipole differences were assessed by calculating the mean of the squared differences between the numbers of black/black, white/white, black/white, and white/black pairs of points for corresponding shifts of the top and middle or the top and bottom regions of a display. The mean square dipole differences were computed for displacements from 1 to 40, 1 to 20, and 1 to 10 pixels respectively in the x and y directions. (A shift of 40 pixels corresponds to a displacement of approximately 43.2'.) Each displacement in the x direction was combined with each displacement in the y direction. Thus, the 40, 20,

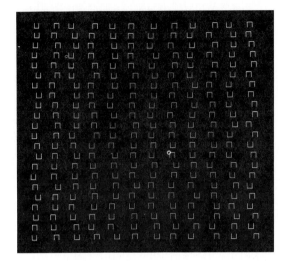

(a)

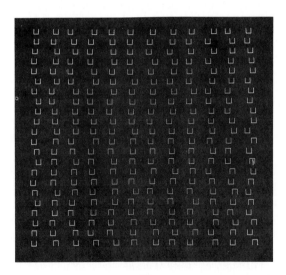

(b)

Figure 16. Strong vertical linking based on proximity interferes with textural segmentation.

and 10 pixel displacements correspond, respectively, to computing the dipole differences over a window 81×81, 41×41, and 21×21 pixels square centered at the origin. The first column in Table 3 identifies a display by its figure number, and and the second, third, and fourth columns present the mean square dipole differences for 40, 20, and 10 pixel displacements respectively. The larger the value in Table 3, the greater is the difference in dipole statistics. A value of zero indicates that the dipole statistics are identical.

Mean square dipole differences fail to correspond to the rated textural segmentations. Table 3 presents the mean square dipole differences for the top ard middle regions of the white squares on a black background in Experiment 1 (stimuli 8a through 8d) for window displacements of 40, 20, and 10 pixels. The mean square dipole differences are largest for stimuli 8b and 8d and smallest for stimuli 8a and 8c for all three displacements. A Tukey test of multiple comparisons, in contrast, revealed that stimuli 8a (mean rating = 3.58) and 8b (mean rating = 4.04) differed significantly at the .01 level from stimuli 8c (mean rating = 1.71) and 8d (mean rating = 2.00). Neither stimuli 8a and 8b nor stimuli 8c and 8d differed significantly from each other. The linear correlations between the mean textural segmentation ratings and the mean square dipole differences are .40, .58, and .20 respectively for window displacements of 40, 20, and 10 pixels. The best fitting logarithmic and power functions in

Table 3

Mean Square Dipole Differences

Stimulus (Identified by Figure Number)	81×81 Window	41×41 Window	21×21 Window
8a	29,620	28,208	16,812
8b	473,704	720,361	219,172
8c	21,434	26,883	19,776
8d	317,268	247,927	209,589
11a	550	466	134
11b	2,208	4,194	608
11c	212	330	116
11d	1,312	566	498
11e	601	502	52
11f	1,925	4,048	172
11g	195	315	80
11h	1,009	261	375
13a	7,789	11,201	6,424
13b	10,526	11,697	8,966
14a	46,260	46,076	24,391
14b	37,364	41,977	16,647
15a	6,062	5,149	3,870
15b	13,641	6,259	8,029
16a	8,442	10,165	8,210
16b	8,677	10,068	7,449
2a	14,134	16,065	10,552
2b	9,661	14,910	6,037

the sense of least squares were also fitted to the data. The correlations between mean textural segmentation ratings and the logarithms of the dipole differences identify the best fitting logarithmic function. They are .32, .39, and .17 for the three window displacements. The correlations between the logarithms of the mean textural segmentation ratings and the logarithms of the dipole differences identify the best fitting power function. They are .32, .37, and .17 for the three window displacements. The dipole statistics for the black squares on a white background (stimuli 8e through 8h) are the same, except for minor variations introduced in making the Polaroid pictures, as the dipole statistics for the white squares on a black background. The textural segmentation of the black squares on a white background is similar to that of the white squares on a black background (cf. Table 1). Thus, the dipole statistics fail in this case also to account for textural segmentation.

Dipole statistics can be generalized to textures consisting of three brightness levels. The dipole statistics consist of the number of pairs of points which are separated by a fixed distance and a fixed direction and which fall on the possible combinations of black, gray, and white points of the texture. The dipole statistics of the stimuli consisting of white and gray squares on a black background (stimuli 9a through 9d) are the same as the dipole statistics of the stimuli consisting of black and gray squares on a white background (stimuli 9e through 9h) except for the interchange of black and white. Thus, the mean square dipole differences of corresponding stimuli in the two sets are essentially equivalent. The textural segmentation produced by the two stimulus sets were, however, completely different. The stimuli with a white background uniformly failed to produce textural segmentation (cf. Table 1).

The mean square dipole differences for the slope stimuli in Experiment 2 consisting of vertical and diagonal lines (stimuli 11a through 11d) and of horizontal and diagonal lines (stimuli 11e through 11h) are shown in Table 3. There are again clear discrepancies between the mean ratings for textural segmentation and the mean square dipole differences. For example, the mean textural segmentation ratings for stimuli 11a, 11b, and 11f were 2.67, 3.33, and 1.83 (see Table 2). A Tukey test of multiple comparisons indicated that stimulus 11b differed significantly at the .01 level from stimulus 11f, but not from stimulus 11a. The mean square dipole differences, however, are more alike between stimuli 11b and 11f than between stimuli 11b and 11a for all three pixel displacements. Table 2 shows that textural segmentation was uniformly better for stimuli with vertical and diagonal lines than for stimuli with horizontal and diagonal lines. Examination of Table 3, however, shows that the dipole statistics with 40 and 20 pixel displacements are quite similar. The linear correlations between the mean textural segmentation ratings and the mean square dipole differences are .64, .50, and .76 respectively for window displacements of 40, 20, and 10 pixels. The correlations between mean textural segmentation ratings and the best fitting logarithmic function of the dipole differences are

.63, .52, and .74 respectively, and between mean textural segmentation ratings and the best fitting power function .66, .50, and .77 respectively for the three pixel displacements.

Table 3 also presents the mean square dipole differences between the top and bottom regions of the U displays. Though no experiments were conducted, there appears to be only a moderate correspondence between differences in dipole statistics and perceived textural segmentation. The textural segmentation of Displays 13b and 14a appears to be similar, but the dipole difference between the top and bottom regions of Display 14a is considerably greater than that of Display 13b for all three pixel displacements. The perceived textural segmentation of Display 14b, moreover, is greater than that of Display 14a though its dipole difference is less for all three displacements. A striking discrepancy with a 20-pixel displacement is that Displays 13a and 13b have similar dipole differences though they differ in their perceived textural segmentation. A similar discrepancy occurs for Displays 13b and 15b with 40 and 10-pixel displacements and with Displays 13b and 16a for all three pixel displacements. The results of Experiments 1 and 2, and the U displays illustrate the insufficiency of global dipole statistics for predicting textural segmentation.

5. Model for Textural Segmentation

The hypothesis proposed is that textural segmentation is a result of (a) local stimulus differences that differentially stimulate receptive fields, and (b) linking operations that form textural elements. Compared to Julesz's original hypothesis that textural segmentation does not occur for textures that have the same first- and second-order statistics, this hypothesis is admittedly imprecise. First, an incomplete list of features is proposed in place of the precisely defined concepts of first- and second-order statistics. Though a great deal is known about the stimulus features which differentially stimulate receptive fields, the list is undoubtedly incomplete. Marr (1976), for example, has pointed out the importance of line ends or terminations for textural segmentation. The fact that there exist elliptical-like receptive fields which would be specifically sensitive to line ends has only recently been pointed out (Frisby & Clatworthy, 1975). Second, an incomplete set of rules for the formation of textural elements is proposed in place of the well-defined points of the retinal array. Textural elements are hypothesized to be formed by proximity, certain kinds of similarity, and good continuation. Others of the Gestalt laws of grouping may play a role in the formation of texture.

The assumptions of the model are: (1) There is linking of the features extracted from the retinal array into textural elements. (2) There is an encoding of the brightness, color, size, slope, and the location of each textural element and its parts. (3) The features belonging to textural elements in neighboring spatial regions are compared and differences encoded. Difference detectors encode the total differences in brightness, color, size, and slope of textural elements in neighboring spatial regions. The regions used by the difference detectors are of a range of sizes. (4) There are decision units which inspect the degrees to which difference detectors are stimulated. The function of the decision units is to segment a display into textural regions on the basis of the difference signals.

Textural elements can differ with respect to more than a single value of a feature. For example, the upright and tilted Ts in Figure 5 differ in two values of a feature (slope). The upright Ts consist of 90° (vertical) and 0° (horizontal) lines; the tilted Ts of 45° and 135° lines. The response of a difference detector, such as for slope, will be assumed to reflect the total difference between textural elements in neighboring spatial regions. The fact that the slopes of the lines of the tilted Ts lie between the slopes of the lines of the upright Ts does not interfere with textural segmentation. It makes little difference if the lines are separated, i.e., half of the display is composed of randomly placed vertical and horizontal lines and the other half of randomly placed 45° and 135° lines. Julesz (1962) found that for brightness, if elements in a display had nonadjacent brightness values, textural segmentation is interfered with. If the number of values of a feature associated with the elements becomes large, textural segmentation is decreased.

What variables affect the magnitude of difference signals? Difference signals are proportional to the degree to which feature analyzers are stimulated by textural elements in one spatial region and are not stimulated by textural elements in a neighboring spatial region. It is further assumed that the difference signals are decreased by shared features that stimulate common analyzers. For example, the slope difference signal is a function of the differences in the distribution of slope analyzers stimulated by textural elements in neighboring spatial regions. The slope difference signal varies directly with the number of slope analyzers that are stimulated by textural elements in one region and not the other and inversely with the number of slope analyzers that are stimulated in common. In a sense, similarity can be regarded as noise which reduces the detectability of the difference signal.

Difference signals arising from two different features, e.g., slope and brightness, will be assumed to summate and strengthen textural segmentation. Beck (1967) found that textural segmentation was increased by increasing the brightness difference between background and disparate figures (the displays were like those in Figure 6). Though the overall textural segmentation increased with an increased brightness difference, the relative strengths of textural segmentation produced by various pattern differences were approximately the same for three different brightness differences. The linear correlations between the mean ratings of textural segmentation for nine disparate figures embedded in a display of upright Ts were .94 (brightness levels 1 and 2), .93 (levels 1 and 3), and .97 (levels 2 and 3). Analogously, the magnitude of the

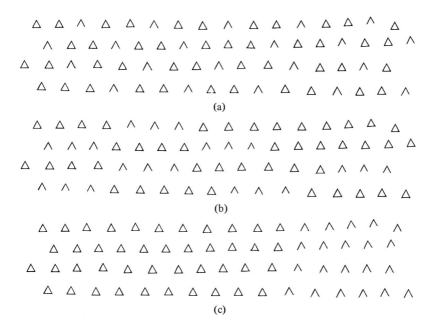

Figure 17. Example of how textural segmentation is a function of the size of the spatial region over which difference signals occur.

final difference signal between two spatial regions will be assumed to be decreased if textural elements are the same with respect to a feature that ordinarily produces textural segmentation. As mentioned earlier, differential stimulation of complex receptive fields by arrangement differences, for example, a T and an L, fail to produce strong textural segmentation. Since the slopes of the component lines of the textural elements are the same, the textural elements would tend to stimulate common slope analyzers that would decrease the difference signal.

We have assumed that differencing operations are taken over a range of field sizes. The strength of a difference signal will also be assumed to be a function of the size of the spatial region over which it is taken. The larger the spatial region for which a difference signal occurs, the stronger the textural segmentation. This is illustrated in Figure 17. In the top display, the difference signal is between a single open triangle and the adjacent closed triangles, in the middle display between three open triangles and the adjacent closed triangles, and in the bottom display between 21 open triangles and the adjacent closed triangles. Another factor affecting textural segmentation may be the spatial concentration of the difference signals. The difference signals become more concentrated in going from the top to the bottom display in Figure 17. In Figure 17a, the difference signals are distributed over the entire display. In Figure 17b the difference signals are spatially more concentrated due to the adjacency of the open triangles. In Figure 17c the difference signals are concentrated in the spatial region on the right. Textural segmentation is also a function of the number of difference signals. The greater the number of display locations over which a difference signal occurs, the stronger will be the textural segmentation.

The function of decision units is to segment a visual pattern into textural components on the basis of the magnitude and distribution of the difference signals. Textural segmentation occurs if the magnitude of difference signals, e.g., in slope, between components in a visual pattern is sufficiently greater than those within component. Textural segmentation is hierarchical. In Figure 5, for example, there is first a segmenting of the backward L, upright T, and tilted T figures from the white paper. The brightness difference between the white paper and the figures presumably produces the largest difference signal. Then, there is a segmenting of the upright Ts from the tilted Ts on the basis of line slope. Textural segmentation may fail to occur because there are too many difference signals. Figure 18 shows vertical, diagonal, and horizontal lines. There is no spontaneous textural segmentation although attention can be directed to either the vertical diagonal, or horizontal lines and they can be held in attention. The presence of strong difference signals uniformly throughout the display interferes with spontaneously segmenting the display. Every line produces a strong slope difference signal because of its difference in slope from neighboring lines. Similarly, no textural segmentation occurs within a texture gradient because the differences between neighboring textural elements are small and approximately equal. There is no distinctive difference signal associated with the texture gradient. As indicated earlier, textural segmentation also occurs in terms of single properties such as slope or color but not in terms of a

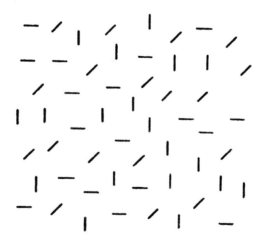

Figure 18. Textural segmentation fails to occur because of the uniform dissimilarity of the lines throughout the display.

conjunction of properties such as slope and color. If competing difference signals are present, alternative segmentations occur which interfere with one another.

6. Julesz's Dipole Model

There is a similarity between the hypothesis that textural segmentation occurs in terms of elementary feature analyzers and the hypothesis that it occurs in terms of nth-order statistics. First- and second-order statistics encode properties relating to the sizes, orientations, and arrangement of the black elements in a black and white textural pattern. Random dot patterns differing in first- and second-order statistics will differentially stimulate cells having receptive fields with different widths and orientations. The use of random dot patterns, therefore, does not in itself preclude discriminations being based on locally defined features produced by statistical constraints (Beck, 1982).

Julesz and his colleagues have constructed textural patterns which have the same global first- and second-order statistics but which give strong textural segmentation. Discrimination in these counterexamples to Julesz's statistical conjecture appears to be based on features which would differentially stimulate concentric and elliptical receptive fields. Two examples taken from Julesz (1980) are shown in Figure 19. The top display has not only the same first- and second-order statistics but also the same third-order statistics. The textural difference, nevertheless, is immediately discriminable. The local features in the left and right halves of the top display would differentially stimulate bar detectors differing in size and orientation. Similarly, the other counterexamples discovered by Julesz and his colleagues would differentially stimulate concentric and elliptical receptive fields. (This is also true of the textures constructed by Gagalowicz [1980] which have the same global, but differing local, first- and second-order statistics.) The bottom display in Figure 19 is of particular interest because the dotted lines of the disparate and background textural elements have the same slopes. Julesz (1980) suggests that textural segmentation occurs because the disparate and background textural elements differ in the number of line ends or terminators. Frisby and Clatworthy (1975) have pointed out that the elliptical-like receptive fields discovered by Rodieck and Stone (1965) would be specifically sensitive to line ends and would respond strongly to their presence.

Julesz (1980) has proposed that two different processes are involved in textural segmentation. First, textural segmentation may be based on the statistical properties of the textures. Textures that have the same first- and second-order statistics do not produce strong textural segmentation. Second, textural segmentation may be based on the differential stimulation of feature analyzers. Julesz hypothesized that textures which differ in elongated blobs (e.g., line segments) or in terminators (e.g., line ends) are perceptually segmented even

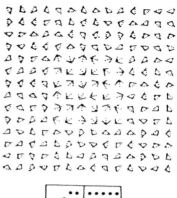

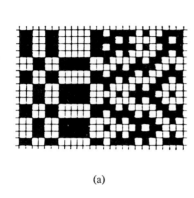

(a)   (b)

Figure 19. Example of discriminable textures which have the same second-order statistics. (After Julesz [1980].)

though they have identical dipole statistics. The counterexamples found by Julesz and his colleagues show that two textural patterns with the same first- and second-order statistics, or, what is equivalent, identical power spectra in a Fourier representation, can be readily segmented. In terms of a Fourier representation, these counterexamples argue that under some conditions phase differences can produce textural segmentation. It is, however, not clear how to characterize the phase differences that produce textural segmentation. On the other hand, as pointed out above, textural patterns differing in first- and second-order statistics will differentially stimulate elementary feature analyzers. In fact, the two primitives proposed by Julesz are, more or less, what one would expect if textural segmentation occurs in terms of simple shape properties such as the slopes and sizes of features and textural elements (Beck, 1972, 1982). As mentioned earlier, Julesz (1981a) has now adopted the view that textural segmentation is solely the result of differences in the first-order statistics of local features. A history of Julesz's second-order statistics conjecture and its disproof can be found in Julesz (1981b).

Acknowledgment

The support of the National Science Foundation under Grant MCS-79-23422 is gratefully acknowledged, as is the help of Janet Salzman in preparing this paper. We would also like to thank David Notley and Peter Hyde for writing the programs that produced the stimuli and the autocorrelation statistics.

References

Beck, J. Perceptual grouping produced by changes in orientation and shape. *Science,* 1966, **154**, 538-540. (a)

Beck, J. Effect of orientation and of shape similarity on perceptual grouping. *Perception and Psychophysics,* 1966, **1**, 300-302. (b)

Beck, J. Perceptual grouping produced by line figures. *Perception and Psychophysics,* 1967, **2**, 491-495.

Beck, J. Similarity grouping and peripheral discriminability under uncertainty. *American Journal of Psychology,* 1972, **85**, 1-19.

Beck, J. Similarity grouping of curves. *Perceptual and Motor Skills,* 1973, **36**, 1331-1341.

Beck, J. Textural Segmentation. In J. Beck (Ed.), *Organization and Representation in Perception,* Hillsdale, NJ: Erlbaum, 1982.

Beck, J., & Ambler, B. Discriminability of differences in line slope and in line arrangement as a function of mask delay. *Perception and Psychophysics,* 1972, **12**, 33-38.

Beck, J., & Ambler, B. The effects of concentrated and distributed attention on peripheral acuity. *Perception and Psychophysics,* 1973, **12**, 482-486.

Caelli, T. M. Facilitative and inhibitory factors in visual texture discrimination. *Biological Cybernetics,* 1980, **39**, 21-26.

Foster, D. H., & Mason, R. J. Irrelevance of local position information in visual adaptation to random arrays of small geometric elements. *Perception,* 1980, **9**, 217-221.

Fox, J., & Mayhew, J. E. W. Texture discrimination and the analysis of proximity. *Perception,* 1979, **8**, 75-91.

Frisby, J. P., & Clatworthy, J. L. Illusory contours: Curious cases of simultaneous brightness contrast? *Perception,* 1975, **4**, 349-357.

Gagalowicz, A. A new method for texture field synthesis: Some applications to the study of human vision. *IEEE Transactions on Pattern Analysis and Machine Intelligence,* 1981, **3**, 520-533.

Green, B. F., Jr., Wolf, A. K., & White, B. W. The detection of statistically defined patterns in a matrix of dots. *American Journal of Psychology,* 1959, **72**, 503-520.

Julesz, B. Visual pattern discrimination. *IRE Transactions on Information Theory,* 1962, **8**, 84-92.

Julesz, B. Experiments in the visual perception of texture. *Scientific American,* 1975, **232** (4), 34-43.

Julesz, B. Spatial nonlinearities in the instantaneous perception of textures with identical power spectra. *Philosophical Transactions of the Royal Society, London,* 1980, **B290**, 83-94.

Julesz, B. A theory of preattentive texture discrimination based on first-order statistics of textons. *Biological Cybernetics,* 1981, **41**, 131-138. (a)

Julesz, B. Textons, the elements of texture perception, and their interactions. *Nature,* 1981, **290**, 91-97. (b)

Julesz, B., Gilbert, E. N., Shepp, L. A., & Frisch, H. L. Inability of humans to discriminate between visual textures that agree in second-order statistics – revisited. *Perception,* 1973, **2**, 391-405.

Marr, D. Early processing of visual information. *Philosophical Transactions of the Royal Society, London,* 1976, **B275**, 483-524.

Olson, R., & Attneave, F. What variables produce similarity grouping? *American Journal of Psychology,* 1970, **83**, 1-21.

Rodieck, R. W., & Stone, J. Analysis of receptive fields of cat retinal ganglion cells. *Journal of Neurophysiology,* 1965, **28**, 833-849.

Schatz, B. R. The computation of immediate texture discrimination. MIT AI Memo 426, 1977.

Wertheimer, M. Untersuchungen zur Lehre von der Gestalt. II. *Psychologische Forschung,* 1923, **4**, 301-350.

Criteria for Representations of Shape

Michael Brady

Massachusetts Institute of Technology
Cambridge, Massachusetts

Abstract

Three criteria for judging representations of shape are proposed. The criteria are illustrated for the Generalized Cylinder representation of three-dimensional volumes and the Smoothed Local Symmetries representation of planar shapes. A representation for the shape of surfaces is proposed, called Curvature Patches. It is used to extend Smoothed Local Symmetries to a novel three-dimensional representation.

1. Introduction

Recent discussions of image understanding (Brady, 1981, 1982a; Marr, 1978, 1982; Ballard & Brown, 1982) chart its progress toward modelling visual perception as the construction of a set of rich representations of the information available in images and their interpretation in terms of existing knowledge structures. Only in few cases, however, are we consciously aware of these representations and the information that they make explicit.

The *ACRONYM* representation of objects (Brooks, 1981), for example, ultimately involves everyday objects of conscious thought, such as airplanes and electric motors. Such representations are reminiscent of Bartlett's (1932) *schemas;* indeed, they may be considered to be a step toward making Bartlett's ideas precise (see Minsky, 1975). They are also closely related to the concept of *frame* developed in Artificial Intelligence (Minsky, 1975; Winston, Katz, Lowry, & Binford, 1982).

We are largely unconscious of the information made explicit in most of the representations that have been proposed in image understanding. These include the *zero-crossing map* (Marr & Hildreth, 1980), the *primal sketch* (Marr, 1976), the *symmetric* or *medial axis transform* (Blum, 1973; Blum & Nagel, 1978), *smoothed local symmetries* (Brady, 1982b, 1982c, and Section 3 below), and the variously named *surface orientation map* (Barrow & Tenenbaum, 1978; Grimson, 1981; Horn, 1982; Marr, 1982; Terzopoulos, 1982).

The processes of visual perception have two associated representations, their input and output. Generally, the input is not the image, but some representation that is the result of earlier processing. Similarly, the output representation normally serves as input representation to subsequent processing. In the Marr-Poggio (1979) theory of stereo, for example, the input representation is the zero-crossing map, the feature points found by a difference-of-Gaussians operator. The output representation marks the visual disparity at each matched pair of zero-crossings.

Dually, any representation has associated processes for which it is the input, and which compute *with* the representation, and processes for which it is the output, and which help *derive* the representation. Generalized cylinders have been developed for modelling objects in a way that facilitates recognition, a process that computes with the representation. It is much less well understood how to derive a generalized cylinder representation from an image. Conversely, the design of the surface orientation map, that associates (for example) the (average) local surface normal with each tessellated portion of a visible surface, has been subordinate to the development of the processes that derive it. Such processes are shape from shading (Ikeuchi & Horn, 1981), shape from contour (Barrow & Tenenbaum, 1981; Binford, 1981; Witkin, 1981; Brady & Yuille, in preparation), and shape from stereo (Grimson, 1981; Marr & Poggio, 1979). It is much less well understood how to compute with the surface orientation map, for example to perform recognition or inspection (Horn, 1982). The curvature patch representation introduced below is intended to repair this deficiency. Computer science suggests that detailed design of a representation can be expected to take into account the needs of all such processes, those that compute with the representation and those that derive it.

We have suggested that representations play a crucial role in the emerging computational theory of vision. But what is a representation, and what criteria can be identified to assess them and guide their design? How can we represent visible surfaces and three-dimensional volumes? These are the issues addressed in this paper. Marr and Nishihara (1979) have proposed three criteria for assessing a representation. The first of these, called *accessibility*, states that the representation should be computable relatively efficiently by some process. This is a minimal requirement. The other two criteria, *scope* and *sensitivity*, are complementary and are a restatement of the idea of representing information at a variety of scales (see for example, Marr & Hildreth, 1980; Witkin, 1981.) We shall return to a discussion of this below.

In this article, we propose the following three additional criteria, and illustrate them in Sections 2 and 3 for generalized cylinders and smoothed local symmetries:

- A representation should have *rich local support*. Rich means that the representation should be information preserving. Local support means that it can be locally computed; this is essential for dealing with occlusion and for performing detailed inspection. We call the local parametric description associated with a representation its *local frame*. Local means the scale at which the representation is being computed.

- Local frames give rise to more global descriptions called *frames* by the processes of *smooth extension* and *subsumption*. Subsumption reduces the set of frames by rejecting locally feasible alternatives that are wholly enclosed by other frames. The frame descriptions are particularly succinct if the integral of one or more of the locally available parameters is "small."

- The frames that correspond to perceptual subparts of a shape can be *propagated* by inheritance or affixment whenever the corresponding frames are "consistent."

In Section 4 we consider the representation of visible surfaces. Finally, we suggest an extension of smoothed local symmetries to three dimensions and consider the determination of grasp positions on objects for a multi-fingered robot hand.

2. An Illustration: Generalized Cones

In this section we illustrate the representational criteria introduced in Section 1 by discussing how they apply to generalized cones. Generalized cones were introduced by Binford as a representation of three-dimensional volumes that emphasizes *elongation*. A shape is described by drawing a cross-section at a fixed angle along an arbitrary three-dimensional "spine" curve, expanding the cross-section according to a "sweeping rule." Figure 1 shows a number of generalized cones. Notice that although elongation is the characteristic property of generalized cones, they are not necessarily perceived as elongated. Nor are generalized cones restricted to having a circular cross-section, though this has been a common simplification in practice.

Consider the wedge depicted in Figure 2a. It admits of many simple descriptions. Suppose objects are represented as joins, intersections, and slices

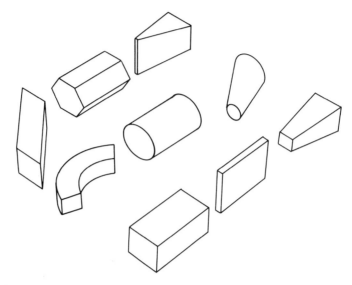

Figure 1. A sampling of three-dimensional objects that can be modeled simply and naturally using generalized cylinders. (After Brooks [1981], Figure 2.1.)

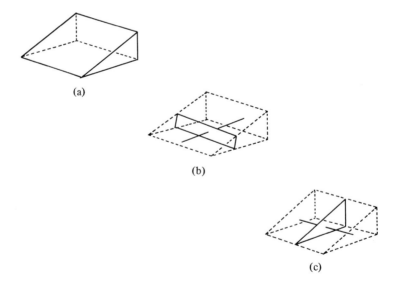

(a)

(b)

(c)

Figure 2. (a) A rectangular wedge; (b) generalized cone representation of the wedge based on drawing an expanding rectangle along a straight line; (c) alternative representation based on drawing a triangle along a straight line.

starting with a repertoire of "basic" shapes (see for example, Braid, 1973). If the wedge is not basic, it might be described as a sliced rectangular box. The generalized cylinder representation, however, encourages a different perspective of the wedge, as a monotonically expanding rectangle (Figure 2b) or a constant triangle (Figure 2c) drawn along a straight axis. Which of these is perceived depends on the dimensions of the wedge.

Generalized cones are well suited to describing objects that have a natural axis. This includes growth structures such as animals (Agin, 1972; Nevatia & Binford, 1977; Marr & Nishihara, 1978). Hollerbach (1975) noted that Greek amphorae are also conveniently described using generalized cones, the spine being the result of producing the amphora on the potter's wheel. Similar representational advantages accrue for objects that result from turning on a lathe or that are produced by extrusion. Objects that are produced by molding, beating, welding, or sculpture tend to be awkwardly described in terms of generalized cones. Brooks (1981, 1982) has used generalized cones to represent airplanes and electric motors, and for robot path planning.

A generalized cone is precisely defined by drawing a (one-parameter) *planar cross-section* curve $c(u)$ at a fixed angle ψ, called the *eccentricity* of the generalized cone, along a space curve $f(s)$, called the *spine* of the generalized cone, expanding the cross-section according to a function $h(s)$, called the *sweeping rule*. Although the spine, cross-section, and sweeping rule are allowed to be arbitrary analytic functions, in practice only simple functions are allowed. Typically, the spine is straight or circular, the sweeping rule is constant or linear, and the cross-section is rectangular or circular. Usually, the eccentricity is a right angle, so that the spine is normal to the cross-section.

Generalized cones satisfy the first of our representational criteria. There is a natural local coordinate frame at all points along the spine. A space curve has a unique local tangent $\mathbf{t}$, and, unless the spine is straight, it has a unique normal $\mathbf{n}$ and binormal $\mathbf{b}$. The vectors $\mathbf{t}$, $\mathbf{n}$ and $\mathbf{b}$ form a local rectangular coordinate system and satisfy the Frenet-Serret equations (Weatherburn, 1927). In general, the direction of the tangent is not uniquely determined; fixing it amounts to reversing the tilt of the cross-section, the "Necker" ambiguity (Figure 3). Often, the spine is normal to the cross-section, so that $\mathbf{b}$ and $\mathbf{n}$ form a basis for a coordinate frame for the cross-section. If the eccentricity of the cone is not ninety degrees, there is still a local, though non-orthogonal, coordinate frame, consisting of the local tangent to the spine and two basis vectors for the cross-section.

According to the second of our representational criteria, local frames become more global by *smooth extension* and *subsumption*. Smoothness is implicit in the requirement that the spine of the generalized cone be smooth. Subsumption isolates the maximal components by rejecting locally feasible alternative descriptions of parts of a shape that are wholly enclosed by other

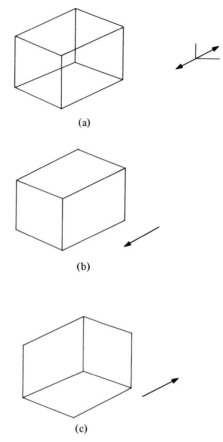

(a)

(b)

(c)

Figure 3. Alternative choices for the sense of the tangent to the spine correspond to tilt-reversal.

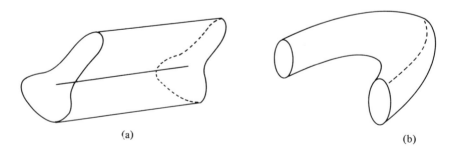

(a)   (b)

Figure 4. (a) A generalized cone with a straight axis; (b) one with a constant cross-section shape.

parts that have a description. Subsumption is implicit in the definition of generalized cones. The summary descriptions are particularly succinct if the integral of (the absolute value of) one or more of the locally available parameters is "small." For example, Figure 4a shows a generalized cone that has a straight axis (the direction of **t** is constant), and Figure 4b shows one whose cross-section is constant.

In general, there is a *partially defined frame* naturally associated with the generalized cone. The frame is partial in that certain choices are underconstrained or arbitrary. The cone in Figure 4a, for example, has a frame that fixes one vector, namely along· the spine of the cone. The circular cylinder shown in Figure 5 has a straight spine whose sense (tilt reversal) is not determined, and a circular cross-section which has a one-parameter family of possible orthogonal coordinate frames. The members of the family correspond to fixing one vector in the cross-section. The origin on the spine is not fixed.

According to the third of our criteria, the frames that correspond to perceptual subparts of a shape can be *propagated* by inheritance or affixment whenever the corresponding frames are "consistent." Consider the robot arm linkage consisting of two right circular cylinders illustrated in Figure 6. The link has an axis of rotation **v**, a vector that lies in the cross-section of both cylinders. Since the coordinate frames for the cross-sections are one-parameter families, the natural choice for both is **v**. This choice allows the coordinate frames of successive linkages to be simply related. It is in fact the representation suggested by Denavit and Hartenberg (1955) and adopted widely in robotics. Marr and Nishihara (1978) propose a restricted form of the Denavit-Hartenberg representation, which is presented here as a special case of the general process of *local frame propagation*. Marr and Nishihara suggest that summarizing the two link frames by a single one amounts to representing an object at a variety of scales. Figure 7 shows a second example of local frame propagation. The hemisphere is formed by drawing a circular cross-section along a straight line; once more the choice of coordinate frame for the cross-section is unspecified. The adjoining rectangle suggests a particular choice.

Brooks and Binford (1980) discuss a hierarchy that derives from class abstraction and that embodies constraints that do not necessarily respect the hierarchy. Nevatia and Binford (1977) developed a multi-level hash coding indexing scheme to ascend and descend the hierarchy to recognize an object in a database of many millions of models. Brooks (1981) shows how the viewpoint may be determined simultaneously with recognition. Marr and Nishihara (1978) do not determine viewpoint explicitly, but assume that a "relaxation" process will line up the model with the image. In their scheme, constraints only propagate down the hierarchy, with recognition proceeding from a "general animal."

Generalized cones have deservedly received considerable attention in image understanding. Marr and Nishihara (1978) and Hinton (1979) even sug-

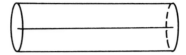

Figure 5. A right circular cylinder. The local frame leaves the sense of the spine and the coordinate frame for the cross-section unspecified.

gest that an analog mechanism called the "spasar" underlies human perception of objects. The evidence cited in support of this claim ranges from data on mental rotations to observations of patients with parietal lesions. Some difficulties with (projections of) generalized cones as a representation of two-dimensional shapes are presented below. Nackman (1982, p. 13) notes that "there is little understanding of the domain of figures that admit generalized cone decompositions. Further, such decompositions are not unique and constraints sufficient to force uniqueness are not, in general, known. Hence all programs that compute generalized cone decompositions are forced to use *ad hoc* rules to choose a single decomposition." Nackman develops a three-dimensional generalization of the symmetric axis transform (q.v.); we shall refer to his work again at the end of Section 4.

3. Two-dimensional Shape Description: Smoothed Local Symmetries

3.1. Introduction

Although the world is three-dimensional, there are important applications of two-dimensional vision. Many objects in stable configurations can be recognized from their bounding contour. Stamped, cast, and forged flat metal parts such as hinges and cogs have this property. So do rotationally symmetric objects such as shafts and springs. Representing two-dimensional shape

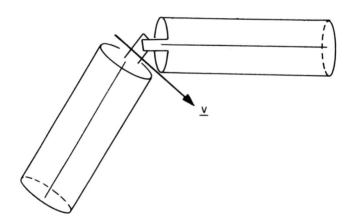

Figure 6. Two robot links inherit the axis of rotation of the joint they form.

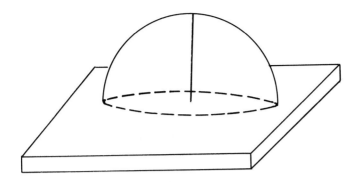

Figure 7. A coordinate frame for a hemisphere on a box.

remains one of the most challenging, ubiquitous and least well understood areas of image understanding. We take descriptive abilities of the sort illustrated in Figure 8 for granted, but they are beyond the abilities of today's vision systems, both in industry and in university laboratories.

Shape is represented in most current robotics or commercially available vision systems (Agin, 1980; Holland, Rossol, & Ward, 1979) in terms of global features such as: the center of area, the number of holes, the aspect ratio of the principal axes, and the ratio of the perimeter squared to the area. Such features can be computed efficiently and they are reasonably insensitive to noise and quantization, even for low resolution binary images. Unfortunately, the computed values of global features for the visible portion of an arbitrarily occluded object bear arbitrary relationships to the values that would be computed for the whole object. It follows that it is difficult or impossible to recognize occluded parts using global features.

Bolles (1982) uses "focus features" to handle limited cases of occlusion. An object model is a graph whose nodes are features such as corners and holes and whose edges are the exact distance and relative orientation of the two nodes related by the edge. In essence, Bolles' representation is a compromise between classical features and representations with rich local support, called for in the first of our criteria.

It is also difficult to inspect parts for other than gross defects using global descriptors. Figure 9b shows a heap of overlapping "sectors" and "nibs" shown in Figure 9a. The convex portion of contour indicated by arrow 1 is consistent with both parts, in the sense that it could form a segment of either a sector or nib in one of several ways. The piece of contour indicated by arrow 2 is, however, only consistent with being one specific segment of a sector. Clearly, although local pieces of contour and shape constrain image-to-model correspondence, not all local pieces are created equal. The "broken" nib shown in Figure 9c is typical of industrial inspection tasks. The fault, indicated by arrow 3, is restricted to a small part of the object's contour and shape.

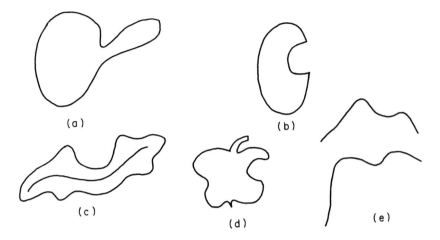

Figure 8. Descriptions of two-dimensional figures. (a) An object that is perceived as the join of two subfigures; (b) an object that is perceived as having a piece cut out; (c) an object that is symmetric about a curved axis; (d) an object that is symmetric apart from a piece that has been cut out; (e) two similar figures, one of which has been rotated and expanded.

In this section, we report progress in the development of a representation of two-dimensional shape and contour that satisfies the criteria set out in the introduction. The representation, called *smoothed local symmetries,* is most closely related to generalized cylinders (Nevatia & Binford, 1977; Brooks, 1981) and the symmetric axis transform (Blum, 1973; Blum & Nagel, 1978). In an implementation of the representation (Brady, 1982b), we isolate the bounding contours of objects, currently using a technique that is based on Marr and Hildreth's (1980) difference-of-Gaussians edge finder. The shape representation has been applied to determine grasp points of laminar objects for a parallel jaw gripper, and is being applied to the recognition of hand-printed characters, the design of character type-faces, and the inspection of industrial parts. There are interesting connections between smoothed local symmetries and the criteria for shape discussed by Palmer in this volume.

3.2. Shape Description

The bibliography on shape description is extensive (see Ballard & Brown, 1982; Pavlidis, 1977; Rosenfeld & Kak, 1976). Proposals for representing shape fall naturally into two classes: those that describe the one-dimensional bounding contour of an object (plus any internal holes) and those that describe the two-dimensional region subtended by an object. Shape is intrinsically two-dimensional, so it would seem that two-dimensional region-based approaches are to be preferred. However, occluded parts can often be identified, even when most subregions, corresponding to objects in a hierarchical description of array element often come from a contiguous portion of contours, but this is not necessary. A lot of sparsely distributed weak evidence can appear to be strong evidence.

The uniform accumulation of evidence in the Hough transform scheme does not meet the objection raised by Figure 9: a small portion of a contour can guarantee the presence of an object even though its "vote" is small. The Hough transform does not provide a means for detecting localized imperfections of the sort shown in Figure 9c. Finally, the shapes shown in Figure 9 are simple; in general a local piece of contour provides too weak a constraint to directly address a single model (Brooks, 1981).

Two-dimensional projections of generalized cones. Generalized cones were discussed in the previous section. Although they were proposed as a representation of three-dimensional shape, projections of generalized cones have been used to represent two-dimensional shape. Brooks (1981) calls two-dimensional projections of generalized cones "ribbons." As we pointed out in the previous section, although the spine $f(s)$ and the sweeping rule $h(s)$ are allowed to be arbitrary analytic functions, in practice the description of shape is only approximate as only simple functions are allowed (f is typically straight or circular, h is typically constant or linear). Generalized cones are information-preserving in the sense that $f(s)$ and $h(s)$ are sufficient to re-draw the original shape.

Generalized cone representations do not seem to be consistent with perceptual judgments in all cases. Figure 10 has a straight axis but subjects typically report that the figure has a curved spine. Generalized cone representations apparently allow the spine to be a bounding contour, as in Figure 11, a special case that should be distinguished. Generalized cones are *defined* analytically; but the defining parameters do not seem to be computed by the human visual system in all cases.

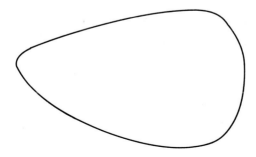

Figure 10. A generalized cylinder with a straight axis $y = 0$, that subjects usually perceive as having a curved axis. The shape was produced as follows: suppose that the contour for $y > 0$ is a function $f(x)$ that does not cross the x-axis; then for $y < 0$, the curve is $-n \cdot f(x)$, for some integer n (here 3).

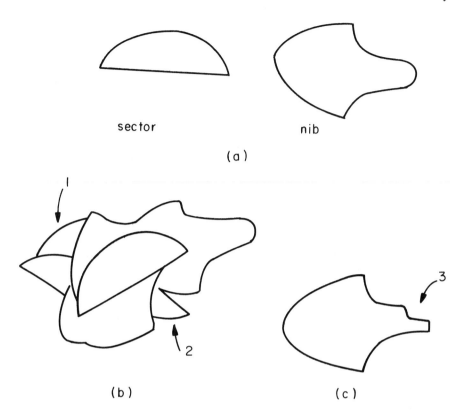

Figure 9. (a) Two hypothetical parts, a sector and a nib; (b) a heap of the parts; (c) a "broken" nib, whose defect is confined to a local portion of the contour.

an object, are partly obscured. Moreover, region-based representations of shape determine their parameters for a given shape from the bounding contour. One needs, for example, to isolate straight line and circular segments of a contour. Descriptions such as "serrated" correspond to determining that the contour tangent orientation $\alpha(s)$ is a periodic function of the contour length s.

The Hough transform. The generalized Hough transform (Ballard, 1981; Sloan & Ballard, 1980) has been proposed as a mechanism that can find instances of occluded shapes. Feature points contribute a vote to those elements of an accumulator array of parameterized object instances that are consistent with the local evidence gathered at the feature point. The visible portion of a contour often contributes enough votes to an accumulator array element to give strong evidence that the corresponding parameterized object is occluded. The inference of a global parameterized object instance from local features is only feasible for a limited class of largely unstructured objects. The accumulation of votes collapses information, a large vote being heuristic evidence that the associated parameterized object is present in the image. The votes in an

Figure 11. A generalized cone whose spine is part of the bounding contour.

Marr (1977) has proved a number of theorems about generalized "cylinders," that is, generalized cones whose spines are straight. The conditions in which the theorems apply are, however, highly restrictive (Bruss & Horn, 1981). No algorithm currently supports Marr's elegant theoretical work. His analysis of joins (Marr, 1977, Theorem 7) does make clear, however, that sub-objects do not necessarily occur at concavities, nor do concavities always signal objects.

The symmetric axis transform. The symmetric axis transform (SAT) (Blum, 1973; Blum & Nagel, 1978) is defined semi-constructively and topologically as the union of the centers of maximal disks that touch at least two points of the bounding contour of an object (Figure 12). A theorem of Calabi-Hartnett (Blum, 1973, p. 216) shows that the SAT is information-preserving. The symmetric axis is the locus of maximal disk centers. Figure 13 shows the SAT for a variety of shapes. The transform is most effective on smoothly curved objects. This is the usual case for biological shapes, the principal application to date (Blum, 1973; Webber & Blum, 1979). Blum and Nagel (1978, p. 169) observe that "the SAT is not the simplest description for rectilinear figures." Figure 14 shows the SAT of a rectangle. In view of this and Figure 11, we conclude that straight portions of contours should be distinguished in shape

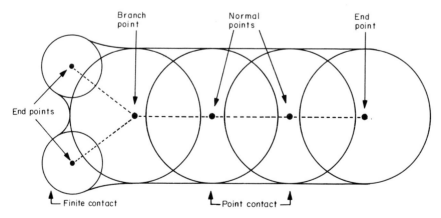

Figure 12. The symmetric axis transform. Maximal disks that touch the contour at more than two points are called branch points (points B_i).

Figure 13. Examples of the symmetric axis transform. (After Blum [1973, Figure 18].)

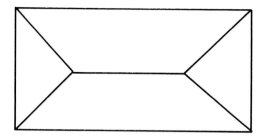

Figure 14. The symmetric axis transform of a rectangle.

descriptions and treated separately. Several algorithms for computing (approximations to) the SAT of a shape have been proposed (Bookstein, 1979; Lee, 1982).

The SAT has been criticized (Agin, 1972) (also cited by Marr [1978] and Hollerbach [1975]) for being too sensitive to small perturbations in the bounding contour. In fact, *exact* generalized cone representations suffer from the same criticism. As we noted earlier, however, generalized cone representations are rarely exact in practice. In any case, the over-sensitivity only arises when there is a single representational scale. The criticism may be countered by providing multiple scales of representation, an idea that has been proposed several times recently (Marr & Hildreth, 1980; Witkin, 1981; Terzopoulos, 1982).

The SAT embodies a definition of local symmetry (see Figure 15). Based on the geometry of Figure 15, Blum and Nagel (1978) define the set of primitive regions shown in Figure 16a. They also distinguish between object and axis curvature, providing the additional descriptive level shown in Figure 16b. Object joins are placed at "ligatures" computed from curvature extrema. A feature vector is then computed for a shape and recognition is based on a conventional pattern recognizer.

Aside from rectangular figures, the SAT produces unintuitive descriptions in many cases. In particular, if the curvature at contour points A and B is small, and the angle between the tangents at A and B is sufficiently obtuse (Figure 17), then the center of the maximal disk that touches the contour at A and B may lie in a perceptually distinct part of the shape.

Rich representations. We have argued that inspection and occlusion demand representations that have local support. We have also proposed that straight line contours, and possible disks, should be treated as a special case. Representations should also be *rich* in the sense of being information-preserving. The SAT and generalized cones are rich, the generalized Hough transform is not. To be useful, a representation requires less storage than the original shape. To achieve this goal it is necessary to represent an object at

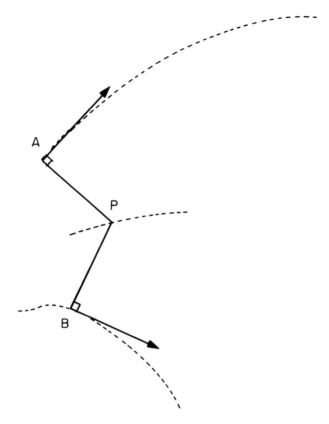

Figure 15. The local geometry of the SAT. The maximal disk centered on *P* touches the contour at *A* and *B*.

different scales that suppress detailed information unless it is required (Hollerbach, 1975). Finally, a shape representation should be invariant under translation, rotation, and magnification. Convexity (Marr, 1977; Hoffman & Richards, 1982) is a property that satisfies this requirement, but it hardly amounts to a rich representation as Figure 18 shows.

3.3. Smoothed Local Symmetries

In this section, we analyze two of the shortcomings of the SAT, discussed in the previous section, and introduce an alternative region-based representation that combines features of the SAT and two-dimensional generalized cones (ribbons). The first problem is illustrated in Figure 17. The maximal disk that touches the contour at *A* and *B* has its center far along the axis of the rectangular portion, at a point determined by the curvature of the pin head. Perceptually, the pin head and shaft are distinct sub-objects, hence the local symmetry between *A* and *B* is more naturally recorded within the pin head. We can overcome this problem very simply. We re-define local symmetry as shown in Figure 19, recording it at the midpoint of the line AB. This reformulation of local symmetry invites a parametric description that is intermediate between those proposed for the SAT and generalized cones. A possible parameterization is developed in the next subsection.

Second, the SAT performs poorly on rectilinear figures (Figure 14). Also, the maximal disk formulation only finds a subset of symmetries, for example the diagonals of a square, yet many objects, especially rectilinear figures, have multiple symmetries. A (side-side) join (Marr, 1977) of two smooth objects is reasonably effectively isolated by a branch point. One source of trou-

Object curvature (k_β)	Object angle (β)		
	Neg	Zero	Pos
Neg	Opening flare		Closing cup
Zero	Opening wedge	Worm	Closing wedge
Pos	Opening cup		Closing flare

The assignment of WIDTH LABELS based on the algebraic signs of object angle and object curvature.

(a)

Axis curvature change (k_α)	Axis curvature (k_α)		
	Pos	Zero	Neg
Pos	Spiral in left		Spiral out right
Zero	Circular left	Straight	Circular right
Neg	Spiral out left		Spiral in right

The assignment of AXIS LABELS based on the algebraic signs of axis curvature and axis curvature change.

(b)

Figure 16. (a) The primitive shapes suggested by Blum and Nagel; (b) the object curvature descriptor. (After Blum & Nagel [1978, Tables 1 and 3].)

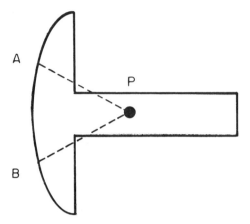

Figure 17. The center of the maximal disk that touches the contour at points A and B is at P, which is perceptually part of the straight line portion of the shape.

ble is that branch points do not always signal object joins, especially for rectilinear figures. Rather, they indicate a lack of smoothness of the contour. Second, the SAT is defined topologically as the locus of maximal disk centers. A pair of contour points, such as A and B in Figure 19, are local evidence for a symmetry. To accumulate such local evidence to produce meaningful axes, we observe that perceived symmetries are along *smooth curves*. The SAT fails to make explicit the important role that smoothness plays in forming axes.

We find all pairs of contour points A and B that satisfy the definition of local symmetry shown in Figure 19, and extract the *maximal smooth loci* of local symmetries as potential axes for the shape. We then delete any axis whose producing region, called the *cover* of the axis, is wholly subsumed by the cover of another axis. This is what is meant by *subsumption* in the statement of the first representational criterion given in the introduction. The surviving axes are called *smoothed local symmetries*. A smoothed local symmetry is a global symmetry if its cover is (a large proportion of) the entire object.

Figure 20 shows the smoothed local symmetries of a rectangle and a square. Notice that both the horizontal and vertical symmetries have been

Figure 18. Convexity is preserved under rotation, translation, and magnification; but it is not a rich representation. It cannot distinguish the shapes shown here.

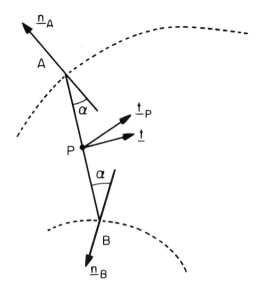

Figure 19. Local symmetry is recorded at the midpoint of *AB* in smoothed local symmetries, to avoid the problem illustrated in Figure 17.

found, irrespective of the aspect ratio of the rectangle, since they always cover the entire shape. Also, the diagonal "mini-axes" that appear in the SAT of a rectangle (Figure 15) have been suppressed by subsumption in all cases except the square, where they too have global cover. The length of the horizontal and vertical axes is equal to the corresponding sides of the rectangle, passing beyond the branch point of the SAT, because of the process of smooth extension.

Figure 21 shows three figures that appear to have the same smoothed local symmetries. The shapes are distinguished by their parametric descriptions (q.v.). Figure 22 shows the smoothed local symmetries of some shapes that have smooth contours. The smoothed local symmetry of Figure 10 is curved, in accordance with human perception. It is easy to show that if an object has an axis of symmetry then that axis is a smoothed local symmetry. This also holds if the object flexes (see Blum, 1973). Clearly, the representation accords more closely with human perception than generalized cones or the SAT.

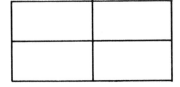

Figure 20. The smoothed local symmetries of a square and a rectangle.

Figure 21. Three shapes that apparently have the same smoothed local symmetries.

A pilot implementation of smoothed local symmetries has been constructed; further examples will be presented when the implementation is complete. The algorithm we have developed is naturally implemented by a parallel processor.

3.4. Parametric Description of Smoothed Local Symmetries

Figure 19 shows a local symmetry at P corresponding to points A and B on the contour. Suppose that the point A (for example) is $\mathbf{r}_A$ in some vector frame. Let the unit vector in the direction BA be $\mathbf{r}(s)$ and let the length of AB be $h(s)$, where s indicates distance along the smoothed local symmetry. By definition of the local symmetry, the angle α between $\mathbf{r}$ and the outward normal at A is equal to that between $\mathbf{r}$ and the inward normal at B. We measure angles counter-clockwise from $\mathbf{r}$. Then $\mathbf{t}_B$ is at angle $\frac{\pi}{2}-\alpha$ and $\mathbf{t}_A$ is at $\alpha-\frac{\pi}{2}$. It follows that

$$\mathbf{r}\cdot(\mathbf{t}_A-\mathbf{t}_B) = 0.$$

Hence $\mathbf{r}$ is perpendicular to the unit vector

$$\mathbf{t} = \frac{(\mathbf{t}_A-\mathbf{t}_B)}{2\cos\alpha}.$$

We assume that $|\alpha| < \pi/2$ for a local symmetry. In general, $\mathbf{t}$ is not the tangent $\mathbf{t}_P$ to the smoothed local symmetry at P since

$$\mathbf{t}_P = \frac{d}{ds}\mathbf{r}_P$$

$$\tag{1}$$

$$= \frac{1}{2}(\mathbf{t}_A\frac{ds_A}{ds}-\mathbf{t}_B\frac{ds_B}{ds}).$$

Suppose that $\mathbf{t}_P$ is at angle ϕ counter-clockwise from $\mathbf{t}$, that is, at angle $\phi-\frac{\pi}{2}$ from $\mathbf{r}$. Then, forming the dot product of Equation (1) with $\mathbf{t}$ gives

$$\cos\phi = (s_A'+s_B')\frac{\cos\alpha}{2}.$$

By forming the cross product of Equation 1 with t_A and t_B, we find

$$\frac{ds_A}{ds} = \frac{2\,|\,t_P \times t_B\,|}{|\,t_A \times t_B\,|}$$

(2a)

$$= 2\frac{\sin(\alpha+\phi)}{\sin 2\alpha}$$

and

$$\frac{ds_B}{ds} = \frac{2\,|\,t_P \times t_A\,|}{|\,t_B \times t_A\,|}$$

(2b)

$$= 2\frac{\sin(\alpha-\phi)}{\sin 2\alpha}.$$

(The case $\alpha = 0$ is discussed below). An incremental step ds along the smoothed local symmetry at P corresponds to a step ds_A along the contour at A and ds_B at B. Bookstein (1979) observes that ds_B has the opposite sign to ds_A. The ratio ds_A/ds_B can be arbitrarily large, as can be seen by drawing a local symmetry of two concentric circles, the ratio of whose radii differ by a large amount.

From $r_A - r_B = h(s)r$, we deduce

$$h'(s) = (t_A \cdot r)s_A' + (t_B \cdot r)s_B'$$

$$= \sin\alpha(s_A' + s_B')$$

(3)

$$= 2\tan\alpha\cos\phi.$$

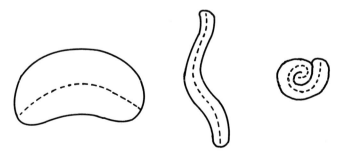

Figure 22. The smoothed local symmetries of some shapes that have smooth bounding contours.

For example, if $\phi = 0$, as it is for a global or flexed symmetry, then $h' = 2\tan\alpha$. Suppose that $\mathbf{t}_A$ is in direction ϕ_A with respect to some base direction and $\mathbf{t}_B$ is in direction ϕ_B. Since $\phi_A - \phi_B = 2\alpha - \pi$, we find

$$2\alpha' = \phi_A' - \phi_B'$$

(4)

$$= \kappa_A s_A' - \kappa_B s_B',$$

where, κ_A, respectively κ_B, is the curvature of the contour at A, respectively B. A useful form of Equations (3) and (4) is

$$2d\alpha = \kappa_A ds_A - \kappa_B ds_B$$

$$dh = \sin\alpha(ds_A + ds_B).$$

This form is used for interpolating local symmetries in an implementation. Solving (3) and (4) we get

$$\begin{bmatrix} s_A' \\ s_B' \end{bmatrix} = \frac{1}{(\kappa_B + \kappa_A)\sin\alpha} \begin{bmatrix} \kappa_B & \sin\alpha \\ \kappa_A & -\sin\alpha \end{bmatrix} \begin{bmatrix} h' \\ 2\alpha' \end{bmatrix}.$$

(5)

From (5) we can find ϕ_A' and ϕ_B'. Since

$$\kappa_P = \phi' + \frac{1}{2}(\phi_A' + \phi_B')$$

(6)

$$= \phi' + \frac{1}{2}(\kappa_A s_A' + \kappa_B s_B'),$$

knowing either of κ_P or ϕ' enables both to be computed. Finally, it is clear from (3) that h'' involves s_A'' and s_B'', which can be computed from (2) if α, ϕ, α', and ϕ' are known.

Consider, as a special case, a (non-flexed) global symmetry. The axis curvature κ_P is everywhere zero, as are ϕ and ϕ'. Also, $\kappa_B = -\kappa_A$. From Equation (6) we find that

$$\phi_A' + \phi_B' = 0,$$

so

$$\alpha' = \kappa_A s_A' = -\kappa_B s_B'.$$

It follows that

$$\alpha' = \frac{\kappa}{\cos\alpha}.$$

Also, extrema of h occur when α is zero. They do not coincide with extrema of κ_A in general but they provide natural places to segment an object into its components.

As a second special case, consider a flexed worm (Blum & Nagel, 1978), for which h is constant. It follows that $h' = h'' = 0$, and that $\alpha = \phi = \phi' = 0$. We find that

$$s_A' = \frac{\kappa_P}{\kappa_A}$$

$$s_B' = \frac{\kappa_P}{\kappa_B}$$

$$\rho_P = \frac{\rho_A + \rho_B}{2}$$

$$\rho_A = \rho_B + h.$$

Not surprisingly, the description of a worm amounts to the description of a plane curve that is the axis, and a statement that the cross-section function is constant. More generally, consider a flexed symmetry (Blum & Nagel, 1978). It is easy to show that a shape is a flexed symmetry if and only if $\phi = 0$. The shape description is neatly factored into (a) a description of the shape of the flexed symmetry axis, (b) a description of the shape orthogonal to the flexed symmetry. At first blush, this appears to be a reaffirmation of generalized cones. However, it is not at all clear that the shape of a non-flexed symmetry, and hence of the shape orthogonal to a flexed symmetry, is well described by the cross-section function h.

Finally, it is clear that ϕ is a measure of the *lack* of symmetry of a shape locally. This is because the angle between $\mathbf{t}_A$ and $\mathbf{t}_P$ is $\alpha - \phi$, while the angle between $\mathbf{t}_B$ and $\mathbf{t}_P$ is $\alpha + \phi$. The difference between these two angles is 2ϕ, and is zero for a flexed symmetry. Generalized cone parameterizations are not well suited to describing asymmetric shapes.

We are currently investigating which of the variables α, ϕ, α', ϕ', κ_A, κ_B, h', κ_P, h'', s_A', s_B', s_A'', and s_B'' should be chosen to parameterize the shape, and which, if any, should be derived? We are basing the investigation on recognition of occluded shapes since occluded portions tend to be local, and the significance of an occluded portion of a shape is related to its use in recognition.

In order to establish a local symmetry at A and B we need to compute $\mathbf{t}_A$, $\mathbf{t}_B$, α, and h. The curvatures κ_A and κ_B can also be computed in any smooth visible portion of a contour (Brady, 1982b). A parametric description involving curvatures can potentially be accessed and checked for consistency

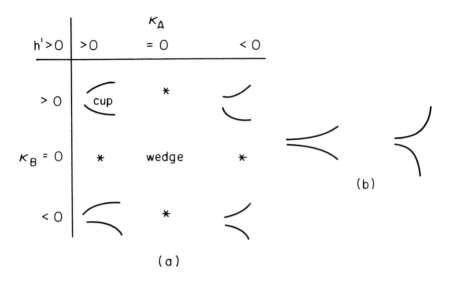

Figure 23. (a) The signs of parameters h', α', κ_A, and κ_B suffice to give basic shapes similar to those proposed by Blum and Nagel (1978). * indicates a distinguished basic shape with a straight edge. (b) Differential rate of flaring is measured by α'.

even for occluded shapes, hence we include the curvatures in the parameterization. We can also compute α' and h' fairly accurately along smoothed local symmetries. Of course, to make the shape description scale invariant, it needs to be scaled by some intrinsic measure such as the average value of h.

Interestingly, just the *signs* of h', κ_A, κ_B, and α' allow us to define more than the basic shapes proposed by Blum and Nagel (1978); see Figure 23a. The rate of flaring shown in Figure 23b can be estimated from α'. It can also be estimated from h'', but we prefer first order parameters (apart from κ_A and κ_B) for reasons of numerical accuracy.

Figure 23a includes a "beak" shape that was not proposed by Blum and Nagel (1978), who describe such a shape as a wedge, cup, or flare with a curved axis. In many cases, it is hard to judge which of their shapes a given beak corresponds to.

The parameters h', α', κ_A, and κ_B enable s_A', s_B', ϕ_A, and ϕ_B to be computed according to equation (5). In general ϕ, ϕ', and κ_P are difficult to compute accurately, though their signs can be computed reliably. Which of them to compute, and how, is currently under investigation.

Our earlier discussion of the symmetric axis transform criticized the definition of branch points. It is easy to show that a perceptual join of two objects corresponds to a discontinuity, or high curvature value of, κ_P. We are currently investigating object joins.

3.5. Propagating Local Frames

We now turn to the second part of our approach to shape description, the iterative refinement of local frames: partly-constrained, parametric descriptions of basic shapes. We consider a shape to be composed of basic shapes joined together in ways like those discussed by Marr (1977). Each basic shape has an associated local frame in which some information is left unconstrained. The local frame is iteratively refined as far as possible to be consistent with, and hence extend, adjacent local frames. We illustrate the idea while describing basic shapes that are distinguished by having a straight or circular contour.

Consider Figure 24a. The curved portion of the contour generates a global smoothed local symmetry. The straight edge PQ also contributes a parametric description of the shape. A straight edge opposite a curve fixes the direction of one vector in a coordinate frame for describing the shape, but leaves the choice of origin, the offset of the vector from the straight edge, and the other vector of the frame unconstrained. There is a set of *default choices* for the shape considered in isolation: the default offset is zero, the second vector is perpendicular to the one given, and the default origin belongs to the set consisting of P, Q, and the midpoint of PQ. The smoothed local symmetry in Figure 24a is consistent with this set of defaults and suggests that the midpoint of PQ should be the frame origin. Consider the shape in Figure 24b. It consists of two components that are globally symmetric shapes of the sort shown in Figure 24a.

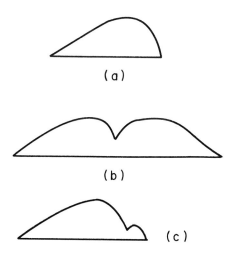

(a)

(b)

(c)

Figure 24. (a) A straight line opposite a curve; (b) local frame propagation; (c) local frame propagation between two shapes whose covers have very different areas.

The straight edge direction is consistent between the shapes, and becomes a vector for a global frame describing the shape as a whole. An offset of zero and a perpendicular vector can be chosen by default. How should the origin be chosen? The intersection of the default sets is the point marked O. It establishes an origin for the shape as a whole and allows points in the two subshapes to be inter-related. The relative sizes of the covers of two subcomponents is important. Figure 24c shows a variation of the shape shown in Figure 24b. The preferred description is that of the larger subcomponent with the smaller component as an adjunct.

A pair of parallel straight lines fixes the vector direction as above, and has the same default second vector. The default offset is that of the line of symmetry midway between the lines. Figure 25 shows two joined subcomponents with the default frame vectors shown. The collinearity of two of the defaults is a special case of smooth extension that underlies the idea of smoothed local symmetry. Choosing the collinear defaults establishes a global frame that is also a global symmetry. Variations of the sort illustrated by Figure 24c are possible and are left for the reader.

Figure 26 illustrates a case in which defaults are parallel but not collinear. The straight edge common to both subcomponents suggests a global frame vector parallel to it. If PQ is longer than QR, the default of the smaller shape is over-ridden and made collinear with the default of the larger sub-component. The extended line is of the larger sub-component. The extended line is treated as a global symmetry, hence the description: a symmetric shape with an adjunct smaller shape.

Two lines that are not parallel form a ''wedge'' shape in Blum and Nagel's (1978) terminology. The frame vector direction is constrained in direction to be between those of the two lines, with the bisector as default. A circular contour is dual to a straight line, in the sense that it fixes the origin, at the

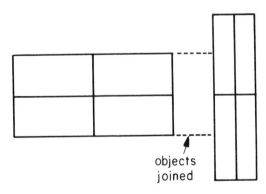

objects
joined

Figure 25. Local frame propagation when a pair of defaults are smoothly joined.

Figure 26. Local frame propagation when a pair of defaults could smoothly extend but for a difference in offset.

center, but leaves the direction of a unit frame vector undetermined. The default is the bisector of the extremal radii on the circular portion of the contour.

Palmer (this volume) has suggested that a "reference frame" is somehow computed for two-dimensional shapes. He suggests that a number of perceptual demonstrations can be accounted for in terms of the choice of different *a priori* competing reference frames. For example, a square is seen as a diamond if it is presented with its axis vertical. The smoothed local symmetries of a square are shown in Figure 20. There are four (global) smoothed local symmetries. It is known that the default image coordinate frame has its axes horizontal and vertical. Presenting the square so that its sides are horizontal and vertical forces the choice of those smoothed local symmetries, with the further default preference for the vertical. If the square is presented so that a diagonal is vertical, then the diagonal smoothed local symmetries are selected. Palmer's Figure 6 presents psychophysical evidence for the computation of axes of a rectangle. The axes of symmetry correspond to high numbers. Note that unlike the SAT both the horizontal and vertical global symmetries show perceptual advantage. Slight perceptual advantage also accrues to points on the SAT mini-axes, that are subsumed in the smoothed local symmetries. We conjecture that the *relative* perceptual advantage of points on the mini-axes would decrease as a rectangle's aspect ratio (ratio of longest to shortest side) increases. It should attain a maximum for squares. There should be no difference for the SAT representation. We hope to perform this experiment.

Brady (1982b) observes that the smooth propagation of local frames provides a suitable context in which sub-objects and nicks can be isolated. We have completed a pilot implementation of smoothed local symmetries and are currently implementing frame propagation for shapes like those discussed in this section. An early application of the shape representation has been to choosing grasp points on an object for a parallel jaw gripper (Brady, 1982b). We use shape information to narrow the search space for an optimum grasp point, even though we risk missing a global minimum. A parallel jaw gripper has a limited repertoire, and can essentially only use symmetry. To date, we have concentrated on grasp positions normal to, or lying along, global symmetries. Figure

27 illustrates this. We are also describing hand printed characters and the design of printed character sets.

So far we have concentrated on single-level descriptions of shape, but we intend to develop eventually hierarchical descriptions like those advocated by Hollerbach (1975). Descriptors that occur at the lower (that is to say, less important) levels of the hierarchy are typically signalled by relatively rapid changes in the curvature of a contour that are smoothed out at higher levels by appropriate Gaussian smoothing (see Brady, 1982b). For example, the shape of a cog is a disk at the highest level. In order to represent the teeth a a higher level it is necessary to recognize the periodicity (texture) of the curvature.

4. Three-dimensional Surfaces: Curvature Patches

4.1. Computing Visible Surfaces

Recent work in image understanding (Ballard & Brown, 1982; Brady, 1982a; Marr, 1982; Nevatia, 1982) has centered on the development of modules that compute three-dimensional depth, or depth gradients. Such modules include: shape from stereo (Baker & Binford, 1981; Grimson, 1981; Marr & Poggio, 1979); shape from shading (Ikeuchi & Horn, 1981) and photometric stereo (Horn, Woodham, & Silver, 1978; Ikeuchi, 1981; Porter & Mundy, 1982); shape from contour (Witkin, 1981; Binford, 1981; Brady & Yuille, in preparation); and shape from motion (Bruss & Horn, 1981; Ullman, 1979; Webb & Aggarwal, 1982). Other work has concentrated on shape from texture (Vilnrotter, Nevatia, & Price, 1981). In applying vision to robotics, direct range finding and structured light have been investigated as techniques for recovering depth (Agin, 1980; Holland, Rossol, & Ward, 1979; Bolles, 1981). Although the work referred to in this paragraph is currently largely experimental, it is clear that robust, efficient, practical three-dimensional vision systems will soon be available.

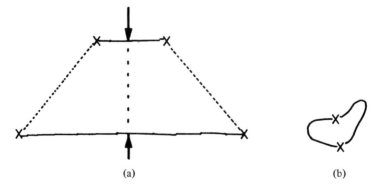

(a)   (b)

Figure 27. Grasp points chosen for a parallel jaw gripper: (a) by minimizing a potential function of the shape; (b) by exploiting the global symmetry of the shape.

Several authors have suggested that the output of the "shape from" processes listed above is a representation variously called the 2-1/2 D sketch (Marr, 1982), the needle map and Gaussian image (Horn, 1982; Ikeuchi, 1981), and intrinsic images (Barrow & Tenenbaum, 1978). It is supposed that this (or these) representation(s) make explicit information such as the local surface normal. There are several unresolved issues, including: the parameterization of the local surface normal; whether depth is made explicit or computed by integration; whether second order quantities, such as the Hessian, the principal curvatures (or some combination of them), or the second fundamental form of the surface, are made explicit or computed by differentiation; how accurately information is recorded in the representation(s) and how susceptible it is to noise; and how many separate representations are maintained. The last of these points bears on the issue of multiple scales of representation of surfaces, discussed in (Terzopoulos, 1982).

Several visual processes, notably stereo and structure from motion, only specify depth and orientation at a discrete subset of the points in an image. The points at which they are specified are typically those where the irradiance changes abruptly, and they usually correspond to edge points. Human perception, however, is of complete, piecewise smooth, surfaces, and such complete surface information is important for most applications of vision. Since, mathematically, the class of surfaces which could pass through the known boundary points provided by stereo, for example, is infinite and contains widely varying surfaces, the visual system must incorporate some additional constraints in order to compute the perceived complete surface.

Using the image irradiance equation formulated by Horn (1977), Grimson (1982) has derived a *surface consistency constraint,* informally known as "no news is good news." The constraint implies that the surface must agree with the information from stereo or motion correspondence, and *not* vary radically between these points. An explicit form of the surface consistency constraint has been derived, by relating the probability of a zero-crossing in a region of an image to the variation in the local surface orientation of the surface, provided that the surface albedo and the illumination are roughly constant.

A second idea is that, in the absence of contrary evidence, the visual system constructs the *most conservative* curve or surface consistent with the given sparse data. This is made precise using the calculus of variations. A crucial aspect of the variational formulation is the choice of performance index to minimize. Grimson (1981) argues that the performance index should be a seminorm, and suggests the quadratic variation $f_{xx}^2 + 2f_{xy}^2 + f_{yy}^2$ of the surface $f(x,y) = z$. Brady and Horn (1981) have noted that any quadratic form in f_{xx}, f_{xy}, and f_{yy} is a seminorm, hence is a plausible performance index. They have shown that the quadratic forms that are, in addition, rotationally invariant form a vector space, which has the square Laplacian and the quadratic variation as a basis. Since the quadratic variation has the smaller null space, it offers the tighter constraint, and is to be preferred.

Brady and Grimson (1981) have suggested that the theory of surface perception developed by Grimson, Brady, Horn, and Terzopoulos underlies human perception of subjective contours. Although subjective contours are ordinarily considered in the context of two-dimensional figure perception, evidence is presented in support of the theory, first advanced by Coren (1972), that the perception is essentially of subjective surfaces. Subjective contours form the boundaries of subjective surfaces that are filled in from a variety of local cues. Surfaces are not only suggested by interposition cues; rather, any cue for surfaces leads to subjective surfaces using that cue. Examples are presented that involve motion, texture, stereo, and even shading. Subjective surfaces can be doubly curved, for example a saddle surface and a subjective version of the sail illusion. Subjective surfaces may contain reflectance patches. Grimson's requirement that surface discontinuities must be made explicit can be used to generate subjective versions of the Mach illusion. Brady and Yuille (in preparation) have developed a variational principle for determining perceived three-dimensional shape from a two-dimensional contour. It is proposed as the basis of the perception of subjective surfaces.

Brady and Horn (1981) suggest that surface interpolation can be posed in terms of a physical model, namely as the variational problem describing the constrained equilibrium state of a thin flexible plate. The variational problem and the physical model have been developed by Terzopoulos (1982). After formulating surface interpolation as an energy minimizing problem over an appropriate Sobolev space, the problem is discretized and approached via the finite element method. In essence, the variational problem is transformed into a large set of linear algebraic equations whose solution is computable by local-support, cooperative, parallel processors.

It has been suggested that visual processes such as edge detection and stereo provide information at a number of distinct scales, spanning a range of resolutions. To exploit the information available at each level of resolution, Terzopoulos (1982) formulates a hierarchy of discrete problems, and a highly efficient multi-level algorithm from the work of Brandt. The algorithm involves both intra-level relaxation processes and bi-directional, inter-level, local, interpolation processes. Intra-level relaxation smooths out high frequency variations, while inter-level interpolation tends to damp out low frequency variations, greatly speeding the overall process. The resulting process is extremely efficient, even on a serial computer, though it is better suited to an array of parallel processors.

Of the names proposed for the representation of visible surfaces, "intrinsic image" is particularly apt, since the representation, completed by interpolation, has a lack of structure similar to an image. Just as applications of two-dimensional vision depend upon the development of rich descriptions, so will applications of three-dimensional vision. In this paper we propose a representation for visible surfaces called *curvature patches*. We show that it satisfies the criteria for a representation listed in the introduction. In particular, we show how curvature patches are defined locally on a surface, how they are smoothly extended, how global coordinate frames can arise, and how surface joins can be described. Curvature patches draw upon work in image understanding, human visual perception, and computer-aided design.

Figure 28. Interpreting line drawings as surfaces. (After Stevens [1981], Figure 10.)

4.2. Curvature Patches

According to the first of our criteria, we need to choose a suitable representation of a *local piece of surface*. Several definitions are possible, including formulations based on the familiar notion of neighborhood in a metric or topological space. The local representation we propose blends ideas from three sources: observations about human perception of surface curves, the definition of parameterized surface patch developed in computer-aided design, and the work on surface interpolation discussed in the previous section.

Stevens (1981) has discussed the perception of drawings, such as that shown in Figure 28, as three-dimensional surfaces. There are many ways of describing surface shape, one of the weakest being the sign of the Gaussian curvature, a descriptor discussed by Stevens. One can, however, clearly distinguish the surface fragments shown in Figure 29, all of which have positive Gaussian curvature. Second, Stevens (1981, p. 64) argues that surface contours are often interpreted as being both planar and geodesic. It is well known (see for example Weatherburn, 1927, p. 103) that a curve that is both planar and a geodesic is necessarily a line of curvature.

In general, it appears that surface contours are most often interpreted as lines of curvature, whether or not they are planar. It is also clear that we can make quantitative estimates of surface curvature, even in impoverished line drawings like those in Figures 28 and 29. As the discussion in the previous section makes clear, however, richer information can be made available in practice or can be computed. In particular, we expect to be able to compute the first and second fundamental forms of a surface and its principal lines of curvature.

It requires two parameters, say u and v, to define a surface $\mathbf{r}(u,v)$. Computer-aided design (CAD) has developed the idea of a surface patch formed by quantizing the parameters u and v. Typically the values of $\mathbf{r}(u_i,v_j)$, $\mathbf{r}_u(u_i,v_j)$, $\mathbf{r}_v(u_i,v_j)$, $\mathbf{r}_{uv}(u_i,v_j)$, and, sometimes $\mathbf{r}_{uu}(u_i,v_j)$ and $\mathbf{r}_{vv}(u_i,v_j)$, are given at the patch corners (u_i,v_j). The surface is then interpolated by a suitable set of blending functions, such as the cubic blending functions of Ferguson, Coons, Forrest, and Bezier (see Forrest, 1972; Faux & Pratt, 1979). Other work uses splines in tension and bicubic splines.

The representations of surfaces proposed for CAD are a compromise between numerical convenience and the desire to achieve perceptual "fairing" of a surface. Although they have been of immense practical value in the design of ships, airplanes, and automobiles, current CAD techniques are not completely satisfactory. Surfaces often have unaesthetic highlights or portions that appear flattened.

The definition of a *curvature patch* has two parts. First, the surface is parameterized so that the webbings $\delta u = 0$ and $\delta v = 0$ correspond to the lines of curvature of the surface. Second, instead of interpolating the surface from

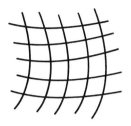 

Figure 29. Distinctive surface fragments, all of which have positive Gaussian curvature.

the values of $\mathbf{r}(u_i, v_j)$ and its first and second directional derivatives at the patch corners using a set of blending functions, we interpolate it using the theory developed by Grimson, Brady, Horn, and Terzopoulos, and discussed in the previous section.

Generally, no restriction is placed on the choice of parameterization u and v of a surface in CAD. Typically, parameterizations are chosen for analytical convenience, or for convenience in design, as a network of points is defined in cross sections or by "lofting." We saw above, however, that the lines in drawings such as Figures 28 and 29 are not interpreted arbitrarily, but as lines of curvature. Choosing the surface parameterization so that the webbings are the lines of curvature has some convenient consequences. First, since the lines of curvature are mutually perpendicular at a point, $\mathbf{r}_u$, $\mathbf{r}_v$, and the surface normal $\mathbf{n}$, which is their normalized cross product, form a local orthogonal coordinate frame. It is easy to show (Weatherburn, 1927, p. 72) that parametric curves are lines of curvature if and only if the first and second fundamental forms of the surface are diagonal. This is if and only if

$$\mathbf{r}_u^T \mathbf{r}_v = \mathbf{n}^T \mathbf{r}_{uv} = 0.$$

It follows that the principal curvatures are especially easy to compute, being

$$\frac{\mathbf{n}^T \mathbf{r}_{uu}}{\mathbf{r}_u^2} \quad \text{and} \quad \frac{\mathbf{n}^T \mathbf{r}_{vv}}{\mathbf{r}_v^2}.$$

Consider tensor-product surface patches (Faux & Pratt, 1979, p. 204), for example Ferguson patches. The surface is defined by a quadratic form that involves the matrix

$$\begin{bmatrix} \mathbf{r}(0,0) & \mathbf{r}(0,1) & \mathbf{r}_v(0,0) & \mathbf{r}_v(0,1) \\ \mathbf{r}(1,0) & \mathbf{r}(1,1) & \mathbf{r}_v(1,0) & \mathbf{r}_v(1,1) \\ \mathbf{r}_u(0,0) & \mathbf{r}_u(0,1) & \mathbf{r}_{uv}(0,0) & \mathbf{r}_{uv}(0,1) \\ \mathbf{r}_u(1,0) & \mathbf{r}_u(1,1) & \mathbf{r}_{uv}(1,0) & \mathbf{r}_{uv}(1,1) \end{bmatrix}$$

A similar matrix arises if the blending functions are chosen to achieve specified cross-boundary second derivatives, in which case the matrix is 6×6. Ferguson's original formulation assumed that the cross derivative r_{uv} was zero. Faux and Pratt (1979, p. 205) observe that "a notable application of this restricted type of patch is in the APT surface-fitting routine FMILL." However, Forrest (1972, Appendix 2) claims that assuming the cross derivative is zero at all webbing intersection points (u_i, v_j) produces the appearance of local flattening of the interpolated surface near patch corners. He notes that the effect is minimized when the webbings *are coincident with lines of curvature*. Since $n^T r_{uv} = 0$ does not imply $r_{uv} = 0$, Ferguson's assumptions are sufficient, though not necessary for a curvature patch webbing.

We remarked earlier that the representations of surfaces proposed for CAD are a compromise between numerical convenience and the desire to achieve perceptual "fairing" of a surface. In fact, this remark essentially applies to the choice of blending functions that define the interpolation of the surface from the information at the patch corners. Since the thin plate theory of surface interpolation discussed in the previous section was developed to model human perception of surfaces (Brady & Grimson, 1981; Grimson, 1982; Terzopoulos, 1982), it is suggested that curvature patch surfaces are perceptually "fairer" than bicubic splines and the other surfaces developed in CAD. This claim is currently being tested. Even assuming that the claim is correct, it may be objected that the perceptual advantage has been bought at the cost of considerable computation. For von Neumann computers this is probably so. However, the algorithms implemented by Grimson (1981) and Terzopoulos (1982) are inherently well suited to being computed efficiently by arrays of microprocessors.

The idea of interpolating a surface by a minimization process has been hinted at, though not to my knowledge stated explicitly, in the CAD literature. Mehlum (1974, p. 175) observes that the nonlinear splines he develops by a variational technique could "serve as patch boundaries." Nielson (1974) observes that splines under tension can be characterized as functions that minimize an appropriate norm (compare Terzopoulos' [1982] use of energy norms), and he proposes an alternative called v-splines. Finally, Pilcher (1974) fits an elastic skin to the given points. The membrane typically only guarantees continuity, however, not smoothness.

The curvature patch representation is rich, or information preserving, in the following sense. Bonnet's "fundamental theorem of surfaces" (see for example Millman & Parker, 1977, p. 151) guarantees that six continuous functions (corresponding to the entries of the first and second fundamental forms) that satisfy the equations of Gauss-Bonnet and Mainardi-Codazzi uniquely determine the surface.

Example 1: A surface of revolution. Consider a surface of revolution, whose axis is aligned with **k**. The surface is formed by rotating a (necessarily one-parameter) plane curve $p(u)\mathbf{i}+z(u)\mathbf{k}$ about **k**. Let θ denote the angle through which the plane has been rotated. Then the surface is

$$\mathbf{r}(u,\theta) = p(u)\cos\theta\mathbf{i}+p(u)\sin\theta\mathbf{j}+z(u)\mathbf{k}.$$

For example, $p(u) = a$, $z(u) = u$ for a right cylinder. It is easy to show (see Millman & Parker, 1977, p. 86, for example) that the lines of curvature of the surface of revolution are the meridians, corresponding to $\delta\theta = 0$, and the parallels, corresponding to $\delta z = 0$. The curvature along the webbing $\delta z = 0$ is $\mathbf{n}^T\mathbf{r}^*/p$, where $\mathbf{r}^* = [\cos\theta \ \sin\theta \ 0]^T$. The foreshortening of the expected curvature $1/p(u)$ exemplifies Meusnier's Theorem (Weatherburn, 1927), since the vector $\mathbf{r}^*$ drawn perpendicular to the **k** axis to the curve is not, in general, a normal section.

Example 2: A generalized cylinder. A generalized cylinder (Marr, 1977) is a generalized cone whose axis is straight. Since the tangent to the spine has constant direction, its use as a vector (say the z axis) in a coordinate frame for the cylinder is suggested (see Section 3). Assume for simplicity that the cross-section function is convex. It is most convenient to use polar coordinates, say $r(\theta)$, θ. If the sweeping rule is $h(z)$ and the eccentricity of the generalized cylinder is ninety degrees, the surface of the generalized cylinder is the set of points

$$\{[r(\theta),\theta,z]:\theta,z\}.$$

We choose θ and z as the parameterization of the cylinder. Observe that $\partial r/\partial\theta$ and $\partial r/\partial z$ are lines of curvature if and only if $h'(z) = 0$ or $r'(\theta) = 0$. That is, the partials r_θ and r_z are lines of curvature if and only if, using the terminology of Marr (1977), the cross section is a skeleton and the contour generator is a fluting. It follows that the curvature patch representation specializes to the skeleton and fluting representation advocated by Marr in the case of generalized cylinders.

4.3. Smooth Extensions of Curvature Patches

In this section, we assume that we have constructed a curvature patch representation of the surface $r(u,v)$ and have quantized the parameters (u_i,v_j) to form the patches. The aim is to isolate the maximal connected sets of curvature patches that admit a rich yet concise description, analogous to the process of finding smoothed local symmetries, described in the previous section.

Differential geometry has developed a hierarchy of descriptors that includes doubly and singly curved surfaces, surfaces of revolution, and even relatively specific types such as cylindrical surfaces. Determining the set of surface types that occur in a practically useful hierarchy is currently an open problem. Probably the weakest statement that can be made is to categorize the Gaussian curvature of a surface patch. The usefulness of this descriptor stems from the "Theorema Egregium," Gauss' modestly named "truly remarkable" theorem, which states that the Gaussian curvature is an intrinsic property of a surface. For complex surfaces such as airfoils, such a weak descriptor may well be the best one can do. The difficulty of machining airfoils attests to their complexity. Many machined surfaces are constructed from parts that are significantly simpler than airfoils, however. For example, many machined parts are surfaces of revolution, which have a relatively simple description, as illustrated in the example above.

Consider the bottle outline shown in Figure 30 (compare the discussion of Greek amphorae, all implicitly surfaces of revolution, in Hollerbach, 1975). We should like to segment the surface into three components corresponding to the cylindrical neck and body, and the shoulder. Suppose that we have discovered a webbing on the surface as a result of stereo processing or using structured light in a system such as the GE IBIS system (Porter & Mundy, 1982). Denote a typical point of the webbing by $\mathbf{r}(u_i, \theta_j)$, for some range of i and j. It is clearly possible to devise a local process that can investigate *local trends* in the principal curvatures by looking at adjacent patch corners. For example, by looking at the curvature along the meridians (although they are not yet known to be meridians, of course) at points (u_{j-1}, θ_0), (u_j, θ_0) and (u_{j+1}, θ_0), local evidence is found for a ruling on the neck and body of the bottle, for all θ_0. By a similar process, the constant curvature can be discovered in each horizontal slice of the bottle, corresponding to a parallel of the surface of revolution. This provides local evidence for a surface of revolution. Significantly, it

Figure 30. Outline shape of a typical surface of revolution.

also provides local evidence for the axis of revolution, given its curvature, suitably foreshortened by Meusnier's theorem. Indeed, the fact that the curvature of the parallel is constant from slice to slice in the neck, then changes smoothly along the shoulder, before becoming constant again in the body, enables the bottle to be segmented into its functional parts.

The processing in the previous paragraph is hypothetical, since it has not yet been implemented (although Hollerbach [1975] has implemented it in two dimensions). We are currently designing an implementation in collaboration with Porter and Mundy of GE CRC, Schenectady using the IBIS structured light system they have developed. Of course, we shall have to deal with noise in the measurements, exactly like one does with an image (it is an *intrinsic* image, as we noted above). It is intended to use the system for robot parts identification and welding tasks. Progress reports will be issued in due course.

4.4. Surface Intersections

Consider Figure 31. It depicts one cylinder drilled or welded to another. How shall the join be described? The first observation is that it is generally very difficult to solve for the intersection curve in closed form. If a parametric surface Σ_1 defined as $r_1(u_1,v_1)$ intersects another Σ_2, defined by $r_2(u_2,v_2)$, then the curve of intersection Γ is given implicitly by the equation

$$r_1(u_1,v_1)-r_2(u_2,v_2) = 0.$$

This gives three equations in four unknowns, the parameters. In general, the solution is computed numerically, typically by moving points $r_1(u_1,v_1)$ and $r_2(u_2,v_2)$ on the surfaces so as to minimize the distance between them. In fact we often do not need to determine the curve Γ precisely. Objects often appear essentially similar from a range of view angles. We are really more concerned with finding significant, reliable features that enable us to identify the object and to determine its orientation relatively precisely.

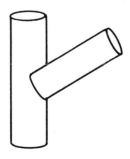

Figure 31. A cylinder drilled into, or welded to, another.

There is an alternative tradition of work that seems relevant, namely the interpretation of line drawings in artificial intelligence. Indeed, Barrow and Tenenbaum (1981) and, particularly, Binford (1981) have suggested extending ideas about line drawing analysis to curved surfaces. The intersection curve Γ might be described as convex, concave, or occluding. Such descriptions only have local validity for curved surfaces, unlike the planar-faced blocks world (Brady, 1982a; Nevatia, 1982), where the convexity or concavity of an edge cannot change along its length. For the same reason, it is not obvious how to develop transform spaces, such as dual space, that have been important for understanding blocks world line drawings. In the case being considered here, there is a richer description available, since we can make an assertion about the surface type flanking the intersection curve. For example, we can determine that the join in Figure 31 is between two cylindrical surfaces. The extra richness yields tighter constraints for identification of the part being viewed, and the viewpoint.

For the surface intersection shown in Figure 31, such a description might suffice to identify the part, but it would hardly constrain its orientation. We propose a richer description of the intersection curve Γ that includes the *angle of intersection* between the surfaces at points along Γ. Suppose that $\mathbf{n}_i$ is the normal to Σ_i "near" a point on Γ where the tangent is $\mathbf{t}$. It is easy to show that the angle of intersection is the triple scalar product $[\mathbf{n}_1, \mathbf{n}_2, \mathbf{t}]$, and that this is

$$\sqrt{1 - (\mathbf{n}_1^T \mathbf{n}_2)^2}.$$

The snag with this formulation is that it assumes that there is a single global coordinate frame in which the vectors $\mathbf{n}_i$ are both represented. In fact, a parametric surface $\mathbf{r}(u, v)$ also sets up a coordinate frame $[\mathbf{i} \ \mathbf{j} \ \mathbf{k}]^T$. Suppose that the coordinate frame for surface Σ_1 is $[\mathbf{i}^* \ \mathbf{j}^* \ \mathbf{k}^*]^T$. In order to compute the angle of intersection, we first need to convert $\mathbf{n}_2$ from $[\mathbf{i} \ \mathbf{j} \ \mathbf{k}]^T$ to $[\mathbf{i}^* \ \mathbf{j}^* \ \mathbf{k}^*]^T$. In general, six degrees of freedom are required to specify the coordinate transform. If the coordinate transformation can be specified by a translation $\mathbf{q}^*$ and then a rotation $\mathbf{R}$, we have

$$\mathbf{r}^*(u, v) = \mathbf{q}^* + \mathbf{R}\mathbf{r}(u, v).$$

It is well known that the transformed normal $\mathbf{n}^*$ is equal to $\mathbf{R}\mathbf{n}_2$.

The computation of the angle of intersection along Γ can be made easier by a suitable choice of coordinate frame. We can discover a suitable choice by a process analogous to the local frame propagation technique illustrated earlier for smoothed local symmetries. A sphere, for example, naturally places the origin of the coordinate frame at its center, but leaves the base angular directions unspecified. Thus 3 degrees of freedom are constrained, 3 are unconstrained. A cylinder fixes the choice of the $\mathbf{k}$ direction, but leaves the position of the origin along the axis and the orientation of the $\mathbf{i}$ direction unspecified.

Suppose that a cylinder intersects a sphere (Figure 32). The idea of local frame propagation is to exploit the unconstrained degrees of freedom of neighboring frames to simplify the interaction between the corresponding surfaces. In particular, we hope to simplify the computation of the angle of intersection, and even determine Γ. Let the frame of the sphere be $[\mathbf{i}^* \ \mathbf{j}^* \ \mathbf{k}^*]^T$ and that of the cylinder be $[\mathbf{i} \ \mathbf{j} \ \mathbf{k}]^T$. Since the choice of $\mathbf{k}^*$ is unspecified, we are free to align $\mathbf{k}^*$ with $\mathbf{k}$. Similarly, we can align $\mathbf{i}^*$ with $\mathbf{i}$. These choices amount to choosing the rotation matrix $\mathbf{R}$ to be the identity. It is then easy to solve for Γ and the angle of intersection along it. In particular, one can show that the angle of intersection either has a unique maximum or is constant. If it is constant, Joachimsthal's theorem (Weatherburn, 1927, p. 68) shows that the intersection is along a line of curvature of *both* the sphere and the cylinder. Finding the maximum angle of intersection suffices to orient the part precisely.

4.5. Applications of Curvature Patches

We close this section with two applications of the curvature patch representation. First, we consider the extension of smoothed local symmetries to three dimensions. Second, we consider the determination of grasp points.

Representing volumes: generalizing smoothed local symmetries to three dimensions. In this subsection we propose an extension of the two-dimensional smoothed local symmetry representation introduced in Section 3, giving a representation of three-dimensional volumes.

Two-dimensional smoothed local symmetries depend upon a definition of local symmetry and are formed by the processes of smooth extension and subsumption. The axis of symmetry of a symmetric planar figure, whether it is straight or curved, is also a smoothed local symmetry. A mathematical mirror symmetry of a volume is a planar surface; hence we expect that the smoothed local symmetries of three-dimensional volumes will, in general, be curved surfaces.

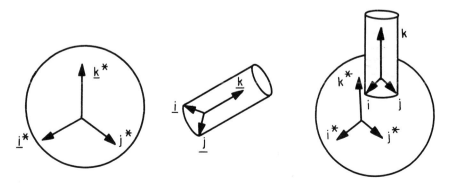

Figure 32. Choosing coordinate frames at an intersection.

The axis (spine) of a generalized cone is a space curve, not a surface. Figure 33 seems to have a flexed symmetry surface, the generalized cone description seeming less natural. It is easy to produce three-dimensional surfaces from planar sections such as that shown in Figure 10; as before, the defining parameters of the generalized cone representation do not appear to be computed by the human visual system.

Nackman (1982) has developed a three-dimensional version of the symmetric axis transform (SAT). The result is a surface that is the locus of all maximal spheres. He proposes a local parametric description of a symmetry based on the Christoffel symbols of the surface. Much of Nackman's thesis is devoted to partitioning objects into what he calls "slope districts." He suggests an extension to three dimensions of Bookstein's (1979) algorithm for computing an approximation to the SAT. The objections raised earlier about the two-dimensional SAT apply also to the three-dimensional version.

The definition of local symmetry underlying smoothed local symmetries applies to an arbitrary planar section of an object. Therefore, we can define the three-dimensional smoothed local symmetries of an object by smooth extension and subsumption of the local symmetries found in an appropriate set of planar sections. What is an appropriate set of planar sections?

Consider a surface that has a mirror symmetry. Without loss of generality we can choose the x-y plane so that it contains the mirror symmetry. Suppose that the surface is $f(x,y)$ for $z \geq 0$ and $-f(x,y)$ for $z \leq 0$. Denote the point $[x,y,f(x,y)]$ by P, and the point $[x,y,-f(x,y)]$ by Q. It is easy to show that if $[l,m,n]$ are the direction cosines of a principal line of curvature at P, then $[l,m,-n]$ are the direction cosines of a principal line of curvature at Q. It is also easy to show that PQ, $[l,m,n]$, and $[l,m,-n]$ are coplanar.

We arrive at our definition of three-dimensional local symmetry: define a *principal slice* of a surface at a point P to be a planar slice that contains a principal line of curvature at P. The principal slices at a point form a one-parameter family, only one of which is a normal section. We say that the point P has a *local symmetry* with a point Q if (i) P and Q are contained in a plane

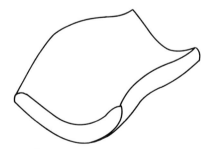

Figure 33. A three-dimensional surface that seems to have a flexed mirror symmetry.

Π that is a principal slice at *both* P and Q, (ii) P and Q form a two-dimensional local symmetry in Π. It follows, in particular, that mirror symmetries, curved or planar, are smoothed local symmetries.

For example, for a surface of revolution, the parallels are lines of curvature and are the boundaries of circular cross-sections that are principal slices at each point of the parallel. The smoothed local symmetry of a circle is its center, so the three-dimensional smoothed local symmetry of a surface of revolution is the (straight) axis of revolution.

Grasping three-dimensional objects. As an important application other than recognition of (two-dimensional) smoothed local symmetries, we have considered (Brady, 1982b) the problem of grasping a two-dimensional lamina with a parallel jaw gripper, without using highly specific information that might be obtained from a part model. One reason for this restriction is that flexibility demands that we be able to grasp sensibly objects we have never seen before. A second reason is that we may not be able to grasp a recognized occluded part in the manner dictated by its model. We seek techniques that rely upon general information.

Hanafusa and Asada (1977a,b) have developed a definition of stability of prehension for a three-fingered, spring-loaded hand. A potential function is defined so that stable grasp configurations occur at potential minima. Asada's approach to grasping is essentially local: he defined a hill-climbing technique to find a minimum of a potential function. His system, and a similar approach developed below, knows essentially nothing about the shape of an object. (Brady, 1982b) uses shape information to narrow the search space for an optimum grasp point, even though there is a risk that a global potential minimum will be missed. A parallel jaw gripper has a limited repertoire of grasps. To date, we have concentrated on grasp positions normal to, or lying along, global symmetries.

Recently, a number of multi-fingered robot hands have been developed (Hanafusa & Asada, 1977a,b; Salisbury & Craig, 1982). Both hands have three fingers, a thumb in opposition to the other two. Salisbury and Craig (1982) have shown how to compute the forces exerted at the fingertips to achieve a given stiffness of grasp and a given center of compliance. The choice of grasp point is assumed given.

Schlesinger (1919) observed people grasping objects and suggested that there are six basic ways of grasping (modes of prehension). Each mode of prehension makes assumptions about the surface geometry and volumetric structure of the object to be grasped: spherical grasp is appropriate, for example, when the grasped surface is approximately spherical and has a curvature which the cupped hand and wrapped fingers are capable of; lateral grasp is more appropriate for laminae that are not too big nor too small. Although we learn, perhaps as a crucial part of their representation, how to pick up objects such as knives, pens, screwdrivers, and cups, we also typically choose sensible grasp positions for objects we have never seen before and for familiar objects whose "default" or preferred grasp position is occluded. One can imagine implementing a system based on Schlesinger's account: a person grasping an object would first choose a mode of prehension, and would then determine a position on the object to apply that mode of prehension.

We are investigating an alternative account. We suggest that, subject to the kinematic constraints imposed by the human hand's limited range of movement, objects are grasped along lines of curvature. This is precisely the information made available in the curvature patch representation. The grasper's problem is to lay his or her fingers along a reachable set of lines of curvature to maximize some performance criterion, for example to grasp as stiffly as possible or to give maximum freedom of movement. We are currently investigating this possibility for a range of objects and a kinematic model of a multi-fingered hand.

As we remarked above, the six modes of prehension isolated by Schlesinger are each associated with a commonly occurring type of surface. Schlesinger's observations may well tell us more about the types of surfaces commonly found or manufactured than they do about the process of grasping.

Acknowledgment

This report describes research done at the Artificial Intelligence Laboratory of the Massachusetts Institute of Technology. Support for the laboratory's Artificial Intelligence research is provided in part by the Advanced Research Projects Agency of the Department of Defense under Office of Naval Research contract N00014-75-C-0643, the Office of Naval Research under contract number N00014-80-C-0505, and the System Development Foundation. I thank Ruzena Bajcsy, Dana Ballard, Tom Binford, Rod Brooks, John Canny, Jerry Feldman, Eric Grimson, Tomas Lozano-Perez, Joe Mundy, Ken Sloan, and Demetri Terzopoulos for their comments on the material presented here. A number of excellent MIT UROP students worked on the smoothed local symmetries, in particular Steve Gander, Victor Inada, and Neil Webber.

References

Agin, G. J. Representation and description of curved objects. Stanford University AI Memo 73, 1972.

Agin, G. J. Computer vision systems for industrial inspection and assembly. *Computer*, 1980, **13** (5), 11-20.

Baker, H. H., & Binford, T. O. Depth from edge and intensity based stereo. Proceedings, 6th International Joint Conference on Artificial Intelligence, 1981, 631-636.

Ballard, D. H. Generalizing the Hough transform to detect arbitrary shapes. *Pattern Recognition,* 1981, **13**, 111-122.

Ballard, D. H., & Brown, C. M. *Computer Vision.* Englewood Cliffs, NJ: Prentice-Hall, 1982.

Barrow, H. G., & Tenenbaum, J. M. Recovering intrinsic scene characteristics from images. In A. R. Hanson & E. M. Riseman (Eds.), *Computer Vision Systems.* New York: Academic Press, 1978.

Barrow, H. G., & Tenenbaum, J. M. Interpreting line drawings as three dimensional surfaces. *Artificial Intelligence,* 1981, **17**, 75-117.

Bartlett, F. C. *Remembering: A Study in Experimental and Social Psychology.* Cambridge, UK: Cambridge University Press, 1932.

Binford, T. O. Inferring surfaces from images. *Artificial Intelligence,* 1981, **17**, 205-245.

Blum, H. Biological shape and visual science (part 1). *Journal of Theoretical Biology,* 1973, **38**, 205-287.

Blum, H., & Nagel, R. N. Shape description using weighted symmetric axis features. *Pattern Recognition,* 1978, **10**, 167-180.

Bolles, R. Three-dimensional locating of industrial parts. Proceedings, 8th NSF Grantees' Conference on Production Research and Technology, 1981, w1-w5.

Bolles, R. C., & Cain, R. A. Recognizing and locating partially visible objects: The local-feature-focus method. *International Journal of Robotics Research,* 1982, **1** (3), 57-82.

Bookstein, F. L. The line-skeleton. *Computer Graphics and Image Processing,* 1979, **11**, 123-137.

Brady, M. *Computer Vision.* Amsterdam: North-Holland, 1981.

Brady, M. Computational approaches to image understanding. *Computing Surveys,* 1982, **14**, 3-71. (a)

Brady, M. Parts description and acquisition using vision. Proceedings, SPIE Conference on Robot Vision, 1982, 20-28. (b)

Brady, M. Smoothed local symmetries and local frame propagation. Proceedings, IEEE Computer Society Conference on Pattern Recognition and Image Processing, 1982, 629-633. (c)

Brady, M., & Grimson, W. E. L. The perception of subjective surfaces. MIT AI Memo 666, 1981.

Brady, M., & Horn, B. K. P. Rotationally symmetric operators for surface interpolation. MIT AI Memo 654, 1981; *Computer Graphics and Image Processing,* in press.

Brady, M., & Yuille, A. A variational principle for shape from contour. MIT AI Laboratory, in preparation.

Braid, I. C. *Designing with Volumes.* Cambridge, UK: Cantab Press, 1973.

Brooks, R. A. Symbolic reasoning among 3-D models and 2-D images. *Artificial Intelligence,* **17**, 1981, 285-348.

Brooks, R. A. Solving the findpath problem by representing free space as generalized cones. MIT AI Memo 674, 1982.

Brooks, R. A., & Binford, T. O. Representing and reasoning about partially specified scenes. Proceedings, DARPA Image Understanding Workshop, 1980, 150-156.

Bruss, A., & Horn, B. K. P. Passive navigation. MIT AI Memo 662, 1981.

Coren, S. Subjective contours and apparent depth. *Psychological Review,* 1972, **79**, 359-367.

Denavit, J., & Hartenberg, R. S. A kinematic notation for lower pair mechanisms based on matrices. *Journal of Applied Mechanics,* 1955, **22**, 215-221.

Faux, I. D., & Pratt, M. J. *Computational Geometry for Design and Manufacture.* Chichester, UK: Ellis Horwood, 1979.

Forrest, A. R. On Coons and other methods for the representation of curved surfaces. *Computer Graphics and Image Processing,* 1972, **1**, 341-359.

Grimson, W. E. L. *From Images to Surfaces: A Computational Study of the Human Early Visual System,* Cambridge, MA: MIT Press, 1981.

Grimson, W. E. L. The implicit constraints of the primal sketch. MIT AI Memo 663, 1982.

Hanafusa, H., & Asada, H. A robotic hand with elastic fingers and its application to assembly process. Proceedings, IFAC Symposium on Information and Control Problems in Manufacturing Technology, 1977. (a)

Hanafusa, H., & Asada, H. Stable prehension of objects by the robot hand with elastic fingers. *Transactions of the Society of Instrument and Control Engineers,* 1977, **13** (4). (b)

Hinton, G. E. Some demonstrations of the effects of structural descriptions in mental imagery. *Cognitive Science,* 1979, **3**, 231-250.

Hoffmann, D. D., & Richards, W. A. Representing plane curves for recognition. MIT AI Memo 630, 1982.

Holland, S. W., Rossol, L., & Ward, M. R. CONSIGHT 1: A vision controlled robot system for transferring parts from belt conveyors. In G. Dodd & L. Rossol (Eds.), *Computer Vision and Sensor Based Robots.* New York: Plenum Press, 1979.

Hollerbach, J. M. Hierarchical shape description of objects by selection and modification of prototypes. MIT AI TR-346, 1975.

Horn, B. K. P. Understanding image intensities. *Artificial Intelligence,* 1977, **8**, 201-231.

Horn, B. K. P. Sequins and quills – representations for surface topography. In R. Bajcsy (Ed.), *Representation of 3-Dimensional Objects.* Berlin: Springer, 1982.

Horn, B. K. P., Woodham, R. J., & Silver, W. M. Determining shape and reflectance using multiple images. MIT AI Memo 490, 1978.

Ikeuchi, K. Determination of surface orientations of specular surfaces by using the photometric stereo method. Unpublished paper, 1981.

Ikeuchi, K., & Horn, B. K. P. Numerical shape from shading and occluding boundaries. *Artificial Intelligence,* 1981, **17**, 141-185.

Lee, D. T. Medial axis transformation of a planar shape. *IEEE Transactions on Pattern Analysis and Machine Intelligence,* 1982, **4**, 363-369.

Marr, D. Early processing of visual information. *Philosophical Transactions of the Royal Society, London,* 1976, **B275**, 483-524.

Marr, D. Analysis of occluding contours. *Proceedings of the Royal Society, London,* 1977, **B197**, 441-475.

Marr, D. Representing visual information. In A. R. Hanson & E. M. Riseman (Eds.), *Computer Vision Systems.* New York: Academic Press, 1978.

Marr, D. *Vision.* San Francisco, CA: Freeman, 1982.

Marr, D., & Hildreth, E. C. Theory of edge detection. *Proceedings of the Royal Society, London,* 1980, **B207**, 187-217.

Marr, D., & Nishihara, H. K. Representation and recognition of the spatial organization of three dimensional structure. *Proceedings of the Royal Society, London,* 1978, **B200**, 269-294.

Marr, D., & Poggio, T. A theory of human stereo vision. *Proceedings of the Royal Society, London,* 1979, **B204**, 301-328.

Mehlum, E. Nonlinear splines. In E. Barnhill & R. F. Riesenfeld (Eds.), *Computer Aided Geometric Design.* New York: Academic Press, 1974.

Millman, R. S., & Parker, G. D. *Elements of Differential Geometry.* Englewood Cliffs, NJ: Prentice-Hall, 1977.

Minsky, M. A framework for representing knowledge. In P. H. Winston (Ed.), *The Psychology of Computer Vision.* New York: McGraw-Hill, 1975.

Nackman, L. R. Three-dimensional shape description using the symmetric axis transform. Doctoral dissertation, University of North Carolina, 1982.

Nevatia, R. *Machine Perception.* Englewood Cliffs, NJ: Prentice-Hall, 1982.

Nevatia, R., & Binford, T. O. Description and recognition of curved objects. *Artificial Intelligence,* 1977, **8**, 77-98.

Nielson, G. M. Some piecewise polynomial alternatives to splines under tension. In R. E. Barnhill & R. F. Riesenfeld (Eds.), *Computer Aided Geometric Design.* New York: Academic Press, 1974.

Pavlidis, T. *Structural Pattern Recognition.* New York: Springer, 1977.

Pilcher, D. T. Smooth parametric surfaces. In R. E. Barnhill & R. F. Riesenfeld (Eds.), *Computer Aided Geometric Design.* New York: Academic Press, 1974.

Porter, G. B. III, & Mundy, J. L. A non-contact profile sensor system for visual inspection. Proceedings, IEEE Computer Society Workshop on Industrial Applications of Machine Vision, 1982, 119-129.

Rosenfeld, A., & Kak, A. C. *Digital Picture Processing.* New York: Academic Press, 1976.

Salisbury, K. J., & Craig, J. J. Articulated hands: Force control and kinematic issues. *International Journal of Robotics Research,* 1982, **1** (1), 1-15.

Schlesinger, G. Der mechanische Aufbau der kunstlichen Glieder. In *Ersatz Glieder und Arbeitshilfen.* Berlin: Springer, 1919.

Sloan, K. R., & Ballard, D. H. Experience with the generalized Hough transform. Proceedings, DARPA Image Understanding Workshop, 1980, 150-156.

Stevens, K. A. The visual interpretation of surface contours. *Artificial Intelligence,* 1981, **17**, 47-75.

Terzopoulos, D. Multi-level reconstruction of visual surfaces. MIT AI Memo 671, 1982.

Ullman, S. *The Interpretation of Visual Motion.* Cambridge, MA: MIT Press, 1979.

Vilnrotter F., Nevatia R., & Price, K. E. Structural analysis of natural textures. Proceedings, DARPA Image Understanding Workshop, 1981, 61-68.

Weatherburn, C. E. *Differential Geometry of Three Dimensions.* Cambridge, UK: Cambridge University Press, 1927.

Webb, J., & Aggarwal, J. K. Structure from the motion of rigid and jointed objects. *Artificial Intelligence,* 1982, **19**, 107-130.

Webber, R. L., & Blum, H. Angular invariants in developing human mandibles. *Science,* 1979, **206**, 689-691.

Winston, P. H., Katz B., Lowry, M., & Binford, T. O. MIT AI Laboratory, 1982.

Witkin, A. P. Recovering surface shape and orientation from texture. *Artificial Intelligence,* 1981, **17**, 17-47.

Contrasts between Human and Machine Vision: Should Technology Recapitulate Phylogeny?

Myron L. Braunstein

University of California
Irvine, California

Abstract

Biological evolution has imposed constraints on the development of the human visual system that do not necessarily apply to machine vision. Three characteristics of human vision illustrate the effects of these special constraints: the primacy of depth perception, the coexistence of perception and contradictory knowledge, and the use of heuristic perceptual processes. One approach that includes a consideration of biological constraints is ecological optics, but the usefulness of that approach is limited by its failure to consider perceptual processes or mechanisms and to deal with conditions of marginal stimulus information. Because of the special constraints imposed by evolution, the effectiveness or lack of effectiveness of a perceptual process in a computational system is not evidence for its presence or absence in human vision. An appreciation of the differences between human and machine vision should facilitate research on problems common to the two fields.

As other papers in this volume clearly demonstrate, there has recently been an increased use of machine vision programs as models of human vision and of data from human vision in developing programs for machine vision. These interactions emphasize the similarities between the processes found in human vision and those that might be useful in programming a machine for intelligent image processing. The purpose of this paper is to suggest that it is also useful to look at differences between human and machine vision. It will be argued that these differences are the result of the special constraints imposed by evolution on the development of the human vision system.

It can be argued naively that because there is no physical similarity between a human being and a computer there is no value in interactions between researchers in human vision and machine vision. This argument is contradicted by the useful interactions that have already taken place. Machine vision programs have tested the plausibility of proposed mechanisms in the human vision system (for example, Marr & Poggio, 1979). The precise specification of stimulus variables and of perceptual mechanisms required in machine vision projects have been useful in clarifying areas of ambiguity in research on human vision (for example, Prazdny, 1981). Data from research on human vision has been used to suggest algorithms for automated image processing and to provide standards for measuring the success of automated image analyzers since the earliest efforts in machine pattern recognition (Selfridge & Neisser, 1960).

Just as it is unreasonable to conclude that there are no similarities between the processes found in human and machine vision, it is also unreasonable to conclude that there are no substantial differences. The processes used by the human visual system are not necessarily the processes most suitable for image analysis programs, and those most suitable for image analysis are not necessarily the processes found in human vision. Although there are natural constraints that apply to both biological and machine systems (Yonas, Thompson, & Granrud, 1981), biological systems develop through an evolutionary process that is subject to constraints that are not applicable to machine vision programs. The latter constraints are likely to be reflected in differences in the processes found in these two types of visual systems.

Each step in the evolution of the human visual system from the first primitive photoreceptor – probably a light sensitive patch on the ectoderm (Polyak, 1957) – was the result of natural selection. The continued development of the visual system through natural selection implies that vision was relevant to the survival of organisms at various stages in the phylogenetic chain from the most primitive system to the human system. Each variant that was a factor in the phylogenetic development of the human system must have provided some advantage to the primitive organism in which it occurred. The development of the human visual system was thus constrained by the necessity of going through stages that contributed to the survival of more primitive organisms. The following three constraints appear plausible on this basis: First, the capability of responding differentially to objects, especially other organisms, on the basis of differences in distance should have been an important criterion in the development of even the most primitive visual systems. The advantage of vision arises from the possibility of responding to features of the environment that are not in physical contact with the organism (Gregory, 1968). This advantage would be severely limited if the visual system did not respond to variations in distance.

Second, primitive organisms using visual information would require automatic mechanisms capable of responding to such stimulation. The ability to cognitively adjust the results of perceptual processes is a much later development in the evolutionary process. Finally, perceptual mechanisms must have developed on the basis of a criterion of sufficiency for survival or of providing an advantage for survival. It cannot be assumed that these mechanisms used optimum solutions to geometric problems from either a mathematical or computer viewpoint. Other criteria, such as the need for rapid response and for responses biased towards more serious threats, may have lead to the development of processes that use heuristic approximations to the geometric problems of reconstructing a three-dimensional scene from two-dimensional projections. The following three sections will consider these suggested differences between evolved biological visual systems and machine systems.

1. The Primacy of Depth Perception

It may be that the photosensitive patches that appeared on the skins of primitive organisms at first provided information only about the orientation of the light source. Even at this stage, distance from the surface in a body of water would have affected the intensity of the stimulation. At a later stage, a change in stimulation – a shadow – may have indicated the direction of a predator. This does not suggest that a recognition and decision process took place at this stage. Instead, there may have been a direct linkage between the photosensitive patch on one side of the body and the mechanisms that propelled the animal in the opposite direction (Polyak, 1957). An increase in the area of a shadow on a photosensitive surface, once such information could be integrated, may have indicated the approach of a potential predator. This integration would then have represented an early form of depth perception. This evolutionary basis is speculative, but there is evidence that mechanisms for detecting "looming" in the human visual system are directly related to the perception of motion in depth. Both psychophysical evidence based on research with human subjects (Beverley & Regan, 1979, 1980) and physiological evidence based on research with cats (Cynader & Regan, 1978) support the existence of such mechanisms. The motion-in-depth perception is apparently triggered by out of phase oscillation of parallel contours, as in a pair of vertical bars moving alternately together and apart or an alternately expanding and contracting square.

This type of mechanism may be contrasted to a process that begins with a two-dimensional representation of the retinal projection and computes a three-dimensional interpretation using one or more intermediate representations (Marr, 1980). This may be an effective approach in machine image analysis, but may not reflect the processes used by the human visual system. Machine vision, unlike animal vision, is not constrained by the need for simple mechanisms that promote survival in a three-dimensional environment.

2. Perception and Contradictory Knowledge

Perception is characterized by processes that are largely automatic and resistant to modification based on contradictory knowledge. A rotating trapezoid, viewed from a sufficient distance, will appear to oscillate regardless of the observer's knowledge that it is rotating. This is true even if the knowledge is based on active exploration of the environment. An observer who reports oscillation when standing ten feet from the trapezoid may report that it is rotating after a closer inspection of the moving object, but when the observer returns to the ten-foot observation distance he or she will again report perceiving the trapezoid as oscillating. Direct perception theorists (Shaw & Bransford, 1977) have suggested that such active exploration of the environment can overcome illusions. In the case of the rotating trapezoid, the claim that the illusion can be overcome by active exploration implies that the observer's knowledge of the actual path is to be regarded as perception but the observer's immediate experience of oscillation need not be considered as a perceptual effect. That interpretation handles the simultaneous occurrence of a perceptual experience and contradictory knowledge by ignoring the perceptual experience and labelling knowledge as perception. Regardless of how it is labeled, immediate experience based on sensory input (or simulated sensory input) is clearly distinguishable from knowledge by psychophysical procedures and cannot be ignored as a category of behavior.

The spiral aftereffect presents another example of the coexistence of a perceptual experience with contradictory knowledge. An observer who has adapted to a contracting spiral, and then looks at another object, will perceive expansion of that object. If, for example, the object observed after the spiral is a person's face, the face will appear to expand. The observer will probably know that the face is not expanding, but this does not prevent the perception of expansion. There are numerous other examples of the coexistence of perception with contradictory knowledge. Most illusions, including virtually all geometric illusions, cannot be overcome by knowledge of the physical characteristics of the distal stimulus.

Although contradictory knowledge may not alter perception, it obviously alters the response to visual stimuli. A person can decide not to flee from a looming shape if there is definite knowledge that it is not a large, approaching object. However, an automatic fear reaction to the looming shape may not be completely absent. The visual system appears to have maintained some capability for automatic reactions to stimulation even as higher level cognitive processes have been added in the course of evolution. Possibly the perceptions that are most critical to survival, e.g. looming, are the most resistant to the effects of contradictory knowledge.

This separation of perception and knowledge may be reasonable from an evolutionary perspective, but would not be useful in a system designed for the accurate analysis of three-dimensional scenes. It would be more appropriate for an image analysis system to remove any artifacts due to its lower level processing mechanics before those artifacts reached higher levels of processing. The human observer, however, experiences many such artifacts.

3. Heuristic Processes in Perception

Human visual processes usually yield perceptions that approximate solutions based on projective geometry and physical constraints, but with unusual stimulation perception may deviate from geometric solutions. Consider, for example, the problem of perceiving a three-dimensional object on the basis of a changing retinal projection. A changing shape projected onto the retina can be the result of an infinite number of distal events. A problem common to human and machine vision is the selection of one, or a small number, of perceptions on the basis of this geometrically ambiguous proximal stimulus. One solution that has been proposed, both as the solution actually used by human observers and as a solution appropriate for machine analysis, is the principle of minimum object change (Jansson & Johansson, 1973) or the rigidity assumption (Ullman, 1979). According to these concepts, the preferred perceptual solution is the one that minimizes perceived changes in the shapes of objects, or put another way, maximizes the perceived rigidity of objects.

It is reasonable to expect that the perceptual processes that developed in the course of evolution would take advantage of what is usual in the environment, and some rigidity is a usual feature of the animal environment. However, taking advantage of the rigidity in the environment does not require that a rigidity assumption be applied to the analysis of each new proximal stimulus. An alternative possibility is that perceptual processes developed that take advantage of some of the specific projective results of the three-dimensional motions of rigid objects. The following three processes are proposed as examples of perceptual processes that appear to serve this purpose in human vision: (1) Converging contours, either explicitly present or implicit in texture density gradients, imply receding surfaces (Braunstein, 1972). (2) A retinal projection that changes in size but remains constant in shape implies motion in depth (Beverly & Regan, 1980; Johansson, 1964). (3) Sinusoidal projected motions imply rotation in depth (Braunstein, 1977).

As processes of the type listed above tend to preserve the rigidity of perceived object motions, it usually is not possible to determine empirically whether the human observer is using such processes or using a mathematically correct rigidity assumption. There are exceptional situations, however, in which it is possible to distinguish between these two alternatives. These situations arise when a perception that results from one of the specific processes suggested above conflicts with the results of a general rigidity solution. The rotating trapezoid illusion is an example of such a situation. With sufficient viewing distance, a trapezoid rotating about an axis parallel to its parallel sides will project converging contours enclosing an approaching side. The principle that converging contours imply recession would cause that side to appear to recede. As this occurs during half of each rotation cycle, the application of this perceptual process will result in perceived oscillation (Braunstein, 1972). As the retinal projections produced by the rotating trapezoid are consistent with rotation of a rigid trapezoid and not with the oscillation of any rigid object, the perception of oscillation violates the rigidity assumption (and requires supplementary assumptions by those who subscribe to it [Ullman, 1979]). The observer in this situation uses a process that usually is consistent with maintaining perceived rigidity – converging contours imply recession – while viewing an unusual distal stimulus for which that process is inappropriate. As a result of using this specific process when it is inappropriate, the rigidity assumption is violated. The specific process is an automatically applied part of the human perceptual system; the rigidity assumption is not.

A second example of a situation that indicates use of specific processes rather than a general rigidity assumption involves the stimulation of size change mechanisms. The perception of motion in depth on the basis of proximal size change usually results in perceptions that are consistent with a rigidity assumption, but unusual stimulation of the size change mechanisms may result in perceptions that violate that assumption. Spirals rotating in the frontal plane produce perceptions of expansion or contraction and sometimes of approach or recession in depth (Gates, 1934). A figure consisting of two spiral-like parts joined together (Figure 1), when rotated in the frontal plane, is universally perceived as a distorting, three-dimensional shape. The retinal projection in this case is consistent with a rigid object in motion, i.e., the actual rotation in the plane of the two-dimensional figure. The pattern of stimulation of motion-sensitive receptive fields apparently produces information for the expansion of part of the figure and the contraction of part of the figure (Braunstein & Andersen, 1980). This apparent simultaneous expansion and contraction cannot occur in a rigid plane figure. Instead, a distorting three-dimensional object is perceived. This perception is not overridden by a rigidity assumption.

The non-veridical perceptions described in this section could be avoided in machine vision through the use of computations that maintain object rigidity. These computations, however, would not match the processes used by the human visual system. Although the approximate processes in human vision sometimes lead to errors (illusions), they have several potential advantages over the use of a rigidity assumption. First, these processes may be less likely to bias the system so heavily towards the perception of rigid motions that the perception of non-rigid motions is adversely affected. When faced with an ambiguous proximal motion, the human visual system is capable of generating compromise perceptions consisting of mostly rigid motion combined with some form change (Braunstein & Andersen, 1981; Marmolin, 1973). Second, the heuristic perceptual processes used by the human observer may be less complex for a biological system than processes incorporating a general rigidity assump-

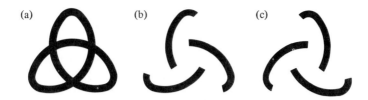

Figure 1. (a) A two-dimensional figure that is perceived as a non-rigid three-dimensional shape when rotated in the frontal plane (designed by Fred Duncan and reproduced by permission). (b,c) Parts of (a) that appear to expand or contract, depending on direction of rotation, when rotated in the frontal plane.

tion. Simplicity and complexity do not always have the same meaning in perceptual analysis as in machine computation (Johansson, 1981). The primitive calculations for machines, e.g., adding and shifting, are probably not the same primitives used by the visual system. The human visual system may use "smart" mechanisms (Runeson, 1977) for functions such as detecting texture convergence or sinusoidal velocities in the retinal projection. These mechanisms may not be decomposable into simpler arithmetic computations. A third reason for the use of perceptual processes that do not provide exact solutions to geometric problems is that these processes may provide more rapid solutions than more precise algorithms. An approximate process that rapidly detected a potential danger would be of more value for the survival of an animal than a more precise process that accurately identified the danger after it was too late to take effective action (Braunstein, 1976, Chapter 7).

4. Ecological Optics

The previous sections argue that an animal's perceptual processes can best be understood in the context of the visual environment in which that animal, and its predecessors, had to survive. This is the same point emphasized by Gibson (1979) in his theory of ecological optics. According to Gibson, an animal perceives the "affordances" of the environment. These are benefits and dangers that are directly perceived on the basis of higher order variables in the reflected light. Drinkable water, dangerous cliffs, and suitable mates are examples of affordances that Gibson argued are perceived in this manner. The merits of this "direct perception" approach have been actively debated, and it has in particular been contrasted to the representational approaches characteristic of research in machine vision (Ullman, 1980). The logical implications of Gibson's approach have been questioned by Fodor and Pylyshyn (1981). For the present discussion, the relevant issue is whether the direct perception approach is useful as a means of adding evolutionary considerations to the study of human visual perception.

On the positive side, Gibson (1979) provided a remarkably insightful analysis of the visual environment of animals, of the features of the environment that relate to survival, and of how these features are represented in reflected light. Many of the features that he describes have been neglected in the perceptual literature. The relationship of the size of the animal to the relevant scale of environmental features is emphasized. A recognition of the importance of kinetic occlusion (the accretion and deletion of texture elements at an edge) to an animal's detection of such environmental features as sharp drop-offs is another unique feature of Gibson's analysis.

On the negative side, the direct perception approach expounded by Gibson is characterized by an unwillingness to consider processes. This creates special problems in attempting to relate perception to evolutionary considerations. One can assume, in accordance with Gibson's approach, that the perception of affordances has evolved as part of the behavior of an animal, or one can assume that perceptual processes have evolved that tend to be helpful in the perception of affordances. If the animal always perceived affordances correctly, it might not be possible to distinguish between these two alternatives on a behavioral level. It could be argued that the question of whether the perception of affordances is direct or whether it is the usual result of certain processes that have evolved is not relevant at a behavioral level, but should be relegated to the domain of neurophysiology. However, if affordances are not perceived perfectly, the nature of the processes is clearly relevant to behavior, as this will determine the conditions under which errors occur and the specific errors that can be expected.

There are many natural situations in which animals do not perceive affordances correctly. Light reflected by dry ground may be perceived as water by a distant observer under certain atmospheric conditions. A predator with effective natural camouflage may be perceived as part of the brush by its intended victim. Proponents of direct perception have argued that these situations are unusual and that perceptual errors are unlikely when an animal is free to explore a rich natural environment (Shaw & Bransford, 1977). This argument has also been applied to laboratory situations in which observers are presented with a few spots of light in the dark and are required to keep their heads stationary. These situations have been criticized as lacking ecological validity and therefore not contributing to an understanding of perception by a mobile animal in a natural environment. The illusions that can be produced under such laboratory conditions have not been considered valid evidence against direct perception.

Limiting the domain of ecological validity to mobile animals in visually rich environments, however, does not take into account the full range of conditions that are important to the survival of an animal. Animals remain stationary under conditions that are clearly relevant to survival. Stationary observation is characteristic of predatory animals (Owen, 1978). An animal may have to respond accurately to spots of light in the dark, as well as to illuminated textured surfaces, in its natural environment. Whether or not two spots of light in the dark are perceived as rigidly connected may determine whether an animal is able to recognize a potential threat (e.g., a tiger at night).

Contrary to the viewpoint associated with ecological optics, it is probably not an animal's visual capabilities in an enriched environment but the boundary conditions of the animal's visual capabilities that are most important for survival. Identifying distant objects that are barely detectable, detecting significant objects in the dark, overcoming natural camouflage, and perceiving prey or predators without moving are examples of capabilities that may be close to the boundaries of an animal's perceptual skills. The animal's performance in these boundary conditions depends on the specific processes that underlie those skills.

In summary, ecological optics is of value in pointing out some of the ways in which perception is related to the needs of animals to survive in their environments but it does not provide a complete account of the effects of evolutionary constraints on human vision. Gibson's emphasis on motion and depth as immediate perceptions, rather than inferences derived from static two-dimensional perceptions, and on the role of higher order variables such as texture gradients, are issues of importance in comparing human and machine vision. His reluctance to consider processes leading from proximal stimulation to perceptual experience and his discounting of research using minimal stimulus conditions limits the usefulness of his approach. A more general theory relating human perceptual processes to the constraints imposed by evolution would be useful in identifying differences that should be expected between human and machine vision.

5. Conclusion

The useful interaction of two scientific disciplines requires attention to the differences as well as to the similarities in the objectives and subject matter of the two disciplines. Machine vision efforts are usually directed towards the preparation of accurate symbolic descriptions of patterns received by photosensitive devices. Algorithms are usually derived on a priori grounds and tested against criteria of accuracy and computational efficiency. The objective of basic research in human visual perception can be characterized as increasing our understanding of relationships between physical stimulation and perceptual experience. (Others may argue that the goal is prediction.) Useful models of human perception may also be derived on a priori grounds, with attention to ecological constraints, but the ultimate test is whether the model matches the results of research with human subjects. A computational demonstration of the effectiveness of a proposed perceptual process in an image analysis system cannot substitute for converging evidence from a body of systematic empirical research with human subjects, as evidence that the process exists in human vision. Conversely, the failure of a process to perform accurately or efficiently in an image analysis system is not sufficient evidence that the process is not used in human vision. Phylogenetic constraints may have led to processes in human vision that appear inefficient or unnecessarily inaccurate from the point of view of machine analysis. These considerations should not detract from the value of increased interaction between researchers in machine vision and human vision. On the contrary, a clearer appreciation of the differences that should be expected between human and machine vision should facilitate research on those problems that are common to the two fields.

Acknowledgment

I thank George J. Andersen for discussions that helped shape the ideas presented here and for valuable comments on preliminary versions of this paper.

References

Beverley, K. I., & Regan, D. Separable aftereffects of changing-size and motion-in-depth: Different neural mechanisms. *Vision Research,* 1979, **19**, 727-732.

Beverley, K. I., & Regan, D. Visual sensitivity to the shape and size of a moving object: Implications for models of object perception. *Perception,* 1980, **9**, 151-160.

Braunstein, M. L. Perception of rotation in depth: A process model. *Psychological Review,* 1972, **79**, 510-524.

Braunstein, M. L. *Depth Perception Through Motion.* New York: Academic Press, 1976.

Braunstein, M. L. Minimal conditions for the perception of rotary motion. *Scandinavian Journal of Psychology,* 1977, **18**, 216-223.

Braunstein, M. L., & Andersen, G. J. Local and global determinants of a stereokinetic illusion. Proceedings, 21st Annual Meeting, Psychonomic Society, 1980.

Braunstein, M. L., & Andersen, G. J. Velocity gradients and relative depth perception. *Perception and Psychophysics,* 1981, **29**, 245-255.

Cynader, M., & Regan, D. Neurons in cat parastriate cortex sensitive to the direction of motion in three-dimensional space. *Journal of Physiology,* 1978, **274**, 549-569.

Fodor, J. A., & Pylyshyn, Z. W. How direct is visual perception? Some reflections on Gibson's "Ecological Approach." *Cognition,* 1981, **9**, 139-196.

Gates, L. W. The aftereffect of visually observed movement. *American Journal of Psychology,* 1934, **46**, 34-46.

Gibson, J. J. *An Ecological Approach to Visual Perception.* Boston, MA: Houghton-Mifflin, 1979.

Gregory, R. L. The evolution of eyes and brains – a hen-and-egg problem. In S. J. Freedman (Ed.), *The Neuropsychology of Spatially Oriented Behavior.* Homewood, IL: Dorsey Press, 1968.

Jansson, G., & Johansson, G. Visual perception of bending motion. *Perception,* 1973, **2**, 321-326.

Johansson, G. Perception of motion and changing form. *Scandinavian Journal of Psychology,* 1964, **5**, 181-208.

Johansson, G. About visual event perception. University of Uppsala, Department of Psychology, Rept. 306, 1981.

Marmolin, H. Visually perceived motion in depth resulting from proximal changes. I. *Perception and Psychophysics,* 1973, **14**, 133-142.

Marr, D. Visual information processing: The structure and creation of visual representations. *Philosophical Transactions of the Royal Society, London,* 1980, **B290**, 199-218.

Marr, D., & Poggio, T. A computational theory of human stereo vision. *Proceedings of the Royal Society, London,* 1979, **B204**, 301-328.

Owen, D. H. The psychophysics of past experience. In P. K. Machamer & R. G. Turnbull (Eds.), *Studies in Perception: Interrelations in the Histories of Science and Philosophy.* Columbus, OH: Ohio State University Press, 1978.

Polyak, S. *The Vertebrate Visual System.* Chicago, IL: University of Chicago Press, 1957.

Prazdny, K. A note on "Optical motions as information for unsigned depth." *Journal of Experimental Psychology: Human Perception and Performance,* 1981, **7**, 286-289.

Runeson, S. On the possibility of "smart" perceptual mechanisms. *Scandinavian Journal of Psychology,* 1977, **18**, 172-179.

Selfridge, O. G., & Neisser, U. Pattern recognition by machine. *Scientific American,* 1960, **203** (3), 60-68.

Shaw, R., & Bransford, J. Introduction: Psychological approaches to the problem of knowledge. In R. Shaw & J. Bransford (Eds.), *Perceiving, Acting, and Knowing.* Hillsdale, NJ: Erlbaum, 1977.

Ullman, S. *The Interpretation of Visual Motion.* Cambridge, MA: MIT Press, 1979.

Ullman, S. Against direct perception. *Behavioral and Brain Sciences,* 1980, **3**, 373-415.

Yonas, A., Thompson, W., & Granrud, C. E. Computer vision: Implications for the psychology of human vision. University of Minnesota, Computer Science Department, TR 81-6, 1981.

Flexibility in Representational Systems

Lynn A. Cooper

University of Pittsburgh
Pittsburgh, Pennsylvania

Abstract

There has been a persisting controversy in the literature on visual information processing and mental imagery as to whether the internal representation
of spatial information is most appropriately characterized as analog or as
discrete and propositional in nature. It is argued that this controversy has been
misdirected, and while succeeding in producing a rich body of experimental
results, has failed to produce definitive theoretical conclusions. The view
advanced in this paper is that the diversity and flexibility of systems for
representing visual information should be explored vigorously, rather than
attempting to isolate invariant properties of such systems. Evidence is reviewed
suggesting that the nature of the representation of spatial information is highly
sensitive to (a) changes in information-processing demands, (b) ability-related
individual differences, and (c) levels of expertise in problem-solving skills.

The spirit of my remarks has much in common with the paper by Posner
and Henik in this volume. These investigators point out that multiple representational systems or "codes" are activated by even simple stimulus events.
These coding systems represent different levels of abstraction, with low-level
codes involving the activation of simple features of visual objects and higher-
level codes involving more integrated aspects of object representation. Furthermore, different levels of coding abstraction can be accessed in accordance with
the nature of the operations that must be performed on such representations.

I am highly sympathetic to this view of representation – one emphasizing
the diversity of available systems for representing visual objects and events –
and I would like to extend Posner and Henik's argument a bit further. My

emphasis will not be on a hierarchy of internal codes for visual/spatial information, but rather on the flexibility of the organism in selecting the representational system most appropriate for the task at hand. The basic idea is that there are no invariant properties of systems for representing visual/spatial information. Rather, the nature of the internal representation of spatial information is highly sensitive to the particulars of task demands, to changes in cognitive organization with learning, and to individual differences in patterns of cognitive ability. I will illustrate these somewhat vague points with several concrete examples from both my own experimental work and that of others. The examples will be drawn from tasks requiring higher-level visual problem-solving skills than those considered by Posner and Henik.

In passing, it should be noted that the idea of multiple, flexible representational systems is hardly a new one in psychology. Indeed, such a notion lies at the heart of Piaget's theory of cognitive development (e.g., Piaget, 1952; Piaget & Inhelder, 1971; see also Mandler, in press, for a review), although Piaget's emphasis was on changes in representational systems with age rather on the interaction between various forms of representation at a given point in development. Nonetheless, this point has been overlooked by many researchers in the area of visual representation. Consider, for example, the controversy that has persisted for the past ten years over whether the internal representation and processing of visual information is analog or discrete and symbolic in nature (see, e.g., Cooper & Shepard, 1973, 1978; Kosslyn, 1980, 1981; Palmer, 1978; Pylyshyn, 1973, 1979, 1981). I will not review the history of this debate here, but suffice it to say that so-called "analog" theorists have pointed to results from studies of mental rotation (e.g., Cooper & Shepard, 1973; Cooper, 1975; Shepard & Metzler, 1971) and imaginal scanning and size scaling (e.g., Kosslyn, 1973, 1975; Kosslyn, Ball, & Reiser, 1978) as support for their position. These experiments appear to demonstrate that the relational structure of external objects and events is preserved in the relational structure of the corresponding internal representations and operations. The so-called propositional theorists have countered by pointing out that the relevant experimental results can be adequately mimicked by appropriately constructed symbolic systems (Anderson, 1978) and have recently suggested that these findings could be attributable to subjects' use of tacit knowledge concerning the nature of physical transformations and their interpretation of experimental situations as inviting the simulation of what they know the appropriate physical transformation to be (Pylyshyn, 1981). This debate has proven extremely difficult to legislate on either empirical or logical grounds, although it has served the function of generating a rich body of data which may place constraints on theories of the nature of representational systems for spatial information. The suggestion I make here is that the focus of the controversy has been misguided. That is, rather than asking whether a representation has properties classifying it as either analog or propositional in nature, it may be more productive to explore what characteristics representational systems assume under different conditions of information-processing demands and for different levels of knowledge and expertise.

As a first example of this notion of flexibility in selecting an appropriate representational system, consider a series of experiments performed by Robert Glushko and myself (Glushko & Cooper, 1978). We were interested in the question of how people compare a verbal description of a visual object with a picture of the object in order to determine whether the two are the same or different. Typical descriptions and visual figures used in our experiment are shown in Figure 1. In an initial study, Glushko and I allowed subjects to study

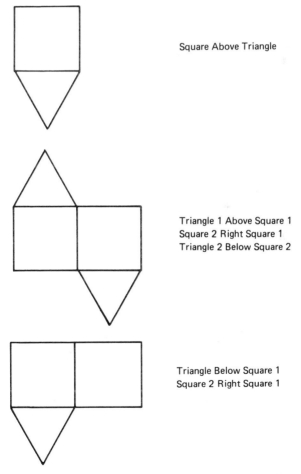

Square Above Triangle

Triangle 1 Above Square 1
Square 2 Right Square 1
Triangle 2 Below Square 2

Triangle Below Square 1
Square 2 Right Square 1

Figure 1. Examples of the stimulus materials (descriptions and figures) used in the Glushko and Cooper (1978) experiment.

the descriptions as long as they wanted before presenting the visual figure to be compared with the representation of the description. Under these conditions, the amount of time taken to process the description increased linearly with the length of the description, and the time required for subsequent description-figure matching was not affected by the complexity of the description of the figure or by the nature of the particular lexical items used in the description.

This finding suggested that subjects were constructing a spatially structured internal representation of the upcoming visual test figure during the comprehension interval and were then rapidly and holistically matching this representation against the test figure when it appeared. However, our findings were at odds with reports in the literature of results from similar experiments (cf. Carpenter & Just, 1975; Clark & Chase, 1972). These investigators have found that when a brief verbal description must be compared with a simple visual picture, the time to make the comparison increases with the linguistic complexity of the description, with times being greater for descriptions containing negation and for descriptions containing linguistically "marked" terms (e.g., "below" and "left" vs. "above" and "right"). These results have been interpreted as indicating that both descriptions and figures are represented internally in a symbolic/propositional manner for purposes of comparison.

Why, then, did Glushko and I find evidence for construction of a spatial internal representation in the description-picture matching paradigm? The answer appears to lie in differences between the temporal structure of our task and tasks studied by other investigators. Specifically, in our version of the description-figure matching task, the comprehension interval (whose duration was controlled by the subject) was sufficiently long to ensure that adequate time was available to construct a spatial representation of the described figure for purposes of rapid matching against the test figure. This situation, then, is optimal for the use of a spatially-based comparison strategy. Other investigators, who have obtained effects of linguistic variables on comparison times, have generally employed simultaneous presentation of descriptions and figures. In this situation, subjects may not have enough time to generate a spatial representation of the figure based on the description. An appropriate strategy might consist of comparing a semi-processed symbolic representation of the description with elements of the visual figure. Thus, temporal constraints imposed by the task structure might induce the use of different representational systems, depending on the precise nature of these constraints.

Glushko and I proceeded to test this idea directly by testing subjects in four experimental conditions. One condition was the same as our previous experiment, in which subjects controlled the duration of the description presentation in order to prepare for the test figure. In two other conditions, the description was presented for a fixed duration – either two or six seconds – before presentation of the test figure. In the final condition, both the description and the figure were presented simultaneously. The pattern of results from this experiment was complex, but one central finding was clear. As the conditions progressed from subject-controlled comprehension times through fixed-duration intervals to simultaneous presentation, effects of linguistic variables on comparison times became more and more pronounced. These results argue strongly against the idea that a uniform type of internal representation is used in this description-figure matching situation. Rather, alternative representational systems are available and can be utilized in a manner consistent with the demands of particular tasks.

A second potential example of the existence of alternative systems for representing spatial information – which may be related to individual differences in spatial aptitude – comes from work that I (in collaboration with Randall Mumaw and Robert Glaser) am currently carrying out on the solution of problems in engineering design graphics. The skill that we are analyzing is the comprehension and comparison of different forms of representing three-dimensional objects in two-dimensional drawings of the sort typically encountered in courses in mechanical engineering. These problems are complex, they seem to require the representation and manipulation of spatial information, and they appear to be capable of being solved by a variety of alternative strategies.

The problem types that we have investigated so far are illustrated in Figures 2, 3, and 4. One sort of problem (shown in Figure 2) involves comparison of an isometric drawing of a three-dimensional object with an orthographic projection to determine whether or not they depict the same structure. In the isometric form of pictorial representation, an object is illustrated in perspective approximately as it would be seen by the eye. In the orthographic form of representation, three separate projections of the front, top, and one side view of the object are displayed as an observer would see each view if looking directly at each of the appropriate surfaces. (Various drawing conventions are used in orthographic projections, such as depicting hidden edges of objects with dashed lines.) A second type of problem, shown in Figure 3, requires the subject to judge the compatibility of various orthographic views, i.e., given two orthographic projections, a judgment of the correctness of a third orthographic view must be made. The third type of problem requires the observer to determine whether or not a verbal description correctly describes an orthographic projection of a three-dimensional object. A typical problem of this sort is illustrated in Figure 4.

Our subjects in this study are students enrolled in a class in mechanical engineering. They have been tested on a battery of psychometric tests of spatial aptitude, and they represent a reasonably wide range of measured ability. In addition, they have been tested on laboratory versions of items typically found on such aptitude tests. We also collect retrospective protocols from the subjects following their participation in the experiment. One of our goals in doing this study is to determine whether individual differences in spatial aptitude are related to differences in the internal representations and processes used

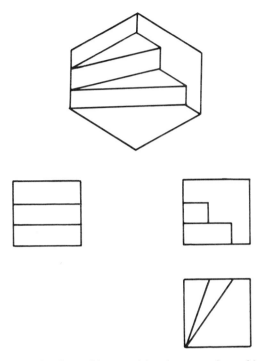

Figure 2. Example of a problem requiring the comparison of isometric and ortho-graphic projections of a visual object.

to solve the design graphics problems. In our experimental situation, the two parts of each problem are presented sequentially on two separate slides, and the subject may alternate between the two slides as frequently as he/she desires. The dependent measures that we have collected include: overall percent correct, total solution time, average number of alternations between the first and second slides, average time of the first look at the first slide, average time of the first look at the second slide, and averages of the times of all other looks at the first and second slides.

We are hoping that individual differences in patterns of performance may provide evidence for the use of alternative representational systems that are related to spatial aptitude. For example, the use of a relatively pure spatial strategy for solving an item requiring the translation between a verbal description and an orthographic drawing of an object might be revealed by relatively long initial inspection of each of the two slides, but relatively few alternations between them. This performance pattern would be consistent with a strategy of mentally constructing a spatial analog of the appropriate isometric drawing from both the verbal description and the orthographic projection and then directly comparing the two generated spatial representations. A more analytic strategy, based on the encoding of lower-level visual features, could be revealed by

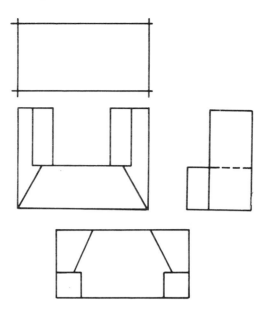

Figure 3. Example of a problem requiring the determination of the compatibility of two orthographic views.

many alternations between the two slides containing the description and the orthographic projection, but relatively brief inspection times. This performance pattern would be consistent with a strategy of sequentially checking for the correspondence between each statement in the verbal description and corresponding features or parts of the orthographic drawing. We have not yet completed the analysis of these data, but we are excited at the prospect that the use of different representational systems for solving these complex visual problems may in fact prove to be a component of spatial aptitude.

A third example of research demonstrating the use of multiple representational systems is work comparing the performance of expert problem solvers with that of novices in particular cognitive domains. The notion emerging from this research is that the nature of the internal representations used in problem solving changes with increasing level of skill. Parenthetically, this is one direction in which we would like to turn in our analysis of the representations and processes used to solve the engineering design graphics problems. That is, we would like to monitor performance of subjects at various stages of skill acquisition in order to see whether there is an orderly pattern of change. One possibility is that subjects continue to use the same representational and processing strategies as they acquire problem solving skill, but that the efficiency with which these different strategies are used increases with practice. The other, more interesting possibility is that individuals come to use qualitatively different representational systems as problem solving skill develops.

A rectangular block has a longitudinal
slot in its top. The front left edge of
the block is cut at an angle. The depth
of the cut is greater than the slot but
is not completly through the bottom
of the block.

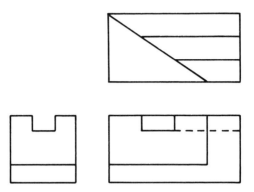

Figure 4. Example of a problem requiring the comparison of a verbal description and an orthographic drawing of an object.

There is evidence for this latter conjecture in other problem-solving domains. Some of the best-known work along these lines is research on expert-novice differences in chess (e.g., Chase & Simon, 1973). This body of work suggests that one dimension of difference between the expert and the novice is in the larger number of well-organized chess configurations that the expert has represented internally and in the expert's ability to directly associate a chess move with a configural representation of the surface features of the board. A second line of research on expert-novice differences in the domain of physics problem solving (e.g., Chi, Feltovich, & Glaser, 1981; Larkin, McDermott, Simon, & Simon, 1980) also suggests that a central component of skill difference lies in differences in the problem representations of experts and novices. This work is too extensive to be reviewed here. Suffice it to say that Larkin et al. (1980) have obtained evidence for four stages of representation in physics problem solving – the first stage being a literal representation of the problem statement, the second stage being a "naive" representation of objects and their spatial relations set up in the problem, the third stage being a "scientific" representation whose entities are abstract concepts such as forces and momenta, and a fourth stage being the algebraic representation of the problem. The Chi et al. work, based on the categorization of physics problems by subjects at different levels of skill acquisition, suggests that novices classify problems according to features explicitly stated in the problem whereas experts use principles derived from these features as a basis for classification.

In summary, I have argued – as have Posner and Henik – that the multiplicity of available systems for representing spatial information is a topic for serious investigation in the field of cognitive science. The flexibility of the organism in selecting an appropriate representation for solving a spatial problem is evident upon consideration of how representations may change as a result of sometimes rather subtle changes in task structure, how individual differences in ability patterns may lead to the selection of alternative representations, and how the nature of representation can change qualitatively as increasing skill in particular problem-solving domains is acquired.

References

Anderson, J. R. Arguments concerning representations for mental imagery. *Psychological Review,* 1978, **85**, 249-277.

Carpenter, P. A., & Just, M. A. Sentence comprehension: A psycholinguistic processing model of verification. *Psychological Review,* 1975, **82**, 45-73.

Chase, W. G., & Simon, H. A. The mind's eye in chess. In W. G. Chase (Ed.), *Visual Information Processing.* New York: Academic Press, 1973.

Chi, M. T. H., Feltovich, P. J., & Glaser, R. Categorization and representation of physics problems by experts and novices. *Cognitive Science,* 1981, **5**, 121-152.

Clark, H. G., & Chase, W. G. On the process of comparing sentences against pictures. *Cognitive Psychology,* 1972, **3**, 472-517.

Cooper, L. A. Mental rotation of random two-dimensional shapes. *Cognitive Psychology,* 1975, **7**, 20-43.

Cooper, L. A., & Shepard, R. N. Chronometric studies of the rotation of mental images. In W. G. Chase (Ed.), *Visual Information Processing.* New York: Academic Press, 1973.

Cooper, L. A., & Shepard, R. N. Transformations on representations of objects in space. In E. C. Carterette & M. P. Friedman (Eds.), *Handbook of Perception VIII: Perceptual Coding.* New York: Academic Press, 1978.

Glushko, R. J., & Cooper, L. A. Spatial comprehension and comparison processes in verification tasks. *Cognitive Psychology,* 1978, **10**, 391-421.

Kosslyn, S. M. Scanning visual images: Some structural implications. *Perception and Psychophysics,* 1973, **14**, 90-94.

Kosslyn, S. M. Information representation in visual images. *Cognitive Psychology,* 1975, **7**, 341-370.

Kosslyn, S. M. *Image and Mind.* Cambridge, MA: Harvard University Press, 1980.

Kosslyn, S. M. The medium and the message in mental imagery. *Psychological Review,* 1981, **88**, 46-66.

Kosslyn, S. M., Ball, T. M., & Reiser, B. J. Visual images preserve metric spatial information: Evidence from studies of information retrieval time. *Journal of Experimental Psychology: Human Perception and Performance,* 1978, **4**, 47-60.

Larkin, J. H., McDermott, J., Simon, D. P., & Simon, H. A. Expert and novice performance in solving physics problems. *Science,* 1980, **208**, 1335-1342.

Mandler, J. M. Representation. In J. H. Flavell & E. M. Markman (Eds.), *Cognitive Development.* (Vol. 2 of P. Mussen (Ed.), *Manual of Child Psychology.)* New York: Wiley, in press.

Palmer, S. E. Fundamental aspects of cognitive representation. In E. Rosch & B. B. Lloyd (Eds.), *Cognition and Categorization.* Hillsdale, NJ: Erlbaum, 1978.

Piaget, J. *The Origins of Intelligence in Children.* New York: International Universities Press, 1952.

Piaget, J., & Inhelder, B. *Mental Imagery in the Child: A Study of the Development of Imaginal Representation.* London: Routledge and Kegan Paul, 1971.

Pylyshyn, Z. W. What the mind's eye tells the mind's brain: A critique of mental imagery. *Psychological Bulletin,* 1973, **80**, 1-24.

Pylyshyn, Z. W. The rate of "mental rotation" of images: A test of a holistic analogue hypothesis. *Memory and Cognition,* 1979, **7**, 19-28.

Pylyshyn, Z. W. The imagery debate: Analogue media versus tacit knowledge. *Psychological Review,* 1981, **88**, 16-45.

Shepard, R. N., & Metzler, J. Mental rotation of three-dimensional objects. *Science,* 1971, **171**, 701-703.

Computing with Connections

J. A. Feldman
D. H. Ballard

University of Rochester
Rochester, New York

Abstract

Much of the progress in the fields constituting cognitive science has been based upon the use of explicit information processing models, almost exclusively patterned after conventional serial computers. An extension of these ideas to massively parallel, connectionist models appears to offer a number of advantages. After a preliminary discussion, this paper introduces a general connectionist model and considers how it might be used in cognitive science. Among the issues addressed are: stability and noise-sensitivity, distributed decision-making, time and sequence problems, and the representation of complex concepts.

1. Introduction

Much of the progress in the fields constituting cognitive science has been based upon the use of concrete information processing models (IPM), almost exclusively patterned after conventional sequential computers. There are several reasons for trying to extend IPM to cases where the computations are carried out by a parallel computational engine with perhaps billions of active units. As an introduction, we will attempt to motivate the current interest in massively parallel models from four different perspectives: anatomy, computational complexity, technology, and the role of formal languages in science. It is the last of these which is of primary concern here. We will focus upon a particular formalism, connectionist models (CM), which is based explicitly on an abstraction of our current understanding of the information processing properties of neurons.

Copyright © 1983 by Academic Press, Inc.

Animal brains do not compute like a conventional computer. Comparatively slow (millisecond) neural computing elements with complex, parallel connections form a structure which is dramatically different from a high-speed, predominantly serial machine. Much of current research in the neurosciences is concerned with tracing out these connections and with discovering how they transfer information. One purpose of this paper is to suggest how connectionist theories of the brain can be used to produce testable, detailed models of interesting behaviors. The distributed nature of information processing in the brain is not a new discovery. The traditional view (which we shared) is that conventional computers and languages were Turing universal and could be made to simulate any parallelism (or analog values) which might be required. Contemporary computer science has sharpened our notions of what is "computable" to include bounds on time, storage, and other resources. It does not seem unreasonable to require that computational models in cognitive science be at least plausible in their postulated resource requirements.

The critical resource that is most obvious is time. Neurons whose basic computational speed is a few milliseconds must be made to account for complex behaviors which are carried out in a few hundred milliseconds (Posner, 1978). This means that *entire complex behaviors are carried out in less than a hundred time steps*. Current AI and simulation programs require millions of time steps. It may appear that the problem posed here is inherently unsolvable and that there is an error in our formulation. But recent results in computational complexity theory (Ja'Ja' & Simon, 1980) suggest that networks of active computing elements can carry out at least simple computations in the required time range. In subsequent sections we present fast solutions to a variety of relevant computing problems. These solutions involve using massive numbers of units and connections, and we also address the questions of limitations on these resources.

Another recent development is the feasibility of building parallel computers. There is currently the capability to produce chips with 100,000 gates at a reproduction cost of a few cents each, and the technology to go to 1,000,000 gates/chip appears to be in hand. This has two important consequences for the study of CM. The obvious consequence is that it is now feasible to fabricate massively parallel computers, although no one has yet done so (Hillis, 1981; Fahlman, 1980). The second consequence of this development is the renewed interest in the basic properties of highly parallel computation. A major reason why there aren't yet any of these CM machines is that we do not yet know how to design, assemble, test, or program such engines. An important motivation for the careful study of CM is to learn more about how to implement parallel computation, but we will say no more about that in this paper.

The most important reason for a serious concern in cognitive science for CM is that they might lead to better science. It is obvious that the choice of technical language that is used for expressing hypotheses has a profound influence on the form in which theories are formulated and experiments undertaken. Artificial intelligence and articulating cognitive sciences have made great progress by employing models based on conventional digital computers as theories of intelligent behavior. But a number of crucial phenomena such as associative memory, priming, perceptual rivalry, and the remarkable recovery ability of animals have not yielded to this treatment. A major goal of this paper is to lay a foundation for the systematic use of massively parallel connectionist models in the cognitive sciences, even where these are not yet reducible to physiology or silicon.

Over the past few years, a number of investigators in different fields have begun to employ highly parallel models (idiosyncratically) in their work. The general idea has been advocated for animal models by Arbib (1979) and for cognitive models by Anderson (Anderson et al., 1977) and Ratcliff (1978). Parallel search of semantic memory and various "spreading activation" theories have become common (though not quite consistent) parts of information processing modeling. In machine perception research, massively parallel, cooperative computational theories have become a dominant paradigm (Marr & Poggio, 1976; Rosenfeld et al., 1976) and many of our examples come from our own work in this area (Sabbah, 1981; Ballard, 1981). Scientists looking at performance errors and other non-repeatable behaviors have not found conventional IPM to be an adequate framework for their efforts. Norman (1981) has recently summarized arguments from cognitive psychology, and Kinsbourne and Hicks (1979) have been led to a similar view from a different perspective. It appears to us that all of these efforts could fit within the CM paradigm outlined here.

One of the most interesting recent studies employing CM techniques is the partial theory of reading developed in (McClelland & Rumelhart, 1981). They were concerned with the word superiority effect and related questions in the perception of printed words, and had a large body of experimental data to explain. One major finding is that the presence of a printed letter in a brief display is easier to determine when the letter is presented in the context of a word that when it is presented alone. The model they developed (cf. Figure 1) explicitly represents three levels of processing: visual features of printed letters, letters, and words. The model assumes that there are positive and negative (circular tipped) connections from visual features to the letters that they can (respectively, cannot) be part of. The connections between letters and words can go in either direction and embody the constraints of English. The model assumes that many units can be simultaneously active, that units form algebraic sums of their inputs and output values proportionally. The activity of a unit is bounded from above and below, has some memory, and decays with time. All of these features, and several more, are captured in the abstract unit described in Section 2.

This idea of simultaneously evaluating many hypotheses (here words) has been successfully used in machine perception for some time (Hanson & Rise-

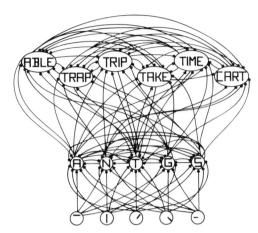

Figure 1. A few of the neighbors of the node for the letter "t" in the first position in a word, and their interconnections (McClelland & Rumelhart, 1981).

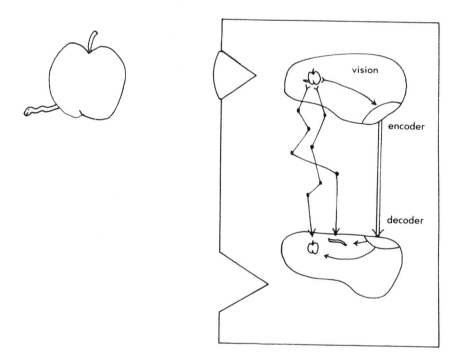

Figure 2. Connectionism vs. symbolic encoding.

=> Assumes some general encoding
-> Assumes individual connections

man, 1978). What has occurred to us relatively recently is that this is a natural mode of computation for widely interconnected networks of active elements like those envisioned in connectionist models. The generalization of these ideas to the connectionist view of brain and behavior is that all important encodings in the brain are in terms of the relative strengths of synaptic connections. The fundamental premise of connectionism is that individual neurons *do not transmit large amounts of symbolic information*. Instead they compute by being *appropriately connected* to large numbers of similar units. This is in sharp contrast to the conventional computer model of intelligence prevalent in computer science and cognitive psychology.

The fundamental distinction between the conventional and connectionist computing models can be conveyed by the following example. When one sees an apple and says the phrase "wormy apple," some information must be transferred, however indirectly, from the visual system to the speech system. Either a sequence of special *symbols* that denote a wormy apple is transmitted to the speech system, or there are special *connections* to the speech command area for the words. Figure 2 is a graphic presentation of the two alternatives. The path on the right described by double-lined arrows depicts the situation (as in a computer) where the information that a wormy apple has been seen is encoded by the visual system and sent as an abstract message (perhaps frequency-coded) to a general receiver in the speech system which decodes the message and initiates the appropriate speech act. Notice that a complex message would presumably have to be transmitted sequentially on this channel, and that each end would have to learn the common code for every new concept. No one has yet produced a biologically and computationally plausible realization of this conventional computer model.

The only alternative that we have been able to uncover is described by the path with single-width arrows. This suggests that there are (indirect) links from the units (cells, columns, centers, or what-have-you) that recognize an apple to some units responsible for speaking the word. The connectionist model requires only very simple messages (e.g. stimulus strength) to cross a channel but puts strong demands on the availability of the right connections. Questions concerning the learning and reinforcement of connections are addressed in (Feldman, 1981b).

For a number of reasons (including redundancy for reliability), it is highly unlikely that there is exactly one neuron for each concept, but the point of view taken here is that the activity of a small number of neurons (say 10) encodes a concept like apple. An alternative view (Hinton & Anderson, 1981) is that concepts are represented by a "pattern of activity" in a much larger set of neurons (say 1,000) which also represent many other concepts. We have not seen how to carry out a program of specific modeling in terms of these diffuse models. One of the major problems with diffuse models as a parallel computation scheme is cross-talk among concepts. For example, if concepts using units

(10, 20, 30, . . .) and (5, 15, 25, . . .) were simultaneously activated, many other concepts, e.g., (20, 25, 30, 35, . . .) would be active as well. In the example of Figure 2, this means that diffuse models would be more like the shared sequential channel. Although a single concept could be transmitted in parallel, complex concepts would have to go one at a time. Simultaneously transmitting multiple concepts that shared units would cause cross-talk. It is still true in our CM that many related units will be triggered by spreading activation, but the representation of each concept is taken to be compact.

Most cognitive scientists believe that the brain appears to be massively parallel and that such structures can compute special functions very well. But massively parallel structures do not seem to be usable for general purpose computing and there is not nearly as much knowledge of how to construct and analyze such models. The common belief (which may well be right) is that there are one or more intermediate levels of computational organization layered on the neuronal structure, and that theories of intelligent behavior should be described in terms of these higher level languages, such as Production Systems, Predicate Calculus, or LISP. We have not seen a reduction (interpreter, if you will) of any higher formalism which has plausible resource requirements, and this is a problem well worth pursuing.

Our attempts to develop cognitive science models directly in neural terms might fail for one of two reasons. It may be that there really is an interpreted symbol system in animal brains. In this case we would hope that our efforts would break down in a way that could shed light on the nature of this symbol system. The other possibility is that CM techniques are directly applicable but we are unable to figure out how to model some important capacity, e.g., planning. Our program is to continue the CM attack on problems of increasing difficulty (and to induce some of you to join us) until we encounter one that is intractable in our terms. There are a number of problems that are known to be difficult for systems without an interpreted symbolic representation, including complex concepts, learning, and natural language understanding. The current paper is mainly concerned with laying out the formalism and showing how it applies in the easy cases, but we do address the problem of complex concepts in Section 4. We have made some progress on the problem of learning in CM systems (Feldman, 1981b) and are beginning to work seriously on natural language processing and on higher-level vision. Our efforts on planning and long-term memory reorganization have not advanced significantly beyond the discursive presentation in (Feldman, 1980).

We will certainly not get very far in this program without developing some systematic methods of attacking CM tasks and some building-block circuits with well-understood properties. A first step towards a systematic development of CM is to define an abstract computing unit. Our unit is rather more general than previous proposals and is intended to capture the current understanding of the information processing capabilities of neurons. Some useful special cases of our general definition and some properties of very simple networks are developed in Section 2. Among the key ideas are local memory, non-homogeneous and non-linear functions, and the notions of mutual inhibition and stable coalitions.

A major purpose of the rest of the paper is to describe building blocks which we have found useful in constructing CM solutions to various tasks. The constructions are intended to be used to make specific models but the examples in this paper are only suggestive. We present a number of CM solutions to general problems arising in intelligent behavior, but *we are not suggesting that any of these are necessarily employed by nature.* Our notion of an adequate model is one that accounts for *all* of the established relevant findings and this is not a task to be undertaken lightly. We are developing some preliminary sketches (Ballard & Sabbah, 1981; Sabbah, 1981; Ballard & Kimball, 1981a,b) for a serious model of low and intermediate level vision. As we develop various building blocks and techniques we will also be trying to bury some of the contaminated debris of past neural modeling efforts. Many of our constructions are intended as answers to known hard problems in CM computation. Among the issues addressed are: stability and noise-sensitivity, distributed decision-making, time and sequence problems, and the representation of complex concepts. The crucial questions of learning and change in CM systems are discussed elsewhere (Feldman, 1981b).

The body of this paper has four sections. Section 2 contains the basic definitions for a tractable and biologically plausible neuron-level computing unit. Although there is a rich tradition of neural modeling research, much of which will be useful to us, our definitions depart from standard ones. A primitive unit can have both symbolic and numerical state, can treat its inputs non-uniformly, and need not compute a linear function. A particularly important construct is the use of groups or "conjunctions" of input connections. Some important special cases and some simple examples, based on lateral inhibition, are presented. Encapsulation techniques are suggested as a basis for simplifying larger problems.

Section 3 is concerned with the general computing abilities of networks of our units. The crucial point is achieving a single coherent action in a diffuse set of units. Winner-take-all (WTA) networks are introduced as our solution to this problem for single layers. More generally, we define and study the idea of a stable coalition of units whose mutual reinforcement has the effect of a single action, perception, etc.

Section 4 concentrates on some specific computations and how they can be effectively performed within the model. We begin with computing simple functions like multiplication and show how general parameters can be treated. Modifiers and mappings are used to show how connections can effectively be treated as dynamic.

In Section 5 we tackle some additional classic problems for connectionism and apply our ideas to some more problems in visual perception. A representation for conjunctive concepts such as "big blue cube" is laid out and applied to the description of complex objects. Finally, as another indication of the way we intend to proceed, sketches of time-varying situations such as speech and movement are discussed.

2. Neuron-like Computing Units

A central part of a generally useful framework for connectionist theories is a standard model of the individual unit. It will turn out that a "unit" may be used to model anything from a small part of a neuron to the external functionality of a major subsystem. But the basic notion of unit is meant to loosely correspond to an information processing model of our current understanding of neurons. The particular definitions here were chosen to make it easy to specify detailed examples of relatively complex behaviors. There is no attempt to be minimal or mathematically elegant. The various numerical values appearing in the definitions are arbitrary, but fixed finite bounds play a crucial role in the development. The presentation of the definitions will be in stages, accompanied by examples. A compact technical specification for reference purposes is given in the Appendix. Each unit will be characterized by a small number of discrete states plus:

p – a continuous value in $(-10,10)$, called *potential* (accuracy of several digits)

v – an *output value*, integers $0 \leq v \leq 9$

$\mathbf{i}$ – a vector of *inputs* $i_l, \ldots , i_n$

2.1. P-units

Some applications can use a particularly simple kind of unit whose output v is proportional to its potential p (rounded) when $p > 0$ and which has only one state. In other words

$$p \leftarrow p + \beta \, \Sigma w_k i_k \qquad\qquad (0 \leq w_k \leq 1)$$

$$v \leftarrow \text{if } p > \theta \text{ then round } (p - \theta) \text{ else } 0 \quad (0, \ldots , 9)$$

where β, θ are constants and w_k are weights on the input values. The weights are the sole locus of change with experience in the current model. Most often, the potential and output of a unit will be encoding its *confidence*, and we will sometimes use this term. The $\leftarrow$ notation is borrowed from the assignment statement of programming languages. This notation covers both continuous and discrete time formulations and allows us to talk about some issues without any explicit mention of time. Of course, certain other questions will inherently involve time and computer simulation of any network of units will raise delicate questions of discretizing time.

The restriction that output take on small integer values is central to our enterprise. The firing frequencies of neurons range from a few to a few hundred impulses per second. In the 1/10 second needed for basic mental events, there can only be a limited amount of information encoded in frequencies. The ten output values are an attempt to capture this idea. A more accurate rendering of neural events would be to allow 100 discrete values with noise on transmission (cf. Sejnowski, 1977). Transmission time is assumed to be negligible; delay units can be added when transit time needs to be taken into account.

The p-unit is somewhat like classical linear threshold elements (Perceptrons; Minsky & Papert, 1972), but there are several differences. The potential, p, is a crude form of memory and is an abstraction of the instantaneous membrane potential that characterizes neurons; it greatly reduces the noise sensitivity of the networks. Without local memory in the unit, one must guarantee that all the inputs required for a computation appear simultaneously at the unit.

A problem with the definition above of a p-unit is that its potential does not decay in the absence of input. This decay is both a physical property of neurons and an important computational feature for our highly parallel models. One computational trick to solve this is to have an inhibitory connection from the unit back to itself. Informally, we identify the negative self-feedback with an exponential decay in potential which is mathematically equivalent. With this addition, p-units can be used for many CM tasks of intermediate difficulty. The Interactive Activation models of McClelland and Rumelhart can be described naturally with p-units, and some of our own work (Ballard, 1981) and that of others (Marr & Poggio, 1976) can be done with p-units. But there are a number of additional features which we have found valuable in more complex modeling tasks.

2.2. Disjunctive Firing Conditions and Conjunctive Connections

In certain cases, the unit should have a more complicated response than the straightforward summation of weighted inputs suggested by our initial characterization of p-units. For example, it is both computationally efficient and biologically realistic to allow a unit to respond to one of a number of alternative conditions. A way of viewing this is to imagine the unit having "dendrites" each of which depicts an alternative enabling condition (Figure 3). For example, one could extend the network of Figure 1 to allow for several different type fonts activating the same letter node, with the higher connections unchanged. Biologically, the firing of a neuron depends, in many cases, on local spatio-temporal summation involving only a small part of the neuron's surface. So-called dendritic spikes transmit the activation to the rest of the cell.

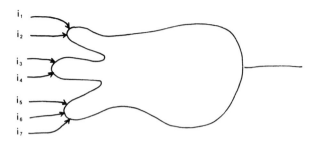

Figure 3. Conjunctive connections and disjunctive input sites.

This could be described in a variety of ways, but one of the simplest is to define the potential in terms of the maximum of the separate computations, e.g.,

$$p \leftarrow p + \beta \mathrm{Max}(i_l + i_2 - \phi, \; i_3 + i_4 - \phi, \; i_5 + i_6 - i_7 - \phi)$$

where β is a scale constant as in the p-unit and ϕ is a constant chosen (usually > 10) to suppress noise and require the presence of multiple active inputs (Sabbah, 1981). The minus sign associated with i_7 corresponds to its being an inhibitory input.

It does not seem unreasonable (given current data; Kuffler & Nicholls, 1976) to model the firing rate of some units as the maximum of the rates at their active sites. Units whose potential is changed according to the maximum of a set of algebraic sums will occur frequently in our specific models. One advantage of keeping the processing power of our abstract unit close to that of a neuron is that it helps inform our counting arguments. When we attempt to model a particular function (e.g. stereopsis), we desire that the number of units and connections as well as the execution time required by the model be plausible.

The max-of-sum unit is the continuous analog of a logical OR-of-AND (disjunctive normal form) unit and it is sometimes convenient to use the latter as an approximate version of the former. The OR-of-AND unit corresponding to Figure 3 is:

$$p \leftarrow p + \alpha \; \mathrm{OR} \; (i_1 \& i_2, \; i_3 \& i_4, \; i_5 \& i_6 \& (\mathrm{not} \; i_7))$$

This formulation stresses the importance that nearby spatial connections *all* be firing before the potential is affected. Hence, in the above example, i_3 and i_4 make a *conjunctive connection* with the unit. The effect of a conjunctive connection can always be simulated with more units but the number of extra units may be very large.

2.3. Q-units and Compound Units

Another useful special case arises when one suppresses the numerical potential, p, and relies upon a finite-state set $\{q\}$ for modeling. If we also identify each input of i with a separate named input signal, we can get classical finite automata. A simple example would be a unit that could be started or stopped from firing.

One could describe the behavior of this unit by a table, with rows corresponding to states in $\{q\}$ and columns to possible inputs, e.g.,

	i_1 (start)	i_2 (stop)
Firing	Firing	Null
Null	Firing	Null

One would also have to specify an output function, giving output values required by the rest of the network, e.g.,

$$v \leftarrow \textbf{if } q \ = \text{Firing } \textbf{then } 6 \textbf{ else } 0.$$

This could also be added to the table above. An equivalent notation would be transition networks with states as nodes and inputs and outputs on the arcs.

In order to build models of interesting behaviors we will need to employ many of the same techniques used by designers of complex computers and programs. One of the most powerful techniques will be encapsulation and abstraction of a subnetwork by an individual unit. For example, a system that had separate motor abilities for turning left and turning right (e.g., fins) could use two start-stop units to model a turn-unit, as shown in Figure 4.

Note that the compound unit here has two distinct outputs, where basic units have only one (which can branch, of course). In general, compound units will differ from basic ones only in that they can have several distinct outputs.

The main point of this example is that the turn-unit can be described abstractly, independent of the details of how it is built. For example, using the tabular conventions described above,

	Left	Right	Values Output
a gauche	a gauche	a droit	$v_1 = 7, \ v_2 = 0$
a droit	a gauche	a droit	$v_1 = 0, \ v_2 = 8$

where the right-going output being larger than the left could mean a right-finned robot. There is a great deal more that must be said about the use of states and symbolic input names, about multiple simultaneous inputs, etc., but the idea of describing the external behavior of a system only in enough detail for the task at hand is of great importance. This is one of the few ways known of coping

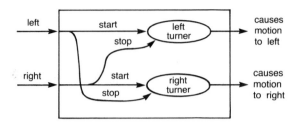

Figure 4. A turn unit.

with the complexity of the magnitude needed for serious modeling of biological functions. It is not strictly necessary that the same formalism be used at each level of functional abstraction and, in the long run, we may need to employ a wide range of models. For example, for certain purposes one might like to expand our units in terms of compartmental models of neurons like those of (Perkel, 1979). The advantage of keeping within the same formalism is that we preserve intuition, mathematics, and the ability to use existing simulation programs. With sufficient care, the units defined above can represent large subsystems without giving up the notion that each unit can stand for an abstract neuron. The crucial point is that a subsystem must be elaborated into its neuron-level units for timing and size calculations, but can (hopefully) be described much more simply when only its effects on other subsystems are of direct concern.

2.4. Units Employing p and q

It will already have occurred to the reader that a numerical value, such as potential, would be useful for modeling the amount of turning to the left or right in the last example. It appears to be generally true that a single numerical value and a small set of discrete states combine to provide a powerful yet tractable modeling unit. This is one reason that the current definitions were chosen. Another reason is that the mixed unit seems to be a particularly convenient way of modeling the information processing behavior of neurons, as generally described. The discrete states enable one to model the effects in neurons of polypeptide modulators, abnormal chemical environments, fatigue, etc. Although these effects are often continuous functions of unit parameters, there are several advantages to using discrete states in models. Scientists and laymen alike often give distinct names (e.g., cool, warm, hot) to parameter ranges that they want to treat differently. We also can exploit a large literature on understanding loosely-coupled systems as finite-state machines (Sunshine, 1979). It is also traditional to break up a function into separate ranges when it is simpler to describe that way. We have already employed all of these uses of discrete states in our detailed work (Sabbah, 1981; Feldman, 1981b). One example of a unit employing both p and q non-trivially is the following crude neuron model.

This model is concerned with saturation and assumes that the output strength, v, is something like average firing frequency. It is not a model of individual action potentials and refractory periods.

As another example of the use of q, suppose the distinct states of the unit $q \in \{normal, recover\}$. In *normal* state the unit behaves like a p-unit, but while it is *recovering* it ignores inputs. The incomplete following table captures almost all of this behavior.

	$-1 < p < 9$	$p > 9$	Output Value
normal	$p \leftarrow p + \Sigma i$	$p \leftarrow -p/$ recover	$v \leftarrow \alpha \, p - \theta$
recover	normal	$<$impossible$>$	$v \leftarrow 0$

Here the change from one state to the other depends on the value of the potential, p, rather than on specific inputs. The recovering state is also characterized by the potential being set negative. The unspecified issue is what determines the duration of the recovering state – there are several possibilities. One is an explicit dishabituation signal like those in Kandel's experiments (Kandel, 1976). Another would be to have the unit sum inputs in the recovering state as well. The reader might want to consider how to add this to the table. A third possibility, which we will use frequently, is to assume that the potential, p, decays toward zero (from both directions) unless explicitly changed. This implicit decay $p \leftarrow p_0 e^{-kt}$ can be modeled by self-inhibition; the decay constant, k, determines the length of the recovery period.

The general definition of our abstract neural computing unit is just a formalization of the ideas presented above. To the previous notions of p, v, and $\mathbf{i}$ we formally add

$$\{q\} - \text{a set of } discrete \ states \, , \, < 10$$

and functions from old to new values of these:

$$p \leftarrow f(\mathbf{i}, p, q)$$
$$q \leftarrow g(\mathbf{i}, p, q)$$
$$v \leftarrow h(\mathbf{i}, p, q)$$

which we assume, for now, to compute continuously. The form of the f, g, and h functions will vary, but will generally be restricted to conditionals and simple functions. There are both biological and computational reasons for allowing units to respond (for example) logarithmically to their inputs and we have already seen important uses of the maximum function.

The only other notion that we will need is modifiers associated with the inputs of a unit. We elaborate the input vector $\mathbf{i}$ in terms of received values, weights, and modifiers:

$$\forall j, i_j = r_j \cdot w_j \cdot m_j \quad j = 1, \ldots, n$$

where r_j is the *value* received from a predecessor $[r = 0, \ldots, 9]$; w_j is a changeable *weight*, unsigned $[0 \leq w_j \leq 1]$ (accuracy of several digits); and m_j is a synapto-syntaptic *modifier* which is either 0 or 1.

The weights are the only thing in the system which can change with experience. They are unsigned because we do not want a connection to change from excitatory to inhibitory. The modifier or gate simplifies many of our detailed models. Learning and change will not be treated technically in this paper, but the definitions are included in the Appendix for completeness (Feldman, 1981b).

We conclude this section with some preliminary examples of networks of our units, illustrating the key idea of mutual (lateral) inhibition (Figure 5). Mutual inhibition is widespread in nature and has been one of the basic computational schemes used in modeling. We will present two examples of how it works to help aid in intuition as well as to illustrate the notation. The basic situation is symmetric configurations of p-units which mutually inhibit one another. Time is broken into discrete intervals for these examples. The examples are too simple to be realistic, but do contain ideas which we will employ repeatedly.

2.5. Two P-units Symmetrically Connected

Suppose $w_1 = 1$, $w_2 = -.5$

$$p(t+l) = p(t) + r_1 - (.5)r_2 \quad\quad r_j = \text{received}$$

$$v = \text{round}(p) \ [0, \ldots, 9]$$

Referring to Figure 5a, suppose the initial input to the unit $A.1$ is 6, then 2 per time step, and the initial input to $B.1$ is 5, then 2 per time step. At each time step, each unit changes its potential by adding the external value (r_1) and subtracting half the output value of its rival. This system will stabilize to the side of the larger of two instantaneous inputs.

2.6. Two Symmetric Coalitions of 2-units

$$w_1 = 1$$
$$w_2 = .5$$
$$w_3 = -.5$$
$$p(t+1) = p(t) + r_1 + .5(r_2 - r_3)$$
$$v = \text{round}(p)$$

A, C start at 6; B, D at 5;

A, B, C, D have no external input for $t > 1$

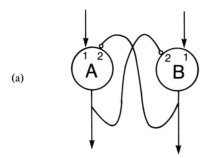

Suppose A₁ received an input of 6 units, then 2 per time step
Suppose B₁ received an input of 5 units, then 2 per time step

t	P(A)	P(B)
1	6	5
2	5.5	4
3	5.5	3.5
4	6	3
5	6.5	2
6	7.5	1
7	9.5	0
8	Sat	0

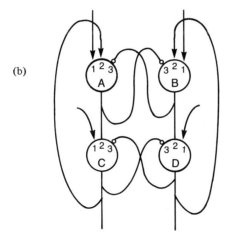

t	P(A)	P(B)	P(C)	P(D)
1	6	5	6	5
2	6.5	4.5	6.5	4.5
3	7.5	3.5	7.5	3.5
4	9.5	1.5	9.5	1.5
	Sat	0	Sat	0

Figure 5. (a) and (b) Small symmetric networks.

The connections for this system are shown in Figure 5b. This system converges faster than the previous example. The idea here is that units A and C form a "coalition" with mutually reinforcing connections. The competing units are A vs. B and C vs. D. The last example is the smallest network depicting what we believe to be the basic mode of operations in connectionist systems. The faster convergence is not an artifact; the *positive feedback* among members of a coalition will generally lead to faster convergence than in separate competitions. It is the amount of positive feedback rather than just the size of the coalition that determines the rate of convergence (Feldman & Ballard, 1982). In terms of Figure 1, this could represent the behavior of the rival letters A and T in conjunction with the rival words ABLE and TRAP, in the absence of other active nodes.

Competing coalitions of units will be the organizing principle behind most of our models. Consider the two alternative readings of the Necker cube shown in Figure 6. At each level of visual processing, there are mutually contradictory units representing alternative possibilities. The dashed lines denote the boundaries of coalitions which embody the alternative interpretations of the image. A number of interesting phenomena (e.g., priming, perceptual rivalry, filling, subjective contour) find natural expression in this formalism. We are engaged in an ongoing effort (Sabbah, 1981; Ballard, 1981) to model as much of visual processing as possible within the connectionist framework. The next section describes in some detail a variety of simple networks which we have found to be useful in this effort.

3. Networks of Units

The main restriction imposed by the connectionist paradigm is that no symbolic information is passed from unit to unit. This restriction makes it difficult to employ standard computational devices like parameterized functions. In this section, we present connectionist solutions to a variety of computational problems. The sections address two principal issues. One is: Can the networks be connected up in a way that is sufficient to represent the problem at hand? The other is: Given these connections, how can the networks exhibit appropriate dynamic behavior, such as making a decision at an appropriate time?

3.1. Using a Unit to Represent a Value

One key to many of our constructions is the dedication of a separate unit to each value of each parameter of interest, which we term the unit/value principle. We will show how to compute using unit/value networks and present arguments that the number of units required is not unreasonable. In this representation the output of a unit may be thought of as a confidence measure. Suppose a network of depth units encodes the distance of some object from the retina. Then if the unit representing depth $= 2$ saturates, the network is expressing confidence that the distance is two units. Similarly, the "G-hidden" node in Figure 6 expresses confidence in its assertion. There is much neuro-physiological evidence to suggest unit/value organizations in less abstract corti-cal maps. Examples are edge sensitive units (Hubel & Wiesel, 1979) and per-ceptual color units (Zeki, 1980), which are relatively insensitive to illumination spectra. Experiments with cortical motor control in the monkey and cat (Wurtz & Albano, 1980) suggest a unit/value organization. Our hypothesis is that the unit/value organization is widespread, and is a fundamental design principle.

Although many physical neurons do seem to follow the unit/value rule and respond according to the reliability of a particular configuration, there are also other neurons whose output represents the range of some parameter, and apparently some units whose firing frequency reflects both range and strength information (Flanagan, 1979). Both of the latter types can be accommodated within our definition of a unit, but we will employ only unit/value networks in the remainder of this paper.

In the unit/value representation, much computation is done by table look-up. As a simple example, let us consider the multiplication of two variables, i.e., $z = xy$. In the unit/value formalism there will be units for *every* value of x and y that is important. Appropriate pairs of these will make a conjunctive connection with another unit cell representing a specific value for the product. Figure 7 shows this for a small set of units representing values for x and y.

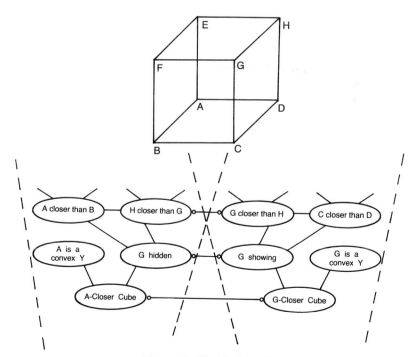

Figure 6. The Necker cube.

Notice that the confidence (expressed as output value) that a particular product is an answer can be a linear function of the maximum of the sums of the confidences of its two inputs. A major problem with function tables (and with CM in general) is the potential combinatorial explosion in the number of units required for a computation. A naive approach would demand N^2 units to represent all products of numbers from 1 to N. The network of Figure 7 requires many fewer units because each product is represented only once, another advantage of conjunctive connections. We could use even fewer units by exploiting positional notation and replacing each output connection with a conjunction of outputs from units representing multiples of 1, 10, 100, etc. The question of efficient ways of building connection networks is treated in detail in Section 4 (cf. also Hinton, 1981a,b).

3.2. Modifiers and Mappings

The idea of function tables (Figure 7) can be extended through the use of *variable mappings*. In our definition of the computational unit, we included a binary modifier, m, as an option on every connection. As the definition specifies, if the modifier associated with a connection is zero, the value v sent along that connection is ignored. Thus the modifier denotes inhibition, or blocking. There is considerable evidence in nature for synapses on synapses (Kandel, 1976) and the modifiers add greatly to the computational simplicity of our networks. Let us start with an initial informal example of the use of modifiers and mappings. Suppose that one has a model of grass as green except in California

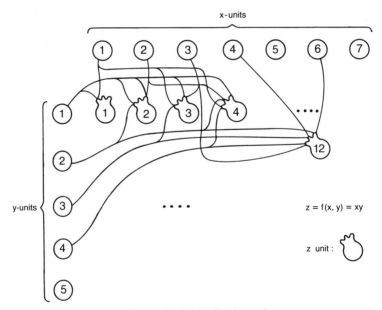

Figure 7. Multiplication units.

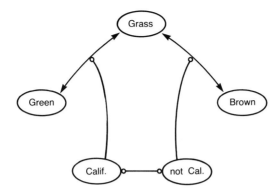

Figure 8. Grass is Green connection modified by California.

where it is brown (golden), as shown in Figure 8. Here we can see that grass and green are potential members of a coalition (can reinforce one another) except when the link is blocked. This use is similar to the cancellation link of (Fahlman, 1979) and gives a crude idea of how context can effect perception in our models. Note that in Figure 8 we are using a shorthand notation. A modifier touching a double-ended arrow actually blocks two connections. (Sometimes we also omit the arrowheads when connection is double-ended.)

Mappings can also be used to select among a number of possible values. Consider the example of the relation between depth, physical size, and retinal size of a circle. (For now, assume that the circle is centered on and orthogonal to the line of sight, that the focus is fixed, etc.) Then there is a fixed relation between the size of retinal image and the size of the physical circle for any given depth. That is, each depth specifies a *mapping* from retinal to physical size (see Figure 9). Here we suppose the scales for depth and the two sizes are chosen so that unit depth means the same numerical size. If we knew the depth of the object (by touch, context, or magic) we could know its physical size. The network above allows retinal size 2 to reinforce physical size 2 when depth = 1 but inhibits this connection for all other depths. Similarly, at depth 3, we should interpret retinal size 2 as physical size 8, and inhibit other interpretations. Several remarks are in order. First, notice that this network implements a function phys = f (ret,dep) that maps from retinal size and depth to physical size, providing an example of how to replace functions with parameters by mappings. For the simple case of looking at one object perpendicular to the line of sight, there will be one consistent coalition of units which will be stable. The network does something more, and this is crucial to our enterprise; the network can represent the consistency relation R among the three quantities: depth, retinal size, and physical size. It embodies not only the function f, but its two inverse functions as well (dep = f_1(ret,phys), and ret = f_2(phys,dep)). (The network as shown does not include the links for f_1 and f_2, but these are similar to those for f.) Most of Section 5 is devoted to laying out networks that embody theories of particular visual consistency relations.

The idea of modifiers is, in a sense, complementary to that of conjunctive connections. For example, the network of Figure 9 could be transformed into the following network (Figure 10). In this network the variables for physical size = 4 and depth = 1 make a *conjunctive connection* with retinal size = 4. Each of the value units in a competing row could be connected to all of its competitors by inhibitory links and this would tend to make the network activate only one value in each category. The general issue of rivalry and coalitions will be discussed in the next two sub-sections.

When should a relation be implemented with modifiers and when should it be implemented with conjunctive connections? A simple, non-rigorous answer to this question can be obtained by examining the size of two sets of units: (1) the number of units that would have to be inhibited by modifiers; and (2) the number of units that would have to be reinforced with conjunctive connections. If (1) is larger that (2), then one should choose modifiers; otherwise choose conjunctive connections. Sometimes the choice is obvious: to implement the brown Californian grass example of Figure 8 with conjunctive connections, one would have to reinforce all units representing places that had green grass! Clearly in this case it is easier to handle the exception with modifiers. On the other hand, the depth relation R (phys,dep,ret) is more cheaply implemented with conjunctive connections. Since our modifiers are strictly binary, conjunctive connections have the additional advantage of continuous modulation.

To see how the conjunctive connection strategy works in general, suppose a constraint relation to be satisfied involves a variable x, e.g., $f(x,y,z,w) = 0$. For a particular value of x, there will be triples of values of y, z, and w that satisfy the relation f. Each of these triples should make a conjunctive connection with the unit representing the x-value. There could also be 3-input conjunctions at each value of y, z, w. Each of these four different kinds of conjunctive connections corresponds to an interpretation of the *relation* $f(x,y,z,w) = 0$ as a *function,* i.e., $x = f_1(y,z,w)$, $y = f_2(x,z,w)$, $z = f_3(x,y,w)$, or $w = f_4(x,y,z)$. Of course, these functions need not be single-valued. This network connection pattern could be extended to more than four variables, but high numbers of variables would tend to increase its sensitivity to noisy inputs. Hinton has suggested a special notation for the situation where a network exactly captures a consistency relation. The mutually consistent values are all shown to be centrally linked (Figure 11). This notation provides an elegant way of presenting the interactions among networks, but must be used with care. Writing down a triangle diagram does not insure that the underlying mappings can be made consistent or computationally well-behaved.

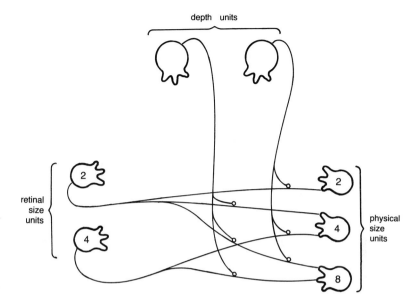

Figure 9. Depth network using modifiers.

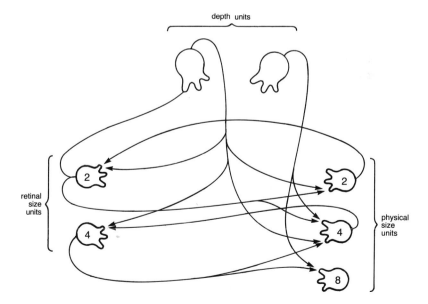

Figure 10. Depth network using conjunctive connections.

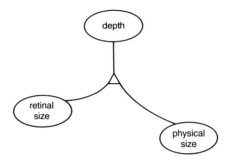

Figure 11. Notation for consistency relations.

3.3. Winner-take-all Networks and Regulated Networks

A very general problem that arises in any distributed computing situation is how to get the entire system to make a decision (or perform a coherent action, etc.). Biologically necessary examples of this behavior abound, ranging from turning left or right, through fight-or-flight responses, to interpretations of ambiguous words and images. Decision-making is a particularly important issue for the current model because of its restrictions on information flow and because of the almost linear nature of the p-units used in many of our specific examples. Decision-making introduces the notions of *stable states* and *convergence* of networks.

One way to deal with the issue of coherent decisions in a connectionist framework is to introduce *winner-take-all* (WTA) networks, which have the property that only the unit with the highest potential (among a set of contenders) will have output above zero after some settling time (Figure 12). There are a number of ways to construct WTA networks from the units described above. For our purposes it is enough to consider one example of a WTA network which will operate in one time step for a set of contenders each of whom can read the potential of all of the others. Each unit in the network computes its new potential according to the rule

$$p \leftarrow \textbf{if } p > \max(i_j, .1) \textbf{ then } p \textbf{ else } 0.$$

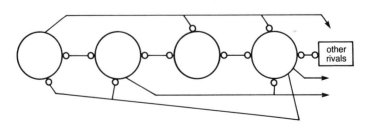

Figure 12. Winner-take-all network. Each unit stops if it sees a higher value.

That is, each unit sets itself to zero if it knows of a higher input. This is fast and simple, but probably a little too complex to be plausible as the behavior of a single neuron. There is a standard trick (apparently widely used by nature) to convert this into a more plausible scheme. Replace each unit above with two units; one computes the maximum of the competitor's inputs and inhibits the other. The circuit above can be strengthened by adding a reverse inhibitory link, or one could use a modifier on the output, etc. Obviously one could have a WTA layer that got inputs from some set of competitors and settled to a winner when triggered to do so by some downstream network. This is an exact analogy of strobing an output buffer in a conventional computer.

One problem with previous neural modeling attempts is that the circuits proposed were often unnaturally delicate (unstable). Small changes in parameter values would cause the networks to oscillate or converge to incorrect answers. We will have to be careful not to fall into this trap, but would like to avoid detailed analysis of each particular model delicacy in this paper. What appears to be required are some building blocks and combination rules that preserve the desired properties. For example, the WTA subnetworks of the last example will not oscillate in the absence of oscillating inputs. This is also true of any symmetric mutually inhibitory subnetwork. This is intuitively clear and could be proven rigorously under a variety of assumptions (cf. Grossberg, 1980). If every unit receives inhibition proportional to the activity (potential) of each of its rivals, the instantaneous leader will receive less inhibition and thus not lose its lead unless the inputs change significantly.

Another useful principle is the employment of lower-bound and upper-bound cells to keep the total activity of a network within bounds (Figure 13). Suppose that we add two extra units, LB and UB, to a network which has coordinated output. The LB cell compares the total (sum) activity of the units of the network with a lower bound and sends positive activation uniformly to all members if the sum is too low. The UB cell inhibits all units equally if the sum of activity is too high. Notice that LB and UB can be parameters set from

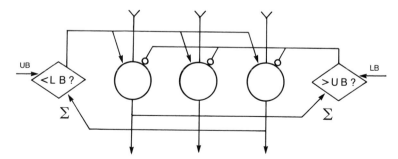

Figure 13. Regulated network using upper-bound and lower-bound units. If sum exceeds UB all units get uniform inhibition.

outside the network. Under a wide range of conditions (but not all), the LB-UB augmented network can be designed to preserve order relationships among the outputs v_j of the original network while keeping the sum between LB and UB.

We will often assume that LB-UB pairs are used to keep the sum of outputs from a network within a given range. This same mechanism also goes far towards eliminating the twin perils of uniform saturation and uniform silence which can easily arise in mutual inhibition networks. Thus we will often be able to reason about the computation of a network assuming that it stays active and bounded.

3.4. Stable Coalitions

For a massively parallel system to actually make a decision (or do something), there will have to be states in which some activity strongly dominates. Such stable, connected, high confidence units are termed *stable coalitions*. A stable coalition is our architecturally-biased term for the psychological notions of percept, action, etc. We have shown some simple instances of stable coalitions, in Figure 5b and the WTA network. In the depth networks of Figures 9 and 10, a stable coalition would be three units representing consistent values of retinal size, depth, and physical size. But the general idea is that a very large complex subsystem must stabilize, e.g., to a fixed interpretation of visual input, as in Figure 1. The way we believe this to happen is through mutually reinforcing coalitions which dominate all rival activity when the decision is required. The simplest case of this is Figure 5b, where the two units A and B form a coalition which suppresses C and D. Formally, *a coalition will be called stable when the output of all its members is non-decreasing.* Notice that a coalition is not a particular anatomical structure, but an instantaneously mutually reinforcing set of units, in the spirit of Hebb's cell assemblies (Jusczyk & Klein, 1980).

What can we say about the conditions under which coalitions will become and remain stable? We will begin informally with an almost trivial condition. Consider a set of units $\{a, b, \ . \ . \ .\}$ which we wish to examine as a possible coalition, π. For now, we assume that the units in π are all p-units and are in the non-saturated range and have no decay. Thus for each u in π,

$$p(u) \leftarrow p(u) + \text{Exc} - \text{Inh},$$

where Exc is the weighted sum of excitatory inputs and Inh is the weighted sum of inhibitory inputs. Now suppose that $\text{Exc} \mid \pi$, the excitation from the coalition π only, were greater than INH, the largest possible inhibition receivable by u, for each unit u in π, i.e.,

$$(\text{SC}) \qquad \forall u \in \pi \ ; \ \text{Exc} \mid \pi > \text{INH}$$

Then it follows that

$$\forall\, u \,\in\, \pi\, ;\, p(u) \leftarrow p(u)+\delta \text{ where } \delta > 0.$$

That is, the potential of every unit in the coalition will increase. This is not only true instantaneously, but remains true as long as nothing external changes (we are ignoring state change, saturation, and decay). This is because $\text{Exc} \mid \pi$ continues to increase as the potential of the members of π increases. Taking saturation into account adds no new problems; if all of the units in π are saturated, the change, δ, will be zero, but the coalition will remain stable.

The condition that the excitation from other coalition members alone, $\text{Exc} \mid \pi$, be greater than any possible inhibition INH for each unit may appear to be too strong to be useful. It is certainly true that coalitions can be stable without condition (SC) being met. The condition (SC) is useful for model building because it may be relatively easy to establish. Notice that INH is directly computable from the description of the unit; it is the largest negative weighted sum possible. If inhibition in our networks is mutual, the upper-bound possible after a fixed time τ, INH_τ, will depend on the current value of potential in each unit u. The simplest case of this is when two units are "deadly rivals" – each gets all its inhibition from the other. In such cases, it may well be feasible to show that after some time τ, the stable coalition condition will hold (in the absence of decay, fatigue, and changes external to the network). Often, it will be enough to show that the coalition has a stable "frontier," the set of units with outputs to some system under investigation.

There are a number of interesting properties of the stable coalition principle. First notice that it does not prohibit multiple stable coalitions nor single coalitions which contain units which mutually inhibit one another (although excessive mutual inhibition is precluded). If the units in the coalition had non-zero decay, the coalition excitation $\text{Exc} \mid \pi$ would have to exceed both INH and decay for the coalition to be stable. We suppose that a stable coalition yields control when its input elements change (fatigue and explicit resets are also feasible). To model coalitions with changeable inputs, we add boundary elements, which also had external "Input" and thus whose condition for being part of a stable coalition, π, would be:

$$\text{Exc} \mid \pi + \text{Input} > \text{INH}.$$

This kind of unit could disrupt the coalition if its Input went too low. The mathematical analysis of CM networks and stable coalitions continues to be a problem of interest. We have achieved some understanding of special cases (see the following example), and these results have been useful in designing CM too complex to analyze in closed form.

3.5. An Artificial Example

The coalitions of units needed to model biologically interesting functions will be large and heterogeneous. We do not yet have mathematical results that enable us to characterize the behavior of general coalitions. What we can do now is develop an artificial example of a coalition and establish the critical aspects of its behavior. This has proven to be useful to us both in aiding intuition and in constraining the choice of weights for real models. (Paul Shields of U. Toledo and Stanford provided the basic analysis.)

The artificial coalition consists of $M+1$ rows, each of which has $N+1$ units which compete by mutual lateral inhibition. We are assuming here that each unit can have potential and output values of unlimited range and accuracy and the output is exactly the potential. This makes it possible to express the competition in a row as a strictly linear rule:

$$X_{mn} \leftarrow X_{mn} - \beta\Sigma_{\substack{j \neq n}} X_{mj} + \text{Coalition Support}.$$

If we further assume that each coalition is exactly a column and provides positive support proportional to the sum of its members, the rule becomes

$$(*) \quad X_{mn} \leftarrow X_{mn} + \alpha\Sigma_{\substack{i \neq m}} X_{in} - \beta\Sigma_{\substack{j \neq n}} X_{mj}.$$

Under all these assumptions (*) defines a linear transformation, T, on the collection of values X_{mn} viewed as a "vector" in the sense of linear algebra. This transformation is sufficiently regular that we can characterize all of its eigenvalues and "eigenvectors." Recall that an eigenvalue, λ, and the associated "vector" X have the property that $TX = \lambda X$. Any such coalition structure, X, will be stable because repeated applications of the relaxation rule (*) will just multiply every element repeatedly by the related λ. What is interesting here is that the configurations of X_{mn} which have this property are easy to discuss in terms of our model.

Suppose that X_{mn} were such that each column had every one of its elements equal. This might be a good resting state for the structure because any row would provide the same answer as to the relative strengths of the various possibilities. The rule (*) becomes:

$$X_{mn} \leftarrow X_{mn} + \alpha \cdot M \cdot X_{mn} - \beta\Sigma_{\substack{j \neq n}} X_{mj}$$

because all M other elements of its column are equal to X_{mn}. If we further assume that each row has the sum of all its elements equal to zero, the remaining summation above must be equal to $-X_{mn}$ and we get:

$$X_{mn} \leftarrow X_{mn} + \alpha \cdot M \cdot X_{mn} + \beta X_{mn}$$

or

$$X_{mn} \leftarrow (1+\alpha M +\beta)X_{mn}$$

which says that $(1+\alpha M +\beta) = \lambda_1$ is an eigenvalue for T, working on "vectors" with constant columns and zero row-sums. The condition of a zero sum for a row captures the idea of competition quite nicely; the fact that this requires negative values to be transmitted is not a serious problem. It is the assumptions of unbounded scale and accuracy that limit the application of these results even in the case of purely row-column coalition structures.

The fact that constant-column, zero-row-sum configurations are stable for this structure is important, but there are several other points to be made. Notice that several columns could have the same constant value; the problem of ties cannot be resolved by such a uniform system. There are also other eigenvalues and "vectors" which do not correspond to desirable states of the system. These are:

Eigenvalue	"Vector" X
$1+\alpha M -\beta N$	matrix of all 1
$1-\alpha-\beta N$	rows equal, column-sums zero
$1-\alpha+\beta$	row-sums and column-sums all zero

By computing the multiplicity of the four eigenvalues, one can show that the total multiplicity is $(N+1)(M+1)$, so that there are no other eigenvalues. The critical point is that powers of a linear system like T will converge to the direction specified by its largest eigenvalue. If we make sure to choose α and β so that $\lambda_+ = 1+\alpha M +\beta$ is the largest eigenvalue, then repetitions of (*) will converge to the desired constant-column zero-row-sum state. This requires (for α, β positive) that

$$1+\alpha M +\beta > \alpha+\beta N -1$$

or

$$(**) \quad 2 > \beta N -\alpha M +(\alpha-\beta).$$

We can ignore $\alpha-\beta$ which is a small fraction. Recall that β is the weight given to the competitors and α the weight given to collaborators. Condition (**) states that if the coalitions are given adequate weight, the system will settle into a state with uniform columns (coalitions). The obvious choice of $\beta = 1/N$ and $\alpha = 1/M$ comfortably meets condition (**). The problem that occurs if β is too small is that mutual inhibition will have no effect and the system will converge to the state where all columns have their initial average value. The relative importance of competition and collaboration will be a crucial part of the detailed specification of any model. There appears to be no reason that discrete values, bounded ranges and overlapping coalitions should change the basic character of this result, but the detailed analysis of a realistic coalition structure for its convergence properties appears to be very difficult. More generally, there will need to be ways of assessing the impact of finite bounds and discrete ranges on systems whose continuous approximation is understood, a classic problem in numerical analysis.

4. Conserving Connections

It is currently estimated that there are about 10^{11} neurons and 10^{14} connections in the human brain and that each neuron receives input from about 10^3-10^4 other neurons. These numbers are quite large, but not so large as to present no problems for connectionist theories. It is also important to remember that neurons are not switching devices; the same signal is propagated along all of the outgoing branches. For example, suppose some model called for a separate, dedicated path between all possible pairs of units in two layers of size N. It is easy to show that this requires N^2 intermediate sites. This means, for example, that there are not enough neurons in the brain to provide such a cross-bar switch for substructures of a million elements each. Similarly, there are not enough neurons to provide one to represent each complex object at every position, orientation, and scale of visual space. Although the development of connectionist models is in its perinatal period, we have been able to accumulate a number of ideas on how some of the required computations can be carried out without excessive resource requirements. Five of the most important of these are described below: (1) functional decomposition; (2) limited precision computation; (3) coarse and coarse-fine coding; (4) tuning; and (5) spatial coherence.

4.1. Functional Decomposition

When the number of variables in the function becomes large, the fan-in or number of input connections could become unrealistically large. For example, the function $t = f(u, v, w, x, y, z)$ implemented with 100 values of t, when each of its arguments can have 100 distinct values, would require an average number of inputs per unit of $10^{12}/10^2$, or 10^{10}! However, there are simple ways of trading units for connections. One is to replicate the number of units with each value. This is a good solution when the inputs can be partitioned in some natural way as in the vision examples in the next section. A more powerful technique is to use intermediate units when the computation can be decomposed in some way. For example, if $f(u, v, w, x, y, z) = g(u, v) \circ h(w, x, y, z)$, where $\circ$ is some composition, then separate networks of value units for $f(g, h)$, $g(u, v)$, and $h(w, x, y, z)$ can be used. The outputs from the g and h units can be combined in conjunctive connections according to the composition operator $\circ$ in a third network representing f. An example is the case of word recognition. Letter-feature units would have to connect to vastly more word units without the imposition of the intermediate level of letter units. The letter units limit the ways letter-feature units can appear in a word.

4.2. Limited Precision Computation

In the multiplication example $z = xy$, the number of z units required is proportional to $N_x N_y$ even when redundant value units are eliminated, and in general the number of units could grow exponentially with the number of arguments. However, there are several refinements which can drastically reduce the number of required units. One way to do this is to fix the number of units at the *precision* required for the computation. Figure 14 shows the network of Figure 7 modified when less computational accuracy is required.

This is the same principle that is incorporated in integer calculations in a sequential computer: computations are rounded to within the machine's accuracy. Accuracy is related to the number of bits and the number representation. The main difference is that since the sequential computer is general purpose, the number representations are conservative, involving large numbers of bits. The neural units need only represent sufficient accuracy for the problem at hand. This will generally vary from network to network, and may involve very inhomogeneous, special purpose number representations.

4.3. Coarse and Coarse-fine Coding

Coarse coding is a general technical device for reducing the number of units needed to represent a range of values with some fixed precision, due to Hinton (1980). As Figure 15a suggests, one can represent a more precise value as the simultaneous activation of several (here 3) overlapping coarse-valued units. In general, D simultaneous activations of coarse cells of diameter D precise units suffice. For a parameter space of dimension k, a range of F values can be captured by only F^k/D^{k-1} units rather than F^k in the naive method. The coarse coding trick and the related coarse-fine trick to be described next both depend on the input at any given time being sparse relative to the set of all values expressible by the network.

The coarse-fine coding technique is useful when the space of values to be represented has a natural structure which can be exploited. Suppose a set of units represents a vector parameter v which can be thought of as partitioned into two components (r, s). Suppose further that the number of units required to represent the subspace r is N_r and that required to represent s is N_s. Then the number of units required to represent v is $N_r N_s$. It is easy to construct examples in vision where the product $N_r N_s$ is too close to the upper bound of 10^{11} units to be realistic. Consider the case of trihedral ("Y") vertices, an important visual cue. Three angles and two position coordinates are necessary to uniquely define every possible trihedral vertex. (Two angles define the type of vertex (arrow, y-joint); the third specifies the rotation of the joint in space.) If we use 5-degree angle sensitivity and 10^5 spatial sample points, the number of units is given by $N_r \approx 3.6 \times 10^5$ and $N_s = 10^5$ so that $N_r N_s \approx 3.6 \times 10^{10}$. How can we achieve the required representation accuracy with fewer units?

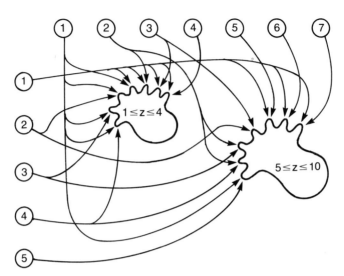

Figure 14. Modified multiplication table using fewer units.

In many instances, one can take advantage of the fact that the *actual occurrence* of parameters is sparse. In terms of trihedral vertices, one assumes that in an image, such vertices will rarely occur in tight spatial clusters. (If they do, they cannot be resolved as individuals simultaneously.) Given that simultaneous proximal values of parameters are unlikely, they can be represented accurately for other computations, without excessive cost.

The solution is to decompose the space v into two subspaces, r and s, each with unilaterally reduced resolution.

Instead of $N_r N_s$ units, we represent v with two spaces, one with $N_{r'} N_s$ units where $N_{r'} \ll N_r$ and another with $N_r N_{s'}$ units where $N_{s'} \ll N_s$.

To illustrate this technique with the example of trihedral vertices we choose $N_{s'} = 0.01 N_s$ and $N_{r'} = 0.01 N_r$. Thus the dimensions of the two sets of units are:

$$N_{s'} N_r = 3.6 \times 10^8$$

and

$$N_s N_{r'} = 3.6 \times 10^8.$$

The choices result in one set of units which accurately represent the angle measurements and fire for a specific trihedral vertex anywhere in a fairly broad visual region, and another set of units which fire only if a general trihedral vertex is present at the precise position. The coarse-fine technique can be viewed as replacing the square coarse-valued covering in Figure 15a with rectangular

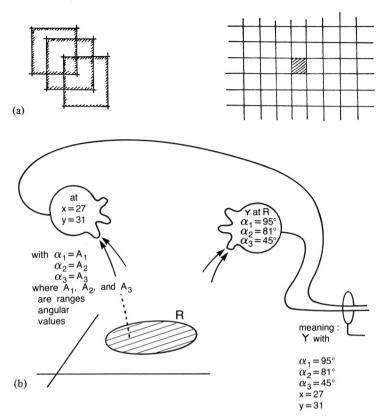

Figure 15. (a) Coarse coding example. In a two-dimensional measurement space, the presence of a measurement can be encoded by making a single unit in the fine resolution space have a high confidence value. The same measurement can be encoded by making overlapping coarse units in three distinct arrays have high confidence values. (b) Coarse angle – fine position and coarse position – fine angle units combine to yield precise values of all five parameters.

(multi-dimensional) coverings, like those shown in Figure 16. In terms of our value units, the coarse-fine representation of trihedral vertices is shown in Figure 15b.

If the trihedral angle enters into another relation, say $R(v, \alpha)$, where both its angle and position are required accurately, one conjunctively connects pairs of appropriate units from each of the reduced resolution spaces to appropriate R-units. The conjunctive connection represents the *intersection* of each of its components' *fields*. Essentially the same mechanism will suffice for conjoining (e.g.) accurate color with coarse velocity information.

An important limitation of these techniques, however, is that the input must be sparse. If inputs are too closely spaced, "ghost" firings will occur. In Figure 16, two sets of overlapping fields are shown, each with unilaterally

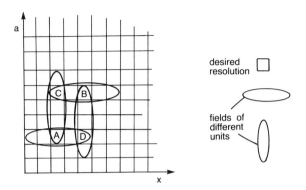

Figure 16. Inputs at A and B cause ghosts at C and D.

reduced resolution. Actual input at points A and B will produce an erroneous indication of an input at C, in addition to the correct signals. The sparseness requirement has been shown to be satisfied in a number of experiments with visual data (Ballard & Kimball, 1981a,b; Ballard & Sabbah, 1981).

The resolution device involves a units/connections tradeoff, but in general, the tradeoff is attractive. To see this, consider a unit that receives input from a network representing a vector parameter v. If n is the number of places where the output is used, and conjunctive connections are used to conjoin the D firing units, then Dn synapses are required. Thus if A is the number of non-coarse coded units to achieve a given acuity, then coarse coding is attractive when $A/D^{k-1} > Dn$, assuming connections and units are equally scarce. This result is optimistic in that, when other uses of conjunctive connections are taken into account, the number of conjunctive units could be unrealistically large.

4.4. Tuning

The idea of tuning further exploits networks composed of coarsely- and finely-grained units. Suppose there are n fine resolution units of a feature A and n fine resolutions for a feature B. To have explicit units for feature values of AB, n^2 units would be required. This is an untenable solution for large feature spaces (the number of units grows exponentially with the number of features), so alternatives must be sought. One solution to this problem is to vary the grain of the AB units so that they are only coarsely represented. This solution has its attendant disadvantages in that separate stimuli within the limits of the coarse resolution grain cannot be distinguished. Also, a set of weak stimuli can be misinterpreted. A better solution is to have a coarse unit that would respond only to a single saturated unit within its input range. In that way a collection of weak inputs is not misinterpreted.

This situation can be achieved by having the units in each finely-tuned network that are in the field of a coarse unit laterally inhibit each other, e.g., in the WTA network of Figure 5a. The outputs of these individual feature units then form disjunctive connections with appropriate coarse resolution multiple feature units. If m is the grain of the coarse resolution units along each feature dimension, the number of disjunctions per coarse unit is $(n/m)^2$. The result of this connection strategy is that a coarse unit responds with a strength that varies as the strength of the largest maximum in the subnetwork of each of the finely-tuned units that correspond to its field. The response of a coarse-tuned unit is the maximum of the sums of the conjunctive inputs from the finely tuned units which connect to it. In terms of Figure 15, a tuned coarse-angle cell would respond only to one high-confidence pair of angles in its range, and not to several weak ones (which couldn't correctly appear all at one position). This is a better property than just having unstructured coarse units and it will be exploited in the next section, when we deal with perceiving complex objects.

4.5. Spatial Coherence

The most serious problem which requires conserving connections is the representation of complex concepts. The obvious way of representing concepts (sets of properties) is to dedicate a separate unit to each conjunction of features. In fact, it first appears that one would need a separate unit for each combination at each location in the visual field. We will present here a simple way around the problem of separate units for each location and deal with the more general problem in the next section.

The basic problem can be readily seen in the example of Figure 17. Suppose there were one unit each for finally recognizing concepts like colored circles and squares. Now consider the case when a red circle (at $x = 7$) and a blue square (at $x = 11$) simultaneously appear in the visual field. If the various "colored figure" units simply summed their inputs, the incorrect "blue circle" unit would see two active inputs, just like the correct "red circle" and "blue square" units. This problem is known as cross-talk, and is always a potential hazard in CM networks. The solution presented in Figure 17 is quite general. Each unit is assumed to have a separate conjunctive connection site for each position of the visual field. In our example, the correct units get dual inputs to a single site (and are activated) while the partially matched units receive separated inputs and are not activated. Only sets of properties which are spatially coherent can serve to activate concept units. This example was meant to show how spatial coherence could be used with conjunctive connections to eliminate cross-talk. There are a number of additional ways of using spatial coherence, each of which involves different tradeoffs. These are discussed in the next section, which considers some sample applications in more detail.

Even given this strategy for managing space, it is still combinatorially implausible to have complex cells such as BTRM(...). However, there is a way

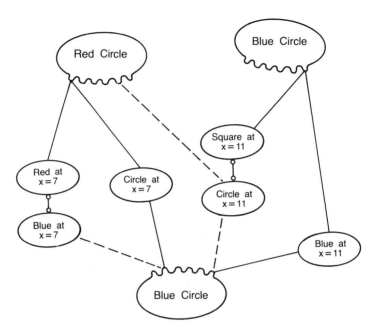

Figure 17. Spatial coherence on inputs can represent complex concepts without crosstalk. Solid lines show active inputs and dashed lines (some of the) inactive inputs.

around this problem using multiple connections from object units. A blue-textured-fast (BTF) object unit can be synthesized from BT and BF units, i.e., BT & BF $\simeq>$ BTF. What the $\simeq>$ symbol means is that the implication is not guaranteed, but very *likely,* given that the BF and BT units are tuned. The BF and BT units detect spatial registration directly via connections like those in Figure 17, but the BTF unit does not. We are saying essentially that, in general, simple (here, pairwise) conjunctions can be kept spatially registered by conjunctive connections, and that more complex property combinations can be synthesized from them. Complex combinations that are important to an individual are presumed to have new units *recruited* (Feldman, 1980; 1981a,b) to represent them explicitly.

5. Applications

This section illustrates the power of the CM paradigm via two groups of examples. The first shows how the various techniques for conserving connections can be used in an idealized form of perception of a complex object. Here the point is that an object has multiple features which are computed in parallel via the transform methodology. The second group of examples starts with a relatively simple problem, that of vergence eye movements, to illustrate motor control using value units. In this example, control is immediate; a visual signal produces an instantaneous output (within the settling time constants of the units). Extensions of this idea use space as a buffer for time. For motor output, space allows the incorporation of more complex motor commands. For speech input, spatial buffering allows for phoneme recognition based on *subsequent* information.

These examples were chosen to show that CM can provide a unified representation for both perception and motor control. This is important since an animal is hardly ever passively responding to its environment. Instead, it seems involved in what Arbib has called a perception-action cycle (Arbib, 1979). Perceptions result in actions which in turn cause new perceptions, and so on. Massive parallelism changes the way the perception-action cycle is viewed. In the traditional view, one would convert the input to a language which uses variables, and then use these variables to direct motor commands. CM suggests that we think of accomplishing the same actions via a transformation: sensory input is transformed (connected to) to abstract representational units, which in turn are transformed (connected to) to motor units. This will obviously work for reflex actions. The examples are intended to suggest how more flexible command and control structures can also be represented by systems of value units.

5.1. Object Recognition

The examples of Figures 1 and 6 are representative of the problem of gestalt perception: that of seeing parts of an image as a single percept (object). An "object" is indicated by the "simultaneous" appearance of a number of "visual features" in the correct relative spatial positions. In any realistic case, this will involve a variety of features at several different levels of abstraction and complex interaction among them. A comprehensive model of this process would be a prototype theory of visual perception and is well beyond the scope of this paper. What we will do here is consider the pre-requisite task of constructing CM solutions to the problems of detecting non-punctate visual features and of forming sets of the features which should help characterize a percept. We will refer throughout to the prototype problem of detecting Fred's frisbee, which is known to be round, baby-blue, and moving fairly fast. The development suppresses many important issues such as hierarchical descriptions, perspective, occlusion, and the integration of separate fixations, not to mention learning. A brief discussion of how these might be tackled follows the technical material.

The first problem is to develop a general CM technique for detecting features and properties of images, given that these features are not usually detectable at a single point in some retinotopic map. The basic idea is to find parameters which characterize the feature in question and connect each retinotopic detector to the parameter values consistent with its detectand.

Consider the problem of detecting lines in an image from short edge segments. Different lines can be represented by units having different discrete parameter values, e.g. in the line equation $\rho = x\cos\theta + y\sin\theta$, the parameters are ρ and θ. Thus edge units at (x, y, θ) could be connected to appropriate line units. Note that this example is analogous to the word recognition example (Figure 1). Edges are analogous to letters and lines to words. As in the words-letter example, "top-down" connections allow the existence of a line to raise the confidence of a local edge. In our line detection example, lines in the image are high potential (confidence) units in a slope-intercept (θ, ρ) parameter space. High confidence edge units produce high confidence line units by virtue of the network connectivity. This general way of describing this relationship between parts of an image (e.g., edges) and the associated parameters (e.g., ρ, θ for a line) is a connectionist interpretation of the *Hough transform* (Duda & Hart, 1972). Since each parameter value is determined by a large number of inputs, the method is inherently noise-resistant and was invented for this purpose. A Hough transform network for circles (like Fred's frisbee) would involve one parameter for size plus two for spatial location, and exactly this method has been used for tumor detection in chest radiographs (Kimme et al., 1975). Notice that the circle parameter space is itself retinotopic in that the centers of circles have specified locations; this will be important in registering multiple features.

The Hough transform is a formalism for specifying excitatory links between units. The general requirements are that part of an image representation can be represented by a parameter vector $\mathbf{a}$ in an image space A and a feature can be represented by a vector $\mathbf{b}$ which in an element of a feature space B. *Physical constraints* $f(\mathbf{a},\mathbf{b}) = 0$ relate $\mathbf{a}$ and $\mathbf{b}$. The space A represents spatially indexed units, and each individual element $\mathbf{a_k}$ is only consistent with certain elements in the space B, owing to the constraint imposed by the relation f. Thus for each $\mathbf{a_k}$ it is possible to compute the set

$$B_k = \{\mathbf{b} \mid f(\mathbf{a_k},\mathbf{b}) \leq \delta_b\}$$

where B_k is the set of units in the feature space network B that the $\mathbf{a_k}$ unit must connect to, and the constant δ_b is related to the quantization in the space $\mathbf{B}$. Let $H(\mathbf{b})$ be the number of active connections the value unit $\mathbf{b}$ receives from units in A. $H(\mathbf{b})$ is the number of image measurements which are consistent with the parameter value $\mathbf{b}$. The potential of units in B is given by $p(\mathbf{b}) \leftarrow H(\mathbf{b})/\Sigma_b H(\mathbf{b})$. The value $p(\mathbf{b})$ can stand for the confidence that the segment with feature value $\mathbf{b}$ is present in the image. If the measurement represented by $\mathbf{a}$ is realized as groups of units, e.g., $\mathbf{a} = (\mathbf{a_1},\mathbf{a_2})$, then conjunctive connections are required to implement the constraint relation.

Implementing these networks often results in a set of *very sparsely distributed* high-confidence feature space units. In implementations of the line detection example, only approximately 1% of the units have maximum confidence values. This figure is also typical of other modalities. In general, each $\mathbf{a_k}$ and the relationship f will not determine a single unit in B_k as in the line detection example, but there still will be isolated high-confidence units. Figure 1 shows why this is the case: different a_k letter-feature units connect to common units in the letter space B.

We have found that parameter spaces combine with the growing body of knowledge on specific physical constraints to provide a powerful and robust model for the simultaneous computation of invariant object properties such as reflectance, curvature, and relative motion (Ballard, 1981).

Of course segmentation must involve ways of associating peaks in several different feature spaces and methods for doing this are discussed presently, but the cornerstone of the techniques are high-confidence units in the individual-modality feature spaces. In extending the single feature case to multiple features, the most serious problem is the immense size of the cross product of the spatial dimensions with those of interesting features such as color, velocity, and texture. Thus to explain how image-like input such as color and optical flow are related to abstract objects such as "a blue, fast-moving thing," it becomes necessary to use all the techniques of the previous sections.

Even if we assume that there is a special unit for recognizing images of Fred's frisbee, it cannot be the case that there is a separate one of these units for each point in the visual field. One weak solution to his kind of problem was given in Figure 17 of the last section. There could conceivably be a separate 3-way conjunctive connection on the Fred's frisbee unit for each position in space. Activation of one conjunct would require the simultaneous activation of circle, baby-blue, and fairly-fast in the same part of the visual field. The solution style with separate conjunctions for every point in space becomes increasingly implausible as we consider more complex objects with hierarchical and multiple descriptions. The spatially registered conjunctions would have to be preserved throughout the structure.

The problem of going from a set of descriptors (features) to the object which is the best match to the set is known in artificial intelligence as the *indexing problem*. The feature set is viewed as an index (as in a data base). There have been several proposed parallel hierarchical network solutions to the indexing problem (Fahlman, 1979; Hillis, 1981) and these can be mapped into CM terms. But these designs assume that the network is presented with sets of descriptors which are already partitioned; precisely the vision problem we are trying to solve. There are three additional mechanisms that seem to be necessary, two of which have already been discussed. Coarse coding and tuning (as discussed in Section 4) makes it much less costly to represent conjunctions. In addition, some general concepts (e.g., blue frisbee) might be indexed more efficiently through less precise units. The new idea is an extension of spatial coherence that exploits the fact that the networks respond to activity that occurs together in time. If there were a way to focus the activity of the network on one area at a time, only properties detected in that area would compete to index objects.

The obvious way to focus attention on one area of the visual field is with eye movements, but there is evidence that focus can also be done within a fixation. The general idea of internal spatial focus is shown in Figure 18. In this network, the general "baby-blue" unit is configured to have separate conjunctive inputs for each point in space, like the blue-square units of Figure 17. The difference is that the second input to the conjunction comes from a "focus" unit, and this makes a much more general network. The idea of making a unit (e.g., baby blue) more responsive to inputs from a given spatial position can be implemented in different ways. The conjunctive connection at the $x = 7$ lobe of the baby-blue unit is the most direct way. But treating this conjunct as a strict AND would mean that all spatial units would have to be active when there was no focus. An alternative would be to have the "focus on 7" unit boost the output of the "baby blue at 7" unit (and all of its rivals) as shown by the dashed line; this would eliminate the need for separate spatial conjunctions on the baby-blue unit, but would alter the potential of all the units at the position being attended. The trade-offs become even trickier when goal-directed input is taken into account, but both methods have the same effect on indexing. If the system has its attention directed only to $x = 7$, then the only feature units activated at all will be those whose local representatives are dominant (in their WTA) at $x = 7$. In such a case, there would be a time when the only concept units active in the entire network would be those for $x = 7$. This does not "solve" the problem of identifying objects in a visual scene, but it does suggest that sequentially focusing attention on separate places can help significantly. There is considerable reason to suppose (Treisman, 1980; Posner, 1978) that people do this even in tasks without eye movement.

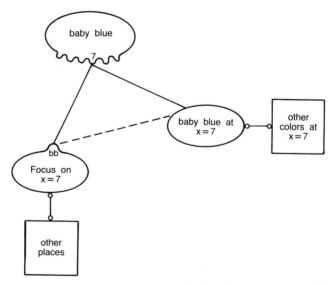

Figure 18. Spatial focus unit can gate only input from attended positions.

There are other ways of looking at the network of Figure 18. Suppose the system had reason to focus on some particular property (e.g., baby-blue). If we make bi-directional the links from "focus on $x = 7$" to "baby-blue" and "baby-blue at 7," a nice possibility arises. The "focus on 7" unit could have a conjunctive connection for each separate property at its position. If, for example, baby-blue was chosen for focus and was the dominant color at $x = 7$, then the "focus on $x = 7$" unit would dominate its rivals. This suggests another way in which the recognition of complex objects could be helped by spatial focus. Figure 19 depicts the fairly general situation.

In Figure 19, the units representing baby-blue, circular, and fairly-fast are assumed to be for the entire visual field and moderately precise. The dotted arrows to the "Fred's frisbee" node suggest that there might be more levels of description in a realistic system. The spatial focus links involving baby-blue are the same as in Figure 18, and are replicated for the other two properties. Notice that the position-specific sensing units do not have their potentials affected by spatial focus units, so that the sensed data can remain intact. The network of Figure 19 can be used in several ways.

If attention has been focused on $x = 7$ for any reason, the various space-independent units whose representatives are most active at $x = 7$ will become most active, presumably leading to the activation (recognition) of Fred's frisbee. If a top-down goal of looking for Fred's frisbee (or even just something baby-blue) is active, then the "focus on $x = 7$" will tend to defeat its WTA rivals, leading to the same result. A third possibility is a little more complicated, but quite powerful. Suppose that a given image, even in context, activates too many property units so that no objects are effectively indexed. One strategy would be to systematically scan each area of the visual field, eliminating confounding activity from other areas. But it is also possible to be more efficient. If some property unit (say baby-blue) were strongly activated, the network could focus attention on all the positions with that property. In this case it is like putting a baby-blue filter in front of the scene, and should often lead to better convergence in the networks for shape, speed, etc.

One should compare the network of Figure 17 with Figures 18 and 19. In the former, parallel co-existing concepts are possible if we assume delicate arrangements of conjunctive connections. The latter networks are more robust but use sequentiality to eliminate cross-talk.

5.2. Time and Sequence

Connectionist models do not initially appear to be well-suited to representing changes with time. The network for computing some function can be made quite fast, but it will be fixed in functionality. There are two quite different aspects of time variability of connectionist structures. One is time-varying responses, i.e., long-term modification of the networks (through changing weights) and short-term changes in the behavior of a fixed network with

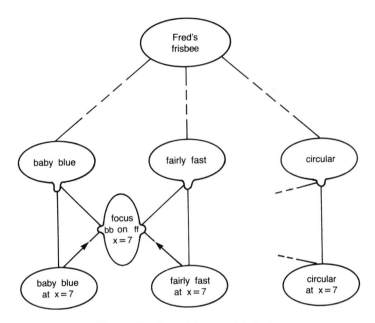

Figure 19. Spatial focus and indexing.

time. The second aspect is sequence: the problem of analyzing inherently sequential input (such as speech) or producing inherently sequential output (such as motor commands) with parallel models. The problem of change will be deferred to (Feldman, 1981b). The problem of sequence is discussed here.

There are a number of biologically suggested mechanisms for changing the weight (w_j) of synaptic connections, but none of them are nearly rapid enough to account for our ability to hear, read, or speak. The ability to perceive a time-varying signal like speech or to integrate the images from successive fixations must be achieved (according to our dogma) by some dynamic (electrical) activity in the networks. As usual, we will present computational solutions to the problems of sequence that appear to be consistent with known structural and performance constraints. These are, again, too crude to be taken literally but do suggest that connectionist models can describe the phenomena.

Motor Control of the Eye. To see how the transform notion of distributed units might work for motor control, we present a simplistic model of vergence eye movements. (The same idea may be valid for fixations, but control probably takes place at higher levels of abstraction.) In this model retinotopic (spatial) units are connected directly to muscle control units. Each retinotopic unit can if saturated cause the appropriate contraction so that the new eye position is centered on that unit. When several retinotopic units saturate, each enables a muscle control unit independently and the muscle itself contracts an average amount.

Figure 20 shows the idea for a one-dimensional retina. For example, with units at positions 2, 4, 5, and 6 saturated, the net result is that the muscle is centered at 17/4 or 4.25. (This idea can be extended to the case where the retinotopic units have overlapping fields.) This kind of organization could be extended to more complex movement models such as that of the organization of the superior colliculus in the monkey (Wurtz & Albano, 1980).

Notice that each retinotopic unit is capable of enabling different muscle control units. The appropriate one is determined by the enabled x-origin unit which inhibits commands to the inappropriate control units via modifiers.

One problem with this simple network arises when disparate groups of retinotopic units are saturated. The present configuration can send the eye to an average position if the features are truly identical. The network can be modified with additional connections so that only a single connected component of saturated units is enabled by using additional object primitives. A version of this WTA motor control idea has already been used in a computer model of the frog tectum (Didday, 1976).

There are still many details to be worked out before this could be considered a realistic model of vergence control, but it does illustrate the basic idea: local spatially separate sensors have *distinct, active* connections which could be averaged at the muscle for fine motor control or be fed to some intermediate network for the control of more complex behaviors.

Converting Space to Time. Consider the problem of controlling a simple physical motion, such as throwing a ball. It is not hard to imagine that in a skilled motor performance unit-groups fire each other in a fixed succession, leading to the motor sequence. The computational problem is that there is a unique set of effector units (say at the spinal level) that must receive input from each group at the right time. Figure 21a depicts a simple case in which there are two effector units (e_1, e_2) that must be activated alternatively. The circles marked 1-4 represent units (or groups of units) which activate their successor and inhibit their predecessor (cf. Delcomyn, 1980). The main point is that a succession of outputs to a single effector set can be modelled as a sequence of time-exclusive groups representing instantaneous coordinated signals. Moving from one time step to the next could be controlled by pure timing for ballistic movements, or by a proprioceptive feedback signal. There is, of course, an enormous amount more than this to motor control, and realistic models would have to model force control, ballistic movements, gravity compensation, etc.

Figure 21b depicts a somewhat fanciful notion of how a variety of output sequences could share a collection of lower level response units. The network shown has a single "Dixie" unit which can start a sequence and which joins in conjunctive connections with each note to specify its successor. At each time step, a WTA network decides what note gets sounded. One can imagine adding the rhythm network and transposition networks to other keys and to other modalities of output.

current eye coordinate

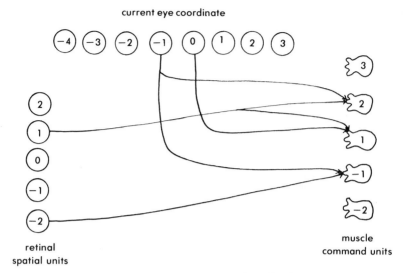

Figure 20. Distributed control of eye fixations.

Converting Time to Space. The sequencer model for skilled movements was greatly simplified by the assumption that the sequence of activities was pre-wired. How could one (still crudely, of course) model a situation like speech perception where there is a largely unpredictable time-varying computation to be carried out? One solution is to combine the sequencer model of Figure 21 with a simple vision-like scheme. We assume that speech is recognized by being sequenced into a buffer of about the length of a phrase and then is relaxed against context in the way described above for vision. For simplicity, assume that there are two identical buffers, each having a pervasive modifier (m_j) innervation so that either one can be switched into or out of its connections. We are particularly concerned with the process of going from a sequence of potential phonetic features into an interpreted phrase. Figure 22 gives an idea of how this might happen.

Assume that there is a separate unit for each potential feature for each time step up to the length of the buffer. The network which analyzes sound is connected identically to each column, but conjunction allows only the connections to the active column to transmit values. Under ideal circumstances, at each time step exactly one feature unit would be active. A phrase would then be layed out on the buffer like an image on the "mind's eye," and the analogous kind of relaxation cones (cf. Figures 1, 6) involving morphemes, words, etc., could be brought to bear. The more realistic case where sounds are locally ambiguous presents no additional problems. We assume that, at each time step, the various competing features get varying activation. Diphone constraints could be captured by (+ or −) links to the next column as suggested by

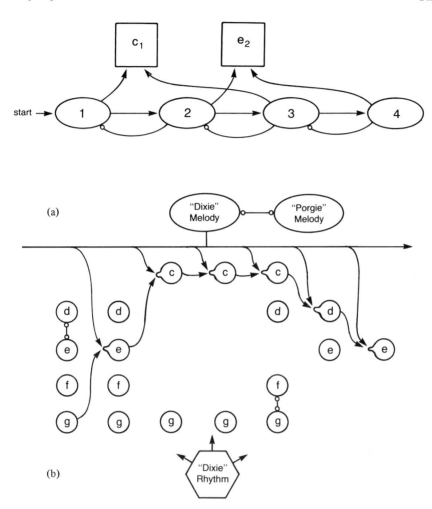

(a)

(b)

Figure 21. Mapping space into time. (a) Sequence and suppression. (b) Whistling Dixie.

Figure 22. The result is a multiple-possibility relaxation problem – again exactly like that in visual perception. The fact that each potential feature could be assigned a row of units is essential to this solution; we do not know how to make an analogous model for a sequence of sounds which cannot be clearly categorized and combined. Recall that the purpose of this example is to indicate how time-varying input could be treated in connectionist models. The problem of actually laying out detailed models for language skills is enormous and our example may or may not be useful in its current form. Some of the considerations that arise in distributed modeling of language skills are presented in (Arbib & Caplan, 1979).

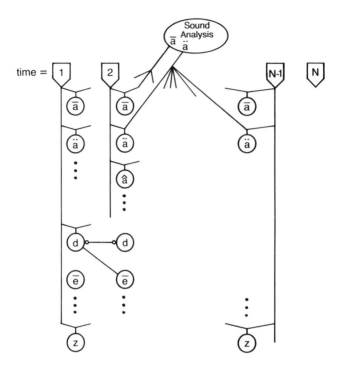

Figure 22. Mapping time into space.

6. Conclusions

The CM paradigm advanced in this paper has been applied successfully only to relatively low-level tasks. There is no reason, as yet, to be confident that an intermediate symbolic representation will not be required for modeling higher cognitive processes. There is, however, the beginning of a collection of efforts which can be interpreted as attempting CM approaches to higher level tasks. These include work which explicitly uses parallelism in planning (Stefik, 1981) and deduction, and work which incorporates more connectionist architectural notions of value units (Forbus, 1981) and coarse coding (Garvey, 1981).

We have now completed six years of intensive effort on the development of connectionist models and their application to the description of complex tasks. While we have only touched the surface, the results to date are very encouraging. Somewhat to our surprise, we have yet to encounter a challenge to the basic formulation. Our attempts to model in detail particular computations (Sabbah, 1981; Ballard & Sabbah, 1981) have led to a number of new insights (for us, at least) into these specific tasks. Attempts like this one to formulate and solve general computational problems in realistic connectionist terms have proven to be difficult, but less so than we would have guessed.

There appear to be a number of interesting technical problems within the theory and a wide range of questions about brains and behavior which might benefit from an approach along the lines suggested in this paper.

Acknowledgment

This research was supported in part by the Defense Advanced Research Projects Agency under Grant N00014-78-C-0164, and in part by the Office of Naval Research under Grant N00014-80-C-0197. Substantial portions of this paper also appeared in *Cognitive Science*, Vol. 6, 1982.

Appendix

Units

A *unit* is a computational entity comprising:

$\{q\}$ – a set of *discrete states,* < 10

p – a continuous value in $[-10,10]$, called *potential* (accuracy of several digits)

v – an *output value,* integers $0 \leqslant v \leqslant 9$

$\mathbf{i}$ – a vector of *inputs* $i_1, \ldots, i_n$ and functions from old to new values of these:

$$p \leftarrow f(\mathbf{i}, p, q)$$
$$q \leftarrow g(\mathbf{i}, p, q)$$
$$v \leftarrow h(\mathbf{i}, p, q)$$

which we assume to compute continuously. The form of the f, g, and h functions will vary, but will generally be restricted to conditionals and simple functions.

P-units

For some applications, we will use a particularly simple kind of unit whose output v is proportional to its potential p (rounded) (when $p > 0$) and which has only one state. In other words

$$p \leftarrow p + \beta \, \Sigma w_k i_k \qquad\qquad (0 \leqslant w_k \leqslant 1)$$
$$v \leftarrow \textbf{if } p > \theta \textbf{ then } \text{round } (p - \theta) \textbf{ else } 0 \quad (v = 0, \ldots, 9)$$

where β, θ are constants and w_k are weights on the input values.

Conjunctive Connections

In terms of our formalism, this could be described in a variety of ways. One of the simplest is to define the potential in terms of the maximum, e.g.,

$$p \leftarrow p + \beta \text{Max}(i_l + i_2 - \phi, \; i_3 + i_4 - \phi, \; i_5 + i_6 - i_7 - \phi)$$

where β is a scale constant as in the p-unit and ϕ is a constant chosen (usually > 10) to suppress noise and require the presence of multiple active inputs. The minus sign associated with i_7 corresponds to its being an inhibitory input. The max-of-sum unit is the continuous analog of a logical OR-of-AND (disjunctive normal form) unit and we will sometimes use the latter as an approximate version of the former. The OR-of-AND unit corresponding to the above is:

$$p \leftarrow p + \alpha \; OR \; (i_1 \& i_2, \; i_3 \& i_4, \; i_5 \& i_6 \& (\text{not } i_7))$$

Winner-take-all (WTA) networks have the property that only the unit with the highest potential (among a set of contenders) will have the output above zero after some settling time.

A *coalition* will be called stable when the output of all of its members is non-decreasing.

References

Anderson, J. A., Silverstein, J. W., Ritz, S. A., & Jones, R. S. Distinctive features, categorical perception, and probability learning: Some applications of a neural model. *Psychological Review,* 1977, **84**, 413-451.

Arbib, M. A. Perceptual structures and distributed motor control. University of Massachusetts, Department of Computer and Information Science, TR 79-11, 1979.

Arbib, M. A., & Caplan, D. Neurolinguistics must be computational. *The Brain and Behavioral Sciences,* 1979, **2**, 449-483.

Ballard, D. H. Parameter networks: Towards a theory of low-level vision. Proceedings, 7th International Joint Conference on Artificial Intelligence, 1981, 1068-1078.

Ballard, D. H., & Kimball, O. A. Rigid body motion from depth and optical flow. University of Rochester, Computer Science Department, TR 70, 1981. (a)

Ballard, D. H., & Kimball, O. A. Shape and light source direction from shading. University of Rochester, Computer Science Department, 1981. (b)

Ballard, D. H., & Sabbah, D. On shapes. Proceedings, 7th International Joint Conference on Artificial Intelligence, 1981, 607-612.

Delcomyn, F. Neural basis of rhythmic behavior in animals. *Science,* 1980, **210**, 492-498.

Didday, R. L. A model of visuomotor mechanisms in the frog optic tectum. *Mathematical Biosciences,* 1976, **30**, 169-180.

Duda, R. O., & Hart, P. E. Use of the Hough transform to detect lines and curves in pictures. *Communications of the ACM,* 1972, **15**, 11-15.

Fahlman, S. E. *NETL, a System for Representing and Using Real Knowledge.* Boston, MA: MIT Press, 1979.

Fahlman, S. E. The Hashnet interconnection scheme. Carnegie-Mellon University, Computer Science Department, 1980.

Feldman, J. A. A distributed information processing model of visual memory. University of Rochester, Computer Science Department, TR 52, 1980.

Feldman, J. A. A connectionist model of visual memory. In G. E. Hinton & J. A. Anderson (Eds.), *Parallel Models of Associative Memory.* Hillsdale, NJ: Erlbaum, 1981. (a)

Feldman, J. A. Memory and change in connection networks. University of Rochester, Computer Science Department, TR 96, 1981. (b)

Flanagan, D. (Ed.). *The Brain. (Scientific American,* September 1979.) San Francisco, CA: Freeman, 1979.

Forbus, K. D. Qualitative reasoning about physical processes. Proceedings, 7th International Joint Conference on Artificial Intelligence, 1981, 326-330.

Garvey, T. D., Lowrance, J. D., & Fischler, M. A. An inference technique for integrating knowledge from disparate sources. Proceedings, 7th International Joint Conference on Artificial Intelligence, 1981, 319-325.

Grossberg, S. Biological competition: Decision rules, pattern formation, and oscillations. *Proceedings, National Academy of Science, USA,* 1980, **77**, 2238-2342.

Hanson, A. R., & Riseman, E. M. (Eds.). *Computer Vision Systems.* New York: Academic Press, 1978.

Hillis, W. D. The connection machine (computer architecture for the new wave). MIT AI Memo 646, 1981.

Hinton, G. E. University of California, San Diego, 1980.

Hinton, G. E. The role of spatial working memory in shape perception. Proceedings, Cognitive Science Conference, 1981, 56-60. (a)

Hinton, G. E. Shape representation in parallel systems. Proceedings, 7th International Joint Conference on Artificial Intelligence, 1981, 1088-1096. (b)

Hinton, G. E., & Anderson, J. A. (Eds.). *Parallel Models of Associative Memory.* Hillsdale, NJ: Erlbaum, 1981.

Hubel, D. H., & Wiesel, T. N. Brain mechanisms of vision. *Scientific American,* 1979, **241** (3), 150-162.

Ja'Ja', J., & Simon, J. Parallel algorithms in graph theory: Planarity testing. Pennsylvania State University, Computer Science Department, TR 80-14, 1980.

Jusczyk, P. W., & Klein, R. M. (Eds.). *The Nature of Thought: Essays in Honor of D. O. Hebb.* Hillsdale, NJ: Erlbaum, 1980.

Kandel, E. R. *The Cellular Basis of Behavior.* San Francisco, CA: Freeman, 1976.

Kimme, C., Sklansky, J., & Ballard, D. Finding circles by an array of accumulators. *Communications of the ACM,* 1975, **18**, 120-122.

Kinsbourne, M., & Hicks, R. E. Functional cerebral space: A model for overflow, transfer and interference effects in human performance: A tutorial review. In J. Requin (Ed.), *Attention and Performance VII.* Hillsdale, NJ: Erlbaum, 1979.

Kuffler, S. W., & Nicholls, J. G. *From Neuron to Brain: A Cellular Approach to the Function of the Nervous System.* Sunderland, MA: Sinauer, 1976.

Marr, D. C., & Poggio, T. Cooperative computation of stereo disparity. *Science,* 1976, **194**, 283-287.

McClelland, J. L., & Rumelhart, D. E. An interactive activation model of the effect of context in perception: Part 1. *Psychological Review,* 1981, **88**, 375-407.

Minsky, M., & Papert, S. *Perceptrons.* Cambridge, MA: MIT Press, 1972.

Norman, D. A. A psychologist views human processing: Human errors and other phenomena suggest processing mechanisms. Proceedings, 7th International Joint Conference on Artificial Intelligence, 1981, 1097-1101.

Perkel, D. H., & Mulloney, B. Calibrating compartmental models of neurons. *American Journal of Physiology,* 1979, **235**, R93-R98.

Posner, M. I. *Chronometric Explorations of Mind.* Hillsdale, NJ: Erlbaum, 1978.

Ratcliff, R. A theory of memory retrieval. *Psychological Review,* 1978, **85**, 59-108.

Rosenfeld, A., Hummel, R. A., & Zucker, S. W. Scene labelling by relaxation operations. *IEEE Transactions on Systems, Man, and Cybernetics,* 1976, **6**, 420-433.

Sabbah, D. Design of a highly parallel visual recognition system. Proceedings, 7th International Joint Conference on Artificial Intelligence, 1981, 722-727.

Sejnowski, T. J. Storing covariance with nonlinearly interacting neurons. *Journal of Mathematical Biology,* 1977, **4**, 303-321.

Stefik, M. Planning with constraints (MOLGEN: Part 1). *Artificial Intelligence,* 1981, **16**, 111-139.

Sunshine, C. A. Formal techniques for protocol specification and verification. *IEEE Computer,* 1979, **12** (9) , 20-27.

Treisman, A. M., & Gelade, G. A feature-integration theory of attention. *Cognitive Psychology,* 1980, **12**, 97-136.

Wurtz, R. H., & Albano, J. E. Visual-motor function of the primate superior colliculus. *Annual Review of Neuroscience,* 1980, **3**, 189-226.

Zeki, S. The representation of colours in the cerebral cortex. *Nature,* 1980, **284**, 412-418.

Stimulus Information and Processing Mechanisms in Visual Space Perception

Ralph Norman Haber

University of Illinois
Chicago, Illinois

Abstract

Most current models of the perception of visual space follow one of two courses. Some begin with a superficial analysis of the visual information contained in the light reflected from surfaces in a visual scene to the eye and then work out highly sophisticated information processing models of how that information is used. Others begin with a detailed analysis of the stimulus information available at the eye, and then provide only a superficial processing description, if they provide one at all. The empiricists, beginning with Helmholtz and followed by many others (e.g., Rock, Epstein, Leibowitz, and perhaps Hochberg) generally pursue the first approach, while Gibson and some of his disciples represent virtually all of the work on the second. It is my thesis that progress on space perception is doomed until we create a thoroughly Gibholtzian approach, in which the stimulus analysis is based on the information available from realistic scenes viewed by two-eyed, moving observers, and the processing analysis is based on how the richness of this stimulus information is extracted or picked up to construct representations of visual space in our heads. The present paper is in two parts. The first describes the major categories of available information contained in light, how those sources of information are related to each other, and how the perceiver acquires or comes into contact with such sources of information. The second part is concerned with how that information is processed or used. Finally, because we are able to perceive the layout of space when looking at photographs and paintings of scenes, and at

157

sequences of pictures that create motion, the same pair of analyses – one for
stimulus information and one for processing – is presented for cases in which
one looks at flat displays of scenes and sequences of motion pictures.

1. Introduction

Normal adult human beings spend nearly all of their waking time gazing
at visual scenes from which they extract information about the layout of space
and the identity of objects and persons. A vastly smaller amount of time is
spent looking at pictures and other two-dimensional representations, most of
which are of scenes containing a layout of space. An even smaller amount of
time (except for intellectuals, scientists, and a few recluses) is spent looking at
nonspatial patterns such as print to be read, butterfly wings to be classified, or
hieroglyphics to be decoded. Humans occasionally interrupt these gazings in
order to look *for* something rather than *at* something, a concern of visual search
and selective attention. Of these four visual activities, I plan to review the first
two, looking at scenes and pictures, which, taken together, account for what I
estimate to be at least 90% of the time our eyes are open. First, I will describe
the kinds of information available to the perceiver from scenes and from pictures: visual information, information from the other senses, information from
stored knowledge about similar scenes, and information based upon general
strategies for scene and picture looking. Second, I will describe how that information is extracted, selected, processed, compared, and stored; that is, how a
perception of the stable visual environment results from it all.

Unfortunately, the topic of space perception is not based entirely on a
well-researched and validated experimental literature, and the competing
theoretical positions have yet to be resolved or amalgamated into a reasonably
agreed upon body of truth and fact. For some of the content of this review, I
shall have to depend almost exclusively on theory, with little support available
from controlled experimentation (see also Haber & Hershenson, 1980). I shall
try to thread my way through the competing theories as best I can. My goal is
not to decide which theory is best, but to try to provide the best description of
what we know about the perception of visual space. This will often be very
difficult because even descriptions of data are given in theoretically loaded
terms. I have covered a few of these theories (Haber, 1983a) in a summary of
the current status of visual perception, including that of space perception. No
scientific enterprise is devoid of theoretical direction and therefore of potential
bias for misdirection. In spite of two millenia of speculation and two centuries
of research, we still have more theories than the available research can support.

1.1. Stimulus Analyses and Process Analyses

A stimulus analysis of scene perception provides a description of the visual information from the scene as it is imaged on the retinal surface. Also included is nonvisual information about the scene acquired at the same time from our other senses (for example, sound, body position, body motion). A process analysis of scene perception provides a description of steps involved in the extraction, combination, processing, storage, and organization of the relevant and necessary features of that information in order to arrive at a perception of the scene. It should be obvious that neither of these analyses by itself is sufficient. Unfortunately, most of the theories of space perception are based on only one or the other. This has often led to dramatically different research strategies that both preclude certain results and compel certain interpretations that do not generalize. The most time honored, widely believed and misleading assumption about nature of the stimulus used for processing I refer to as the *The Eye-Is-a-Camera Analogy*. When scientists in the seventeenth century first began to appreciate the image-forming properties of the eye, they rather naturally assumed this image in the eye is the source of visual data on which perception is based. With the invention first of the camera obscura and then especially of the photographic camera 150 years ago, this notion was elaborated until the starting point for all analysis became the picture on the retinal surface, in the same way that the image on the film surface in a camera is the beginning of the photographic picture. This implicit assumption is strengthened by the similarity of the optical mechanisms of the eye and the camera, so that the outcome in each case is a negative image obtained through the interaction of the quantal and wavelength properties of light acting upon a two-dimensional photosensitive surface.

While I have no quarrel with the similarities of the optical mechanisms (but see Boynton, 1974), the analogy must be rejected as a model of perception. We should never think of the retinal image as a picture, because then we have to ask how that picture resembles or differs from the real scene on the one hand and from our perception of the scene on the other. The retinal image is not a template which must be matched to the scene or else tampered with until it does match. Any attempt to use the retinal image as a picture to be perceived is doomed to lead all theorists and researchers into an acrimonious and futile debate and into silly experiments as well. This happens whenever someone attempts, for example, to account for how a square object in space, which projects a trapezoidal retinal image, could yet be perceived as square. While it is true that squares viewed obliquely have trapezoidal retinal images, no little man sits in the head looking at a trapezoid and trying to determine what its distal shape had to have been. Braunstein (1975) has the best discussion of the theoretical and practical pitfalls of the eye-camera analogy.

It is more accurate to think of the retinal image as simply information. Pretend that we never heard of a camera. Think of light reflected from a scene and imaged on a photosensitive surface as containing information about the nature of the surfaces in that scene. One step in the study of scene perception is to work out the transformations that relate the information on the retinal surface to the structure of the scene. These transformations are related to the laws of perspective dating to Euclid and were described in more detail by the early Renaissance painters, especially Leonardo da Vinci. Once we know the transformation rules, then we can go to the next step: identification of the information extraction processes that complete the processing of that information into perception. Perspective rules always transform the images so that individual parts do not match (in the sense of a template) even though the transformation from scene to image is entirely lawful. The intermediate step of an image with pictorial properties that more or less match the scene is unnecessary. Unfortunately, we shall see the mischief of the retinal picture concept in our consideration of both stimulus and processing analyses.

The best example of a good stimulus analysis comes from Gibson (1950, 1966, 1979). He has, more than anyone else, appreciated the complex nature of the perspective and motion transformations that describe the relationship between the surfaces reflecting light to the eye and the pattern of light imaged on the retinal surface. I shall depend heavily on his analysis, with some additions as necessary, in the next section of this paper. But according to Gibson, space is directly perceived, which implies that the stimulus analysis can be substituted for the process analysis. All the stimulus analysis has described is the pattern on the retina. While Gibson knows the retinal image is not a picture and is not looked at as a picture, he fails to account for how the information in the retinal image becomes a perception of a layout of space. The rules that relate the scene to the image are only the beginning, not the end, of space perception. While I liberally borrow from Gibson's stimulus analysis, he provides only the beginnings of a model about processing.

In my opinion, there is no "best example" of a good processing analysis of visual space perception, mainly because most of the analyses available start either with very restricted inputs rather than with natural scenes, or from very diverse theoretical positions. Epstein (1977) has brought together a number of major process models of space perception, all of which follow in various ways the Helmholtzian tradition of "unconscious inference." While Helmholtz is still the father to most processing models, I shall consider one or two that he might not wish to recognize as his children.

1.2. Rich Versus Impoverished Scenes

Gibson starts with natural scenes and tries to describe all of the information available at the eye. In contrast, most of the processing models presented in Epstein (1977) start with very simple line drawings or simple objects. This permits the less complicated stimulus information to be described in terms of a complex set of processing steps. When studying a complex information system, it is reasonable to reduce the input to its fundamental components, treating each in isolation so its properties can be analyzed. But this "good science" runs the risk of removing from the analysis just that combination of information typically used by perceivers in their perceptions. When that happens, well-controlled experiments often find that additional information has to be added before perceivers can properly process the scenic information. Therefore, until we know the kind of stimulus information that is routinely used by perceivers, we run serious risk of misinterpretation when experiments concentrate on grossly impoverished stimuli. It may be, in the final analysis, that some of the nonvisual sources of information about the layout of visual space are unnecessary to account for visual perception, except when the display is impoverished.

In addition, I suspect the dynamics of processing change as the environmental information is restricted. There are many sources of information, some scaled over the entire retinal surface and some very local. Following a general assumption about all information processing, I believe that processing focus is on the highest level of information first and only turns to more local sources if they are needed to resolve ambiguities or to provide extra assurance. Thus, the study of the processing of impoverished scenes, such as simple line-drawn objects, may reveal a fundamentally different set of processing strategies than those found for richly informative natural scenes. For example, Gregory (1966) demonstrated that convergence is used like a rangefinder device for specifying the distance of a single point of light in the dark (Gregory, 1970), but it rarely, if ever, contributes to the perception of the layout of natural scene (Ogle, 1962). It is empirically correct to call convergence a cue for depth, but only for impoverished displays. It seems to be ignored in processing rich scenes.

As a second example, Gilchrist (1977) has now convincingly shown that the perceived lightness of different parts of a scene has to be computed after the layout of space is determined, not concurrently or before (also see Hochberg & Beck, 1954). The perception of brightness contrast between two areas on the retinal image depends entirely upon whether those areas are perceived to be adjacent to each other in space. If they are seen at different distances, then no contrast occurs. Thus, perceiving the layout of a scene in depth has to be completed before lightness (and presumably also color) values are assigned to each surface in the scene. This fact is obscured when only simple coplanar displays are used to demonstrate contrast.

There undoubtedly are many other examples lurking in the psychophysical literature in which some function, computed for a single case, has little relevance for natural perception. One candidate is measurement of color discriminability. Boynton (1973) describes a number of experiments in which subjects are asked to adjust some value of half of a small field so that the edge separating it from the other half disappears or is as minimal as possible. The field used is always relatively small, seen against a black background, and is located either at some fixed or indeterminate distance from the perceiver. What would happen, however, if the left half were seen in front of the right half? Generalizing from Gilchrist's results, the edge would never disappear, because it was formed by a depth difference. Consequently, the values found for color discrimination could be quite different. If the generalization of the functional relationship is to be applied to the natural scenes, which procedure is best?

2. Stimulus Analysis of Space in Scenes

The explanation of how we maintain visual contact with the environment around us requires a careful description of the sources of visual information and the transformations that map that information from scene to retinal image. I shall begin with reflectance and with imaging of light on an image plane, such as the retina. Then I will discuss the laws of perspective that describe the transformations that occur in the patterns of light from scene to eye. Finally, I conclude with the nature of these transformations on a single stationary retina, on a movable retina, and on two retinas.

2.1. Reflectance as a Property of Surfaces

Visible light radiates from a light source (such as the sun, light bulbs, or fires), diverging out from the source in all directions. Some of the rays strike surfaces and are either absorbed or are reflected by them. Most surfaces in the natural world reflect light diffusely, so that regardless of the angle of incidence, the light rays diverge in all directions again. In this sense, such surfaces act like specialized light sources themselves. Some of these reflected rays may happen to impinge on the cornea of an eye, passing through the cornea, the lens, and the optical media inside the eyeball, strike one of the photoreceptors in the retina, and be absorbed by one of its photopigment molecules. It is only those particular rays of light which have any potential visual significance. All others bound about in the universe unnoticed by human beings and therefore without visual effect on human beings.

Two properties of light, as it interacts with surfaces, provide perceivers with information about those surfaces – its intensity and its wavelength.

The intensity of a ray of light is a function of the number of quanta it contains. For practical purposes, there is no loss of quanta in each ray until it encounters something, such as a surface. All surfaces absorb some of the quanta in the ray and reflect others. What proportion is reflected is a property of the surface itself, with some surfaces reflecting nearly 100% of those quanta that are incident to them and others almost none.

The other relevant property of light is wavelength – the physical dimension that we perceive as color, or more technically, as hue. While most light sources emit light of all visible wavelengths, surfaces absorb light selectively as a function of wavelength. Thus, some surfaces absorb all short wavelengths (ones that normally produce perceptions of blue) and reflect only long wavelengths (ones that look red). This results in a perception of a surface being red since only long wavelength light reaches the retina.

Therefore, the pattern of brightnesses and of hues that we perceive in a scene is initially the result of the selective reflectances of the number of quanta and their wavelengths from each part of each surface reflecting light to the eye. In a technical sense, then, surfaces of objects are not light or dark, nor are they red, green or yellow. These are properties of our perceptions alone. Surfaces differ only in the number of quanta and the wavelength of the light they reflect. However, because of the correspondence between the patterning of the quanta and wavelength being reflected and the pattern on the retina, even visual scientists find themselves referring to a red book or a black piece of paper as if the property of the hue or brightness belonged to the object and not to our perception of it.

When looking at a typical scene, wide variation in the pattern of reflectances is present. Some result from variation within a single extended surface – the light and dark grain lines in a piece of wood, for example. Others arise because two adjacent surfaces have different reflectances, such as a white piece of paper resting on a desk. In general, physical edges between surfaces or objects almost always create visual edges – that is, abrupt changes in reflectances of both quanta and wavelength. Thus, if a photometer, sensitive to both aspects of reflectance, is moved over a scene, it registers nearly all of the physical discontinuities in that scene.

The above is a description of a *reflectance edge* – a difference in the quanta or wavelength between adjacent rays of light (or areas in the retinal image of that light) resulting from differences in reflectances of surfaces. In contrast, an *illumination edge* results from different amounts of illumination falling on adjacent parts of a surface in space. Even if there is no difference in reflectance, different amounts of light are reflected from either side of an illumination edge. Illumination edges are produced whenever there is more than one light source in a scene, or more typically, whenever objects cast shadows over each other from a single source. In terms of properties in the pattern of light as imaged on the retinal surface, these two types of edges cannot be distinguished by local cues along their border, but require higher level processing. I will return to this problem in the section on processing later.

We have started with light from a source diverging in all directions in space. From every point on every surface encountered by each ray, the light is reflected in diverging paths again. Now consider what happens when an optical system such as an eyeball is placed in the path of some of these flying photons.

2.2. The Imaging of Light

When light is diffusely reflected on diverging paths from a single point on a surface, some of those rays may strike the cornea and lens of an eye, if one is around. Both of these optical surfaces refract the light so that it is again imaged as a single point on the retinal surface. Since the focal length of the eye is fixed at about 17 mm, proper accommodation of objects at different distances is accomplished by the lens changing its shape. If the light is being reflected from several places on an extended surface, then the divergent rays from each of these points are recombined to project points behind the lens. The image behind the lens is inverted and mirror imaged in relation to the points reflecting the light in space.

As long as the corneal-lens optical system is properly accommodated, the pattern of light reflected from surfaces in space is reproduced on an image plane inside the eye coincident with the retinal surface. Thus, we have surfaces reflecting patterns of light in all directions in space and an optical system which images one slice of that pattern on the photoreceptive surface. The image is inverted with respect to the reflecting surface, but that is the least interesting of the transformations.

2.3. The Laws of Perspective

While the inversion transformation is invariant, there is another transformation which is not invariant, a transformation of size as a function of distance. This transformation is described by the laws of perspective using the principles of Euclidean geometry. Regarding perspective of size, Euclid said that an object of fixed physical size takes up a smaller part of the retinal image if it is far away than if it is near: the size of the retinal image of the object in space is a direct function of how close the object is to the eye. Retinal size is specified by the angle made between the reflected rays of light from the extremities of the object. If an object is itself big or is nearby, the angle made by the light is large, thereby creating a large retinal image. The retinal size of projected objects is usually specified in terms of this visual angle in degrees. The sun is about $1/2°$, a typical thumbnail at arm's length is nearly $2°$ (about the area of the fovea), and a 50-foot flagpole at 200 feet is nearly $15°$ of visual angle, as is a 30×30 inch picture at ten feet viewing distance. The entire retinal image for one eye is about $150°$ (or larger if you have a very small nose).

As the observer backs away from the scene, the visual angle made by each object decreases, making the retinal image of each object smaller. If every surface of every object in a scene is directly facing the observer, and at the same distance, then as he backs up, the compression in the size of the image of each surface on the retinal image is uniform – no changes in the shapes of any parts of the retinal images occur.

However, the natural environment is never arranged so that all surfaces are equidistant from you. Because the size of an image is related simultaneously to the size of the object and to its distance, if some parts of the surface are nearer than other parts (that is, it is slanted with respect to your line of sight), then the nearer parts project larger images than the farther parts.

The most important surface of all, the ground on which we stand, never is equidistant. That surface stretches away from us at a slant so that some parts of the surface are much farther away than other parts.

Imagine you are standing on a railway looking down the track. The physical length of each crosstie is, of course, physically constant, but those farther into the distance subtend progressively smaller and smaller visual angles on the retina. Given this, the light reflected from the physically parallel tracks produces converging retinal images – converging toward a vanishing point in the distance. For the same reason, a checkerboard pattern of squares on the ground produces a pattern of trapezoids on the retina. The degree of compression of the top relative to the bottom of each trapezoid depends upon the slant of the ground away from the observer. Similarly, each variation in reflectance of the surface of the ground produces smaller patterns of discontinuities on the retina the farther away the variation on the surface is from the eye.

One of the most important variations in reflectances of surfaces is referred to as its texture density. All surfaces have a physical graininess – variation in the uniformity of the physical surface elements. Most such graininess is randomly regular. For example, a grass lawn in any one small area has some tall and short blades which vary somewhat in width and color, but about the same variability appears in any area of the lawn. Therefore, we can consider the graininess or texture density of the entire lawn as uniform or regular. Of course, the texture density of a lawn is quite different than that of a highway or the surface of a lake, but in each case, their respective texture density is fairly constant over the entire surface. Texture density is one of the most distinctive characteristics of the surfaces of objects.

If the observer looks at a surface straight on, then the reflected light reproduces the texture density in the retinal image as uniformly as it is on the original surface. If the surface is oblique to the line of sight, as it always is from the the ground to the eye, then the density of the texture from nearby points of the surface is coarser on the retina than the density from far away parts of the surface. While the physical texture density may have been uniform and constant, the projected density on the retina has a gradient – coarse to fine – as as a function of distance of the surface to the observer. In this sense, then, texture density undergoes the same transformation with distance as do other discontinuities in light reflected from surfaces to the eyes of observers. Texture is a particularly important source of information about distance, since the textured ground on which we stand stretches away in all directions from us and thereby projects a gradient on the retina that accurately reflects the slope of the ground and the relative distance of its elements from the observer.

These transformations of perspective are regular and mathematically specified. They represent a set of rules which relate the arrangement of light-reflecting surfaces to the pattern of reflected light imaged on the retinal surface.

There is only one pattern on the retina that can be produced by any given natural scene viewed from a particular position, and as long as the ground on which the perceiver stands is visible and attached to both the scene and the observer, there is only one scene that can produce any particular retinal pattern. It is, of course, possible for purposes of research to create ambiguous scenes or provide impoverished information in which the reversibility of the laws of perspective do not hold, but these do not occur in the natural world. See Stevens (1980, 1981) for a discussion of the mathematical properties of the specification of texture, and how those relate to representation of three-dimensional information about textured surfaces.

The above discussion of how reflectances of light from distal surfaces form a retinal image describes the kind of information available from a three-dimensional scene to a stationary eye, equivalent to what an observer would see if he used only one eye and did not move. I shall refer to these transformation rules as those of *stationary or surface perspective*. Of course, in the normal perception of scenes, we use two eyes and we move both our eyes and heads extensively.

2.4. Effects of Eye Movements on the Retinal Image

Two changes of importance occur when an eye moves (without any head movement): first, the retinal image is displaced with respect to the retinal surface, and second, the part of the scene seen with maximum clarity and contrast shifts. The displacement produces no transformation in the pattern of the image – no compression or expansion, nor any transposition among the visual edges produced by luminance discontinuities, irrespective of variation in distance away from the objects in the scene. Of course, the part of the image farthest to the side opposite to the direction of movement no longer falls on any receptive part of the retina and therefore is occluded. Conversely, part of the scene on the other side becomes visible as the eye rotates toward it.

The effects of clarity are different. While there is no shape transformation in the retinal image produced by eye rotation, the part of the scene represented with sharpest focus shifts, since that always is the part in a direct line of sight straight ahead. Because of this, eye movements are an important component in visual search, since without moving the eye, not all of the scene can be seen clearly. However, eye movements without head movements do not provide any additional direct information about the spatial arrangements of the scenes (but see Hochberg, 1981).

2.5. Effect of Head Movements on the Retinal Image

Unlike eye movements, head movements create changes which provide major sources of information about the layout of space in three-dimensional scenes. In general, whenever the entire eye is displaced, the pattern of discontinuities in the light on the retinal image undergoes a transformation: light arising from nearby surfaces is shifted in the retinal image more than light arising from far-away surfaces. In watching the countryside from a moving car, the light reflected from nearby trees sweeps across the retina more quickly than the light reflected from trees in the distance. The relationship between the relative rates of movement of the light patterns across the retina is an exact function of the differences in distances of the objects. The visual edges created by the physical discontinuities in reflectance are transposed, crossing each other as the observer moves.

The different ways in which the observer moves within a scene affect the particular kinds of transformations that occur in the retinal image (see Ullman [1979] for a more general discussion). Each pattern of transformation is unique to each type of movement. In the above example, movement is laterally across a scene, with fixation moving as well. However, if while moving laterally you fixate some point in the scene, then light reflected from that point does not change on the retina. All other parts of the image change, with those parts nearer than the point of fixation changing in one direction, while those points farther than the point of fixation change in the other. This is one aspect of the familiar case of motion parallax.

Some movements are radial rather than lateral. If you move into a scene directly toward what you are looking at, the retinal image expands from the center out, with light reflected from objects on the side moving toward the edge of the image and being occluded from the image altogether as you pass the object by. The relative rate of movement from the center of the image outward is again a direct function of how close these objects are to you.

In addition to displacements across the retina, the shapes of the light pattern change, too, when you move in or out of a scene. A rectangular window slightly to the side of straight ahead in the distance projects a nearly rectangular retinal image, but as you walk straight ahead, that window's image is displaced to the side of the retina and at the same time becomes more and more trapezoidal in shape on the retina.

These different transformations with movement provide a set of laws of *motion perspective* that also mathematically specify the relation between the light reflecting surfaces and the continuous changes in the retinal pattern produced by displacements of our eyes in space. Not all of the mathematics have been worked out, and undoubtedly some of them will be more complex than we presently know. See Koenderink and van Doorn (1976) for one example of specification of the information available from motion parallax. Because both of these sets of rules, stationary perspective and motion perspective, refer to the same scene, they must provide the same information about that scene, though, of course, the particular content of that content of that information on the retina differs vastly.

2.6. Effects of Using Two Eyes

The above discussion has assumed that the observer had an eye patch over one eye. Technically, removing the eye patch produces no changes in the uncovered eye, but because the two eyes are slightly separated in the head, the two retinal images of a scene can be slightly different, depending upon whether the scene has any variation in depth. If you look at a scene which has no depth to it, where all parts are equidistant from your eyes, such as a wall or picture, then your two retinal images correspond exactly over their entire extent. However, if there are parts of the scene closer or farther away than the point in the scene at which your two eyes are fixated, then your two images do not overlap perfectly but are out of registration. Specifically, the part of the image which corresponds to parts closer than the point of your fixation is displaced outward in the images relative to those parts perfectly aligned, whereas parts farther away are displaced inward. The degree of displacement or noncorrespondence is exactly proportional to the difference in distance (nearer or farther) from the distance of the object at which you are fixated. If you now change fixation to some other object at a different distance, a new pattern of correspondence and noncorrespondence is created, but again, the disparity between the parts of the two images is proportional to the relative distances. In this sense, then, the amount and location of disparities between the entire two retinal images can be described by the amount of *binocular disparity perspective* present between each location on the two retinal surfaces. This disparity perspective is the basis for stereoscopic depth perception (see Julesz, 1971); it also provides another fundamental set of mathematical perspective relationships between light-reflecting surfaces in space and patterns in our images, though in this case their perspective is based on a comparison of the two images – one from each eye.

The relationship of the pattern on the two eyes also is informative of the arrangements of space when you move into a scene or when an object comes toward (or away from) you. When you walk directly toward some object, the image of that object enlarges on each retina (as described above as part of motion perspective). In addition, the location of that image on each retina shifts by an equal amount in opposite directions across each retina – in this case, by each shifting outward as you get closer to the object. The equal and opposite movement is a consequence of the collision course. If you walk toward but to the right of an object, then its image slides across both retinas in the same direction, though at a greater rate on the left retina than on the right. Same-direction motion is indicative of a noncollision course, with the relative differences indicating the amount of miss. If the image moves on one retina and merely expands without movement on the other, then the object is on a collision course with the latter eye only.

These relationships define information about motion in depth; that is, movement toward or away from the perceiver. The geometry provides a precise description of the paths of moving objects, and even more completely, on the path of the perceiver while moving. These relationships depend upon the use of two eyes, but are quite different from the binocular disparity information that defines stereopsis. Further, Regan, Beverly, and Cynader (1979) argue that useful motion in depth information is available from parts of a scene several thousand feet away, two or three times farther than that available from stereopsis (Julesz, 1971).

2.7. Scales of Space vs. Local Depth Cues

The preceding treatment is in contrast to the conventional one, which first describes the monocular depth cues of familiar size, familiar shape, texture, accommodation, interposition, adjacency, aerial perspective, motion parallax, linear perspective, and the like, and then the binocular depth cues of convergence and binocular disparity. While conventional, the organization of information into monocular vs. binocular has not proved to be particularly useful. Rather, what I have done so far is describe three general sources of information that are specified across the entire retinal image rather than over just a portion of it. I have referred to stationary, motion, and disparity perspective as specifying three *scales of space* (Haber, 1978) to differentiate them from the local sources of information about space. Each of these three perspective laws specifies a scale because the transformation rules apply to the entire image. Thus, standing on a grassy expanse, the stationary perspective specifies a continuous change over the entire retinal surface stretching away from you, irrespective of where you look. The gradient of texture is a metric specifying rate of change: that is, distance. Similarly, motion perspective provides rates of displacement across the entire scene. Unlike the stationary scale, where the reference point is the observer's feet, the reference point on the motion perspective scale always is the point of fixation. This is also true for the binocular disparity scale, where zero disparity is at the point of converged fixation in the scene. Each of these specifies the spatial arrangement simultaneously for every visual edge over the entire retinal surface, and each has scalar properties, acting as rules relating variations across the two-dimensional retinal image to information in the third dimension extending away from the observer.

In contrast, local depth cues provide information for spatial arrangements only among adjacent areas on the retinal image. Interposition, for example, is considered a depth cue because it specifies the retinal image relations when one surface occludes another one. While interposition is a powerful source of information about the relative distances of two objects (see Hochberg, 1971), it only specifies adjacent discontinuities and does not specify the relation of these two surfaces to any other ones that are not adjacent to each other in the line of sight. Thus, interposition has no scalar properties. The same is true for adjacency (Gogel, 1977), in which areas in the retinal image that share a common contour are seen as being at the same distance from the perceiver. In a sense, familiar size or shape is also a local depth cue – if you know the true size and shape of an object, then its retinal size is a clue too how far away it is, and its retinal shape is a clue to its orientation or slant.

I have had less to say about the local depth cues, partly because I feel they are of less importance than those that possess scalar properties, and also because they have been so well analyzed in conventional treatments of depth perception (e.g., Hochberg, 1971; Kaufman, 1975). I will return to them in the consideration of processing because the local sources of information may be processed quite differently from the scalar ones.

Since all of these sources of information arise from a common scene, they must be each be informative of that scene. They cannot be contradictory, even if any particular one may be insufficient by itself to uniquely specify only that one scene as its source. The three scales of space substantially overlap in their information content because they each represent scalar information from the entire scene. Normally, the three scales of space, as well as all local sources, are always available simultaneously during perception. It is up to processing analyses to show how they are used singly or in unison. Even so, the three scales are not completely redundant among themselves because each alone has a small range in which it fails to provide complete determinancy of a scene. For example, Ullman (1979) has shown the conditions for indeterminancy for motion perspective, along with Koenderink and van Doorn (1976) and Nakayama and Loomis (1974). It remains to be worked out in terms of the geometry how the three scales together provide sufficient information by themselves for uniqueness.

Gibson's analysis of visual information from scenes, from which the scales of space are derived, stresses the notion of invariance of information. What Gibson means is that some pattern in the light always turns out to be invariant or unchanging with distance or slope or orientation. For example, the surface scale of space specifies an invariant relationship between texture density gradient and distance along the ground. I have described the relationships in term of scales or metrics rather than invariants as Gibson does, but the two descriptions have the same properties.

2.8. Nonvisual Sources of Information

Full retinal scales of space plus local depth cues comprise sources of visual information about the arrangements of surfaces and objects in visual space. While it is still a matter of theoretical debate, it is possible that we also receive and use information from nonvisual sources that provide information about visually perceived space. These include information about the changes in the positions of our eyes and head; information about changes in our overall locomotion through the scene; information in the form of general assumptions about the nature of the content of the scene (for example, in man-made scenes, we might assume that intersections of edges occur at right angles, or that elliptically-shaped objects are really circular); and information in the form of specific expectancies about the contents or arrangements of the particular scene in view. I shall save discussion of these nonvisual sources of information until the section on processing, since they have received most attention in the context of different approaches to processing.

2.9. The Role of the Retinal Clarity Gradient

Most stimulus analyses are usually carried out on a hypothetical plane between the eyes and the scene, through which the reflected light from the scene passes. In theory, if the plane is perpendicular to the line of sight, the pattern of reflectances from the scene is exactly reproduced on the plane and on the retinal surface (except for the inversion of the latter). But such a theoretical analysis fails to account for the poor optical properties of the eye. Even with high light intensity and the smallest possible pupil opening, we have little depth of field in human vision. Further, only the central area of the light pattern projected on the retina, that which passes through the center of the cornea and lens, comes anywhere near being sharply focused. Blur becomes pronounced even a few degrees off center (out of a nearly 160° field of view). Contrast between adjacent areas of luminance discontinuity loses sharpness, and thin lines are spread out and are reduced in contrast against their backgrounds. These facts constitute one important reason why a stimulus analysis cannot substitute a projection plane in front of the eyes for the proximal stimulus. It means that every momentary retinal image is clear and sharp only in its center and progressively more fuzzy the further from that center.

Decreasing clarity of the retinal image from center outward has several important implications for both the stimulus analyses and for subsequent processing. On the stimulus side, it means that the three full-retinal scales of space do not provide the same specificity of information over the entire retinal surface. This does not matter too much for the motion perspective scale because current evidence suggests that motion information can be extracted equally from low as compared to high contrast and spatial frequency stimuli (Sekuler, Pantle, & Levinson, 1978). The scaling of stationary perspective would seem to be much more dependent upon good contrast and fine detail, both of which are concentrated in central vision and deficient in the periphery.

These facts and their consequences have been incorporated into several theories proposing that we effectively have two visual systems (see Schneider, 1969; Held, 1970; but also Campion, Latto, & Smith, 1983; Haber, 1983c). For example, Leibowitz, Rodemer, and Dichgans (1979) argue that central vision, including the fovea and a few degrees around it, is concerned with recognition and identification processes requiring fine detail. They call central vision the "WHAT" system. In contrast, the entire retina, excluding the small central part, is concerned with detecting and locating stimulation – a "WHERE" system – irrespective of what that stimulation might be. The

WHERE system locates a stimulus and then directs an eye movement so that the WHAT system can get a good look at it. The detection and location process does not require high resolution but only good motion and edge detection, activities that we know do not diminish from fovea to periphery to the retina.

There are a number of processing consequences of this division of responsibility between these two systems and I will discuss some of them in the processing sections below.

This, then, provides an overview of the visual stimulus information available to perceivers about the layout of real three-dimensional scenes. Before turning to how this information is processed, let me first consider a parallel stimulus analysis of the information available to us from pictures of scenes.

3. A Stimulus Analysis of Pictures

When you look at a photograph of a scene or a landscape painting you have no difficulty perceiving the correct layout of space: the photograph looks like the scene. Yet a photograph is flat, printed on paper. We can bend it, put it in a frame, and even look behind it. We can perceive it as a flat piece of paper and not a scene at all. These two ways of perceiving pictures are always available to us, and they make the perception of pictures a more complex process than the perception of three-dimensional scenes, which normally lack this duality. However, the principles are identical. In this section I examine the stimulus information projected to the eye from photographs and paintings of scenes. I am most concerned with the three scales of space because it is through an analysis of the information they provide that we can see how picture perception differs from scene perception.

Start with a flat, empty visual world, such as a flat wall perpendicular to the line of sight. All three scales of space indicate the flatness. The stationary scale is uniform – it contains no perspective changes, no vanishing points, and no texture gradients. The motion scale is also uniform; no matter how you move, no transformation indicating depth occurs. Rather, all parts of the image of the wall surface move together. Further, the binocular scale shows no disparity anywhere – both retinal images are the same because no part of the scene is different in distance from any other part. Finally, all the local sources of information about depth indicate flatness. In short, the overwhelming perceptual experience is of a flat surface.

3.1. Stationary Perspective Information of Flatness

What differences in the pattern of light reaching the eye occur when, instead of looking at a wall, you look at a flat surface, such as a photograph or a representational painting of a three-dimensional scene?

Consider first a photograph of a landscape. If taken with a 50 mm lens on a 35 mm camera, then the negative or a print made from the negative contains all of the stationary perspective transformations of light coming from the scene, as if the negative had been a retinal image of the scene itself with the eye placed where the camera had been. This means that if you look with one stationary eye at a scene or if you look in the same way at a picture of that same scene (as long as the picture is viewed from the same angle and distance as the camera viewed the scene when it took the picture), the pattern of light imaged on the retina is the same in each case. The laws of linear perspective that describe the relationship of the patterns of light reflected from surfaces of objects in a scene, to the pattern of light imaged on the retina, describe the same relationship when the light is reflected instead from a picture of the scene. Changes in these laws can be introduced if the photograph is taken with a lens that does not correspond to the focal length of the human eye. A telescopic lens foreshortens the photographic image so that near and far surfaces appear to be closer together than they really are. Even so, the proper perspective can be achieved if the properties of the lenses are taken into account. Rosinski and Farber (1980) describe the geometrical effects of magnification and minification in some detail.

If the surface perspective defined by the laws of linear perspective demonstrates that the pattern of light from a scene and from a picture of a scene are the same, why don't we always confuse pictures with scenes? We rarely confuse them because the surface perspective is only part of the information we receive from a picture and from a scene: the rest of the information from pictures and scenes is dramatically different. The net result is that we can perceive a picture of a scene in two ways – as the scene it represents (as if it were a window opening out onto the scene) and as merely a flat piece of paper.

Before we can understand why pictures can be perceived as both flat and as the scenes they represent, the differences in available information from pictures and scenes must be described. For pictures but not for scenes, information is provided that the surface is flat.

There are three different sources of information from the surface of the picture that lead to perceptions of flatness: the flat frame and surround of the picture, the picture itself, and failures of the surface perspective information to match that coming from the real scene.

The surround of a picture – its frame and the flat wall on which it is hung – provides strong indications that the picture is flat, since all of the surface information about distance and orientation from the frame and wall are perfectly correlated with the information from the picture surface itself.

The picture surface contributes massive information for flatness. We can usually see the texture of the canvas or photographic paper. This projects a zero texture gradient over the surface if you stand directly in front of the picture. If you view the picture from an angle, the gradient exactly matches that of the wall surface. These effects of flatness from texture are minimized if viewing is sufficiently far, so that the textural surface is below acuity threshold.

Finally, even though the linear perspective information from a photograph matches that of the linear perspective from the scene itself, illumination and wave length levels do not. In a picture made up of pigments applied to canvas, or black marks on paper, or in the image on a photographic print, no matter what the source of illumination had been over the scene, the ratio of the highest reflectance of the surface of the print to the lowest rarely exceeds 30 to 1. This is a limitation imposed by the nature of flat reflecting surfaces.

In a natural scene, however, specular reflectances – the light reflected from water, mirrors, metal, or narrow edges of almost any object – may be hundreds, thousands, or even millions of times more intense than the light emitted from the same source reflected from other surfaces in the scene. One of the important functions of these specular reflectances is to define changes in orientation, such as folds in a curtain or a gown, or the presence of a knife edge catching the sun.

More important than spectral reflectances is the luminance contrast produced by shadows. Even with a single light source, such as the sun, the amount of light directly reaching the ground may be thousands or millions of times greater than that indirectly reaching a shadowed part of the ground, say under a tree or behind a rock. Therefore, while the difference in light reaching the eye from directly and indirectly illuminated areas of a scene can be enormous, it can never be so in a photograph (or a painting) of a scene. Following similar logic, a photograph or painting never matches the range of hue or saturation found in a scene.

The perceptual impact of these contrast restrictions is to make pictures look flatter than the scenes they represent. Natural scenes have a large gradient of saturation and of brightness – surfaces in the foreground appear brighter and have more saturated colors, whereas distant objects appear less bright and less saturated. When this gradient is restricted, as in a picture, then the difference in distance between objects in the nearground and farground is lessened.

Thus, the stationary scale of space provides simultaneous information about the spatial relations in a scene depicted by picture and that the picture surface itself is flat. The two other perspective scales of space are unambiguous, however. They specify flatness no matter how a normal photograph or picture is made or displayed.

3.2. Motion Information of Flatness

In the previous section on the stimulus analysis of scenes, a number of different types and combinations of eye and head movements were described. Those same movements can occur when looking at a picture of a scene, but for pictures, motion perspective information is consistent with flatness and not depth. Every variety of observer motions before a flat picture reveals its flatness because all visual edges in the pattern shift together without transposition – each discontinuity remains in the same relation to every other one.

3.3. Binocular Disparity Information of Flatness

Here the story is the same as for motion. When you look at a flat picture straight on, there is no dispaity between the two retinal images corresponding to any part of the picture. This happens only when all parts of the scene are equidistant from the two eyes – hence, the scene must be flat.

If the picture is large relative to the viewing distance, then the edges of the picture are farther away from the eyes than is the center, producing a slight difference in the distances from the center to the eye and from the edge to the eye. This produces a disparity between the two images regardless of where the observer is fixating on the picture surface. However, this disparity is exactly consistent with the disparity of the wall and of the frame; hence, the surface of the picture and the surface of the wall are flush. Further, there are no local disparities over the picture surface – no part of the picture is farther away or closer than any of its neighboring parts – only a continuously uniform gradient of disparity indicating a flat, though extended, surface. The same is true if the picture is viewed from an angle; this produces a gradient of disparity indicative of the slant, but, without any local disparities, it still has to be a flat surface.

3.4. Local Depth Cues

In the previous section on visual information from scenes, I listed several local sources, such as interposition, adjacency, and familiar size and shape. These are present in pictures of scenes in exactly the same way they are in scenes themselves. Thus, the local sources provide some information of three-dimensionality, just as does the surface perspective information.

3.5. Differences Between Photographs and Paintings

The stationary perspective scale of space is reproduced in a photograph simply as a result of the properties of geometrical optics. A painter, however, has a free choice in how to paint a landscape. Consequently, linear perspective cannot be taken for granted. Art historians still disagree on how perspective was discovered or invented in painting. However, by the late fifteenth century, a number of treatises had been published to train painters in the use of perspective, e.g., Dürer (Panofsky, 1948), Alberti (Grayson, 1972), and most important of all, the notebooks of Leonardo da Vinci (Richter, 1970). Each of these describes rules for producing patterns on canvas which project the same pattern of light to the eye as does the scene being portrayed. See Hagen (1980a,b) for the best discussion of these.

The rules are based upon what we now know as geometrical optics. When an artist paints a landscape while viewing the scene, he chooses a station point, the position of his eye as it receives light reflected from the surfaces in the scene. Leonardo recommended that initially the novice use a plate of glass instead of a canvas on which to sketch. If the glass is placed between his eye and the scene so that he sees the scene though the glass, then while holding his head still, the novice can sketch the scene on the glass so that each line on the glass exactly overlaps the visual edges in the scene. In this way, the artist can create on the glass a reproduction of all of the patterns in the light – discontinuities in luminance and wavelength – that reach his eye and are focused on his retina. In theory, then by careful drawing and application of pigment on the glass, the pattern of light reflected from the painting of a scene to the eye and from the scene itself are identical.

The fifteenth century treatises were procedures or rules for the apprentice. Leonardo never claimed this was how a painter should paint a picture. Once a novice mastered the rules for making pictures look exactly like scenes, then he could learn how to create art through controlled variations and violations of these rules.

Leonardo knew that while the linear perspective could be reproduced on the glass and transferred to a canvas, the contrasts of brightness and hue could not. His notebooks contain many instructions to painters on how to increase the apparent contrast of adjacent areas of the canvas so as to better mimic the contrast contained in a real scene. Hochberg (1979) describes a number of different solutions worked out by painters to solve this problem; that is, to put the depth back into their pictures. The major techniques involve contrast induction, use of shadows, pairing of complementary colors on adjacent parts of the picture surface, and the use of additive rather than subtractive mixtures of colors. Ratliff (1965) also describes a set of devices based upon lateral inhibitory processes between adjacent areas, which produce larger apparent contrasts than are present on the canvas.

Thus, for the painter to produce a picture that reflects the same pattern of light to the eye as does the scene it represents, he must distort the local color and brightness relationships. In other words, to paint a picture that looks like a scene requires some cheating on the laws of perspective. The apprenticeship of a painter is, in part, learning how to cheat. Failure to cheat in painting means the resulting picture looks flatter than the original scene. I will have much more to say on this in the section on processing information from pictures.

3.6. Summary of the Stimulus Analyses of Pictures

The light reflected from a picture of a scene contains information that the picture is flat, while at the same time providing some of the same information that comes from the actual scene portrayed in the picture. The local depth cues and the stationary perspective scale of space provide both two- and three-dimensional information, whereas the motion and binocular disparity perspective scales provide only information for flatness.

So far I have focused on stimulus analyses of information reaching the eye from scenes and pictures. But before I turn to the much more difficult half

– how we process this information to arrive at perceptions of scenes and pictures – I shall consider the stimulus information available in viewing motion pictures.

4. The Stimulus Information in Motion Pictures

In considering information from scenes, I included conditions in which objects *in* the scene moved, or in which the observer moved, and described the visual information so generated by those movements. The section on information from pictures included only still pictures, in which there is no internal movement. But a rapidly viewed sequence of still pictures produces a perception of motion, as well as a number of other perceptual changes that would not have been predicted by anything discussed so far. Unfortunately, there has been little perceptual research on either the stimulus information provided by viewing sequences of pictures, or how such information is processed. Hochberg and Brooks (1978) have one of the few analyses of the perception of motion pictures, stressing both the stimulus information and the perceptual processing, and I will make frequent reference to their work. I will also include some discussion of one of the more recent uses of motion picture sequences for dynamic closed loop displays, such are used in some automobile driver education simulators and nearly all visual pilot training simulators. In these cases, the control movements of the perceiver, such as turning the wheel or pressing the brake pedal, are programmed to produce the corresponding perceptual changes on the display that would be perceived as if the perceiver actually made those movements in a real scene. (See Haber [1983b] for a detailed discussion of visual simulation for piloting high speed aircraft.)

Motion pictures (and television images) are produced as a succession of still pictures, in which a small displacement of an edge or object from one view to the next is perceived, not as a sequence of separate static views, but as the smooth, continuous movement of a single object from one location to another. The visual system translates a succession of static views into a single view moving continuously in time. This stroboscopic or beta movement is in the class of apparent movement phenomena, "apparent" because no smooth displacement from one location to another on the retina occurs at all. This is the entire basis for the perception of motion pictures.

We know a fair amount about the exact stimulus parameters for stroboscopic movement in simple displays, and virtually nothing about them for rich pictures, even though it is the latter that we inevitably view when looking at motion pictures and television. In the laboratory, stroboscopic movement is studied by presenting a single two-dimensional test form, such as a circle, first in one location and then in another, separated by some distance D, alternating the two presentations back and forth, with some interval T between them in which nothing is on view. Smooth continuous movement is perceived for reasonable values of D when T is in the range of 10-20 milliseconds (Hochberg, 1971; Kaufman, 1975). If T is much shorter than this, then two objects are visible simultaneously, and if T is much longer, each object is seen as pulsing in place on and off, but with no movement back and forth. Within this range, the single object is seen as moving smoothly across the empty space. Further, if the two objects presented are not identical, but differ in size or shape, then not only are they seen as moving, but during the flight across empty space, the size or the shape transforms. A small circle on the left gradually enlarges into a large one as it moves to the right, only to then shrink as it moves back again. With respect to shape, as a triangle on the left begins to move, it transforms into the circle on the right by a continuous rounding, and then by regrowing its vertices as it goes back to the left. More relevant and interesting are examples in which the two objects are three-dimensional transformations of one another, differing in their orientation, such as an outline of a house viewed from two different perspectives. When these are flashed alternately, what is seen is a single house moving and also rotating smoothly in three-dimensional space as it goes. Johansson (1974) and Kolers and von Grunaw (1976) have examined these parameters in great detail, as they are more relevant to motion picture perception.

The variable of distance of displacement (D) from one static presentation to the next is more complicated to specify. Hochberg and Brooks (1978) note evidence that holding distance across the retina constant, stroboscopic motion varies with the apparent location of the objects in space. Hence, the value of D cannot be specified merely as retinal distance. Two objects projected in close proximity on the retina but appearing to be a great distance away from the viewer, and hence a great distance apart from each other in the scene, may not be seen in apparent motion, no matter what temporal alternation rate is used.

The above stimulus values have applied to very simple displays of an alternation of a pair of objects. Analysis of this condition has produced quite stable results. Transfer to motion pictures or television is more complex. For motion pictures, the presentation rate for the succession of still images is usually 24 frames per second (though to reduce the perception of flicker, the viewing of each frame is further interrupted three or four times). A 24 frame per second rate provides an offset to onset interval (T) of about 20 milliseconds between successive frames. For American television, the rate is 30 frames per second, but the computation of the interval between frames is more complex because each frame is generated by displaying all of the alternative lines of the television raster 60 times per second.

4.1. Types of Camera Motion

The critical feature is not how the frames are generated, but what they contain. When looking at a motion picture or television sequence, each successive still picture displays many objects, each statically fixed at a specific size, shape, orientation, color, shading, and so forth. The next frame can differ from the last for a number of reasons. In the simplest case, a stationary camera can produce a sequence of pictures of a scene in which some of the objects moved during the sequence. When a viewer then looks at this sequence of still pictures, the displacement of each one of these moved objects from one frame to the next is perceived as a smooth and continuous movement of those objects. Since the stationary objects are in identical locations in each successive frame, apparent motion needs to be explained only for the displaced ones. As we will see in the processing section on motion pictures, a Gibsonian model based on invariant transformations of information from one frame to the next fits the data for this task quite well. It is closely analogous to what occurs when a perceiver watches a scene containing some objects in motion while he maintains his head and eyes motionless. The only difference is that the object is actually creating a continuously moving retinal image in the latter case, but is only apparently so in the former. The two look alike as a result of stroboscopic motion.

Just as a perfectly stationary observer is unusual, so is a motion picture sequence made with a stationary camera. More typically, the difference in the succession of frames of a motion picture sequence results from changes in the position of the camera from frame to frame. These changes themselves can be continuous, or successive frames can be discontinuous, as in a cut from one view or scene to the next. Four kinds of continuous camera movements can be distinguished, and for three of these, there is an analogous characteristic movement made by perceivers when looking at an actual scene.

First, the camera can *pan* or turn on a stationary tripod, analogous to a saccadic or pursuit eye movement without any head movement. As with the eye movements, a pan camera movement produces no motion perspective information since there is no relative transposition of visual edges in the scene.

Second, the camera can *track* perpendicular to the direction of the lens laterally across the front of the scene. As with body motion, camera tracking does produce motion perspective information because near objects are displaced more than far ones from frame to frame.

Third, while mounted on a stationary tripod, the camera can *zoom* into a scene by changing the focal length of the lens. Zooming has no direct analogy to human vision. While we can shift our gaze from near to far objects, this typically changes the depth of field and sharpness of focus, but not the magnification of the image. However, Roscoe (1980) suggests that some accommodation changes are accompanied by magnification changes. In any event, zooming produces no motion perspective information because there is no relative movement of edges with magnification or minification.

Finally, the camera can move into the scene on a *dolly,* analogous to a perceiver walking ahead. Dolly movements do produce motion perspective.

For all four of these camera movements, the resulting displacement of objects from one frame to the next is perceived as continuous smooth movement attributed to the appropriate movement of the camera. What is seen by the viewer watching the motion picture sequence is a coherent scene through which the camera is moving continuously, as if the perceiver himself were making the same movements of tracking, panning, or dollying. Both the perception of motion and the perception of the invariant scene have to be explained by mechanisms of information processing.

All four of these camera movements typically produce perception of both the coherent scene through which the camera is moving, and perception of the pattern of motion made by the camera itself. Perceivers rarely if ever are uncertain about whether the camera is on a dolly or a track, or is panning or zooming. This is the case even though there is no direct link between the perceiver and the camera's locomotion mechanisms. Thus, this perception need depend upon visual information only, since there are no afferent or efferent motor connections as when the perceiver is moving under his own control.

Automobile and visual flight simulators go one step further, by making the perceiver into the camera operator – in a sense making the camera the eyes of the perceiver directly. A simulator still displays a succession of static pictures on a flat screen, but the content of the pictures is modified by the control movements made by the perceiver. If he presses a control (such as an accelerator pedal in a simulated automobile), which makes the camera dolly faster into the scene, the succession of pictures display a rapid dolly sequence. Obviously, a computer is required to create such a closed loop control (Haber, 1983b). The result is not only an improvement in the apparent three dimensionality of the scene, but a clear perception that the camera and the perceiver are one and the same. It is now the perceiver that is moving in the scene, not merely an abstract camera.

The final type of camera motion is where the change from one frame to the next is not gradual, and there may be no overlapping features at all. Such discontinuous motion is called a *cut*. Perceivers can determine the arrangements of objects in a scene across a cut, and can even perceive the direction that the camera must have moved during the cut. As we will see in the section on processing, being able to explain perception of scenes involving cuts poses the greatest difficulty for our processing models.

5. Processing of Stimulus Information about the Layout of Space

Different theorists have concentrated on different aspects of processing. Thus, some have focused on how we can determine, from an ambiguous retinal image, how far away is the object or what is its physical shape. Others ask how we come to have an integrated panoramic perception of the entire scene and do not ask questions about individual cues to depth. A third approach begins with an analysis of visual edges and uses these to construct a layout of space, paying less attention to the traditional sources of information about space. These approaches do not necessarily contradict one another; but they do not speak to the same goals, and hence are often difficult to compare at all.

In this section I describe several of the different approaches to the processing of information of scenes. My interest is not so much to contrast the approaches or comment on them as correct or incorrect, but rather to suggest that with their focuses on different aspects of perception, all of them have something to contribute. I shall suggest a sequence to processing of information about spatial layout and that the different approaches focus on different steps or stages in this sequence.

A recurrent problem in this section concerns the relation of form to space perception. The older view, which coincides with most of the modern Helmholtzian models, takes a bottom-up approach in which we must first perceive the forms or shapes of patterns, then the surfaces enclosed by the forms, then the objects defined by those surfaces, and only then the arrangements of the objects in space. Form perspective is primary and prior to space perception. I have argued elsewhere (Haber, 1983a) that this view has to be incorrect. Treatment of the visual edges on the retinal images as patterns, without consideration of where they were reflected from in space, is silly and compounds all the problems in understanding how perception works. This is another version of the eye-as-a-camera metaphor, treating the retinal image as a picture to be looked at and interpreted.

5.1. Perceptual Constancy

Probably the most important goal in natural space perception is to maintain the constancy or the visual world. The physical properties of the world around us are constant and invariant, irrespective of our viewing orientation, our viewing distance, our viewing stability, and the momentary direction and content of the incident illumination present in the scene. Constancy refers to the perception of those invariant or intrinsic aspects of the world – the sizes, shapes, locations, orientations, and surface features of the objects and surfaces in the scene.

All theories of perception treat the achievement of constancy as fundamental. Beyond that statement, however, there is little agreement. Even the definition of the problem produces argument. Most of the work reported in Epstein (1977) takes the momentary retinal image, considered as a picture, to be the starting point and asks what has to be added, deleted, or transformed in it so that it can resemble or match the scene that gave rise to it. Thus, such researchers search for compensatory processes that act on the retinal image. Gibson, on the other hand, says that all the information about the intrinsic properties of visual space is already contained in the information imaged on the retina so that no compensation is needed. Accordingly, it is misleading to view the retinal image as a distorted picture which must then be transformed in order to reveal the true scene. In this sense, for Gibson, explaining the constancies is no different from explaining perception itself, and the constancies do not represent any special problem or difficulty as they do to the non-Gibsonians. These two positions are explicitly contrasted several times in later sections.

Most of the attention in the theoretical literature goes to the constancies of size and shape, but these two are really the least interesting. Once you stop thinking of the pattern on the retinal image as the size or shape to be adjusted to remain constant, and start paying attention instead to the appropriate information in the light that does actually remain constant with the intrinsic size and shape of objects, the problem is simplified. As we move around and view an object from different directions, orientations, and distances, the pattern of light reflected from the object to our eyes maintains constant values on the three scales of space.

Besides size and shape, a number of other examples of constancies or invariances of perception have intrigued theorists. While I will consider a few of these in detail as I go through different aspects of processing, I have made a list of most of them here. For each, we perceive the world as it is, in a coordinate system and with a stability independent of ourselves as perceivers. Each therefore represents a different instance in which what we perceive is quite different from the momentary "picture" on the retina. The gyrations of that "picture" as a picture go unnoticed.

Movement Constancy. When the pattern of light refelected from objects in space and imaged on the retina is displaced, we do not perceive movement in those objects if that retinal displacement is caused by eye movements, head movements, or body movement. In some fashion, therefore, movement on the retina is cancelled by information or knowledge that the movement was the result of the perceiver's motion and not of the object's motion in space. This compensation may be the result of feedback from the eye muscles, or efferent information coming from the eye movement control systems, or afferent information coming from the patterning of changes on the retina, or from combinations of all of these. Whatever the cause, however, constancy clearly results. It is one of the most prominent examples where changes in the retinal image are not reflected in perceptual experience (see MacKay [1973] for a review).

Direction Constancy. We know the direction of straight-ahead, regardless of the orientation of the eyes. The layout of space remains oriented in relation to our "cyclopean" eye irrespective of where our movable eyes are looking. This constancy probably depends upon processes similar to those for motion constancy (see Matin [1976] for a review).

Saccadic Blur Compensation. Scenes are examined by means of continuous successive samplings in which the eye makes discrete fixations separated by saccadic movements of tremendous velocities. We do not see the blur of the retinal pattern during these violent eye movement velocities. Rather, what we perceive is a continuousness of a stable scene, and we are unaware of the successive sampling that our eye takes by brief fixations on it. Thus, while the visual scene is broken up into small retinal slices several times a second, separated by a massive blur as the eye shifts position, none of that is reflected or registered in our perceptual experience (see Volkmann [1976] for a review).

Depth of Field Compensation. In addition to lateral excursions of the eye, we continuously adjust the optical power of our eyes through changes in accommodation, enabling us to focus objects at different distances. This is necessary because even with a small pupil opening, as in very bright daylight, we have a relatively small depth of field, yet we rarely, if ever, notice our shifting depth of focus or even that the entire scene is not in clear focus all of the time.

Binocular Disparity Compensation. When we look at a scene, the two retinal images are not the same because the two eyes are some distance apart. If we saw our images, that is, if they were the basis of our perception, then we would continually have rivalry between them or suppression of one of them (see Blake & Camisa, 1978). Rather, however, we have a cyclopean image. We construct a perception that uses the disparities from the two retinal images to create a third dimension over a third image. The cyclopean image does not duplicate either retinal image alone but requires a comparison of the two of them. There again, our perceptual experience does not correspond to what is on either retina.

Tunnel Vision Compensation. Optical factors produce a sharply focused image with high contrast only over the center of the retina. Further, neural coding is concentrated in this area, producing coding of fine detail, color and high contrast only in central vision. Thus, the visual system acts as if our eyes are like a telescopic lens. Yet the appearance of typical scenes is not of clear centers with fuzzy surrounds, but of a sharply focused panoramic "wide angle" construction of the visual world. In order to construct this panorama, multiple eye movements are required over the scene, with a substantial overlap between a large number of fixations (Hochberg, 1981). The panorama that we perceive does not resemble any of the retinal images, but is constructed from a sequence of them.

Stationpoint Compensation. We can view flat pictures or photographs from different orientations and distances without producing any change in the layout of space we perceive. Different oblique views create different retinal images in which the perspective information about space is quite different, yet the perception stays the same. I will discuss below in some detail how this compensation is made possible from the processing of two-dimensional information that specifies our orientation and distance in relation to the picture. Further, from the frame, and from the surface texture of the picture on the wall and from motion perspective and binocular disparity, we can determine that the picture surface is flat. From that two-dimensional information we can process our orientation of the picture surface and thereby account for the perspective changes produced by oblique viewing (see Pirenne, 1970; Haber, 1979).

5.2. Gibson's Direct Model of Scene Processing

Gibson does not really have a processing model of space perception, but there is a sense in which not only does he have one but that it is the easiest model to describe. Gibson argues that the pattern of reflectances in the light reaching the eye both is unique to the scene that gave rise to that pattern and contains all of the information about the spatial arrangements of the surfaces in that scene. In the terms used here, the three scales of space, plus the local depth cues, are sufficient to uniquely specify the scene. No further information is needed according to Gibson. Since all of the information is in the light, there is no need for any processing mechanisms at all. As perceivers we merely have to pick the information up. Processing implies for Gibson integrating, interpreting, combining, mediating, or transforming – all activities he says are unnecessary. In this sense, perception is direct.

Borrowing a metaphor used by Runeson (1977), Gibson can be said to argue that human beings have evolved "smart devices" that respond to the complex relationships in the visual information on the retina directly without having first to process the individual components of the information separately. Gibson demands that theorists recognize the richness of visual information reaching the eyes. His theorizing has focused upon how much that information uniquely specifies the scene that gave rise to it. The information Gibson has focused most on is the information about interrelationships – what I have called the three full-retinal scales of space, and not the individual local depth cues.

In theory, it is easy to think of a matrix of detectors that responds to a scale of space providing a depth value for every location on the retinal surface at each instant in time. We already have evidence of cortical cells whose response is proportional to the amount of disparity in excitation falling on corresponding regions of the two retinas (Richards, 1975). A matrix of such cells shows resting level activity for regions around the converged fixation and for all other retinal areas that correspond to surfaces at the same distance as that of the fixation surface in space. Excitatory activity levels are present in all cells that correspond to surfaces farther from fixation, with a degree of excitatory activity proportional to the distance, and inhibitory activity levels for all cells that correspond to surfaces nearer than fixation. The matrix of binocular disparity detectors in the cortex is therefore, in theory, capable of assigning for every region of stimulation on the retinas a corresponding relative distance in space from which the stimulation was reflected. I say "in theory" because no one has actually measured the extent to which such a matrix might work in this way. Electrophysiological recordings have not been made in the presence of surfaces in depth stimulating large areas of the retina.

The same type of smart device can be imagined for motion perspective, built up of differences among motion detectors corresponding to different retinal regions. Similarly, variation in texture densities or outline convergences of visual edges could be specified by a matrix of detectors tuned to luminance differences produced by these variations across the retinal surfaces. Again, neither Gibson nor anyone else has collected the relevant data that document the existence of such integrated matrices of detectors, but enough evidence does exist about the component parts that the rest of the system is far from fantasy.

The recent work by Regan and his associates (e.g., Regan, 1982; Regan, Beverly, & Cynader, 1979) has uncovered one candidate for a smart device that processes complex stimulus information, meeting the requirements for Gibsonian direct perception. Regan has focused on how we can perceive motion in depth, such as when an object is moving toward or away from the perceiver or when the perceiver is moving toward or away from objects. The former case is a bit simpler to describe, but it is logically equivalent to the latter, which is the far more typical and important.

First, consider the stimulus analysis for radial motion of an object toward (or away) from the perceiver. If a ball is thrown directly toward one eye, the retinal image of the ball expands, from moment to moment, with its rate of expansion proportional to its velocity. When the path is directly toward the eye, then the successively larger optical patterns are concentric, whereas if the path of the ball misses the eye, then the expansion patterns do not share the same center but drift across the retina as the ball gets closer. Hence, expansion itself is a source of information about an approaching object and the concentricity of the expansion is information about its path. But what happens in the other eye? If the ball heads toward one eye, it must eventually miss the other. Hence, there is no drift of the expanding pattern in the eye to be hit, whereas in the other, the pattern drifts outward. If the ball is aimed toward the nose, each expansion pattern drifts outward across each retina, with the rate of drift exactly equal and opposite. Finally, consider what happens if the path misses the head altogether: now there is a drift across both retinas, but they always are in the same direction, with the path of the object given by the ratio of rates. If their drift rates are exactly equal, the ball is on a path perpendicular to the line of sight.

Up to this point, this is a stimulus analysis. But Regan extends it into a processing analysis as well. He argues that as perceivers, we care most about objects that might collide with us, so we should be most sensitive to patterns that drift oppositely across the two retinas, as compared to those where both patterns are in the same direction. This concern specifies only a few degrees of radial motion, but these degrees are the only ones we cannot afford to misinterpret.

Therefore, Regan argues, because we are so good and so fast at detecting and responding to collision motion, we must have specialized detectors (what I am calling a smart device) just for those stimulus features – in this case, the relative drift in opposite directions of the retinal images across the two retinas. Regan set up an experimental program to look for this smart device, and it is exactly what he has found. First, using a variation of a stereoscopic display, he projected an image on each eye which he could then move across the retina under his control. When he varied the patterns according to the above stimulus analysis, each of his subjects reported perception of the appropriate movement. This showed the stimulus information is sufficient for perception of radial motion. Second, he measured the subject's sensitivity to the detection of an object in oscillatory motion in different directions, including collision, near collision, and miss paths. After assessing these different sensitivities, he then adapted one of them by having the subject watch a simulated object move back and forth along one pathway for 20 minutes. Following this adaptation, the subject's sensitivity was tested again. Regan found that there was a specific loss in sensitivity for radial motion along the pathway adapted. No loss in sensitivity occurred for nonadapted pathways. Finally, Regan repeated the experiments in cats, measuring the electrical activity of certain cells in the visual cortex thought to be responsive to motion. He had no trouble finding cells that were responsive only to particular combinations of drift across the two retinas. These cells were not responsive to drift over only one retina, so they were responding only to the specific stimulus information about collision. Regan found most sensitivity for motion that would either hit the head or just narrowly miss the head, but relatively little sensitivity for all other directions.

This is an example of an excellent stimulus analysis that guided the search for specific kinds of processing – in this particular case, processing at a relatively peripheral site in the visual nervous system. It seems likely to me that most aspects of the motion perspective scale of space are processed in similar ways. We need only look.

Regan's work is very elegant, but it is only one example of what may be a number of comparable smart devices. The power in his work is the combination of a careful stimulus analysis which leads to a specific examination of how that stimulus information might be extracted. I consider this an instance of direct coding (of motion in depth) in the sense that the stimulus information is extracted by a dedicated neural structure without the need for other information to be added, or for inferences to be made, or for some logical process to be involved to combine different sources of information – all of which are aspects of other processing models, to be considered below. Regan was fortunate in being able to demonstrate evidence of the neural structures themselves. For most of the smart devices that can be proposed, we may not have such evidence for a long time to come. The approach is still very useful.

A direct coding of space as outlined above has some special properties quite different from any of the other approaches to the processing of space.

First, because what is represented in the code is relative distance of all the reflecting surfaces in space, information about the layout of space is available to the perceiver without his having to process any of the traditional cues to space separately. Thus, the distance and the slant of a surface are specified without an analysis of its retinal image size or shape nor any of its other local properties on the retina. As we shall see below, the "taking into account" theory places heavy reliance on relationships such as shape-slant invariance and size-distance invariance, something completely unnecessary in a direct coding theory.

Second, since direct coding requires inputs from the entire retina and requires that the normally rich information from space be available, such a system would not function very efficiently when exposed to impoverished displays. Thus, for example, the unimpressive results of the experiments reviewed by Epstein and Parks (1963) are to be expected. Their review showed that outline convergence information provided by a single object was not a very good source of information about space. Adding texture to the object's surface did not help much either. A direct coding system can only function well when both entire retinas are richly stimulated, because only then are the three scales of space available to provide stimulus information. I expect that it will turn out that the smart devices are all sensitive primarily to full-retinal sources of information and not at all to the traditionally described local depth cues. If the input is limited to just such local features, alternative and probably not very efficient or accurate methods to perceive them properly would be needed.

Third, Gibson's claim that all of the information about the layout of visual space is contained in the patterning of light reaching the eye has been open to attack from two quarters. Some have questioned the adequacy of his stimulus analysis itself and have argued that visual information is not capable of uniquely specifying each scene (see Ullman [1980] for a review and critique). The other attack on Gibson's claim ignores the uniqueness issue and concentrates on evidence that perceivers do use nonvisual information to perceive the arrangement of surfaces and objects in space. This is a more interesting argument and I will discuss it in some detail in a section on sensory motion interactions below.

Finally, there is nothing in such a direct coding scheme that describes how we perceive objects. It only accounts for the representation of the distance of reflecting surfaces, but distance representation is far from trivial. A direct model provides an account of the major source of information lacking about space: the distance from the observer. Once distance is available, and this model makes it potentially available at quite an early stage in processing, then the rest of the properties of space can be processed – where are the boundaries of surfaces that delineate one object from another and what are the surface properties of those objects (lightness, color, size, shape, position, and motion, for example). Thus, the other models of space perception are not necessarily opposed to a direct coding one, but simply account for stages of processing after the distance information is available.

Gibson has attempted to include the perception of objects within his concept of direct perception. In his most recent book (Gibson, 1979) he refers to this as the direct perception of the affordances of objects. I shall not describe this here, as it is one of the most controversial properties of his theory and one that I expect to drop by the wayside, even as other aspects are shown to be crucial.

There is a further aspect of Gibson's approach to perception that has profound processing implications, even if he did not describe it that way. Gibson argued that perceivers initiate actions in order to produce perceptual consequences. For example, we move our heads in order to generate motion perspective, information which is informative of the arrangements of objects in the scene in front of us. It is not just that our movements produce changes in stimulation, but that part of the motivation for making those movements is to gain access to the information inherent in the change in stimulation. Thus, we have an interdependence of action and perception, in which perception is informative of the environment upon which we can act, and those actions then produce new information, and so on, back and forth. For Gibson, a major goal of perceiving is to guide actions, and a major goal of action is to guide perceiving.

One obvious implication of this interdependence is that perceiving and acting have to be considered together, so that we can never fully understand perception when we study a stationary observer, or an observer who is not free to act upon the environment he is perceiving (see Owen & Warren, in preparation; Fitch & Turvey, 1977). The classical psychophysical procedure, of presenting a stimulus to which the observer reacts, is an inadequate model for the study of space perception. It does not include any account of the ways in which perceivers seek information that is potentially available to them. The information in a single static slice of stimulation may be insufficient to specify all of the arrangements of surfaces in a scene, but a mere flip of the head or a step forward by the perceiver may yield a pattern of changes in stimulation that unambiguously specifies every object. Such a flip or step has to be considered part of the perceptual process (see also Lee, 1976).

A second implication of the interdependence of action and perception is stressed by Owen and Warren (in preparation). The traditional criterion for specification of some perceptual sensitivity is that the observer's responses should covary with stimulation. They note that the converse should also be true: variation in stimulation should be related to the observer's actions, if those actions were motivated to produce or control stimulation. Thus, the observer's responses can be used to predict the pattern of stimulation. If the observer is controlling stimulation through his actions, then stimulation can become a dependent variable in perceptual research, with motor behavior being the independent variable. Owen and Warren illustrate this notion through an analysis of the visual stimulation reaching a pilot during several types of landing tasks. They argue that if the pilot is guiding his actions (through the controls of the plane) in order to obtain or maintain certain kinds of visual information, then it should be possible to plot some index if stimulation (dependent variable) as a function of the position and locomotion of the observer (independent variable). If such plots are regular, they suggest that the pilot is moving in such a way as to acquire potential kinds of visual information. That analysis can isolate the relevant visual information being used for that locomotor task.

These comments about action and perception together imply that the acting upon the environment in order to produce stimulation is part of the processing that perceivers carry out on stimulation.

5.3. Processing of Objects in Space from Visual Edge Analysis

A visual edge can be defined as a locus of points that separates retinal image regions of different luminances. We know that if there are no edges, there is no vision; nothing is perceived in a uniform field (Ganzfeld). A revolution began in vision research 25 years ago, stemming from the work by Barlow (1953) and independently by Kuffler (1953). They laid the groundwork for the discovery of receptive field structures as peripheral as the ganglion level that code the presence of luminance discontinuities – visual edges – on the retina. We now have a relatively complete specification of how edge information is initially sorted out by analyses of luminance differences between adjacent regions (see Thomas [1975] and Robson [1975] for reviews). All of the neurophysiological coding mechanisms studied are heavily based upon retinal and cortical spatial inhibition processes.

While this work has revolutionized our thinking about neural coding, it has had virtually no impact on shaping theories about space and object perception. The reason is that local luminance discontinuities in the retinal image do not uniquely specify the spatial arrangements that give rise to them. Treating all visual edges as the same produces fundamental errors in the processing of spatial information about space. Nothing in the way visual edge detectors work, as currently described in the neurophysiological literature, can make the necessary discriminations. Neurophysiology has not even acknowledged that different types of visual edges may exist.

Four different kinds of spatial arrangements that create visual edges can be distinguished. If a single extended uniformly reflecting surface is illuminated by two different light sources (or a single source that is occluded over part of the surface, placing that part in shadow), then the visual edge along the boundary of the luminance difference is due to differing amounts of illumination only. This is called an *illumination edge*. On the other hand, if a single light source uniformly illuminates two coplanar adjacent surfaces of different reflectances, the resulting visual edge at the boundary of the two surfaces is a *reflectance edge*. Both of these types of edges are coplanar in the sense that the surfaces on either side of the edge are at the same distance and orientation from the eye.

Two further types of edges can be described as being in depth. They arise whenever one surface in space occludes a more distance one or when two surfaces meet at a common boundary but at different angles, as when walls of a room meet at a corner. An edge in depth can be a *depth illumination edge* if the two surfaces have identical reflectances (as usually do walls in a room) but differing illumination. Finally, a *depth reflectance edge* along an occluding boundary occurs when one surface of one reflectance occludes another surface of a different reflectance. Of course, many visual edges combine several of these, as when surfaces at different distances differ both in reflectance and receive different amounts of illumination.

Since a reflectance edge arises at the boundary of two surfaces of different reflectances, the luminance differences correspond to properties of the two surfaces, such as lightness, color, texture, and the like. They tell us something about surfaces of objects in space. However, if the edge is due to different luminances falling on the same surface, then any differences are accidentally caused by the momentary illumination and are not intrinsic properties of the surfaces themselves. In other words, we have to ignore illumination edges when processing information about the layout of space.

To illustrate the importance of this distinction about edges, consider how a viewer perceives lightness constancy. A piece of white paper looks white because of its high reflectance, black coal looks black because of its low reflectance, irrespective of the amount of luminance reaching the eye from those surfaces. The fact that lightness is independent of luminance is what is meant by lightness constancy. The luminance reaching the eye from a surface is the joint product of the illumination incident to the surface and the reflectance of the surface. How can we determine just the reflectance unconfounded by illumination?

A solution to this processing problem was worked out by Wallach (1948) in terms of the contrast across luminance discontinuities. If the same source of light illuminates a highly reflecting piece of white paper and a darker background, the luminance reaching the eye is the result of the common illumination multiplied by the two reflectances. If a visual system takes the ratio of luminance across the edge, the common illumination term in the numerator and denominator of the ratio cancels out so that the resulting ratio is of the reflectances of the two surfaces alone. Thus, by knowing or assuming a common light source, the ratio of luminances provides an unconfounded value of the reflectances of the two surfaces, a property of the two surfaces. In this way, lightness constancy is theoretically possible. Wallach's proposal has worked well for simple displays, but there is substantial evidence (see Beck [1972] for a review) that when the spatial arrangements are made more complex than a disk on a circular surround, equal ratios need not always produce equal lightness.

Even so, this ratio taking process makes sense only for coplanar reflectance edges, but never for illumination edges nor for reflectance edges in depth. For example, if an edge is an illumination edge, whether coplanar or in depth, the basic assumption of equal illumination on either side of the edge is violated, and taking a ratio yields the wrong information about reflectances. Rather, instead of taking a ratio, the visual system can simply assume equal reflectances and ignore the luminance difference.

Recent work has documented these distinctions. In a typical contrast experiment (see Kaufman [1975] for a review), the viewer is asked to make judgments about the lightness of a patch under conditions which the amount of light reflected from surrounding patches is varied. In general, the judged lightness of a patch decreases as the amount of light reflected by its surround increases. Such experiments invariably present all of the patches at the same distance on the same background surface, so that the adjacent areas on the retina are, in fact, presumed to correspond to adjacent surfaces in space as well. Thus, the edges studied are always coplanar reflectance edges.

Gilchrist (1977) has now shown that if the edge is not coplanar, no ratio is taken and the lightness judgment is based on total luminance. To do this, he manipulated the apparent distances of the patch and surround. He found that the traditional lightness contrast effects are reproduced whenever the patch and surround are *perceived* to be adjacent in space and coplanar with each other. However, if the patch is presumed to be in front of the surround or behind it, then even though the retinal extents are still adjacent, no contrast occurs between them. The surround now does not modify the appearance of the patch. This result has been reproduced in several quite different experimental procedures.

Gilchrist's results make it impossible for inhibition among adjacent retinal areas to be the underlying mechanism of contrast. In all of Gilchrist's conditions, the retinal patterns are identical. Rather, whether interaction occurs depends upon whether the observer perceives the retinally adjacent surfaces to be spatially adjacent as well. That is, the observer has to construct the layout of space *first,* and only then can he assign the lightness (and presumably color) values to the surfaces he has already segmented in depth. Gilchrist's result certainly makes sense, so much so that it is strange that we accepted the earlier "conventional wisdom" for so long. Perceptual decisions cannot be made simply on the basis of adjacent retinal areas. The perceiver has to know what is adjacent in space; that is, he has to achieve the layout of space before he can reasonably know how to treat any of the lower-level features on the retina.

The above analysis calls into question all of the theories and models of form and object perception that begin with the pattern of edges on the retina and use it to determine the objects in space. Beck (1972) also notes that early encoding of luminance differences (contrast) cannot be the basis for subsequent surface and object perception. Rather, what seems absolutely necessary is that the nature of the visual edges be determined first, particularly how far away the surfaces are that reflected them, before any patterning among those edges is examined.

A number of sources of information have been suggested (Gilchrist, 1978) by which visual edges can be classified. Some of these are potentially very powerful, while others seem unlikely to be useful at all. The two most powerful stem from depth and intersection information.

Given a likelihood of direct coding of distances from the various perspective transformations, perceivers should be able to determine whether or not the two surfaces on either side of a visual edge are coplanar. In this way, the two kinds of depth edges can be distinguished from the two kinds of coplanar edges directly. In light of the earlier comments about ratio taking, if there is a difference in depth on either side of an edge, lightness can be determined from the luminances directly without taking the ratio across the edge. If the two surfaces on either side of the visual edge are parallel and coplanar, the edge may still be either a reflectance or illumination edge, so some additional information is needed.

The second source of information comes from the nature of the intersections of visual edges. Gilchrist (1977) has described in some detail how the intersection of coplanar edges differs, depending on whether they are both illumination edges, both reflectance edges, or mixed. While the entire analysis has not been worked out yet, it seems like that whenever intersections among several edges occur in a scene, no ambiguity would remain over which kind was which. If so, then given both depth and intersection information, a complete classification of edges would be possible for all fully articulated scenes presented to the eyes. This would seem a natural place for computer vision research to provide the proper analyses and, in fact, most such programs of research focus heavily on edges. Guzman (1968) provided the first fully developed description, but while he concentrated on interaction of edges, he did not distinguish between illumination and reflectance edges in this way. A more recent example is described below.

Gilchrist suggests several other probabilistic sources of information from which edges can be sorted out, but these seem to be of less power. The first is the range of luminance. From a single light source, the maximum ratio between luminances over different parts of the retina (or scene) can rarely exceed 30 to 1. This restriction is due to the fact that few surfaces reflect more than 90% or less than 3% of the light that falls on them. Therefore, reflectance edges are rarely formed by luminance ratios greater than 30 to 1 and most are far less than that. Since many typical scenes have only one light source (the sun when outdoors or a single area of light when inside), using the presence of low contrast to identify reflectance edges is reasonable.

On the other hand, illumination edges have no limit on their luminance ratios and it can often be hundreds, thousands, or even millions to one. Hence, just statistically, edges with large luminance ratios tend to be illumination edges, while those with small ones are reflectance edges.

Another source of information that could help distinguish illumination from reflectance edges is discontinuity in wave length. Again, statistically, if an edge separates two different colors, the likelihood is that the color differences arose from selective spectral absorption rather than selective spectral illumination. If the colors are equal on either side of an edge, no information is available as to the type of edge.

A final property is edge sharpness. In the natural world, illumination edges produced by shadows are not very sharp (a gradually changing luminance profile), whereas reflectance edges are generally much steeper, being the result of adjacency of physical edges of objects in space. While this relationship between sharpness and type of edge is not a necessary one, so that exceptions can be found, it is certainly a statistical one of nearly the same weight as the luminance and wavelength differences mentioned above.

A substantial research program is badly needed to demonstrate the ways in which the visual system does sort out these different kinds of edges. It is likely that all of the ways Gilchrist described are used, and perhaps while none is perfectly reliable as a predictor, all in combination might be.

5.4. Computer Edge Processing of Scenes

Numerous programs of research have attempted to have noncognitive and nonpurposive systems (such as computers) look at a scene with some type of photosensitive sensor (such as a TV camera) which produces an image on an image plane (such as a TV or a CRT screen) from which the spatial layout of the scene and the nature of the objects present are determined. Winston (1977) provides an introductory presentation of the different approaches to the problem, and the collections by Winston (1975) and by Hanson and Riseman (1978) contain a set of high level papers describing specific systems. While the goals of research on computer vision systems and on human vision systems need not be and usually are not very similar, I want to single out two proposals that focus on some of the many processing problems discussed here.

Barrow and Tenenbaum (1978) argue that the proper goal of the visual system is to recover; that is, to extract the intrinsic scene characteristics from the image generated by the photo sensor. By "intrinsic," they mean those characteristics of the scene that are independent of momentary illumination, momentary viewer orientation or position, and familiarity of this scene to the viewer. Thus, the visual system should recover the distance, the orientation, the reflectance, and the incident illumination of every surface in the scene; in other words, the three-dimensional layout of the scene. More importantly, they propose that this recovery is the first step rather than the last. Processing begins by locating edges in the image (discontinuities in intensity value on the image surface) and on the basis of the relational properties around those edges, plus reasonable constraints, the surfaces bounded by the edges are immediately and directly classified in terms of the veridical properties of space. This means that the first step arrives at a three-dimensional representation without first working on all of the lower steps.

The constraints that Barrow and Tenenbaum use are important ones, and some have already been discussed. For example, surfaces are considered to be continuous in space and usually have a uniform reflectance over their entire extent. From this, distance and orientation can also be assumed to be continuous and the reflectance is constant over the entire image except for edges that correspond to surface boundaries. It is also reasonable to assume that incident illumination is constant over the scene. Therefore, step changes in intensity occur at edges marking shadows or at surface boundaries. If the scene has man-made elements in it, then straight edges usually correspond to boundaries of flat planar surfaces. It is even possible to make assumptions about shapes as well, so that ellipses are circles viewed from an oblique position.

As an example, consider a nearby painted wall on which the intensity across the wall is not uniform, but varies. In theory, the variation could be due to variation of the reflectance of the surface, or to variation in illumination, or to variation in surface orientation, or to any combination of these sources. However, the few constraints suggested remove all ambiguity. Since the reflectance of the surface within edge boundaries is assumed to be uniform, the intensity variation must be due to smooth variation in illumination or to variation in the orientation of the surface. The straight edge of the wall implies that it is a planar surface, so that leaves only variation in the illumination as the source of the intensity variation. If the wall is viewed through a reduction tube so the edge cannot be seen, perceivers have trouble telling whether the intensity variation is due to a uniformly illuminated surface curving away from the viewer or a flat surface illuminated unevenly. Thus, when all parts of the scene can be evaluated together, what might be inherently ambiguous over one part is at once resolved into a single unique solution of the scene seen in three dimensions.

All the hard work by the system is done on edges, changes in intensity in the image. The system is designed to classify edges as to whether they arose from a boundary between regions of different reflectances, from marks or printing on an otherwise continuous region (both of these being reflectance edges, though of two different sorts), from illumination differences produced by a shadow falling across the region, or from a surface shadowing itself (both being illumination edges but of two different sorts).

Edges enclose regions which represent surfaces. Surfaces can vary in orientation, in degree of curvature, in reflectance, in texture, and in marking. Each of these intrinsic characteristics affect the uniformity of the intensity across the image so bounded. Again, the constraints imposed by the recovery system based upon intensity variation and edge type allow for virtually unique solutions for the intrinsic characteristics of the regions. The nature of the junction that an edge makes when it intersects another edge in the image also provides information in this system. Intersection information has been heavily exploited by computer vision systems (for example, Guzman, 1968; Waltz, 1975) and while Barrow and Tenenbaum do not treat junctions as primary sources of information, they are often used as final arbitrators when a unique solution has not been found by other constraints on processing.

I have described a few of the basic notions behind the Barrow and Tenenbaum system because it focuses upon deriving spatial layout as the first step, rather than as derived from earlier partial solutions. Further, it focuses on edge information as the major source to carry out these analyses. Some of their constraints are inconsistent with what human perceivers seem to do, but the approach certainly is in the same direction.

Probably the most complete attempt to work out a computational approach to human vision has been by Marr and his colleagues at MIT (e.g., Marr, 1978; Marr, 1982; Marr & Hildreth, 1980; Marr & Poggio, 1979; Crick, Marr, & Poggio, 1981). He proposes three stages to processing, in which the first locates edges, the second constructs a sketch of the scene from the momentary viewpoint of the observer, and the third a sketch independent of the observer. These stages are treated as sequential, so that the second stage requires the representation constructed by the first stage, and so forth.

Marr called the first stage construction of a primal sketch, a primitive but rich description of a momentary visual image that is to be used to determine the reflectance and illumination of each visible surface and its orientation and distance from the viewer. This is basically a task of locating all of the visual edges through a filtering and combining process, and then using the constraints of the local geometrical relationships to make explicit the two-dimensional relations among the significant edges in the image.

The next stage Marr calls the 2-1/2 D sketch, suggesting that it is more than two-dimensional, but not quite the three-dimensional scene that exists independent of the viewer. It is viewer-centered, that is, constructed in coordinates defined by the position, orientation and discontinuity of each surface in the scene relative to the observer. Depth is given by scalar distance of each surface from the observer, and orientation is determined by variation in depth. These values are achieved by the use of motion, stereopsis, texture, occlusion, shading, lighting – the same sources of information I have described.

The final stage, the 3-D sketch, uses a coordinate system of the natural scene itself, independent of the observer's position in that scene. Much of

Marr's work on the third stage has been to show how a natural coordinate system can be derived from an image of the scene.

I have provided even less than a 2-D sketch of Marr's approach, partly because most of the details are computational in nature. Thus, while Marr continuously refers to the mechanisms and the goals of natural human vision, he is basically building an artificially intelligent vision machine that *may* see like a human being sees. As evidence to support that likelihood, he argues that processes seen in the early stages of neural coding in the retina, the LGN, and the visual cortex seem functionally similar to construction of a primal sketch, that is, of finding and locating edges in a two-dimensional representation. These relationships are certainly suggestive, but are by no means compelling.

A further reason for my reticence in treating Marr's approach as a full-scale human processing model is the absence of any evidence that the sequential stages actually are distinct in human perception. For example, if the processing of a stage could be stopped in mid-flight, say by a visual mask or some other interruption, would the perceiver report the scene in terms of his own coordinates (the 2-1/2 D sketch) rather than natural coordinates (the 3-D sketch)? There is no hint of such a report in the literature, even though there have been many masking experiments. As will be noted in the picture processing section below, masking might even make pictures look more 3-D, not less. Thus, I see the three stages are computationally relevant, but as yet only non-functional metaphors for human vision. This is especially relevant given Gilchrist's evidence (reviewed earlier) suggesting that the depth arrangements in natural coordinates may have to be extracted before edges can be defined and precisely located. This may not be a problem for Marr, but it certainly is for human vision.

5.5. Inference and Expectancy Processing Models of Space

A computer vision model, such as the one by Barrow and Tenenbaum, combines edge processing with some assumptions about what the surfaces and scene might be. These assumptions are described in an intuitive and reasonable fashion but are not dictated by any theoretical analysis. Two rather different models, both grown out of the Helmholtz (1925) notion of unconscious inference, do offer a theoretical basis for the assumptions in terms of inference processes and expectancies. The first of these is best exemplified by Rock's "taking into account" or combinatorial model, and the second by Hochberg's mental structure model of expectancy. I shall consider each of these in turn.

5.6. Rock's Combinatorial Model of Scene Processing

"The perceptual system takes account of factors such as distance or illumination in arriving at a perceptual judgment of what the retinal image represents in the world . . . . The information taken into account derives from a source separate from the retinal image of the object: for example, cues to distance or illumination" (Rock, 1977, pp. 321-322). What is taken into account need not be limited to current visual stimulation but can be efferent signals directing eye or body motion or stored information about comparable visual scenes viewed at some prior time. For example, the stability of the visual world occurs during head or body movement because we take those movements, or rather the efferent intentions to make those movements, into account along with the sweep of the retinal image across the retinal surface. In essence, for Rock, perception is always a result of a combinatorial process in which what we see results from some property of the retinal image of an object in combination with other information about the location or orientation of the object or the state of the perceiver.

I have quoted Rock, although a number of related positions (for example, Epstein, 1977; Ebenholtz, 1977; Gogel, 1977) could have been used as models. All of these have three major factors. First, information about the layout of space is divided up, according to the task, into two basic components – the retinal feature to be presented and all other information. If you are perceiving the size of something, then you use information from retinal size on the one hand, and all other information regarding direction, distance, illumination, prior knowledge, and the like on the other hand. If you are perceiving the lightness of something, then you use the pattern of retinal illumination, to be combined with size, distance, and the other information about the scene being viewed. Second, these two sets of components, especially the first one in each case, has the potential status of perception in its own right. Thus, you can be aware of the pattern on the retina as it is being transformed by the operation of the various perspective rules. And third, the resulting perception of a stable, constant, accurate visual scene requires these various sources of information be combined by an inferential, thoughtful, albeit unconscious procedure. While Rock refers to this model as a cognitive one, presumably because of this third inferential feature, the inferences he draws on are logical or mathematical, rather than guided by the perceiver's expectations.

The combinatorial process is described as analogous to syllogistic reasoning. For example, an extent in space is perceived based on the size of the retinal image. This is combined with information from whatever source that the object producing that retinal image size is at a particular distance. Therefore, according to the invariance of size to distance in perspective geometry, the object must be of a particular spatial size. Rock, like Helmholtz, argues that the inference process does not have to be accessible to awareness nor do the separate components yet to be combined have to be perceived as such. The process is considered automatic, especially after much practice with similar perceptions over one's lifetime.

The combinatorial models have focused on the various constancies as perceptions to be explained. They examine the kinds of information that have to be taken into account, along with the retinal information pertaining to the object so that the object's intrinsic size, shape, color, lightness, position, and motion are perceived. These models have not concerned themselves very much with panoramic perception of complete natural scenes, nor with the recognition or identification of objects and scenes of varying familiarity, nor with problems of how successive glances are combined into an integrated perception.

It is difficult to evaluate the combinatorial models without an extensive review of experimental literature. Much of that literature clearly supports the notions proposed by Rock. He treats the various sources of information as separate and independent and therefore in need of combination. As a result of this assumption, Rock rarely considers how we might perceive all aspects of a scene in its entirety at the same time. However, to me it seems unreasonable that we first work out all the sizes of objects by taking their distances into account, then work out all the distances by taking something else into account, and then all the colors and lightness, and so forth. As I have tried to suggest in the stimulus analysis, much of the information about these different aspects is available simultaneously in the retinal projection (or from other nonretinal sources). This conclusion does not imply that a combination, or integration, or taking into account is not needed; only that Rock has focused on the task at too low a level.

The taking-into-account or combinatorial models were developed primarily by considering very simple displays, often flat line drawings. It is therefore not surprising that in such models so little stimulus information is available to the perceiver. To interpret perception of such displays, it seems parsimonious that the perceiver first tries to deal with it piecemeal, and in doing so, includes so much nonvisual information to make up for what is lacking in the normally rich visual scene he is used to perceiving. This is not to say that the kind of inferences being made are never made normally, but it seems unlikely that this approach provides a reasonable or complete description of typical perception.

5.7. Sensory Motor Interactions

An example of a combinatorial model can be illustrated by a more detailed examination of the role of self-produced motion in perception. Such an example also illustrates one of the difficulties with such models.

In Gibson's attempt to describe a theory of direct visual perception of visual space, he denied that the observer ever needs access to nonvisual information to perceive the layout of space. Specifically, he argued that we had no need to know the position of our eyes or bodies. We can determine by visual means alone whether displacement of part or all of the retinal image over the retinal surface is caused by our own eye or head movements or by objective movement of part of the scene. If we can discriminate these from visual information alone, why involve nonvisual efferent mechanisms to do it as well?

Gyr (1972; Gyr, Willey, & Henry, 1979) criticized this position, arguing in part on logical grounds, but mainly on the basis of experimental evidence, that perceivers have to access information about their own movements in order to properly attribute the source of the retinal image motion. In other words, information about the initiation of our own motion has to be combined with visual information in order to perceive the layout of space correctly.

The most powerful part of Gyr's argument is that perceivers are capable of using information from efferent signals, in combination with the visual information on the retina, to achieve a correct perception of the presence or absence of objective motion. A number of studies have shown (see Lackner [1977] and Welsh [1978] for recent reviews) that when the perceiver has little visual information, he depends upon and uses efferent information. What is missing in this research is evidence that perceivers rely on efferent information even when sufficient visual information is present – a problem that plagues many of the combinatorial models (see Haber, 1979b). It is reasonable but unfortunate that this research has used such simple displays for the most part, nearly always flat two-dimensional displays and not full scenes. Gibson's claim and Gyr's attack can never be settled properly unless the experiments are done while looking at natural scenes that provide all of the visual information Gibson says is sufficient.

A number of processing tasks appear to require nonvisual information. Our judgment of what is upright (an orientation constancy) has been assumed to be partially dependent on the gravitation detectors in our vestibular system (see Ebenholtz, 1977, for a detailed discussion). However, Lee (1974; Lee & Lishman, 1977) has shown that in rich visual environments, visual information overwhelms vestibular inputs so that a person uses the visual frame of reference irrespective of what his own gravitational upright detectors tell him. Even in a less rich visual environment, Dichgans and Brandt (1978) have shown that when the eyes are open, perception is more likely to match the visual information than vestibular signals if these conflict. So again, it may be that perceivers use nonvisual information when the visual data are impoverished, but do not depend upon it in an enriched visual scene.

Another task often used against Gibson is perceiving where straight ahead is. To tell whether an object being fixated is straight ahead of our nose seems to require that we know the orientation of our eyes at the same time. If they are also straight ahead, then so is the object. So, visual direction constancy – that objects remain in the same place straight ahead of our nose, regardless of the direction of our gaze – seems dependent upon combination of information from vision and from eye position (Matin, 1976). While this seems reasonable, it is not directly relevant to Gibson's claims. Gibson says we perceive the layout of space irrespective of the momentary orientation of our eyes and head. However, if we want to know where we are in relation to objects and surfaces in that scene, then it seems obvious that *we* need information about our own orientation. This is the case of knowing the relation of our nose to some object in the scene. Similarly, maintaining fixation on an object in space as our head and body move (such as trying to catch a fly ball on the run) must require a complex integration of vestibular inputs and visual inputs to the eye movement motor control centers (Melvill-Jones, 1976).

Another area of research used in support of the need for nonvisual sources of information has been studies on adaptation to visual rearrangement produced by prisms, lenses, or mirrors. Welsh (1978) has an excellent review of this work. For example, if a perceiver is given a pair of prisms to wear in front of the eyes that displace the light 10° to the side, then when first looking through them, everything appears to be displaced by that amount. When the perceiver points to an object or attempts to walk toward it, he makes a 10° mistake in direction. This is exactly what should happen, since the visual information itself has been altered. However, what is surprising is that after some period of time of wearing the prisms, all of the perceiver's perceptual motor behavior becomes accurate. No longer are errors of pointing being made, even though the prisms are still altering the information in the light.

The evidence is overwhelming that to adapt to such rearrangements, the perceiver has to recalibrate the relationships between the visual information he receives as a consequence of his movements and actions with the location and control of the various parts of his body. For example, Harris (1965) showed that a subject adapted (became accurate) to displacing prisms when he pointed at visible targets with his hand that he could also see through the prisms. There was no improvement in accuracy of pointing with a hand that was not visible through the prisms. From this and other evidence, Harris argued that the adaptation process involved having to learn again the relationship between where the hand is seen and where it is felt to be. Correct perception and moving occur when these are in agreement, and the relearning is needed because the prisms alter the relationship that had been previously in effect. As this argument stands, it suggests that perceivers do use nonvisual information (felt position of their body parts, including eyes, hands, and limbs) to combine with the visual information to arrive at a correct perception, a position contrary to Gibson's.

Unfortunately, in spite of now voluminous data on optical transformations, the issue of the necessity of nonvisual information is still unsettled. Take the following argument. The normal adult perceiver knows (acquired by whatever means) that making a particular kind of movement produces a retinal displacement of the pattern of light of a certain kind. Rotating the eye causes the entire image to slide across the retinal surface without distortion or transformation except at the occluding edges. Rocking the head sideways back and forth produces a different pattern of changes. Once these relationships are known, then given the change on the retina, the perceiver can determine what pattern of self-motion caused them, and if none, then the movement must be in the scene. He does not need information about the movement itself at that moment because he can determine what the movement had to have been from the visual information alone.

But now the devious psychologist comes along and alters that knowledge by changing the familiar and predictable correlation between one's movements and the optical consequences of these motions. Now the perceiver cannot predict and in order to perceive correctly is going to have to relearn the relationships anew. So he attends to both the visual information *and* to the pattern of his own movements until he again can predict correctly. This is the process of adaptation. Once completed, he need no longer monitor his movements because he can predict them from the visual changes alone. The motion information is redundant.

According to this argument, nonvisual information (about self-motion) is normally unnecessary for visual perception because it is fully redundant with visual information. However, anything that destroys or alters that redundancy (such as the slow normal growth process of arms getting longer, eyes growing further apart or the very unnatural optical devices used in experiments) forces the perceiver to use nonvisual self motion information in order to correctly perceive his position in the visual scene. Since all of the experimental evidence suggesting that motion information is required was found in tasks in which the redundancy of the motion consequences was destroyed, this evidence is not sufficient to decide whether motion information is always necessary.

We just do not have enough data to clarify the role of sensory-motor interactions in the normal perception of space. It is easy to show that such interactions occur, but it has not been easy to show that such interactions are necessary for stable perception (see Haber, 1978a). I do not mean to imply that visual information produced by motion of the observer is unimportant. Quite the contrary: one of the most powerful ways to resolve the layout of space in a scene is to move your head. Head movements generate the motion perspective scale of space, which provides information available for processing. But neither the generation nor the processing of this scale requires information about the source of the motion.

5.8. Hochberg's Expectancy Guidance Model of Processing

Combinatorial models provide little room for the perceiver to be an active guide to his perception, in spite of their inferential appeal. Hochberg (1968, 1981) redresses this imbalance. He argues that repeated encounters with the structure of the physical world provide us with *mental structures*. These are learned expectations about the outcome of possible perceptual motor acts, such as the sensory consequences of looking at an object from different distances and durations, or of touching it in different ways. Using these mental structures, we perceive just that scene, object, or event that would be most likely to have produced any given pattern of sensory stimulation. Hochberg also refers to these mental structures as schematic maps, again as a set of contingent expectancies as to what the perceiver will see as a result of particular sensory-motor movements, such as moving the fovea to another place presently in peripheral vision and of changes of occlusion and of motion parallax that will occur if he moves his head. The mental structures are continually added to or modified as the perceiver tests his expectations about what he will see if he moves so as to bring presently hidden or blurred parts of the scene into clear vision.

Hochberg is most concerned with the integration of successive fixations. He describes three facts about single glances. First, the retinal image is not uniformly clear; rather, only its center is represented in sharp focus with maximally available contrast. Second, many of the critical processing tasks can only be carried out on the information in central vision, so that what is seen peripherally can be processed only if subsequently shifted over the fovea by an eye movement. And third, in order to be able to perceive an entire scene, the information picked up by each of the successive glances over the scene has to be integrated. The mental structure is this integration, a construction of the scene that goes beyond any particular sample of it that the perceiver received. While Hochberg's ideas are couched in quite different terms than the WHAT-WHERE distinction offered by Leibowitz, their concerns are identical and Hochberg provides a description of the sharing of processing by these two systems.

Hochberg dwells extensively on local depth cues as sources of information about the layout of space because he argues that the information contained in the three scales of space is not available at any one instant. Thus, in any one glance, information is available from only a local region, and not from an entire full retinal scale of space. He especially focuses on the surface scale, the one containing high-detail information from edges and texture. In a number of experiments he has shown that the nature of the objects in the scene layout cannot be determined from a single glance, but requires that the eyes look at different parts of the scene to pick up all of the surface feature information. Given this argument, then perception requires some integration of the information picked up from the successive glances at different parts of a scene.

While Hochberg describes the mental structures primarily as a means to integrate successive glances into a coherent scene (see also recent work by Jonides, Irwin, & Yantis, 1982), it seems equally plausible that such expectations are useful at all stages of processing. Except perhaps upon first opening one's eyes in the morning, each particular moment of perception is a continuation of the previous one. The perceiver's general assumption that the scene is still the same as the one he was just viewing is almost always correct. Further, he assumes that the particular objects he has already identified are still present and stable. It is difficult to explain the continuity of perception if each momentary act of perception is treated as unique and without an immediately preceding history. Such uniqueness (so typical of most experiments) is so much the exception that we are forced to build into our processing model some way to account for our continuous memory of scenes.

This is a much more cognitive theory than Rock's, since Rock has no place for expectations, nor for an internally constructed representation that goes beyond any particular sensory event. Hochberg's mental structures are hypotheses, making perception a process of sampling, testing, and construction, based upon knowledge, past experience, and sheer guesswork on the part of the perceiver.

Hochberg depends more upon past experience and knowledge than does Rock, not because the stimulus information is insufficient but because it is not sufficiently available in any one glance. Thus, the perceiver is always required, even for single glances, to make reference to what he expects to see. The simpler or more familiar the scene, the less stimulus information is needed to create the mental structure; that is, the perception of the scene.

Hochberg also provides a basis for incorporating some of the evidence that perceivers make assumptions about the properties of scenes they look at. Stevens (1981) has shown that perceivers interpret most intersecting edges as if they meet at a right angle. Consequently, whenever perspective produces a non-right angle on the retinal image, perceivers treat it as if it were a right angle intersection viewed at some slant. Misperceptions occur whenever the natural environment has non-right angles. Further, perceivers can use shadow information to help with object arrangements. Therefore, they assume that the source of illumination is overhead (Yonas & Hagen, 1973). Whenever that assumption is false, misperceptions occur. These are but two examples of general assumptions. Several more are considered in the section on picture processing, because there the misperceptions are especially obvious. Given this kind of evidence, we need a processing model that has a way to incorporate assumptions about the scene into the perception. Hochberg's model is the most explicit on this score.

The function of peripheral vision, Hochberg argues, is primarily for guidance about where to look next. Peripheral vision provides the basis for deriving hypotheses that then can be tested by subsequent eye movements. He recognizes the kind of information contained in the three perspective scales of space, as well as the local depth cues. He argues that the retinal image is not sufficiently clear to represent that information explicitly and that that processing system only works upon centrally-provided information.

While there are still many psychologists who reject mental structure as mentalism, Hochberg's conclusions seem inescapable to me. If a single glance cannot provide the information and we cannot process all of it in a glance, then to see an entire scene requires multiple glances. Perception of the entire scene has to be constructed in our head, since it is not present at any one time in the retinal image.

Hochberg (1972) proposed a concept of canonical form to suggest that perceivers need not fully process an entire pattern in order to determine what it represents. He leaned on the results of Ryan and Schwartz (1956), who showed that subjects are often quicker to identify familiar objects from outline drawings or cartoons than from photographs or more detailed drawings. I have suggested elsewhere (Haber, 1979) that the notion of canonical form can equally be applied to perceiving the layout of space. Thus, it may not be necessary to process all of the stimulus information coming from a scene, but rather, the perceiver can search out some small set of features that are sufficient to specify the space. This notion is particularly intriguing given Hochberg's demonstrations (e.g., Hochberg, Brooks, & Roule, 1977) that the stimulus information about a scene is never completely available at any one time. Canonical space features would enable the perceiver to find the prototypic aspects of the depth arrangement at a glance before he has completed the integration process of combining several successive glances (see Berbaum, Tharp, & Mroczek [1983] for some experiments).

While this idea is not especially new, we do not know very much about what some of these canonical depth features might be. One way to find out would be through an analysis of line drawings of scenes in depth. What kind of minimum marks are necessary for perceivers to agree upon the layout of space? It would seem likely that a horizontal line two-thirds of the way up the page is treated as a horizon, especially if two other lines are shown converging toward a vanishing point on that same horizon. Specification of the vanishing point may be a canonical depth feature which then at least minimally specifies the slope of the ground and the location of the viewer vis-a-vis that ground. If any object is then added to the sketch, the presence of the vanishing point helps locate the object in that scene as well.

A second candidate for a canonical depth feature, this time of more local origin, might be the presence of a T intersection (as specified by Guzman [1968]) among luminance discontinuities in the retinal image. Whenever two visual edges intersect so that they form a T, it is usually safe to interpret the continuous line as the edge of the surface that is lying in front of an occluded surface. In a sense, the T intersection is a minimal component of interposition. Guzman (1968), Waltz (1975), and others have shown that such an interpretation of the T intersection is invariably correct and the basis for a powerful hypothesis about the position, relative distance, and orientation of two surfaces.

Without elaborating a potential list of canonical depth features further, one aspect of a processing model for perceiving space might be the initial identification of canonical depth features in the information on the retina in any particular glance. These might then be a guide for Hochberg's expectancy generation underlying the mental structure we construct of a scene.

Canonical depth features might also be a basis for the integration across successive glances. Hochberg, Brooks, and Roule (1977) have shown that integration is not possible without significant landmarks that reoccur in each overlapping glance. In one experiment they presented successive views of a pattern, each view displaced slightly with respect to the previous one, simulating the changes in the retina pattern produced by saccadic eye movements. When the patterns were dense and devoid of any distinctive features, then even with maximal overlap, integration did not occur and viewers could not even tell in which direction the pattern was moving. Successive overlap by itself was not sufficient. While the patterns used in his experiments were all two-dimensional, one type of significant landmark that may permit the integration to occur are the features of canonical depth.

While I find much to support in Hochberg's theorizing, one aspect of his work has been too narrow. His argument is that because of the fall-off of sharpness or clarity in the retinal image from the fovea outward, we have to move our eyes to pick up clear information from the entire scene. This argument applies, however, only to information from the surface scale of space. As noted earlier, motion perspective can probably be perceived over all parts of the retina, because it does not require high spatial frequencies or high acuity detection. Hence, if the perceiver is actively looking at a scene, he is getting information from the entire scene, even though he can see details clearly from only one part of it. The same applies to the binocular disparity scale, since disparity can be registered over the entire retina, and is not limited to only the high resolution foveal regions. Thus, Hochberg's processing model is concerned primarily with the WHAT system, and does not apply to WHERE processing. However, it is the natural complement to at least my interpretation of the Gibson model, which tells us primarily about how we know WHERE objects are, and little about WHAT they are.

5.9. Summary of Information Processing of Scenes

Drawing on the different approaches to the processing of information about the layout of space, I have suggested a rough processing sequence. At the first, and probably most peripheral level, a correspondence is assigned between the pattern of optical stimulation at each point of each retinal surface and the distance from the perceiver of each of the surfaces in space that had reflected the light to the eye. Such assignments are done by neural coding networks organized to respond according to the geometrical rules that describe the various perspective transformations that the optical ray of light undergoes when focused on a retina.

Following this, each luminance discontinuity on the retinal surface is located and categorized according to the type of visual edge that gave rise to it in space. This categorization is probably based on depth and intersection analyses, plus other local features along the edge. With edge classification, intrinsic lightness and color values of each surface can be processed according to the appropriate rules, objects assembled from the surfaces, and the full scene articulated. Throughout this sequence of processing, each step is guided by the perceiver's expectations and knowledge of the structure of the scene.

I have written this section in a sequential fashion, and I have used sequential metaphors to describe the steps: first this stage, then the next, and so forth. Such description should be treated metaphorically, and not as implying a serial model of processing. I see no reason in theory why all of the steps could not be done in parallel, or, if sequential, why the order always has to be fixed and unvarying. We are going to need much more knowledge about space perception before it makes sense to take categorical stands on the sequential nature of such processing. The above is intended only to point out the kinds of steps needed.

6. Information Processing of Pictures

As the earlier stimulus analysis makes clear, pictures contain information that can be perceived with a dual reality. They can be perceived as flat objects in their own right, with certain shapes, brightnesses, contours, and colors, often a frame around them, and so forth. They can also be perceived as representative of some three-dimensional reality in which the flatness and frame are ignored and the edges are used to perceive a three-dimensional layout of space. Perceivers can apparently extract either of these realities of pictures and go back and forth between the two easily. The dual reality of pictures is clearest for so-called representational pictures, especially photographs of natural scenes. Perceivers have little difficulty recognizing or matching such two-dimensional representations to the natural scenes that gave rise to them, nor do they have much trouble correctly interpreting the object information and the spatial layout in such pictures.

I shall consider a number of factors that contribute to the processing of both of these perceptual realities. For the three-dimensional reality, I have argued in detail elsewhere (Haber, 1979a, 1980) that perceivers treat the stationary perspective information they receive from pictures in the same way as they treat stationary perspective from scenes. This means part of the time perceivers simply disregard the flatness information. The evidence from young children's perception of picture suggests they may always disregard the flatness and have to learn to take flatness into account, both to see the flat reality of pictures and to maintain a constant perception of the layout of space regardless of viewing angle and distance. Evidence from visual illusions and from the ease with which we can perceive the layout of space from simple line drawings suggests that even with learning and extensive practice, the three-dimensional reality dominates wherever possible. In this section, I shall review the kinds of evidence that support these assertions as ways of describing how we process the information from pictures. The best resource on pictorial perception is the two-volume collection edited by Hagen (1980a,b). Many of the articles referred to in this section are discussed in detail in one or more chapters of her books.

6.1. Developmental Evidence of the Dual Reality of Pictures

With the assumption that the perception of depth in pictures involves the same processes as those for perceiving the depth in real scenes, then it follows that the two develop and improve together. Similarly, the ability to perceive the flatness of pictures also develops with age and experience. If these two abilities develop together, then children at all ages have access to the dual reality of pictures. This has been a reasonable hypothesis and accepted uncritically without much examination. One pervasive, though silly alternative has been that pictures, because they have only two dimensions, are somehow easier to perceive, so that young children might be able to perceive the flatness of pictures more easily than the depth in real scenes. A third, and much more interesting hypothesis is the one that appears to be true, namely that children do not perceive the flatness of pictures until relatively late, and until then they treat pictures and scenes in the same way and as the same thing. Such equivalence leads to quite predictable types of errors in perceiving pictures. Evidence suggests this is, in fact, what happens.

Cooper (1975), as noted in Hagen (1976b), studied the ability of children to compensate for perspective distortion in pictures viewed obliquely. He found that three-year-olds treated the projections on the retina from pictures as if they had come from real objects and not pictured objects. Thus, the children's responses indicated a perception of three-dimensionality in a scene without recognition that the scenes are only two-dimensional. Similarly, Benson and Yonas (1973) investigated the development of the utilization of direction of illumination as a source of information about depth in pictures. They found that adults assume that illumination is always at the top of the picture, but three year olds apparently assume that illumination is overhead in space regardless of the picture orientation. Young children make many errors in distinguishing convexity from concavity whenever the picture is turned on its side, but not when it is oriented properly. Benson and Yonas interpreted these data to mean that the ability to see a picture as a flat object in its own right develops later than the ability to perceive the depth relation in pictures (or in real scenes).

Hagen (1976b) also found a developmental improvement from ages five to twenty in the utilization of shadow information to determine the direction of the source of illumination in pictures. Since Piaget and Inhelder (1967) show that this process is fully developed for scenes in real space by ages seven to eight, Hagen argued there is a lag in perceiving the surface information in pictures beyond that of perceiving the depth information in pictures.

The nature of this lag is not clear in these experiments. Young children must have some access to the sources of information about flatness because these same sources must be used by them to determine depth. It appears as if they do not apply such cues to pictures, but prefer to treat all projections on the retina as if they came from three-dimensional scenes. In doing so, young children make the kinds of mistakes in their perception described above.

Another series of developmental experiments manipulated the effects of flat surface information on the perception of depth relationships in pictures. Yonas and Hagen (1973) attempted to remove flatness information from pictures by eliminating motion parallax and using rear-projected slides. When flatness information was removed, perceivers at all ages (three, seven, and twenty years) improved in their ability to perceive the depth relations in pictures. Conversely, Hochberg (1962) tried to make real scenes look like pictures by placing cellophane with a frame around it between the viewer and the scene and restricting viewing to one motionless eye. He found that when this kind of flatness information was provided, viewers could not distinguish the real scene from the picture which represented it. Even more critically, Hagen, Glick, and Morse (1978) showed that if you make a viewer think he is looking at a picture, by adding irrelevant cues to flatness (a plate of glass, for example, placed between the viewer and the scene), the perceiver treats the new scene as if it were a picture of the scene, and he makes the same type of errors that he makes in looking at an actual print of the scene. Thus, not only does he confuse the modified scene with a picture of it (Hochberg's findings), but he perceives the depth relations in the modified scene and in the picture of the scene in the same way.

This line of evidence suggests children develop the ability to perceive the three-dimensional relationships in real scenes and in pictures before they also perceive the surface qualities, especially the flatness of pictures. Thus, for some period in normal development, pictures do not have a dual reality for children. This does not mean that pictures look more vivid or are seen in plastic depth (Schlosberg's [1941] term). Rather, being less sensitive to the surface quality and the flatness of pictures leads to qualitatively different perceptions from pictures. The most obvious error to expect is a confusion of pictures with scenes. It is possible that very young children do have such confusion, but it is unlikely to persist for long because of the content differences. The more likely misperceptions occur when the correct perception of the flatness information is necessary in order to perceive the depth in the picture correctly. As will be discussed below, the surface perspective information from a scene and a picture of that scene match only when the picture is made (photographed or drawn) from the same viewing point in the scene as the perceiver is now standing; and when the picture is looked at from the same distance and orientation relative to the scene as the camera or artist was when the picture was made. Violation of either of these requirements distorts the surface perspective scale of space, a distortion which could lead the viewer to misperceive the depth relationships. This misperception can be avoided only by correctly perceiving the flatness information of the picture surface, so that a compensation can be made for the distortion. If you cannot see that the picture is tilted relative to your line of sight, you cannot correct for the distortion in perspective. This is what young children apparently cannot do (see below).

There are several important processing implications to the hypothesis that processing of flatness information develops later and independently of depth information. One is that all perceivers, and especially children, treat all stimulation reaching their eyes as if it arose from a scene in depth and not from a picture of a scene. Thus, while pictures create two perceptual realities, the depth reality takes precedence, even in adults who do have access to both. It appears as if the human visual system is designed to perceive scenes in depth and constructs all stimulation as if it came from a scene. Archaeological evidence (Jaynes, 1976) suggests that pictures are a relatively recent invention in the evolution of human beings, so that for most of our evolution, we never had two-dimensional representations to look at.

A second processing implication for pictures is that they are more difficult to perceive. They take extra processing capacity because the flatness information has to be included with the depth information in order to properly perceive the spatial arrangements. Far from being easier to see because they have fewer dimensions, they are harder for that same reason. This suggests that under high task loadings or brief viewing time, or for immature viewers, the spatial arrangements in pictures are more likely to be misperceived than the spatial arrangements in real scenes. It is also possible that under such conditions, the flatness information in pictures may not be adequately processed, so that pictures both look more three-dimensional than when normally viewed, and the spatial arrangements are more misperceived compared to normal viewing. None of these implications regarding difficulty of perceiving pictures has been tested yet, as far as I know.

A third implication is that perceivers rarely, if ever, misperceive the spatial arrangements in natural scenes, but often are fooled by pictures. Nature creates few visual illusions, but psychologists have created a myriad of them using pictures.

6.2. Station Point Correction

The stationary perspective information available from a scene and from a photograph or a Leonardo-executed picture of that scene is identical only for a one-eyed, motionless observer whose single eye is in the same place vis-a-vis the picture, as is the camera lens or the artist's eye when he outlined the painting on the plate of glass. We have already examined why an observer using two eyes and moving picks up information that the picture is flat, but the proper viewing point requirement is more complex and puzzling. Failure to view from the proper station point results in a pattern on the retinal surface (ignoring blur for a moment) that is no longer isomorphic with the one on the hypothetical plate of glass. It does not contain the proper stationary scale of space and the local cues for the correct three-dimensional scene. Accordingly, failure of the observer to have his eye where the painter had his (or where the camera's lens was for a photograph) should lead to errors in the perception of the layout of the three-dimensionality perceived from the picture.

Farber and Rosinski (1978) have described the geometrical changes that occur under a variety of incorrect station point observations. Their review suggests that whenever the viewer is restricted to a static monocular view, what he sees depends entirely on the geometrical distortions created by the incorrect station point. However, whenever the viewer can use two eyes or move, then the resulting perception often seems to be independent of the viewing point and cannot be predicted by the geometrical distortions at all. If the correct perception of the third dimension in a picture depends upon the picture projecting the same pattern of light as did the scene represented by the picture, why is the viewer's position or orientation irrelevant?

The most developed answer to this problem comes from Pirenne (1970). He argues that as long as the viewer can register both realities of a picture, the three-dimensional scene it represents and the two-dimensional surface of the picture, then he can determine what should be the correct station point and compensate for any distortion on the retina produced by an incorrect station point. It is being able to see that flatness that allows the viewer to align himself before a picture properly and see its depth correctly. Pirenne's argument coincides closely with the pattern of results reviewed by Farber and Rosinski (1978). Since it is only under free binocular viewing of pictures that both realities can be perceived, it is only under such conditions that incorrect station point distortions are irrelevant.

Hagen (1976a) tested Pirenne's explanation by asking observers to look at pictures shown with reduced information for flatness. Subjects viewed a picture containing two unlike-sized objects, one placed near and one far. They had to point to the larger. A correct response required correct perception of the depth relationships. Viewing was monocular through a peephole, either from the correct station point or obliquely from 40° to the side. To manipulate surface information, Hagen used photographic prints which have high surface information, and rear-projected transparencies, which have low surface information. The subjects were four years old, seven years old, and adults. She included children to see if they are less sensitive to the surface information in pictures from which to compensate, and further, if the compensatory mechanisms require experience and practice with viewing pictures and so should be less developed in children.

The results are entirely consistent with the Pirenne hypothesis. Adults use surface information to compensate for the wrong station point in order to correctly perceive the depth relation of pictures. When that surface information is removed, as when viewing transparencies, then depth is distorted when viewed from the wrong station point. Such surface information is not available to children, so viewing from the wrong station point reduces the accuracy of their depth perception regardless of condition.

Pirenne argued that station point compensation is learned on the basis of experience with viewing pictures. He would expect young children to show poor compensation, and therefore, that pictures would have a distorted layout of space when viewed by children from the wrong station point. Pirenne would attribute this to young children's lack of experience with pictures. However, evidence already presented suggests that this is due to their poor perception of the two-dimensional reality, which they need in order to determine the orientation of the picture surface. Specific experience with pictures may be unnecessary for station point compensation to occur.

6.3. Pictorial Illusions and Dual Reality

A perception is called an illusion when its content does not agree with some physical measurement of the surface reflecting light to the eyes (Coren & Girgus, 1977, 1978). The lengths of the lines in the Müller-Lyer illusion do not correspond to the way nearly all perceivers report seeing them. This definition of illusion, however, as Coren and Girgus emphasize, does not make sense when applied to any two-dimensional picture, photograph, or even a line drawing. Consider the problem first with respect to representational paintings or photographs. Such pictures already have undergone the perspective transformation from the three-dimensional scene they represent to a two-dimensional picture surface. To whatever extent a viewer perceives the three-dimensional scene represented in the picture, there has to be a mismatch between the picture surface and his perception. Thus, a photograph of a receding railroad track contains vastly *unequal* line lengths of the various railway ties (as measured with a ruler on the picture surface), but these are perceived *as equal*. In fact, every aspect, feature, or element of the picture surface would produce an illusory perception when defined this way.

But no one calls such mismatches illusions! Rather, these are examples of appropriate applications of size constancy, in which the perceiver uses the information in the picture to construct a correct representation of the three-dimensional scene. Perceivers might be aware or might be made aware that the picture is flat and that the railway ties are of all different sizes on the picture surface. If that occurs, then the perceiver has two different perceptions, the two realities of pictures. However, since the three-dimensionality of representational pictures and photographs is so powerful, even the cues to flatness are not sufficient to prevent the construction of the three-dimensional layout of space with its attendant constancy scaling. Thus, normally the three-dimensional reality overrides the two-dimensional reality.

In the experimental literature, the term "illusion" is reserved, not for these instances of representational pictures or photographs, but for certain kinds of line drawings – drawings not intended by their creators to be representative of three-dimensional scenes at all. They are drawings in which the balance between the two realities has swung so far toward flatness that perceivers are not expected to perceive their three-dimensionality. So, perceivers act as if they are looking at a flat drawing, but they still perceive enough of its three-dimensional reality as if were a representation of some three-dimensional scene. Since the artist had no intention of drawing a three-dimensional scene, such drawings are said to produce a misapplication or inappropriate constancy scaling.

This explanation of line-drawn illusions was first proposed by Thiery (1896), but is more typically identified with Gregory (1966, 1973). The theory states that perceivers use the two-dimensional information in outline drawings to construct a representation of a three-dimensional scene in the same way they can construct three-dimensional space from fully representational pictures or photographs. It is a misapplication of constancy only because there is no three-dimensional scene supposedly being denoted by the illusion-producing drawings. That is, while we as theorists insist the drawing should be treated as flat, perceivers treat it like any other picture and construct a perception of it as if it represented a scene.

One problem with this otherwise reasonable theory is that perceivers often do not report perceiving the apparent depth relationships that the theory says are responsible for the apparent size, shape, or orientation changes. Thus, for the Müller-Lyer illusion, viewers should perceive the apparently longer line as farther away than the apparently shorter line in order to account for why they see the size difference. Yet, many observers, even those with strong illusions of size, see the distances as equal, or even occasionally, that the apparently longer line is closer (Worrall, 1974).

It is easy to see why this might happen, at least with respect to the vast majority of two-dimensional line illusions. The drawings are so impoverished in detail that they appear flat, with no depth in them at all. The two-dimensional information that specifies three-dimensionality may be registered, but it is not perceived as such (see Epstein, 1973; Rock, 1975, 1977), because it is at variance with the clear flatness of the picture given by other sources of information. The registration, or taking into account, is sufficient to produce the apparent size (or shape or directionality) changes, but it itself is not perceived.

Gregory (1966) has shown that if the cues to flatness are reduced in the drawing, then not only is the illusory effect greatly increased, but the depth becomes apparent to the perceiver as well. To demonstrate this, Gregory made the Müller-Lyer elements out of wire which he then treated with phosphorescent paint so that they glowed in the dark. When viewed in the dark, observers had no two-dimensional reality of the flat drawing to contrast with the perspective information induced by the glowing arrowheads. Consequently, the observers reported that the wire figure appeared to be three-dimensional, and in accordance with Gregory's explanation, the illusory effect was greatly enhanced.

Coren and Girgus (1978) provide a detailed review of nearly all of the visual illusions created by artists or psychologists, as well as the few that occur in nature – see below. While I have stressed a dual reality interpretation of these illusions, their review makes it clear that such an interpretation cannot account for all illusions, and may even have problems with a few of them it seems to cover well. They end by stressing that the illusions are not all the same, and we should not expect to find a common explanation. Many may even be due to very peripheral interactions that arise from the way line elements are encoded on the retina (e.g., Ginsburg, 1975) and have nothing to do with space perception at all.

It should be noted that nature does provide a few "three-dimensional" illusions directly, the most famous of which is the size illusion of the moon at various orientations in the sky. The moon illusion has been explained by inappropriate constancy scaling (Rock & Kaufman, 1962). The relative absence of depth information from the sky from such far-away objects leads the zenith moon to be interpreted as nearer than the horizon moon. Since the visual angle of the moon is constant regardless of its elevation, the registered difference in distance results in a perceived difference in size. This explanation is not the only one, and it is not even entirely in agreement with all of the facts (see Hershenson, 1982; Roscoe, 1980). Rather than argue over this illusion, which has resisted explanation for nearly two millenia, it is easier to note that while the moon in the sky is physically part of a three-dimensional scene, it is one of the few natural examples in which there is virtually no rich three-dimensional information available to perceivers. You might as well be looking at a picture of the moon for all the three-dimensionality you get from the evening sky. All three scales of space are compromised and no local depth cues occur except when the moon is near the horizon. Nature tricked us only by visually impoverishing that scene.

6.4. Subjective Contours and Depth

Coren (1972) discusses a set of illusory drawings that create subjective contours. These are illusory because most perceivers report seeing a contour where no intensity or wavelength discontinuity exists. Three examples are the edges perceived in stereoscopically viewed random dot stereograms, the edges defining objects when only the shadows cast by the objects are drawn and the edges of an unshown object seen overlapping a more articulated one (Kanizsa, 1974, 1976).

Coren argues that in all three instances, the subjective contour is perceived because there is depth information in the drawings – their three-dimensional reality – that leads perceivers to stratify the entire configuration on several depth levels. Even if no contour is drawn to separate the different levels, one is perceived because one part of the figure is seen in front of another part. In support of his argument, Coren presents data about each of the three types of subjective contour examples to show that perceivers do see them in depth. Further, when they are redrawn so as to eliminate the depth information, the subjective contours also disappear.

The dual reality of pictures cannot account for all subjective contours or for all of the visual illusions. Other factors besides stratification in depth can induce contours in perception where none existed in the luminance discontinuities. For example, many of the Gestalt laws of organization are statements about where to expect perceptual structure even in the absence of physical structure. While some of these laws concern depth, others do not. The law of closure is most explicitly a statement predicting that subjective contours connecting physically unconnected elements occur. Ware and Kennedy (1977) and Kennedy (1979) report subjective contour demonstrations which are much more akin to closure than to induced stratification.

6.5. Perceiving Incomplete Pictures

One of the traditional objections to a perspective theory of picture perception is that it does not account for the perception of the layout of space in outline drawings, sketches, cartoons, and caricatures or in pictures in which the perspective information is incompletely rendered. This objection constitutes a problem any theory of picture perception must answer.

Hochberg (1972, 1979) has proposed the notion of canonical form to account for the easy recognition of objects from outline drawings, sketches, and cartoons. Recognition does not depend upon the full articulation of all of the features and properties of an object, but can occur if only the object's essential characteristics or distinctive features, i.e., its canonical form, are provided. All ambiguous features, all features that the form shares with other forms, so they are not definitional or distinctive, and all features which are irrelevant or noninformative are unnecessary for recognition. Outline drawings or cartoons are natural examples since their purpose is to use pictures to promote identification or recognition of a specific object quickly. Ryan and Schwartz (1956) tested this notion on the recognition of single objects and found that photographs, which supplied the most visual information, often required the longest exposure duration, whereas cartoons or outline drawings often needed the shortest duration.

Perkins (1975) and Goldman and Hagen (1978) have empirically addressed the problem of feature variation or distortions in caricatures. With such drawings, recognizability is essential, so the selection of features to distort cannot be haphazard. Goldman and Hagen analyzed 100 caricatures of Richard Nixon created by 17 different artists. They found great consistency, both within and between artists, and across time, in the features selected for exaggeration, especially jowls, box-like jaw, and length of nose. Different public figures have different features exaggerated; for Mr. Nixon, these features must be part of his canonical form.

Canonical depth, as described earlier, refers to perspective features that define all the possible three-dimensional relationships in a picture without requiring a full articulation of all perspectives and other depth information. Indication of a vanishing point is one example of a possible canonical depth feature. Such canonical depth features would permit the correct perception of the layout of space from sketches, outline drawings, cartoons, and other incomplete figures. The presence of a canonical depth feature may also account for why some two-dimensional drawings create visual size and orientation illusions. If a few depth features are present, just enough to register the three-dimensionality but insufficient to cast the whole drawing into depth, then the two-dimensional features in the drawing could be misperceived.

The usefulness of canonical form and canonical depth features is only a hypothesis. We need research on what these features might be, and we need to know how perceivers develop the ability to extract and generalize from them. This may involve a process of perceptual learning, or it may turn out that our perceptual systems extract such features without prior practice and experience with them.

Canonical depth features can be considered a kind of convention or symbolization in pictures that observers use to perceive the layout of space properly. While these kinds of features have not been described before as conventions, a number of others have been.

One of the most important yet usually overlooked conventions is the use of lines to represent edges or boundaries in space. Objects in scenes are not surrounded by lines, yet we have no trouble representing and perceiving objects in drawings simply by outlining their boundaries. Kennedy (1974) has argued that this convention is unlearned, and certainly the data of Hochberg and Brooks (1962) (see below) support him. It seems quite likely that the information coming to the eye from the edges of objects in a scene and from the lines bounding objects in a drawing are both coded by the visual system in the same way before that information even reaches the cortex. The spatial frequencies represented at the eye from luminance and spectral discontinuities and those represented by an outline drawing of the same scene might be sufficiently similar to permit the use of lines in a picture to stand for boundaries or edges in space.

A second potentially unlearned convention has been suggested by the work of Benson and Yonas (1973). They showed that even young children assume the source of illumination in a picture comes from the top. Since illumination defines shadows in relation to objects and therefore helps to locate those objects in space, this conventional assumption is an important way of specifying the layout of space.

Other conventions are clearly learned through interaction with pictures. Movements of objects in a picture can be indicated by multiple representations of legs, by a smear drawn behind the object indicating speed, or by blurring the edges, to name just a few conventions. Friedman and Stevenson (1975, 1980) have provided a detailed historical and developmental analysis of the different conventions used to represent movement in pictures. They report evidence that these movement conventions are learned during preschool and early primary school ages. We badly need comparably good analyses for spatial and for sequential conventions, such as these used in reading cartoon strips. Examples of other types of conventions are drawing stars above someone's head in a cartoon to show that he is "seeing stars" or using big print in a cartoon balloon to mean the speaker is talking in a loud voice. The analysis of caricature by Goldman and Hagen (1978) also involves conventionality.

Developmental sequences should be found for those conventions that are learned. There also should be cultural differences for any of the learned conventions, and, in addition, there probably are individual differences within age and culture, depending upon the amount of experience with pictorial conventions.

A research program by Biederman (e.g., 1972, 1977) has examined a different aspect of convention, specifically whether objects in a scene violate relationships assumed to be true of those objects. These assumptions include the need for support (objects cannot float freely but must rest on surfaces), familiar size (a telephone cannot be drawn to appear larger than a car), appropriate location (gasoline pumps are not found inside a kitchen), appropriate orientation, and the like. He found that subjects took longer to identify the objects in question or to identify the overall theme of the picture when one or more of these assumptions are violated. Further, subjects made proportionately more errors during brief presentations of the picture when more violations were present. These results suggest that perceivers have expectations about the organization of objects in a scene, assumptions that aid or speed up the processing of the scene. When these assumptions are violated, processing is slowed down or misguided. Biederman's goal is to find the grammar or syntax of scenes, the rules that describe what objects can or do go together. From such a description, he wants to know how much that grammar guides perception. While he is working with pictures, his conclusions would apply equally to scenes.

Our discussion, so far, has described pictures in terms of the layout of space they represent. It is also possible to describe pictures by organizational principles within the two-dimensional reality without respect to depth. Beginning nearly seventy years ago, the Gestalt psychologists used two-dimensional line drawings for their research because relationships could easily be specified and studied. They said very little about three-dimensional organization; their focus was entirely on identifying the properties in the retinal projection that lead the perceiver to construct one type of perception rather than another. Gestalt psychologists proposed a set of principles that describes these properties. Principle of proximity: elements in a drawing that are near each other tend to be organized together. Principle of similarity: visually similar elements tend to be organized together. Principle of common fate (a derivative of similarity): elements that change or move as a group become organized together by their common fate. This principle explains how we see patterns in dancing; when several dancers move together, they are seen as a group. This is an example of similarity of movement rather than of shape. Principle of continuity: elements that form interesting patterns tend to be organized as a continuous figure rather than as disconnected units. Principle of closure: gaps between elements of a continuous pattern are filled in perceptually. Principle of good figure: some organizations are more stable, more complete, or more satisfying than others. According to this principle, circles are better figures than ellipses, so a figure has to be quite elliptical before it fails to be seen as a circle. This is perhaps the most controversial principle because of the great difficulty in providing any independent definition of what is good. For most applications, what is good is what is seen, and this principle therefore becomes circular.

These principles describe what features are organized or grouped together in flat drawings. However, the Gestalt rules are generally not useful for pictures that can be organized in depth. Depth organization follows principles of perspective, not principles of proximity, similarity, and so forth. Therefore, for most drawings or pictures, these Gestalt principles have had very limited application. In fact, they are most often incorrect when applied to scenic pictures. The Gestalt principles were constructed in an era when treating the retinal image itself as a picture was the accepted metaphor. But visual edges near each other on the retina are perceived as related together only if they also seem to be the same distance away. This fact makes any analysis of pattern on the retina or other two-dimensional surface irrelevant for most kinds of pictures. Gestalt rules, in their traditional form, are useful only for decorative art.

6.6. Do We Have to Learn to Perceive Pictures?

Pictures and other two-dimensional representations are invented or constructed by human beings. Because of this, some theorists have assumed that perceivers would not have been able to perceive the layout of space in pictures or recognize the objects portrayed in pictures without prior visual training or experience with pictures. Two aspects of picture perception have been discussed in the context of learning: perceiving the layout of space represented by the picture and recognizing the objects portrayed in the picture. Unfortunately, these are not always distinguished, so that one is tested while the other is discussed.

Given the interrelationship of scene and picture perception, experimental tests to determine the learning or innateness of picture perception alone are very difficult to interpret. With the reasonable likelihood that scene perception comes first, then as soon as one aspect of scene perception can be accomplished (whether with or without the need of experience), that same aspect can be performed on pictures without further experience. The classic study by Hochberg and Brooks (1962) tested an eighteen-month-old child who had never been exposed to any two-dimensional representations. When asked to identify familiar objects, he was as accurate in doing so from pictures and from drawings of the objects as he was when seeing the objects themselves. Hochberg and Brooks did not report any data about the perception of the layout of space in pictures, but at least with respect to object identification, no prior experience *with pictures* seemed necessary. This is a powerful result, perhaps even an instance in which we can generalize from one American child to all children for all time.

Such logic can also be applied to the cross-cultural literature on picture perception (see Hagen & Jones [1978] for a recent review). While there are cultural groups who do not produce or utilize two-dimensional representations, finding that aspects of their picture perception are accurate, even upon first exposure, does not necessarily specify the acquisition process. On the other hand, finding differences in picture perception between people with no experience and those with lots of experience does suggest that experience is necessary.

Unfortunately, the choice of stimuli (e.g., those of Hudson, 1960) used in the cross-cultural research has been poor enough to weaken most reported findings. While the evidence for naming objects from such pictures is generally consistent with the conclusions of Hochberg and Brooks, the data on extracting the layout of space suggests that many persons who have never seen pictures before having trouble doing it form Hudson's pictures. Unfortunately, this is true both for African subjects (e.g., Hudson, 1960; Derogowski, 1972) and for Westerners. Hagen and Jones (1978) examined some of the stimuli in detail and then used them to test subjects who were thoroughly familiar with pictures. They found that many of the Hudson pictures, as well as others used in the cross-cultural research, did not produce reliable responses, even from experienced observers. This casts suspicion on all the African research with these pictures.

Until we have better data, we cannot interpret any of the reported research as supportive of a learned component of the perception of the layout of space in pictures and certainly not supportive of a learned process independent of space perception in general. The theoretical difficulty in determining the role of learning in picture perception, even after we solve the methodological and materials problems, is that picture perception and scene perception are so linked and intertwined. It makes no sense to look for evidence of learned or unlearned aspects of picture perception without also knowing what aspects of scene perception are learned or innate. Given the argument presented here, if perceiving the layout of space in pictures follows the same processing as it does in real scenes, but develops later for pictures, then if some aspect of perceiving space in scenes is learned, that same aspect is likely to be learned in picture processing as well. But it may be difficult to demonstrate that learning with pictures because the learned part occurred earlier with respect to scenes. This logic suggests the learning question for pictures is not likely to be settled by itself, but only as part of a more general analysis of learning in perception.

6.7. Perceived Structure and Organization of Pictures and Scenes

Conventional wisdom holds that a picture is worth a thousand words (or maybe only 600 – see Haber [1981a] for the exact number). This wisdom has been examined in several different ways, including the massive research and theoretical literature on memory for pictures. Knowing the ways pictures, as compared to words or other more abstract symbols, represent information is obviously important for the understanding of picture perception; this research area has concentrated on pictures only because they are easier to manipulate and expose to subjects than are scenes. Standing (1973) presented his subjects with 10,000 pictures of scenes. While that alone seems herculean, imagine trying to do this with 10,000 different actual scenes! Aside from methodological ease, the conventional wisdom applies to scenes and to pictures of scenes interchangeably and is not meant to imply some special property of pictures as distinct from scenes. However, since virtually all of the research has been done on pictures, I have saved the discussion of perceived structure until this section.

There is substantial evidence that recognition memory for pictures exceeds that of any other symbolic representation, and even more interestingly, pictures can be recognized with virtually no errors. For example, in one experiment (Standing, Conezio, & Haber, 1970), college students were briefly shown over 2,500 candid photographs taken by amateur photographers. The students were later shown photographic pairs, each of which contained one photograph from the set they had seen and one photograph they had not seen before. The students were able to pick out the familiar photograph from each of many hundreds of pairs more than 90 percent of the time; for some students, this number was 98 percent. They were able to recognize the familiar photographs whether they were tested on the same day that they saw the last of the 2,500 photographs or on the following day, whether they had originally been given one second or ten seconds to look at each photograph, and even whether they saw the familiar photograph in its original or in its mirror-imaged orientation.

Standing (1973) has extended these results for a memory load of 10,000 pictures. The comparable studies with words, with numbers, with sentences, or even with brief passages (see Shepard [1967] for an example) show lower accuracies of recognition memory. One striking difference, besides accuracy, is the general lack of effort needed by subjects when trying to remember pictures, as compared to the effort expended to remember more abstract or language-based symbols.

According to current cognitive theory, such memory and effort differences must reflect differences in the way in which perceivers organize or represent the information contained in scenes, as compared to abstract symbols of language. I have described this difference (Haber, 1981a,b; Haber & Wilkinson, 1982), both in terms of the three-dimensionality of scenes and of the way in which meaningfulness provides a structure to scenes.

Because the human visual system can respond so easily to the three-dimensionality of scenes and pictures of scenes, that structure in the stimuli provides a structure in the perception as well. The three scales of space especially organize the scene. The thousands of visual edges in a typical scene do not exist as independent elements to be somehow learned and combined. Rather, they are organized and grouped into surfaces and objects located in space by the operation of these scales. No mnemonics are needed such as are used for unrelated rote items to be learned. Random arrangements of lines are difficult to learn and impossible to remember. Make those lines the projections from a natural scene, even a totally unfamiliar one, and the lines are grouped (chunked) together so that no separate learning is necessary.

The same ideas can be illustrated with respect to the role of meaningfulness as an organizer of scenes and pictures. As one example, Freedman and Haber (1974) showed subjects a number of very high contrast face pictures (originally developed by Mooney [1954]) in which each face is rendered only as white and black areas corresponding to the highlight and low-light reflections from the face when photographed. While a few of the faces were very easy to see as faces, most required many seconds of examination before the face "popped out." For all faces, some subjects saw a face and some did not. The drawings themselves did not change, so the change in perception must have occurred in the perceiver who had sometimes organized a percept and sometimes not.

The subjects were then shown a larger set of drawings that contained all of the original ones, plus an equal number they had never seen. They were asked to indicate for each one whether they saw a face and whether the drawing was familiar; that is, had they seen it before. For the drawings in which they could see a face, the subjects were accurate in reporting whether it was new or old, but if they did not see a face in the drawing, then they were at chance in being able to tell whether it was new or old. Recognition memory was excellent for the drawings the perceiver had been able to organize, but the same drawing perceived as a random assortment of blobs was not remembered at all. Clearly, it is easier to remember a unified perception such as the face of a young girl rather than nine irregular white blobs on a black background. The faceness organizes the otherwise unrelated elements into a unity that is easier to remember, a unity based upon familiarity of meaningfulness. This unity seems to be different from the kinds of organizations covered by the Gestalt principles and is also different from organizations based on perspective principles underlying a three-dimensional layout of space.

6.8. Summary of Picture Processing

To perceive the layout of space represented in pictures requires the same processing of the stationary perspective scale of space and of local depth cues as does processing three-dimensional scenes. Because so much additional information is available to show that picture surfaces also are flat, perceivers must have access to both a three-dimensional and two-dimensional reality to pictures. I have reviewed evidence that this is not always the case with young children who, because they are less sensitive to the flatness information in pictures, make errors when looking at pictures. It is also not the case with certain kinds of pictures which produce illusions of size or distance because the two realities are not balanced properly. The causes of these errors, especially of station point corrections, were discussed in some detail. The role of canonical features and pictorial convention was considered as providing means by which we can perceive the spatial arrangements and recognize the objects in incompletely rendered pictures. The relationship of Gestalt organizational principles to depth organizations and meaningful organizations was considered, with some discussion of why pictures and scenes are better organized than more abstract representations such as language.

All of these aspects were described in the context of three basic ideas: perceivers treat and organize all stimulation reaching their eyes as if it were reflected by a three-dimensional scene; perceivers have more difficulty perceiving pictures than scenes; and perceivers rarely, if ever, misperceive a natural scene, whereas they easily can a picture. All of my discussion has focused on the primacy of scene perception, and that we cannot properly understand picture perception except in that light. Pictures cannot be studied independently.

7. The Processing of Motion Pictures

Hochberg and Brooks (1978) list several perceptual characteristics of motion pictures that are absent when looking at still pictures. The most important of these are: (1) motion pictures provide motion-dependent information about the three-dimensional layout of a scene that is unavailable when looking at still pictures; (2) the perceived scene displayed by motion pictures can be very much larger than the screen, suggesting a spatial storage of visual information; and (3) the perceived scene can cover a much longer (or shorter) time frame than the actual duration of the presentation, suggesting a fluidity of temporal coding. To this list I add an obvious fourth characteristic: (4) viewing motion pictures creates the perception of movement.

When perception of apparent movement is investigated in the laboratory, using an alternating display of two simple forms, the rules that describe the processing seem straightforward. These are based on an analysis of stroboscopic movement, much of which dates back to Korte (1915), but with much added from relatively recent research reviewed by Hochberg and Brooks (1978). They suggest two general factors that operate in stroboscopic movement and therefore are also relevant to motion picture perception. Following a rapid change of view, a fast transient response of apparent movement occurs between regions of similar luminance that are in similar retinal locations across the successive views. They refer to this as an early and fast response to general "blobs" in the field of view, thus a response dependent only in low spatial frequencies, independent of the specific forms of the blobs undergoing perceived movement. At the same time, a slower form-dependent response begins as a result of the displacement of objects between successive views, but this takes longer to develop and it does take into account the overall shapes of objects and their finer identifying details. Hochberg and Brooks suggest that it is these slower responses that account for the apparent movement of objects in perceived space. Consequently, the slow and the fast responses have to produce the same perceptions or else the observer misaligns the objects across successive views. Such misalignment is usually not relevant in simple stroboscopic displays, containing as they do relatively few choices of objects, but it easily occurs in motion picture sequences when the camera motion results in a cut rather than a continuous displacement. Thus, we shall see that the explanation of continuous perception across discontinuous cuts poses severe problems for all theorists.

While there are still unsettled issues about stroboscopic motion, these problems seem trivial when compared to explaining how we perceive the movement in richly generated motion picture sequences. Consider just two observations. First, whatever the role played by nonvisual sources of information in the perception of real motion, these factors can have no role at all in perceiving the motion generated by a moving camera. The viewer provides no control inputs to the camera nor receives any control feedback from the camera regarding its movements. Everything knowable about those movements must be determined by the resulting visual information alone. Afferent, efferent, reafferent, or other compensatory signals do not exist. Yet perceivers are normally never confused about the pattern of motion of the camera nor about the spatial arrangements of objects in the scene.

Second, the scene displayed in motion pictures is seen on a TV or projection screen, a surface that is fixed in size and in location, both in absolute space and in relation to the viewer. The viewer's distance from the screen is normally fixed, and no matter how the sizes of objects and the orientation of surfaces displayed on the screen change, nothing in these changes can fool the perceiver into thinking he is moving or the screen is moving. As Hochberg and Brooks (1978, p. 278) say: ". . . the camera may point in different directions at different times, but the scenes are still all viewed only from straight ahead." So, knowing where our gaze is directed cannot be of much use in helping us perceive the motion and the scene generated by motion pictures.

There have been two considered attempts to explain how we perceive motion picture sequences. They represent the basic dichotomy I have already described for scenes and still pictures – a Gibsonian versus a Helmholtzian theoretical approach. Neither is fully adequate to the task, but I will briefly discuss the successes and failures of each. Gibson (e.g., 1954; see also Johansson, 1950; 1974) applies a similar analysis to sequences of discontinuous successive views generated by motion pictures to that of continuous sequences generated by eye and body motion. Thus, we respond to the successive views of the visual world that reach our eyes by differentiation and extraction of the invariant structure – that structure which mirrors the intrinsic properties of the stationary surfaces – as it undergoes continuous transformation in the pattern of light reaching the eyes. This model was developed to apply to viewing a stationary scene in which the discontinuous succession was created primarily by saccadic eye movements. It has also been extended to viewing a scene in which some objects are themselves in motion against a stable background of the rest of the scene. In both cases, the magnitude of each saccade in a sequence of saccades is usually quite small in relation to the full field of view, so that only a small displacement of each object occurs from one fixation to the next. The small leap applied here by Gibson is the suggestion that we also can and do extract invariants between successive views when the succession is created by sequences of discontinuous motion pictures. The leap may be reasonable when the succession is of substantially overlapping shots but seems on the surface to be unable to account for smooth perception across discontinuous cuts in which there is no overlap at all. It is difficult to see in the absence of overlap how an invariant under transformation can be defined without some referent to prior memory or future expectation (the basis of the Helmholtzian approach to be explored in a moment).

Hochberg, Brooks, and Roule (1977) point out another limitation of a transformational analysis, even for overlapping successions of scenes. They used visual mazes as stimuli and presented successive views in which they varied the amount of overlap from one view to the next. The viewer was asked to report which way the maze shifted from view to view – a task that required that each view be related to the next. With very small displacements, viewers were able to do this, but as soon as the displacement exceeded a few degrees, performance was at chance. What they needed was a landmark or prominent feature that could be used as an anchor. While Hochberg et al. agreed that the mathematical specification of the mazes is as rigorous and precise as for a natural scene and hence as useful to the extraction of invariants under translation, the absence of landmarks is sufficient to render each momentary view independent and hence unrelatable. In their argument, the perceiver needs landmarks to guide the integration; otherwise, the invariants are not evident or sufficient.

But it is with the nonoverlapping cuts between successive views that the Gibsonian approach has its greatest difficulty, and no one has yet been able to bridge this gap. The alternative approach, best represented by Hochberg and

Brooks (1978), adds to Gibson's ideas the cognitive factors that seemingly have to fill in for the missing perceptual information.

Hochberg (1968) has offered a simple demonstration. A viewer looks through a small peephole in which is visible only a small amount of a figure at a time. If the figure is a Maltese cross, all that can be seen is a single right-angle intersection between a pair of adjacent lines. The observer is shown a discrete succession of such views, as if each successive corner of the cross were placed under the peephole. There is no overlap between views. When viewed in this way, whether with long or short viewing times, the observer usually cannot determine what the form is (it could just as easily be two hands of a clock moving about erratically), cannot tell whether there is any coherence to the sequence or to the direction of movement, and is at a loss whenever a view is skipped or a shortcut is taken across part of the figure. These results all change if the observer is first shown the complete cross (or is told verbally in some detail about the shape of the cross). After this initial "establishing shot," a cross is perceived beneath the peephole (even though only a small corner of it is visually available at any one time), moving continuously, and any skipped views are perceived as shortcuts, not as confusion. For Hochberg, the establishing shot provides a perceptual schema or cognitive map of what it is that will be shown, so that the nonoverlapping views can be tied together.

Hochberg and Brooks (1978) suggest that a complete theory of motion picture perception requires three components: fast stimulus factors, slow stimulus factors, and slow cognitive factors. The fast stimulus factors are those that create apparent motion between the nearest objects that occupy adjacent position at successive times. The transients are short, probably on the order of 30-300 milliseconds, and occur only over quite local regions, probably on edges no more than 4° apart from one view to the next. Since they depend upon only low frequency information, these transients can produce apparent motion between objects that are only barely similar to each other. Such transients therefore create substantial misperception with large displacements between very rapidly presented sequence of views when there are many contours available for alignment.

The slow stimulus factors use the specific object shapes for determining the direction of apparent change in location but are available only with relatively long viewing times – nearer to 500 milliseconds or more per view. Hochberg and Brooks accept that a Gibsonian transformational model can be applied to these slower factors, so that direction and stability of the scene can be extracted from the invariants between overlapping views, but only if the notion of landmark features is added to the model.

The slow cognitive factors are needed in nonoverlapping cutting, in which successive views do not contain any object or landmark in common. There, the relative location of the successive views has to be determined by nonstimulus means. The viewer must be ready to assume that it is the same general spatial domain still being viewed and not a cut to some other completely different scene. Further, the viewer has to have some expectation of where in the scene the new shot is in relation to the previous one or have some overall knowledge about the content of the entire scene. As long as the viewer already has a specific schematic map in mind about the scene and is shown successive glimpses at an appropriate rate and order, only a very brief set of such glimpses is necessary (as long as the fast stimulus factors listed above are arranged so they do not interfere).

Hochberg and Brooks (1978) describe some research that bears on each of these three sets of factors, but they note that we know far too little about each of them as they apply to motion picture sequences of rich normal scenes. They make a convincing case that at least with nonoverlapping cutting, a purely stimulus approach is insufficient. Whether their demand for a schematic map addition is the solution is not yet obvious, though it is difficult for me to see what other alternatives can work.

More importantly, their critique of a Gibsonian processing approach for nonoverlapped views might be extended to all views, even those with substantial overlap and prominent landmarks. Do we only use and need cognitive maps for the extreme case of nonoverlapping cuts or do we always need them? The same question can be asked about the role of afferents and efferents. Since they are irrelevant to perception of motion picture sequences, maybe they are irrelevant to the perception of continuously viewed scenes as well. Hochberg and Brooks do not answer these questions, but future research will have to do so.

While there is a paucity of research on motion picture sequences of normal rich scenes, there is even less information available on how closed loop visual simulators work. That they work is not in question. I have described elsewhere (Haber, 1983b) the major properties of visual flight simulators and how they are used.

The important difference between the appearance of a screen on which is projected a sequence of motion pictures and a screen displaying a simulated flight (or automobile ride) is that in the latter, there is no perceived camera as distinct from the viewer himself. It is the viewer who feels and sees himself moving though the scene, not a camera that is merely projecting an image of the scene on the screen. It is a nearly complete destruction of the two-dimensional reality and as complete and overpowering a control of the three-dimensional reality as is possible with a flat two-dimensional display surface.

Not all visual flight simulators are the same, with most of the differences occurring in the fidelity of the visual information projected on the screen. In theory, a perfect simulator displays exactly that pattern of stimulation to the viewer that would have been seen if the viewer were actually moving through the scene being depicted. Thus, nearby objects have coarser texture and move across the screen faster than more distant ones. Further, with attached head position sensors, movement of the viewer's head displaces objects on the screen as would occur with motion parallax. In addition, the image on the screen seen with the left eye differs from that seen by the right eye for those parts of the scene near enough to produce binocular disparity. This is in theory, except that there are no simulators that do all of these things. For most, both eyes see the same view, even for very nearby scenes, and none at present shift parts of the scene to compensate for changes in the viewer's head position. But even without stereopsis and motion parallax, the texture and relative motion changes provide a powerful sense of self-motion through the scene being displayed.

Commercial airlines and equipment manufacturers, and especially the United States military research laboratories, are presently engaged in substantial research programs to improve the fidelity of visual information available in simulators, including the development of binocular displays and ones capable of representing motion parallax effects. Most of this work is directly oriented toward improving the displays, rather than trying to work out theoretically the ways in which such displays exceed the perceptual effects of ordinary motion pictures. Here applied work is fast outstripping its theoretical underpinnings.

8. Conclusion

I shall not attempt a summary, as this paper has been a very eclectic treatment of the perception of space portrayed in natural scenes, in still pictures, and in motion pictures. The major novelty in this presentation, other than perhaps scope, is the explicit separation and description of stimulus information available to the eye from that of the processing of that information. I have argued throughout the paper that a failure to adequately describe the stimulus, and to separate stimulus variables from processing variables, has hindered almost all of the theoretical attempts to explain space perception.

I have tried to be even-handed in my presentation, though I am sure some of my biases are evident. On the stimulus analysis side, I have stressed the full retinal scales over local depth cues, contrary to most conventional treatments. On the processing side, I have been unable to provide a neat structure because the different theories of processing are stated in such disparate terms. Consequently, while I have given as many examples of models and theories as I could, I may have left out a few. I have not tried to resolve the major theoretical battles that still underlie most of the processing discussion, though I could not refrain occasionally from noting what I think are major advantages owing to one side or the other. My preference has always been toward models or theories that have testable properties, as have most of those I presented. Hopefully, this should reduce the number of theories that will need discussion when someone comes to write this review again.

References

Barlow, H. B. Summation and inhibition in the frog's retina. *Journal of Physiology,* 1953, **119**, 69-88.

Barrow, H. G., & Tenenbaum, J. M. Recovering intrinsic scene characteristics from images. In A. R. Hanson & E. M. Riseman (Eds.), *Computer Vision Systems.* New York: Academic Press, 1978.

Beck, J. *Surface Color Perception.* Ithaca, NY: Cornell University Press, 1972.

Benson, C. W., & Yonas, A. Development of sensitivity to static pictorial depth information. *Perception and Psychophysics,* 1973, **13**, 361-366.

Berbaum, K., Tharp, D., & Mroczek, K. Surfaces depicted with and without illumination: Looking for canonical depth in Pandora's box. *Perception,* 1983, in press.

Biederman, I. Perceiving real world scenes. *Science,* 1972, **177**, 77-79.

Biederman, I. On processing information from a glance at a scene: Some implications for a syntax and semantics of visual processing. In S. Treu (Ed.), *User-Oriented Design of Interactive Graphic Systems.* New York: Association for Computing Machinery, 1977.

Blake, R., & Camisa, J. Is binocular vision always monocular? *Science,* 1978, **200**, 1497-1499.

Boynton, R. M. Implications of the minimally distinct border. *Journal of the Optical Society of America,* 1973, **63**, 1037-1043.

Boynton, R. M. The visual system: Environmental information. In E. C. Carterette & H. P. Friedman (Eds.), *Handbook of Perception I: Historical and Philosophical Roots of Perception.* New York: Academic Press, 1974.

Braunstein, D. R. *Depth Through Motion.* New York: Academic Press, 1975.

Campion, J., Latto, R., & Smith, Y. M. Is blindsight due to scattered light, spared cortex and near-threshold effects? *Behavioral and Brain Sciences,* 1983, **6**, in press.

Cooper, R. The development of recursion in pictorial perception. Presented at Conference on Picture Perception, Center for Research on Human Learning, University of Minnesota, 1975.

Coren, S. Subjective contours and apparent depth. *Psychological Review,* 1972, **79**, 339-367.

Coren, S., & Girgus, J. S. Illusions and constancies. In W. Epstein (Ed.), *Stability and Constancy in Visual Perception.* New York: Wiley, 1977.

Coren, S., & Girgus, J. S. *Seeing is Deceiving: The Perception of Visual Illusions.* Hillsdale, NJ: Erlbaum, 1978.

Crick, F. H. C., Marr, D. C., & Poggio, T. An information processing approach to understanding the visual cortex. In F. O. Schmitt (Ed.), *The Cortex.* Cambridge, MA: MIT Press, 1981.

Deregowski, J. B. Pictorial perception and culture. *Scientific American,* 1972, **227** (5), 82-88.

Dichgans, J., & Brandt, T. Visual-vestibular interaction: Effects of self-motion perception and postural control. In R. Held, H. W. Leibowitz & H. L. Teuber (Eds.), *Handbook of Sensory Physiology VIII: Perception.* Berlin: Springer, 1978.

Ebenholtz, S. The constancies in object orientation: An algorithm processing approach. In W. Epstein (Ed.), *Stability and Constancy in Visual Perception.* New York: Wiley, 1977.

Epstein, W. The process of "taking into account" in visual perception. *Perception,* 1973, **2**, 267-285.

Epstein, W. (Ed.). *Stability and Constancy in Visual Perception.* New York: Wiley, 1977.

Epstein, W., & Parks, J. Shape constancy: Functional relationships and theoretical formulations. *Psychological Bulletin,* 1963, **60**, 265-288.

Farber, J., & Rosinski, R. R. Geometric transformation of pictured space. *Perception,* 1978, **7**, 269-282.

Fitch, H. L., & Turvey, M. T. On the control of activity: Some remarks from an ecological point of view. In D. M. Landers & R. W. Christina (Eds.), *Psychology of Motion Behavior and Sport.* Champaign, IL: Human Kinetics, 1977.

Freedman, J., & Haber, R. N. Why we rarely forget a face: The role of organization in perceptual memory. *Bulletin of the Psychonomic Society,* 1974, **3**, 107-109.

Friedman, S. L., & Stevenson, M. B. Developmental changes in the understanding of implied motion in two-dimensional pictures. *Child Development,* 1975, **46**, 773-778.

Friedman, S. L., & Stevenson, M. Perception of movement in pictures. In M. A. Hagen (Ed.), *The Perception of Pictures I: Alberti's Window: The Projective Model of Pictorial Information.* New York: Academic Press, 1980.

Gibson, J. J. *The Perception of the Visual World.* Boston, MA: Houghton-Mifflin, 1950.

Gibson, J. J. The visual perception of objective motion and subjective motion. *Psychological Review,* 1954, **61**, 303-314.

Gibson, J. J. *The Senses Considered as Perceptual Systems.* Boston, MA: Houghton-Mifflin, 1966.

Gibson, J. J. *The Ecological Approach to Visual Perception.* Boston, MA: Houghton-Mifflin, 1979.

Gilchrist, A. L. Perceived lightness depends on perceived spatial arrangement. *Science,* 1977, **195**, 185-187.

Gilchrist, A. L. Color perception without contrast. Presented at the Lake Ontario Visionary Establishment (LOVE), 1978.

Ginsburg, A. P. Is the illusory triangle physical or imaginary? *Nature,* 1975, **257**, 219-220.

Gogel, W. C. The metrics of visual space. In W. Epstein (Ed.), *Stability and Constancy in Visual Perception.* New York: Wiley, 1977.

Goldman, M., & Hagen, M. A. The forms of caricature: Physiognomy and political bias. *Studies in the Anthropology of Visual Communication,* 1978, **5**, 30-36.

Grayson, C. *L. B. Alberti, On Painting and On Sculpture.* (Translation from the Latin.) London: Phaidon Press, 1972.

Gregory, R. L. Visual illusions. In B. Foss (Ed.), *New Horizons in Psychology.* Baltimore, MD: Penguin, 1966.

Gregory, R. L. *The Intelligent Eye.* London: Weidenfeld and Nicholson, 1970.

Gregory, R. L. *Eye and Brain* (second edition). New York: McGraw-Hill, 1973.

Guzman, A. Computer recognition of three-dimensional objects in a visual scene. Doctoral dissertation, MIT, 1968.

Gyr, J. W. Is a theory of direct visual perception adequate? *Psychological Bulletin,* 1972, **77**, 246-261.

Gyr, J., Willey, R., & Henry, A. Motor-sensory feedback and geometry of visual space: A replication. *Behavioral and Brain Sciences,* 1979, **2**, 59-64.

Haber, R. N. Visual perception. In M. R. Rosenzweig (Ed.), *Annual Review of Psychology,* 1978, **29**, 31-59.

Haber, R. N. Leonardo: Our first theorist on the perception of pictures. In D. F. Fisher & C. Nodine (Eds.), *What is a Painting?* New York: Praeger Press, 1979. (a)

Haber, R. N. When is sensory-motor information necessary, when only useful, and when superfluous? *Behavioral and Brain Sciences,* 1979, **2**, 68-70. (b)

Haber, R. N. How we perceive depth from flat pictures. *American Scientist,* 1980, **68**, 370-380.

Haber, R. N. The power of visual perceiving. *Journal of Mental Imagery,* 1981, **5**, 1-16. (a)

Haber, R. N. Visualizing and remembering. *Journal of Mental Imagery,* 1981, **5**, 32-40. (b)

Haber, R. N. One hundred years of visual perceptions: An ahistorical perspective. In S. Koch & D. E. Leary (Eds.), *A Century of Psychology as Science: Retrospectives and Assessments.* New York: McGraw-Hill, 1983, in press. (a)

Haber, R. N. The simulation of high speed aircraft flight. *Scientific American,* 1983, **248**, in press. (b)

Haber, R. N. The two visual system hypothesis loses a supporter. *Behavioral and Brain Sciences,* 1983, **6**, in press. (c)

Haber, R. N., & Hershenson, M. *The Psychology of Visual Perception* (second edition). New York: Holt, Rinehart, and Winston, 1980.

Haber, R. N., & Wilkinson, L. The perceptual components of computer graphic displays. *Computer Graphics and Applications,* 1982, **2** (3), 23-25.

Hagen, M. A. The development of sensitivity to cast and attached shadows in pictures as information for the direction of the source of illumination. *Perception and Psychophysics,* 1976, **20**, 25-28. (a)

Hagen, M. A. Influence of picture surface and station point on the ability to compensate for oblique view in pictorial perception. *Developmental Psychology,* 1976, **12**, 57-63. (b)

Hagen, M. A. (Ed.), *The Perception of Pictures I: Alberti's Window: The Projective Model of Pictorial Information.* New York: Academic Press, 1980. (a)

Hagen, M. A. (Ed.), *The Perception of Pictures II: Dürer's Devices: Beyond the Projective Model of Pictures.* New York: Academic Press, 1980 (b).

Hagen, M. A., Glick, R., & Morse, B. The role of two-dimensional surface characteristics in pictured depth perception. *Perceptual and Motor Skills,* 1978, **46**, 875-881.

Hagen, M. A., & Jones, R. K. Cultural effects in pictorial perception: How many words is one picture really worth? In R. Walk & H. Pick (Eds.), *Perception and Experience.* New York: Plenum Press, 1978.

Hanson, A. R., & Riseman, E. M. (Eds.), *Computer Vision Systems.* New York: Academic Press, 1978.

Harris, C. S. Perceptual adaptation to inverted, reversed, and displaced vision. *Psychological Review,* 1965, **72**, 419-444.

Held, R. Two modes of processing spatially distributed visual stimulation. In F. O. Schmidt (Ed.), *The Neurosciences: Second Study Program.* New York: Rockefeller University Press, 1970.

Helmholtz, H. von. *Handbook of Physiological Optics.* New York: Optical Society of America, 1925.

Hershenson, M. Moon illusion and spiral aftereffect: Illusion cue to the loom-zoom system. Unpublished paper, 1982.

Hochberg, J. E. The psychophysics of pictorial perception. *Audiovisual Communication Review,* 1962, **10**, 22-54.

Hochberg, J. E. In the mind's eye. In R. N. Haber (Ed.), *Contemporary Research and Tneory in Visual Perception.* New York: Holt, Rinehart and Winston, 1968.

Hochberg, J. E. Perception: I. Color and Shape. In J. Kling & L. A. Riggs (Eds.), *Handbook of Experimental Psychology.* New York: Holt, Rinehart and Winston, 1971.

Hochberg, J. E. The representation of things and people. In E. H. Gombrich, J. Hochberg & M. Black (Eds.), *Art, Perception, and Reality.* Baltimore, MD: Johns Hopkins University Press, 1972.

Hochberg, J. E. Art and Perception. In E. C. Carterette & M. P. Friedman (Eds.), *Handbook of Perception X: Perceptual Ecology.* New York: Academic Press, 1978.

Hochberg, J. E. Some of the things that paintings are. In D. F. Fisher & C. Nodine (Eds.), *What is a Painting?* New York: Praeger Press, 1979.

Hochberg, J. E. Levels of perceptual organization. In M. Kubovy & J. R. Pomerantz (Eds.), *Perceptual Organization,* Hillsdale, NJ: Erlbaum, 1981.

Hochberg, J. E., & Beck, J. Apparent spatial arrangement and perceived brightness. *Journal of Experimental Psychology,* 1954, **47**, 263-266.

Hochberg, J. E., & Brooks, V. Pictorial recognition as an unlearned ability. A study of one child's performance. *American Journal of Psychology,* 1962, **75**, 624-628.

Hochberg. J. E., & Brooks, V. The perception of motion pictures. In E. C. Carterette & M. P. Friedman (Eds.), *Handbook of Perception X: Perceptual Ecology.* New York: Academic Press, 1978.

Hochberg, J. E., Brooks, V., & Roule, P. Movies of mazes and wallpaper (abstract). Proceedings, Eastern Psychological Association, 1977, 179.

Hudson, W. Pictorial depth perception in subcultural groups in Africa. *Journal of Social Psychology,* 1960, **52**, 183-208.

Jaynes, J. *The Origin of Consciousness in the Breakdown of the Bicameral Mind.* Boston, MA: Houghton-Mifflin, 1976.

Johansson, G. *Configurations in Event Perception.* Uppsala, Sweden: Almquist and Wiksell, 1950.

Johansson, G. Spatio-temporal Differentiation and Integration in Visual Motion Perception. Report No. 160, Department of Psychology, Uppsala University, Sweden, 1974.

Jonides, J., Irwin, D. E., & Yantis, S. Integrating visual information from successive fixations. *Science,* 1982, **215**, 192-194.

Julesz, B. *Foundations of Cyclopean Perception.* Chicago, IL: University of Chicago Press, 1971.

Kanizsa, G. Contours without gradients or cognitive contours? *Italian Journal of Psychology,* 1974, **1**, 93-113.

Kanizsa, G. Subjective contours. *Scientific American,* 1976, **234** (4), 48-52.

Kaufman, L. *Sight and Mind: An Introduction to Visual Perception.* New York: Oxford University Press, 1975.

Kennedy, J. M. *A Psychology of Picture Perception: Information and Images.* San Francisco, CA: Jossey-Bass, 1974.

Kennedy, J. M. Picture perception and phantom contours. In D. F. Fisher & C. Nodine (Eds.), *What is a Painting?* New York: Praeger Press, 1979.

Koenderink, J. J. & Van Doorn, A. J. Local structure and movement parallax of the plane. *Journal of the Optical Society of America,* 1976, **66**, 717-723.

Kolers, P. A., & von Grunaw, M. Shape and color in apparent motion. *Vision Research,* 1976, **16**, 329-335.

Korte, A. Kinematoskopische Untersuchungen. *Zeitschrift für Psychologie,* 1915, **72**, 193-296.

Kuffler, S. W. Discharge patterns and functional organization of mammalian retinas. *Journal of Neurophysiology,* 1953, **16**, 37-68.

Lackner, J. R. Adaptation to visual and proprioceptive rearrangement: Origins of the differential effectiveness of active and passive movements. *Perception and Psychophysics,* 1977, **21**, 55-59.

Lee, D. N. Visual information during locomotion. In R. E. MacLeod & H. L. Pick, Jr. (Eds.), *Perception: Essays in Honor of James J. Gibson.* Ithaca, NY: Cornell University Press, 1974.

Lee, D. N. A theory of visual control of braking based upon information about time-to-collision. *Perception,* 1976, **5**, 437-459.

Lee, D. N., & Lishman, R. Visual control of locomotion. *Scandinavian Journal of Psychology,* 1977, **18**, 224-230.

Leibowitz, H. W., Rodemer, G. S., & Dichgans, J. The independence of dynamic spatial orientation from dominance and refractive error. *Perception and Psychophysics*, 1979, **25**, 75-79.

MacKay, D. M. Visual stability and voluntary eye movements. In R. Jung (Ed.), *Handbook of Sensory Physiology VIII (3)*. Berlin: Springer, 1973.

Marr, D. C. Representing visual information. *Lectures in the Life Sciences*, 1978, **10**, 108-180.

Marr, D. C. *Vision*. San Francisco, CA: Freeman, 1982.

Marr, D. C., & Hildreth, E. Theory of edge detection. *Proceedings of the Royal Society, London*, 1980, **B207**, 187-217.

Marr, D. C., & Poggio, T. A theory of human stereo vision. *Proceedings of the Royal Society, London*, 1979, **B204**, 301-328.

Matin, L. Saccades and extraretinal signals for visual direction. In R. A. Monty & J. W. Senders (Eds.), *Eye Movements and Psychological Processes*. Hillsdale, NJ: Erlbaum, 1976.

Melvill-Jones, G. The vestibular system for eye movement control. In R. A. Monty & J. W. Senders (Eds.), *Eye Movements and Psychological Processes*. Hillsdale, NJ: Erlbaum, 1976.

Mooney, C. M. Age in the development of closure ability in children. *Canadian Journal of Psychology*, 1954, **11**, 219-226.

Nakayama, K., & Loomis, J. M. Optical velocity patterns, velocity-sensitive neurons, and space perception: A hypothesis. *Perception*, 1974, **3**, 63- 80.

Ogle, K. N. The optical space sense. In H. Davson (Ed.), *The Eye* (vol. 4). New York: Academic Press, 1962.

Owen, D. H., & Warren, R. The reciprocity of perception and action: Actions as perceptual reports and variables of stimulation as measures of performance. In preparation.

Panofksy, E. *Albrecht Dürer* (vol. 1). Princeton, NJ: Princeton University Press, 1948.

Perkins, D. N. A definition of caricature and recognition. *Studies in the Anthropology of Visual Communication*, 1975, **2**, 1-20.

Piaget, J., & Inhelder, B. *The Child's Conception of Space*. New York: Norton, 1967.

Pirenne, M. H. *Optics, Painting, and Photography*. New York: Cambridge University Press, 1970.

Ratliff, F. *Mach Bands: Quantitative Studies on Neural Networks in the Retina*. San Francisco, CA: Holden-Day, 1965.

Regan, D. M. Visual information channeling in normal and disordered vision. *Psychological Review*, 1982, **89**, 407-444.

Regan, D., Beverly, K. I., & Cynader, M. The visual perception of motion in depth. *Scientific American,* 1979, **241** (1), 136-151.

Richards, W. Visual space perception. In E. C. Carterette & H. P. Friedman (Eds.), *Handbook of Perception V: Seeing.* New York: Academic Press, 1975.

Richter, J. P. *The Notebooks of Leonardo da Vinci* (vol. 1). New York: Dover, 1970.

Riggs, L. A. Saccadic suppression of phosphenes: Proof of a neural basis for saccadic suppression. In R. A. Monty & J. W. Senders (Eds.), *Eye Movements and Psychological Processes.* Hillsdale, NJ: Erlbaum, 1976.

Robson, J. G. Receptive fields: Neural representation of spatial and intensive attributes of the visual image. In E. C. Carterette & M. P. Friedman (Eds.), *Handbook of Perception V: Seeing.* New York: Academic Press, 1975.

Rock, I. *An Introduction to Perception.* New York: Macmillan, 1975.

Rock, I. In defense of unconscious inference. In W. Epstein (Ed.), *Stability and Constancy in Visual Perception.* New York: Wiley, 1977.

Rock, I., & Kaufman, L. The moon illusion II. *Science,* 1962, **136**, 1023-1031.

Roscoe, S. N. Visual judgments of size and distance. In S. N. Roscoe, *Aviation Psychology.* Ames, IA: Iowa State University Press, 1980.

Rosinski, R. R., & Farber, J. Compensations for viewing point in the perception of pictured space. In M. A. Hagen (Ed.), *The Perception of Pictures I: Alberti's Window: The Projective Model of Pictorial Information.* New York: Academic Press, 1980.

Runeson, S. On the possibility of "smart" perceptual mechanisms. *Scandinavian Journal of Psychology,* 1977, **18**, 172-179.

Ryan, T. A., & Schwartz, C. Speed of perception as a function of mode of presentation. *American Journal of Psychology,* 1956, **69**, 60-69.

Schlosberg, J. Stereoscopic depth from single pictures. *American Journal of Psychology,* 1941, **54**, 601-605.

Schneider, G. E. Two visual systems. *Science,* 1969, **163**, 895-902.

Sekuler, R., Pantle, A., & Levinson, E. Physiological basis for motion perception. In R. Held, H. W. Leibowitz & H. L. Teuber (Eds.), *Handbook of Sensory Physiology VIII: Perception.* Berlin: Springer, 1978.

Shepard, R. N. Recognition memory of words, sentences, and pictures. *Journal of Verbal Learning and Verbal Behavior,* 1967, **6**, 156-163.

Standing, L. S. Learning 10,000 pictures. *Quarterly Journal of Experimental Psychology,* 1973, **25**, 207-222.

Standing, L., Conezio, J., & Haber, R. N. Perception and memory for pictures: Single trial learning of 2500 visual stimuli. *Psychonomic Science,* 1970, **19**, 73-74.

Stevens, K. A. The information content of texture gradients. *Biological Cybernetics,* 1980, **42**, 95-105.

Stevens, K. A. The visual interpretation of surface contours. *Artificial Intelligence,* 1981, **17**, 47-73.

Thiery, A. Ueber geometrisch-optische Tauschungen. *Philosophische Studien,* 1896, **12**, 67-126.

Thomas, J. P. Spatial resolution and spatial interaction. In E. C. Carterette & M. P. Friedman (Eds.), *Handbook of Perception V: Seeing.* New York: Academic Press, 1975.

Ullman, S. *The Interpretation of Visual Motion.* Cambridge, MA: MIT Press, 1979.

Ullman, S. Against direct perception. *Behavioral and Brain Sciences,* 1980, **3**, 373-415.

Volkmann, F. C. Saccadic suppression: A brief review. In R. A. Monty & J. W. Senders (Eds.), *Eye Movements and Psychological Processes.* Hillsdale, NJ: Erlbaum, 1976.

Wallach, H. Brightness constancy and the nature of achromatic colors. *Journal of Experimental Psychology,* 1948, **38**, 310-324.

Waltz, D. Understanding line drawings of scenes with shadows. In P. H. Winston (Ed.), *The Psychology of Computer Vision.* New York: McGraw-Hill, 1975.

Ware, C., & Kennedy, J. M. Illusory line linking solid rods. *Perception,* 1977, **6**, 601-602.

Welsh, R. B. *Perceptual Modification: Adapting to Altered Sensory Environments.* New York: Academic Press, 1978.

Winston, P. H. *The Psychology of Computer Vision.* New York: McGraw-Hill, 1975.

Winston, P. H. *Artificial Intelligence.* Reading, MA: Addision-Wesley, 1977.

Worrall, N. A test of Gregory's theory of primary constancy scaling. *American Journal of Psychology,* 1974, **34**, 505-510.

Yonas, A., & Hagen, M. A. Effects of static and kinetic depth information on the perception of size by children and adults. *Journal of Experimental Child Psychology,* 1973, **15**, 254-265.

Mapping Image Properties into Shape Constraints: Skewed Symmetry, Affine-Transformable Patterns, and the Shape-from-Texture Paradigm

Takeo Kanade
John R. Kender

Carnegie-Mellon University
Pittsburgh, Pennsylvania

Abstract

In this paper we demonstrate two new approaches to deriving three-dimensional surface orientation information ("shape") from two-dimensional image cues. The two approaches are the method of affine-transformable patterns and the shape-from-texture paradigm. They are introduced by a specific application common to both: the concept of skewed symmetry. Skewed symmetry is shown to constrain the relationship of observed distortions in a known object regularity to a small subset of possible underlying surface orientations. Besides this constraint, valuable in its own right, the two methods are shown to generate other surface constraints as well. Some applications are presented of skewed symmetry to line drawing analysis, to the use of gravity in shape understanding, and to global shape recovery.

1. Introduction

Certain image properties, such as parallelisms, symmetries, and repeated patterns, provide cues for perceiving 3-D shape from a 2-D picture. This paper demonstrates how we can map these image properties into 3-D shape constraints by associating appropriate assumptions with them and by using appropriate computational and representational tools.

We begin with the exploration of how one specific image property, "skewed symmetry," can be defined and formulated to serve as a cue to the determination of surface orientations. Then we will discuss the issue from two new, broader viewpoints. One is the class of affine-transformable patterns. It has various interesting properties, and includes skewed symmetry as a special case. The other is the computational paradigm of shape-from-texture. Skewed symmetry is derived in a second, independent way, as an instance of the application of the paradigm. Also, it is proven that the same skewed-symmetry constraint can arise from greatly different image conditions.

This paper further claims that the ideas and techniques presented here are applicable to many other properties under a general framework of the shape-from-texture paradigm with the underlying meta-heuristic of non-accidental image properties.

2. Skewed Symmetry

In this section we assume the standard orthographic projection from scene to image, and a knowledge of the gradient space (see Mackworth, 1973).

2.1. Definition, Assumption and Constraints

Symmetry in a 2-D picture has an axis for which the opposite sides are reflective; in other words, the symmetrical properties are found along the transverse lines perpendicular to the symmetry axis. The concept *skewed symmetry* was introduced by Kanade (1979) by relaxing this condition a little. It means a class of 2-D shapes in which the symmetry is found along lines not necessarily perpendicular to the axis, but at a fixed angle to it. Formally, such shapes can be defined as 2-D affine transforms of real symmetries. Figures 1a-c show a few examples.[1]

Stevens (1980) presents a number of psychological experiments which suggest that human observers can perceive surface orientations from figures with this property. This is probably because such qualitative symmetry in the image is often due to real symmetry in the scene. Thus let us associate the following assumption with this image property:

A skewed symmetry depicts a real symmetry viewed from some unknown view angle.

Note that the converse of this assumption is always true under orthographic projection.

We can transform this assumption into constraints in the gradient space. As shown in Figure 1, a skewed symmetry defines two directions; let us call them the skewed-symmetry axis and the skewed-transverse axis, and denote

---

[1]The mouse hole example of Figure 1c is due to K. Stevens (1980).

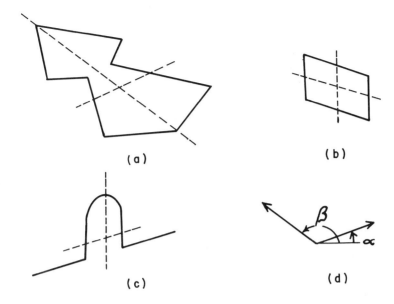

Figure 1. Skewed symmetry.

their directional angles in the picture by α and β, respectively (Figure 1d). Let $G = (p, q)$ be the gradient of the plane which includes the skewed symmetry. The 3-D vectors on the plane corresponding to the directions α and β are

$$(\cos\alpha, \sin\alpha, -p\cos\alpha - q\sin\alpha) \text{ and } (\cos\beta, \sin\beta, -p\cos\beta - q\sin\beta).$$

The assumption demands that these two vectors be perpendicular; their inner product vanishes:

$$\cos(\alpha - \beta) + (p\cos\alpha + q\sin\alpha)(p\cos\beta + q\sin\beta) = 0. \tag{1}$$

By rotating the p-q coordinates into the p'-q' coordinates so that the new p'-q' axes are the bisectors of the angle made by the skewed symmetry and skewed-transverse axes, it is easy to show that

$$p'^2\cos^2(\frac{\alpha-\beta}{2}) - q'^2\sin^2(\frac{\alpha-\beta}{2}) = -\cos(\alpha - \beta) \tag{2}$$

where

$$p' = p\cos\lambda + q\sin\lambda$$
$$q' = -p\sin\lambda + q\cos\lambda$$
$$\lambda = (\alpha + \beta)/2.$$

Thus, the (p, q)'s are on the hyperbola shown in Figure 2. That is, the skewed symmetry defined by α and β in the picture can be a projection of a real symmetry if and only if the gradient is on this hyperbola. The skewed symmetry thus imposes a one-dimensional family of constraints on the underlying surface orientation (p,q). As we will see in Section 5, other constraints can be exploited for the unique determination of surface orientation.

The tips or vertices G_T and G_T' of the hyperbola represent special orientations with interesting properties. First, since they are closest to the origin of the gradient space, and since the distance from the origin to a gradient represents the magnitude of the surface slant, G_T and G_T' correspond to the least slanted orientations that can produce the skewed symmetry in the picture from a real symmetry in the scene.

Second, since they are on the line (the axis of the hyperbola) which bisects the obtuse angle made by α and β, they correspond to the orientations for which the rates of depth change along the directions of α and β in the picture are the same. In other words, the apparent ratio of length to width of the object in the picture represents the real ratio in the scene (see Kanade [1979] for the proof.)

2.2. Rationale and Justification

Skewed symmetry has straightforward applications to scenes containing objects that have been manufactured, whether naturally or artificially. Many constructed items exhibit symmetry, occasionally about many axes.

Some symmetries are introduced due to economies of the manufacturing process: an object is often composed of identically formed component parts (fibers, cells, bricks, etc.). The symmetries result from the three-dimensional tessellation of the components into the whole. Often the tessellation is effectively two-dimensional, in laminae (cloth, honeycombs, walls, etc.), and the application of the skewed symmetry method is straightforward. Further, the requirement for a close symmetric packing of the components occasionally imposes a local symmetry on the individual components, too. The method can then be applied to individual parts (such as the bricks themselves). Notice the method does *not* assume 3-D symmetry of the whole object; what is assumed is *local* 2-D symmetry.

A further source of symmetry is the bilateral symmetry that results from biological manufacture (growth). It not only contributes symmetric objects to the environment; it may also be responsible for an imitative esthetic bias in human manufacture. If the extent of a bilaterally symmetric pattern into the third dimension is not too great (a face, a leaf, an airplane), the skewed symmetry method can be approximately applied also.

3. Affine-Transformable Patterns

In texture analysis we often consider small patterns (texels=texture elements) whose repetition constitutes "texture." Suppose we have a pair of texel

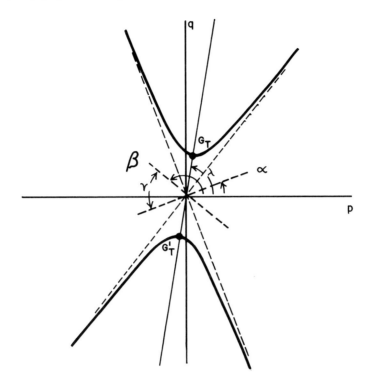

Figure 2. The hyperbola determined by a skewed symmetry defined by α and β.

patterns in which one is a 2-D affine transform of the other; we call them a pair of *affine-transformable* patterns. Let us assume:

> A pair of affine-transformable patterns in the picture are projections of similar patterns in the 3-D space (i.e., they can be overlapped by scale change, rotation, and translation).

Note that, as in the case of skewed symmetry, the converse of this assumption is always true under orthographic projection. The above assumption can be schematized by Figure 3. Consider two texel patterns P_1 and P_2 in the picture, and place the origins of the x-y coordinates at their centers, respectively. The transform from P_2 to P_1 can be now expressed by a regular 2×2 matrix $A = (a_{ij})$. P_1 and P_2 are projections of patterns P_1' and P_2' which are drawn on the 3-D surfaces. We assume that P_1' and P_2' are small enough so that we can regard them as being drawn on small planes. Let us denote the gradients of those small planes by $G_1 = (p_1, q_1)$ and $G_2 = (p_2, q_2)$, respectively; i.e., P_1' is drawn on a plane $-z = p_1 x + q_1 y$ and P_2' on $-z = p_2 x + q_2 y$.

Now, our assumption amounts to saying that P_1' is transformable from

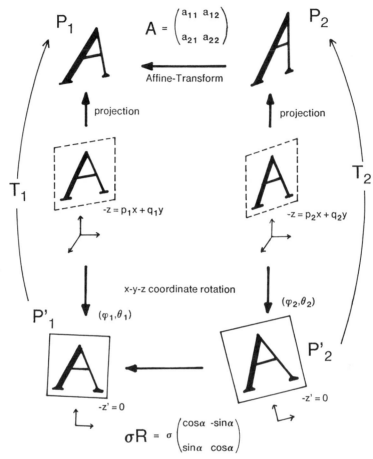

Figure 3. A schematic diagram showing the assumption about the affine-transformable patterns.

P_2' by a scalar scale factor σ and a rotation matrix:

$$R = \begin{pmatrix} \cos\alpha & -\sin\alpha \\ \sin\alpha & \cos\alpha \end{pmatrix}.$$

(We can omit the translation from our consideration, since for each pattern the origin of the coordinates is placed at its gravity center, which is preserved under the affine transform.) Thinking about a pattern drawn on a small plane, $-z = px + qy$, is equivalent to viewing the pattern from directly overhead; that is, rotating the x-y-z coordinates so that the normal vector of the plane is along the new z-axis (line of sight). For this purpose we rotate the coordinates first by ϕ around the y-axis and then by θ around the x'-axis (Figure 4). We have

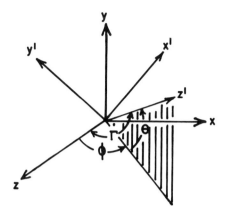

Figure 4. Rotation of the x-y-z coordinates.

the following relations among ϕ, θ, p, and q:

$$\sin\phi = p/\sqrt{p^2+1}, \qquad \cos\phi = 1/\sqrt{p^2+1},$$
$$\sin\theta = q/\sqrt{p^2+q^2+1}, \qquad \cos\theta = \sqrt{p^2+1}/\sqrt{p^2+q^2+1}. \tag{3}$$

Further, let Γ denote the angle of slant of the pattern, i.e., the angle between the old and the new z axes. Then

$$\cos\Gamma = 1/\sqrt{p^2+q^2+1}. \tag{4}$$

The plane which was represented as $-z = px + qy$ in the old coordinates is, of course, now represented as $-z' = 0$ in the new coordinates.

Let us denote the angles of the coordinate rotations to obtain P_1' and P_2' in Figure 3 by (ϕ_1,θ_1) and (ϕ_2,θ_2), respectively. The 2-D mapping from P_i' (x'-y' plane) to P_i (x-y plane) can be conveniently represented by the following 2×2 matrix T_i which is actually a submatrix of the usual 3-D rotation matrix:

$$T_i = \begin{pmatrix} \cos\phi & -\sin\phi\sin\theta \\ 0 & \cos\theta \end{pmatrix}.$$

Now, in order for the schematic diagram of Figure 3 to hold, what relationships have to be satisfied among the matrix $A = (a_{ij})$, the gradients $G_i = (p_i, q_i)$ for $i = 1,2$, the angles (ϕ_i, θ_i) for $i = 1,2$, the scale factor σ, and the matrix R? We equate the two transforms that start from P_2' to reach P_1: one following the diagram counter-clockwise, $P_2' \rightarrow P_2 \rightarrow P_1$, and the other clockwise, $P_2' \rightarrow P_1' \rightarrow P_1$. We obtain

$$AT_2 = T_1\sigma R.$$

That is,

$$
\begin{aligned}
a_{11}\cos\phi_2 &= \sigma(\cos\alpha\cos\phi_1 - \sin\alpha\sin\phi_1\sin\theta_1) \\
a_{12}\cos\theta_2 - a_{11}\sin\phi_2\sin\theta_2 &= -\sigma(\sin\alpha\cos\phi_1 + \cos\alpha\sin\phi_1\sin\theta_1) \\
a_{21}\cos\phi_2 &= \sigma\sin\alpha\cos\theta_1 \\
a_{22}\cos\theta_2 - a_{21}\sin\phi_2\sin\theta_2 &= \sigma\cos\alpha\cos\theta_1.
\end{aligned}
\tag{5}
$$

By eliminating σ and α and substituting for $\sin\phi_i$, $\cos\phi_i$, $\sin\theta_i$, and $\cos\theta_i$ from (3), we have the following equations in p_1, q_1, p_2, and q_2:

$$
\sqrt{p_2^2 + q_2^2 + 1}\,(a_{11}(p_1^2 + 1) + a_{21}p_1 q_1)
$$
$$
= \sqrt{p_1^2 + q_1^2 + 1}\,(a_{22}(p_2^2 + 1) - a_{21}p_2 q_2)
\tag{6}
$$

$$
(-a_{12}(p_2^2 + 1) + a_{11}p_2 q_2)(p_1^2 + 1) - (a_{22}(p_2^2 + 1) - a_{21}p_2 q_2)p_1 q_1
$$
$$
= a_{21}\sqrt{p_1^2 + q_1^2 + 1}\sqrt{p_2^2 + q_2^2 + 1}.
$$

We thus find that the assumption of affine-transformable patterns yields the constraint represented by (6) on surface orientations. The constraint is determined solely by the matrix $A = (a_{ij})$, which is determined by the relation between P_2 and P_1 *observable* in the picture without knowing either the original patterns (P_1' and P_2') or their relationships (σ and R) in the 3-D space.

In order to have an idea about the degree of the constraint represented by (6), if we assume that the orientation of P_2' is known (i.e., $G_2 = (p_2, q_2)$ is known), then (6) gives two simultaneous equations for $G_1 = (p_1, q_1)$. The system appears to be of degree 4, but it can be shown that there are only two solutions; they are of the form (p_0, q_0) and $(-p_0, -q_0)$, which are symmetrical around the origin of the gradient space (see the Appendix).

From (5) we can also derive the following relationship:[2]

$$
\frac{\det(A)}{\sigma^2} = \frac{\sqrt{p_2^2 + q_2^2 + 1}}{\sqrt{p_1^2 + q_1^2 + 1}} = \frac{\cos\Gamma_1}{\cos\Gamma_2}.
\tag{7}
$$

This means that the ratio of cosines of the slant angles of the patterns is equal to the ratio $\det(A)/\sigma^2$. If we assume $\sigma = 1$ (the original patterns are of the same size) or that σ is known, (7) shows that we can order the texel patterns according to the magnitude of slant, Γ_i or $\sqrt{p_i^2 + q_i^2}$, using the values of $\det(A)$.

---

[2]This indicates that $\det(A)$ should be positive. But if it is negative, then we can assume that P_1' and P_2' are mirrored patterns, and put $R = \left(\begin{smallmatrix} \cos\alpha & \sin\alpha \\ \sin\alpha & -\cos\alpha \end{smallmatrix}\right)$.

3.1. Skewed Symmetry from Affine-transformable Patterns

The affine transform from P_2 to P_1 is more intuitively understood by how a pair of perpendicular unit-length vectors (typically along the x and y coordinate axes) are mapped into their transformed vectors. As shown in Figure 5, two angles (α and β) and two lengths (τ and ρ) can characterize the transform. The components of the transformation matrix $A = (a_{ij})$ are represented by

$$a_{11} = \tau\cos\alpha \qquad a_{12} = \rho\cos\beta$$
$$a_{21} = \tau\sin\alpha \qquad a_{22} = \rho\sin\beta. \tag{8}$$

Suppose, for simplicity, the orientation of P_2 in Figure 3 is known to be $(p_2, q_2) = (0,0)$. This simplifies equation (6) to

$$a_{11}(p_1^2+1)+a_{21}p_1q_1 = a_{22}\sqrt{p_1^2+q_1^2+1} \tag{9}$$
$$-a_{12}(p_1^2+1)-a_{22}p_1q_1 = a_{21}\sqrt{p_1^2+q_1^2+1}.$$

If we assume that α, β, τ, and ρ are known, then (p_1, q_1) has two possible solutions. This is essentially the case which Ikeuchi (1980b) investigated in his shape recovery method by assuming a known standard pattern, even though he used the constraint only partially.

Let us consider the case where α and β are known, but τ and ρ are not. One can substitute a_{ij} in (9) by (8), and eliminate τ and ρ. Then we obtain

$$(p_1\cos\alpha+q_1\sin\alpha)(p_1\cos\beta+q_1\sin\beta)+\cos(\alpha-\beta) = 0$$

which reduces to the same as the hyperbola (1). This can be interpreted as follows.

As was noted in the previous subsection, a pair of affine-transformable patterns impose the constraints (6) between their surface orientations, in which, if one is fixed, the other has only two possible orientations. However, if we loosen the transform in such a way that the angular (rotational) correspondence (α and β) is known while the length relationship is not known (or arbitrary), then the one-dimensional constraint of the skewed-symmetry hyperbola is obtained.

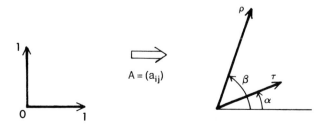

Figure 5. An affine transform (without translation) as characterized by two angles and two lengths.

4. The Shape-from-texture Paradigm

This section derives the same skewed-symmetry constraints from a second theory, different from that of the affine-transformable patterns. The shape-from-texture paradigm is a method of relating image texture properties to scene object properties, by explicitly incorporating assumptions about the imaging phenomenon into a computational framework. The paradigm is briefly presented here, but a fuller discussion can be found in (Kender, 1980).

The paradigm has two major portions. In the first, a given image textural property is "normalized" to give a general class of surface orientation constraints. In the second, the normalized values are used in conjunction with assumed scene relations to refine the constraints. If there are sufficiently many textural elements ("texels") in the image to be normalized, and if enough assumptions are made about their scene counterparts, then the underlying surface's orientation can be specified uniquely. Somewhat more weakly, only two texels are required, and only one assumption (equality of scene textural objects, or some other simple relation), to generate a well-behaved one-dimensional family of possible surface orientations. The method of skewed symmetry – the use of qualitative symmetries in the image to create a perspectively distorted right angle – is an example of such a weak method.

The first step in the paradigm is the normalization of a given texel property. The goal is to create a normalized texture property map (NTPM), which is a representational and computational tool relating image properties to scene properties. The NTPM summarizes the many different conditions that may have occurred in the scene leading to the formation of the given textural element. In general, the NTPM of a certain property is a scalar-valued function of two variables. The two input variables describe the postulated surface orientation in the scene (top-bottom and left-right slants: (p,q) when we use the gradient space). The NTPM for a horizontal unit line length in the image summarizes the lengths of lines that would have been necessary in 3-D space under various orientations: at surface orientation (p,q), it would have to be $\sqrt{p^2+1}$.

More specifically, the NTPM is formed by selecting a texel and a texel property, back-projecting the texel through the known imaging geometry onto all conceivable surface orientations, and measuring the texel property there. The representation chosen for the two-dimensional space of orientations is important; we will, however, only use the gradient space here.

In the second phase of the paradigm, the NTPM is refined in the following way. Texels usually have various orientations in the image, and there are many different texel types. Each texel generates its own image-scene relationships, summarized in its NTPM. If, however, assumptions can be made to relate one texel to another, then their NTPMs can also be related; in most cases only a few scene surface orientations can satisfy *both* texels' requirements. Some examples of the assumptions that relate texels are: both lie in the same plane, both are equal in textural measure (length, area, etc.), one is k times the other in measure, etc. Relating texels in the manner forces more stringent demands on the scene. If enough relations are invoked, the orientation of the local surface supporting two or more related texels can be very precisely determined.

4.1. Skewed Symmetry from the Paradigm Applied to Slope

What we now show is that the skewed symmetry method is a special case of the shape-from-texture paradigm; it can be derived from considerations of texel slope.

To normalize the slope of a texel, it is back-projected onto a plane with the postulated orientation. The back-projected texel now has a new shape on this new surface. Its exact value, however, depends upon the coordinate system on this surface plane. Many coordinate systems are possible; we chose here a coordinate system whose x-axis lies along the gradient direction. The normalized slope is then the angle that the back-projected texel makes with respect to the surface coordinate system x-axis. The calculation is a bit involved, especially under perspective, which requires a knowledge of both the location of the center of focus and the length of the focal distance.

Using the construction in Figure 6, together with several lemmas relating surfaces in perspective to their local vanishing lines, slope is normalized as follows. Assume a slope is parallel to the p-axis; the image and gradient space can always be rotated into such a position. (If rotation is necessary, the resulting NTPM can be de-rotated into the original position using the standard two-by-two orthonormal matrix.) Also assume that the slope is somewhere along the line $y = y_s$, where the unit of measurement in the image is equal to one focal length. The normalized value of the slope is equal to the tangent of the 3-D space angle η, whose base (of length R) is parallel to the surface plane, and is in the direction of the gradient. R is determined from the focal distance, and from the point of the nearest approach of the vanishing line of the plane. This line has equation $px + qy = 1$ (or $G \cdot P = 1$) and its nearest approach is $G / \|G\|^2$. The distance d is given by the intersection of the line $y = y_s$ with the vanishing line. Then, the normalized slope value – the Normalized Texture Property Map – is given by

$$\frac{q - y_s(p^2 + q^2)}{p\sqrt{1 + p^2 + q^2}}. \tag{10}$$

This normalized value can be exploited in several ways. Most important is the result that is obtained when one has two slopes in the image that are assumed to arise from equal slopes in the scene. Under this assumption, their normalized property maps can be equated. The resulting constraint, surprisingly, is a simple straight line in the gradient space. It is intimately related to the vanishing

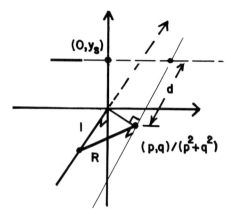

Figure 6. Back-projecting an image slope onto a plane with gradient (p, q).

point formed by the intersection of the extensions of the two image slopes (Kender, 1980).

The constraint equations resulting from assuming that the two slopes arose from perpendicular lines in the scene is, however, enormously complex. It unfortunately does not appear to have many tractable forms or special cases.

Under orthography, nearly everything simplifies. The normalized slope of a texel becomes

$$\frac{q}{p\sqrt{1+p^2+q^2}}. \tag{11}$$

It is independent of y_s; in effect, all slopes are at the focal point.

Considering two image slopes to have arisen from parallel lines in the scene has a trivial solution. If the image slopes are parallel, the entire gradient space is a solution. If they are not, there is no solution at all. This corresponds to the projective geometry theorem that under orthography, parallels are taken into parallels regardless of surface orientation.

In the case where the scene slopes are assumed to be perpendicular, we again get a simplification, but this time a useful one. Not only is the solution tractable, it is the skewed symmetry method of Section 2. We derive it as follows.

Consider Figure 7. Note that under orthography, texels can be translated arbitrarily, since the focal length is infinite and the focal point is effectively everywhere; there is no information in image position. Given the angle that the two texels form, rotate the gradient space so that the positive p-axis bisects the angle. Call this adjustment angle λ; we will use it to de-adjust our results into the original position after they have been computed.

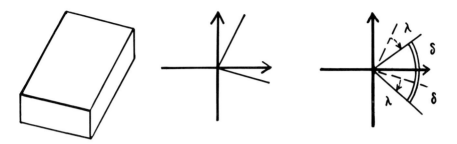

Figure 7. Two image texels assumed to be perpendicular in the scene.

Let the angle that is bisected be 2δ. The normalized value of either slope is obtained directly from the standard normalized slope formula, corrected for the displacement of $+\delta$ and $-\delta$ respectively. That is, for the slope at the positive δ orientation, instead of formula (11), we use the formula under the substitution $p\cos\delta+q\sin\delta$ for p, $-p\sin\delta+q\cos\delta$ for q. We proceed similarly for the slope at $-\delta$. Note that the factor $\sqrt{1+p^2+q^2}$ is invariant under this transformation (it is the length of the normal vector of the surface).

The fact that the normalized slopes are assumed to be perpendicular in the scene allows us to set one of the normalized values equal to the negative reciprocal of the other. The resultant equation becomes

$$p^2\cos^2\delta-q^2\sin^2\delta = \sin^2\delta-\cos^2\delta = -\cos2\delta. \tag{12}$$

This is exactly the hyperbola in Section 2 with $2\delta = |\alpha-\beta|$.

4.2. Skewed Symmetry from the Paradigm Applied to Length and Angle

The paradigm is similarly applicable to other texture measures. Using texel length as the property to be normalized, we find that under perspective, lengths must lie on the same line in order for the resultant equations to be simpler than the fourth order. If they are collinear, again the resultant gradient space constraint is a simple straight line.

Under orthography and the assumption that image lengths have arisen from equal scene lengths, the constraint equation is again a hyperbola – the skewed-symmetry hyperbola, somewhat offset. In fact, the geometric construction in Figure 8 shows that the assumption of equal length can be made equivalent to skewed symmetry.

First, a triangle is formed by translating one or the other of the lengths so that they meet at a common endpoint. Under orthography, such translations do not affect the resulting constraints. Connecting the remaining endpoints creates a triangle which must be isosceles in the scene. Further, under orthography,

Figure 8. Assuming lengths are equal generates the skewed symmetry constraint.

midpoints of lines are preserved (the midpoint of the base of the scene triangle is imaged as the midpoint of the base of the image triangle). The line connecting the vertex and this midpoint has the property that, in the scene, it must form a right angle with the base. Its distortion to something other than a right angle in the image – the induced angle 2δ – is precisely the distortion which characterizes skewed symmetry. Therefore, the same methods apply.

One other case is worth mentioning. Suppose the image has two angles such that one leg of the first is parallel to one leg of the second. See Figure 9. In this case, again the constraint is equivalent to skewed symmetry, as the construction shows. Choosing one of the angles, extend its non-parallel leg until it intersects both legs of the other angle. (If it cannot do so, then first translate the angle before extending.) The resulting triangle must be isosceles in the scene, since the angles are assumed equal in the scene. However, this is the same situation encountered above with the construction involving lengths. Therefore, the altitude from the midpoint of the base (here, the midpoint of the parallel side) to the vertex must form a right angle. Again, the distortion observed in the image is the skewed symmetry distortion.

5. Applications of Skewed Symmetry and Affine-transformable Patterns

5.1. Quantitative Shape Recovery from Line Drawings

Given the line drawing of Figure 10a, we usually perceive a right-angled parallelepiped. The Huffman-Clowes-Waltz labeling scheme for the trihedral

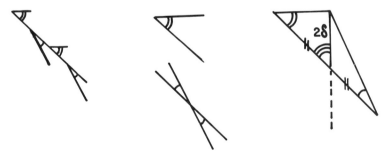

Figure 9. Assuming angles are equal generates the skewed symmetry constraint.

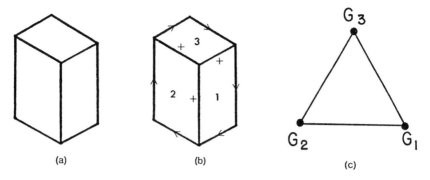

Figure 10. (a) A line drawing of a block; (b) Huffman-Clowes-Waltz labeling; (c) constraints in the gradient space.

world gives the labeling shown in Figure 10b, which signifies that the three edges meeting at the central FORK vertex are all convex, i.e., the object is a convex corner of a block. However, it does *not* specify a particular quantitative shape. In fact, the labeling indicates only that the gradients of the three surfaces should be placed in the gradient space so as to form the triangle shown in Figure 10c. The edges of the triangle should be perpendicular to the picture edges separating the corresponding regions, but the location and size of the triangle are arbitrary in the gradient space. Therefore, the object is not necessarily right-angled.

We can use skewed symmetry here to provide additional constraints. The three regions are skewed-symmetrical with the axes shown in Figure 11a. The hyperbolas corresponding to these regions are shown in Figure 11b. Thus the problem is now how to place the triangle of Figure 10c in Figure 11b so that each vertex is on the corresponding hyperbola. Kanade (1979) proves that the combination of locations shown in Figure 11b is the only possibility, and that the resultant shape is a right-angled block.

It is interesting to note that if we apply the same procedure to the line drawing of Figure 12, we find that there is no way for all the three regions to satisfy the skewed symmetry assumptions. That is, at least one of them has to be non-symmetrical (skewed) in the 3-D space; in other words, the object *cannot* be right angled, but should be rhomboid (a prism). Remember that Figure 10a *can* be either right-angled or rhomboid, but it is usually perceived as right-angled.

Figure 13 demonstrates how the above procedure results in the interpretation of the drawing as a trapezoidal block in this case.

5.2. Skewed Symmetry under Gravity

One principal influence toward symmetry seems to be an object's structural necessity to oppose the gravitational field. Objects that must support

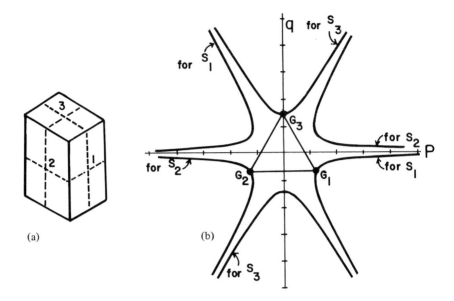

(a)   (b)

Figure 11. (a) Axes of the skewed symmetry of the regions of Figure 10a; (b) corresponding hyperbolas and allocations of the gradients.

themselves tend to have structural members aligned parallel to the direction of force, that is, vertically. Such members are mutually parallel – a type of symmetry. The base of such an object is often perpendicular to gravity to distribute weight and provide balance. Together, then, the base and structural members provide a local symmetry frame that can also be exploited by the skewed-symmetry method. One can show that in this last case it is usually possible to specify surface orientation uniquely.

We will assume that the direction of the gravity field is known, say the top-to-bottom lines in the image frame are assumed to be true projections of a line of gravity force. The gradient space is also considered to be aligned in the direction of the gravity field; $-q$ is also "down."

Under such conditions, suppose we do find a portion of the image that is assumed to be a vertical, symmetric surface: say, a building face as in Figure

Figure 12. A line drawing of a rhomboid: this *cannot* be a right-angled block. Notice that Figure 10a can be a rhomboid.

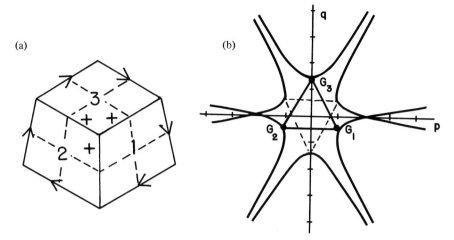

Figure 13. Shape recovery of a trapezoid block: (a) axes; (b) gradient allocations.

14. Using skewed symmetry (or even direct observation), it is not hard to obtain an angle in the image that corresponds to a right angle in the scene. Suppose one of the legs of the angle is parallel to the known gravity field as in Figure 14. The skewed-symmetry method generates the following constraint hyperbola:

$$p = -(q + 1/q)\cot\gamma. \tag{13}$$

This constraint is somewhat interesting: it expresses p (left-right slant) as a *function* of q (top-bottom slant). The value of q itself is easily obtained.

If gravity points in the $-q$ direction, the ground plane must have as its orientation $(0, q_g)$, for a value of q_g determinable through sensing. Since all

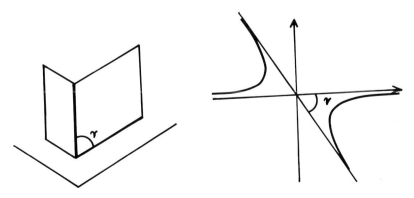

Figure 14. Assumptions about gravity can uniquely specify surface orientations.

vertical planes are perpendicular to the ground plane, all vertical planes must have the orientation $(p_v, -1/q_g)$, for variable p_v. (A quick check shows that the dot product of the corresponding normals is zero: $(0, q_g, 1) \cdot (p_v, -1/q_g, 1) = 0$.) Note that the value of q for *any* vertical plane is fixed at $-1/q_g$. Thus, in our example, p is also determined: it is $-(q_g + 1/q_g)\cot\gamma$. Since q_g is a constant, p varies simply with γ. Figure 14 shows the constraints graphically.

5.3. Shape Recovery of an Object with Many Patterns Stamped

Consider the problem of recovering the shape from a picture of a ball with a number of patterns stamped on it (see Figure 15). For each pair of texel patterns, if they are affine-transformable, we compute a transformation matrix A. Thus we obtain many constraints on the gradients of texels. From these, however, we cannot uniquely determine the surface orientation of each texel.

We need more assumptions or data. We will suppose we know the gradients of some particular texels, and assume that the surface is smooth (together, maybe, with an assumption of global concavity or convexity). Then a relaxation or cooperative technique similar to the one for shape-from shading (Woodham, 1977; Ikeuchi, 1980a) will allow us to determine consistent assignments of gradients to the texels which satisfy those many constraints. Notice that we need not assume that the original pattern is known, nor that the patterns are stamped in a particular manner. Even other patterns can be mixed together with them.

One of the plausible methods of determining the gradient of one particular texel is to use equation (7). Assuming $\sigma = 1$, we order the texels by the magnitude of $\sqrt{p_i^2 + q_i^2}$, and assign $(p, q) = (0, 0)$ (the orientation that is directly facing the viewer) to the least slanted texel. This is analogous to a similar hypothesis in shape from shading. That is, we tend to assign to the brightest point the orientation directly facing the light source, even though under the assumptions of parallel lights and a matte surface, one can only say that the brightest pixels have the minimum incident angle of light, not necessarily $0°$.

6. Conclusion

The assumptions we used for skewed symmetry, affine-transformable patterns, and texture analysis can be generalized as:

Properties observable in the picture are not accidental, but are projections of some preferred corresponding 3-D properties.

This provides a useful meta-heuristic for exploiting image properties: we can call it the meta-heuristic of *non-accidental image properties*. It can be regarded as a generalization of general view directions, often used in the blocks world, to exclude the cases of accidental line alignments.

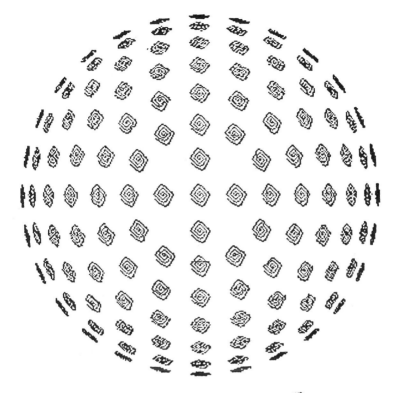

Figure 15. A picture of a ball with a number of 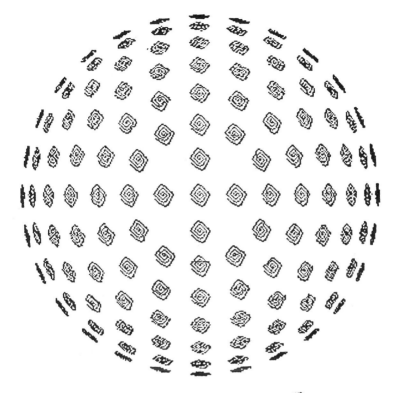 s stamped.

Instances that can fall within this meta-heuristic include: parallel lines in the picture vs. parallel lines in the scene, texture gradients due to distance, and sets of lines convergent to a vanishing point.

The most essential point of our technique is that we relate certain image properties to certain 3-D space properties, and that we map the relationships into convenient representations of shape constraints. We explicitly incorporate assumptions based either on the meta-heuristic or on *a priori* knowledge of the world. The shape-from-texture paradigm provides a computational framework for our technique. In most of our discussion we assumed orthography. Similar (though more involved and less intuitive) results can be obtained under perspective projection.

Acknowledgment

This research was sponsored by the Defense Advanced Research Projects Agency (DOD) under ARPA Order No. 3597, and monitored by the Air Force

Avionics Laboratory under Contract F33615-78-C-1551. The views and conclusions in this document are those of the author and should not be interpreted as representing the official policies, either expressed or implied, of the Defense Advanced Research Projects Agency or the U.S. Government.

Appendix

Proof that (6) has two symmetrical solutions:

We will try to solve (6) for p_1 and q_1, assuming that p_2, q_2, and $A = (a_{ij})$ are known. We assume $\det(A) > 0$. Let us put $\gamma = \sqrt{p_1^2 + q_1^2 + 1}$. Then (6) can be rewritten as

$$Ca_{11}(p_1^2 + 1) + Ca_{21}p_1q_1 = B\gamma$$
$$A(p_1^2 + 1) - Bp_1q_1 = Ca_{21}\gamma \tag{14}$$

where

$$A = a_{11}p_2q_2 - a_{12}(p_2^2 + 1)$$

$$B = a_{22}(p_2^2 + 1) - a_{21}p_2q_2$$
$$C = \sqrt{p_2^2 + q_2^2 + 1}.$$

We can derive a quadratic equation on γ from (14):

$$f(\gamma) = DE\gamma^2 - (D^2 + E^2 + F^2)\gamma + DE = 0, \tag{15}$$

where

$$D = C(Ba_{11} + Aa_{21}) = C \det(A)(p_2^2 + 1) > 0$$
$$E = B^2 + (Ca_{21})^2 > 0$$
$$F = -C^2a_{11}a_{21} + AB.$$

The discriminant of (15) is

$$\begin{aligned} \text{disc} &= (D^2 + E^2 + F^2)^2 - 4(DE)^2 \\ &= F^4 + 2F^2(D^2 + E^2) + (D - E)^2 \\ &\geq 0. \end{aligned}$$

Thus, $f(\gamma)$ has real roots. Now, notice that $\gamma \geq 1$ and thus we are interested in the root greater than or equal to 1. Let us check the sign of $f(1)$ multiplied by the coefficient of γ^2:

$$f(1)DE = (2DE - (D^2 + E^2 + F^2))DE$$
$$= -(F^2 + (D - E)^2)DE$$
$$\leqslant 0.$$

This means that one and only root of $f(\gamma)$ is greater than or equal to 1. Let us denote this root by γ_0. By substituting γ_0 into (14), we can solve it as a simultaneous quadratic equation on p_1 and q_1, and know that (p_1, q_1) has two solutions in the form of (p_0, q_0) and $(-p_0, -q_0)$, which are symmetrical to the gradient space origin.

References

Ikeuchi, K. Numerical shape from shading and occluding contours in a single view. MIT AI Memo 566, 1980. (a)

Ikeuchi, K. Shape from regular patterns (an example of constraint propagation in vision). MIT AI Memo 567, 1980. (b)

Kanade, T. Recovery of the 3-dimensional shape of an object from a single view. *Artificial Intelligence,* 1981, **17**, 409-460.

Kender, J. R. Shape from texture. Doctoral dissertation, Carnegie-Mellon University, 1980.

Mackworth, A. K. Interpreting pictures of polyhedral scenes. *Artificial Intelligence,* 1973, **4**, 121-137.

Stevens, K. A. Surface perception from local analysis of texture and contour. MIT AI Memo 512, 1980.

Woodham, R.J. A cooperative algorithm for determining surface orientation from a single view. Proceedings, 5th International Joint Conference on Artificial Intelligence, 1977, 635-641.

Visual Computation

Paul A. Kolers

University of Toronto
Toronto, Canada

Abstract

A theory of vision that can be implemented on a computer requires specification of the elements of vision. Some experiments on human perception of shape and motion suggest that the human visual system does not work in terms of visual elements that it combines according to rule, but works rather in terms of skills and dispositions. It is not clear how the latter would be implemented on a machine.

The requirements for a computational theory of perception are not always defined; some writers in fact do not discriminate between descriptions of human and machine operation, treating the two commonly. In what follows I will describe briefly some of the formal requirements of a computational system, and then I will show that some attributes of human perceiving do not subscribe to those characteristics.

The principal requirement on a computational system is a set of elements and a set of rules that manipulate them. The preeminent examples are mathematical systems where, as in arithmetic or algebra, characters are defined independently of the operations performed on them, such as addition and subtraction and their associative and commutative combinations. Modern physics and chemistry, similarly, distinguish between the items manipulated and the rules governing their manipulation, as in the rules of combination or dissociation of high energy particles or of chemical compounds. The rules and operations are context-free, moreover, allowing for the same predictions and outcomes not only throughout all human laboratories, but also, in Einstein's assertion, throughout the universe.

The case is not quite the same in human perception. In fact, a very great deal of work has gone forward trying to identify elements of perception. At one time it was thought that discrete sensory experiences might be taken as the elements (Titchener, 1902). During the 1930's, Clarence Graham and his associates (Graham, Brown, & Mote, 1939; Graham & Margaria, 1935) tried to relate the stimulation of local retinal regions to the notion of elements of form. Subsequently, it was thought that a small alphabet or vocabulary of elements might be identified from neurophysiological researches on frog, cat, or monkey (Hubel & Wiesel, 1962; Lettvin, Maturana, McCulloch, & Pitts, 1959). The idea generally developed that an edge or a brightness might be taken as the element out of which the visual system computed perceptible objects.

1. Stability and Motion

It is important to define the temporal limits over which the discussion is to range. When we speak of the object of perception, we usually mean something lasting several seconds, if not minutes. The reason is that we usually remain in visual contact with a single object or scene for relatively long intervals – the person we are talking to, the road on which we are walking, the object at which we are looking – and much of the discussion takes those objects as the basic facts of perception. In laboratory circumstances we find, however, that a similar stability of forms is absent in the brief intervals of time that characterize single fixations of the eyes.

The eye is in near-continual motion, fixating for about 250 msec and moving for about 50 msec, making three to four fixations per second. If shapes are presented to the eye for part of a fixation interval they are seen to transform in interesting ways. A rectangle alternated with a neighboring triangle, for example, is seen neither as rectangle nor triangle, but as a continuously changing shape; indeterminacy of the forms is increased when the two presentations are of irregular polygons (Kolers, 1972a). Moreover, if their duration or intensity is increased they become, not more stable, but exactly the opposite: the greater the intensity or duration of the first flash, the more likely is the configuration to transform (Kolers & Pomerantz, 1971).

Alternately, if the configuration is modified slightly, one shape now followed by a circumscribing second shape, proper selection of stimulus values can so interfere with the perception of the shapes that the first or the second can be made to disappear (Kolers & Rosner, 1960). If the edge of either shape were an element in a perceptual computation, or if either shape were a coded unit, it is unlikely that one could eliminate the other from perception merely as a function of their temporal separation.

2. Control

Two forms of model mark contemporary conjectures. One proposes that stimulation reaches sensory surfaces which then transform the energy according to prescription and pass it on for further processing. This is called the bottom-up approach, and it supposes that the visual system is organized according to levels of operation. These levels may be the nodes in a serial processing model (Hubel & Wiesel, 1962), or the results of a set of operations which are then operated upon again (Marr, 1976). In these views, as in Graham's of a generation earlier, the system need have neither knowledge nor the ability to acquire it, beyond the ability to react to and transform stimulation according to prescription.

A second form of model, called top-down, supposes that prior knowledge, expectations, inferences, or like cognitive processes characterize perceiving. On this view antecedent conditions modify or modulate the operation of the transducers, setting them to respond in particular ways (Bruner, 1957). People's ability to identify objects on the basis of a sharply limited set of features, but by utilizing contextual or other information extrinsic to the stimulus itself, is evidence for such processing (Kolers, 1972b).

In two demonstrations of apparent motion, both the feature analytic and the semantically oriented views were questioned. In one case a Necker cube was alternated with one of its many components, those being either its logical or its neurological features, such as vertical, horizontal, or diagonal lines, outlines, or the like (Figure 1). If the human observer's visual system performed a feature analysis, it would abstract the local feature from the cube and move that, allowing the remainder of the cube to flicker on and off. Rather, people saw a single configuration, a feature growing into a cube in one direction of apparent motion, or a cube shrinking to the feature in the other direction (Kolers, 1972a). Thus the visual system seemed to deal with whole objects that changed their characteristics in time, but not with features that it isolated or extracted and operated upon.

In the second case, the word CAT or the word BOX was alternated with a simple containing outline that looked like a box. If any sort of semantic relatedness bound the linguistic and the graphemic ("pictorial") representations of box, then BOX and the contour should exchange identity more readily or over a greater length of time than CAT, or any other three-letter group, alternated with the contour. The finding was that all moved readily, implying that the top-down or semantic aspect was of little significance in affecting the relation between word and outline shape. Rather, at the level of processing at which the interaction occurred, all of the words were treated merely as shapes. The "wordness" of a word – its semantic, orthographic, or other linguistic features – was not influential in these brief presentations.

3. Shape and Color

Color would certainly qualify as one of the elements or basic attributes of a visual object, hence a difference in the way that the visual system treats edges

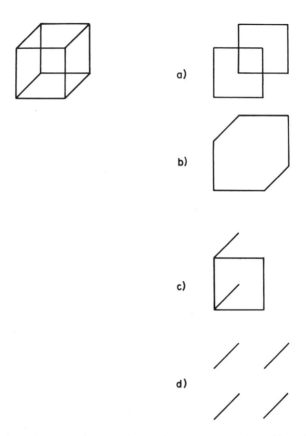

Figure 1. A Necker cube was alternated in apparent motion with each of several of its features.

and colors warrants consideration. If two shapes are alternated, say a right triangle and a square, the perceiver sees one change smoothly into the other at appropriate timing. The change is smooth and continuous as measured (Kolers & von Grunau, 1976). One sees a triangle change smoothly into a square, typically by billowing out along the hypotenuse to square off the fourth corner; or one sees the square change equally smoothly into a triangle. If, however, two colors are alternated, no equivalent smoothness of change occurs.

Suppose that a pair of reds or a pair of greens is alternated. Either pair yields the perception of a color moving across the screen. Suppose that a red is now alternated with a green. How would the disparity of color be resolved? Red could change through orange and yellow to green, or red could desaturate to gray and gray become green, as two alternatives allowing for smooth resolution of the disparity. In fact, neither occurred, but red moved smoothly across the screen to about the midpoint, where it suddenly changed to green; or green moved smoothly across the screen and suddenly changed to red. The same abruptness of change occurred with other colors as well, including achromatic colors. Thus, not only are we uncertain of what might make up the elements of perception, we find that the operations that are performed on two possible elements, color and form, actually differ, analog operations directed at form, digital operations directed at color.

With this small set of examples I have attempted to show that shapes, those fundamental aspects of visual experience, cannot be functioning as elements in a general calculus of perception. Rather, shapes or forms are the results of visual processing, not its elements. Form is an achievement of the visual system, an outcome of its operations, and not the basis out of which other events are made.

4. Elements and Skills

The notion of a compound as made up out of elements operated upon according to rule is fundamental to much of science; indeed, elements plus rules constitute the basis for most scientific paradigms, including the notion of visual perception as based on computation. From some recent experiments, not very well known, a modification of that view emerges, which is that notions of skill and performance may provide a better platform from which to describe human perception than notions of elements and rules do.

I will describe two experiments, briefly, in which the principle seems to be manifested. In the first, college students were presented with pages of alphabetic characters and were required to name them one at a time. They named some pages rightwards, in their normal reading direction, and some pages leftwards. The pages appeared as characters separated by spaces, as illustrated in example N of Figure 2. Even though the task was to name characters singly, people named them in the rightwards direction more rapidly than leftwards. The suggestion that springs immediately to mind is, of course, that the well inculcated skill of reading English rightwards influenced performance. Hence, as a check, people were required to name characters one at a time, rightwards or leftwards, when they appeared as in R in Figure 2. If the right-goingness were the basis of the difference in performance on N, then the characters of R should also be named faster in the rightwards direction.

Exactly the opposite result was found: people named the characters of R faster leftwards than rightwards (Kolers & Perkins, 1969a,b). Thus it was not direction of naming, by itself, that accounted for the prior performance, but direction of naming interacting with the orientation of the characters. The account given was that people develop skills in acquiring the clues by which to identify objects, and practiced search routines pay off in superior performance. When the stimuli were transformed geometrically, as for R, people had either to transform their search routines to accommodate the transformed sample, or to transform an "image" of the sample that they operated on with stereotyped search routines. If they transformed the sample and applied fixed routines, then

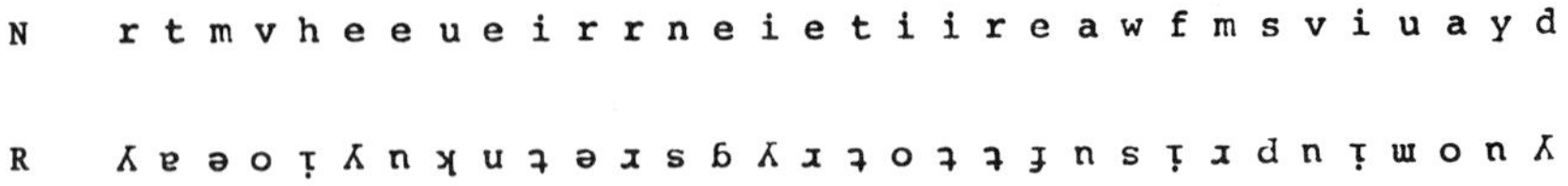

N   r t m v h e e u e i r r n e i e t i i r e a w f m s v i u a y d

R   ʌ ɐ ǝ o ʇ ʌ u ʞ u ʇ ǝ ɹ s ƃ ʌ ɹ ʇ o ʇ ʇ ɟ n s ʇ ɹ d n ʇ ɯ o u ʌ

Figure 2. Two orientations of alphabetic characters that were named rightwards or leftwards.

R would have been named more rapidly in the rightwards direction. The data, in coming up otherwise, suggest that people actually performed the task by transforming their search routines, and it was the searching or clue-selecting that was the skilled action rather than some wholistic naming of characters. The critical point, however, is that clue-finding was integrated with a visuo-motor sampling of the text. People found the clues within a motor sequence of operations; clue finding was not separated from the direction of naming.

In a second experiment people read pages of text in the transformations of Figure 3, separate groups of students each reading 24 pages in a single transformation (Kolers & Perkins, 1975). In reading that many pages, people acquired considerable skill at the task. The students then each read two pages in each of the eight samples illustrated, in properly counterbalanced order. The idea of the experiment was to find how practice at one transformation transferred to performance on the others.

If reading transformed text exercised merely some mental capability to carry out a geometric transformation, then practice with one transformation should be as good as practice with any other. The finding, however, was that the transformations affected each other to different degrees. Table 1 shows in the leftmost column the transformation that a group was practiced on, and in subsequent columns rightwards, the percentage transfer from that training to the test named in the column. The rightmost column, in turn, averages performance within rows. (Transfer of training is calculated as $100(a-x)/(a-b)$ where a is initial performance on a training task, b is final performance on the task, and x is performance on the test task. Thus, performance on the test task is measured as a fraction of final performance on the training task. If the test task is performed more poorly, transfer ranges downward from 100% toward zero; and if performance on the test task is better than final performance on the training task, transfer exceeds 100%. The latter is sometimes called "anomalous transfer.")

Table 1 shows that the degree of transfer varied greatly among the transformations, the same training transformation affecting tests differently (rows), and the same test affected differently by the different training transformations (columns). Without going into the many details of analysis, the main point is that training and test transformations interacted. Performance was not due to some general adaptation or some general learning of transformations; rather, performance was due to the specific interactions between skills acquired

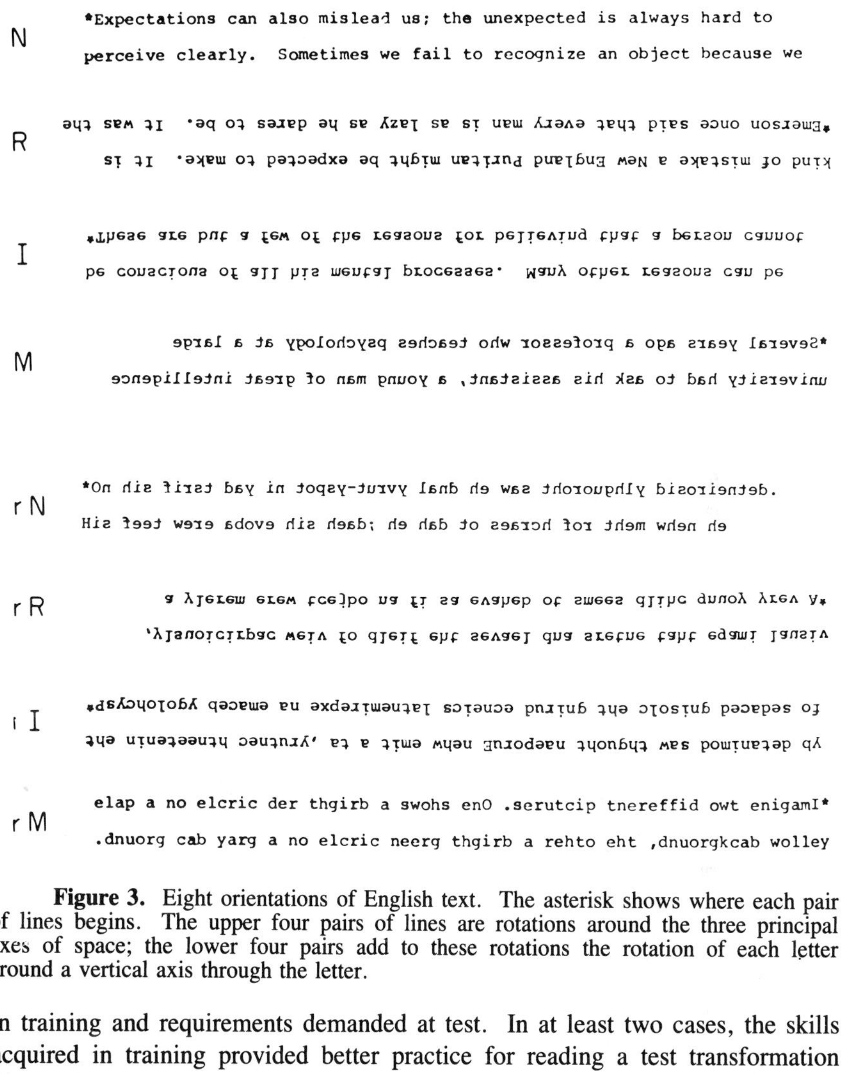

N
*Expectations can also mislead us; the unexpected is always hard to
perceive clearly. Sometimes we fail to recognize an object because we

R
*Emerson once said that every man is as lazy as he dares to be. It was the
kind of mistake a New England Puritan might be expected to make. It is

I
*There are but a few of the reasons for believing that a person cannot
be conscious of all his mental processes. Many other reasons can be

M
*Several years ago a professor who teaches psychology at a large
university had to ask his assistant, a young man of great intelligence

rN
*On his first day in topsy-turvy Land he was thoroughly disoriented.
His feet were above his head; he had to hold for dear life when he

rR
*A very young child of some months as it lay in an old cradle saw objects that were merely a
visual images that surface and leave the field of view sequentially,

rI
*Psychology became an experimental science during the decades to
the nineteenth century, at a time when thought was detained by

rM
*Imagine two different pictures. One shows a bright red circle on a pale
yellow background, the other a bright green circle on a gray background.

Figure 3. Eight orientations of English text. The asterisk shows where each pair of lines begins. The upper four pairs of lines are rotations around the three principal axes of space; the lower four pairs add to these rotations the rotation of each letter around a vertical axis through the letter.

in training and requirements demanded at test. In at least two cases, the skills acquired in training provided better practice for reading a test transformation than training on the test transformation itself did. I refer to the anomalous transfer of training on I and rI and test on R.

The account offered is that tasks are carried out by execution of skills, and that skills are composites. At least three components enter into the composition for geometrically transformed text, two concerned with the orientation of the text and a third concerned with the sequence of letters. The relative contribution of the three seems to vary with the degree of training, and to be exercised differently by the various transformations (Kolers & Perkins, 1975). Thus the exact composition of components will vary from task to task. Skills made up of components seem to be at the heart of a number of perceptual activities

Table 1

Percentage Transfer from Training to Test

Training	Test							Training average
	M	R	rR	rN	rI	I	rM	
I	95	118	97	80	84	100	52	89
rI	92	112	90	82	100	78	72	89
rR	81	84	100	68	56	79	56	74
rM	88	67	70	57	37	44	100	66
R	87	100	70	37	58	44	41	62
M	100	69	71	72	38	28	52	61
rN	82	3	57	100	36	34	20	47
Test average	89	79	79	71	58	58	56	

related to manipulation of symbols. Such a notion of skills does not lend itself readily to the sort of notating of atomic elements and their rules of combination that a computational theory typically requires.

5. People and Machines

There is little doubt that people are not born with but acquire skills. The fundamental aspect of a skill is that an action executed now depends for its accomplishment on the execution of prior actions: activities are carried out through transfer of components acquired in prior actions, carried out on the basis of things done before. A machine's actions, however, are largely independent of what has gone before; its ability to do something is due to its design, not to its past experience. People acquire skills, but machines have their skills built in. In a new situation the human operates by applying all of the skills it has acquired, reinforcing those that are appropriate and sloughing off the irrelevant; the human thereby acquires the specialization that is skill. The machine, in contrast, is specialized from the outset, built and functioning according to a formalism. Unlike a machine's, a human's skill is never complete, never fixed, but is always undergoing modification and change; a skill is not an element but a performance. (Some efforts have been made to notate skill, of course, as in the production system of Newell [1980]. Characteristically, procedures are altered by tying new productions to them. Complicating productions to accommodate changes in performance provides a mathematical solution to the problem, but not an esthetically pleasing or psychologically persuasive one.)

Computational approaches to vision should be construed as falling within a branch of engineering, where a machine's successes are the justification of the machine. A machine's failures, in contrast, may be taken as challenges – both to the engineering and the psychological student of perception; for if successes are their own reward, failures are stimuli to action, to better understanding. I am sure that the psychological and the computational approaches to understanding perception can be mutually beneficial, illuminating for each other critical aspects of their common problems. The nature of their solutions, and so the structure of their theories, will be different, however, for the objects of their study are different. Computational approaches to perception may satisfy the requirements of developing task-specific machines, but human perception does not seem to subscribe to a computational approach in the same way. Armchair solutions in which mechanisms are invented and rules prescribed for their operation may be appropriate to modeling at the machine level, but they do not provide a useful account of the greater complexity of human performance, some of which I described above.

Acknowledgment

These remarks have benefitted from discussion with W. E. Smythe. Their preparation was aided by Grant A7655 from the National Science and Engineering Research Council, Canada.

References

Bruner, J. S. On perceptual readiness. *Psychological Review,* 1957, **64**, 123-152.

Graham, C. H., Brown, R. H., & Mote, F. A. The relation of size of stimulus and intensity in the human eye. *Journal of Experimental Psychology,* 1939, **24**, 555-573.

Graham, C. H., & Margaria, R. Area and the intensity-time relation in the peripheral retina. *American Journal of Physiology,* 1935, **113**, 299-305.

Hubel, D. H., & Wiesel, T. N. Receptive fields, binocular interaction, and functional architecture in the cat's visual cortex. *Journal of Physiology,* 1962, **160**, 106-154.

Kolers, P. A. *Aspects of Motion Perception.* Oxford, UK: Pergamon Press, 1972. (a)

Kolers, P. A. Reading pictures: Some cognitive aspects of visual perception. In T. S. Huang & O. J. Tretiak (Eds.), *Picture Bandwidth Compression.* New York: Gordon and Breach, 1972. (b)

Kolers, P. A., & Perkins, D. N. Orientation of letters and errors in their recognition. *Perception and Psychophysics,* 1969, **5**, 265-269. (a)

Kolers, P. A., & Perkins, D. N. Orientation of letters and their speed of recognition. *Perception and Psychophysics,* 1969, **5**, 275-285. (b)

Kolers, P. A., & Perkins, D. N. Spatial and ordinal components of form perception and literacy. *Cognitive Psychology,* 1975, **7**, 228-267.

Kolers, P. A., & Pomerantz, J. R. Figural change in apparent motion. *Journal of Experimental Psychology,* 1971, **87**, 99-108.

Kolers, P. A., & Rosner, B. S. On visual masking (metacontrast): Dichoptic observation. *American Journal of Psychology,* 1960, **73**, 2-21.

Kolers, P. A., & von Grunau, M. Shape and color in apparent motion. *Vision Research,* 1976, **16**, 329-335.

Lettvin, J. Y., Maturana, H. R., McCulloch, W. S., & Pitts, W. H. What the frog's eye tells the frog's brain. *Proceedings of the IRE,* 1959, **47**, 1940-1951.

Marr, D. Early processing of visual information. *Philosophical Transactions of the Royal Society, London,* 1976, **B275**, 483-524.

Newell, A. HARPY, production systems, and human cognition. In R. A. Cole (Ed.), *Perception and Production of Fluent Speech.* Hillsdale, NJ: Erlbaum, 1980.

Titchener, E. B. *An Outline of Psychology.* New York: Macmillan, 1902.

The Psychology of Perceptual Organization: A Transformational Approach

Stephen E. Palmer

University of California
Berkeley, California

Abstract

Psychological research on perceptual organization is reviewed, and a theoretical framework is presented to account for it. The review focusses on the organizational phenomena of shape constancy, motion perception, figural goodness, perceptual grouping, and reference frame effects. It is argued that the key to understanding them within a unified framework lies in the concept of local invariance over the group of Euclidean similarity transformations. The theory offered to account for these phenomena is based on a parallel processing system constructed from many *spatial analyzers* that are related to each other by similarity transformations. They are compared for output equivalence by *invariance analyzers* and structured more globally by *frame analyzers*. The latter are used to select the maximally informative (i.e., simplest) organization of sensory data for further processing and shape description.

1. Introduction

Most psychologists believe that certain molar phenomena of perception are central to understanding the structure of the visual system. Perceptionists differ among themselves in terms of which ones they take to be most important – naturally, the ones they themselves study are always high on the list – but there does seem to be fairly general agreement on a set of phenomena that are thought to provide important clues to the organization of the visual system. In fact, the study of these phenomena has become a fairly stable subarea within psychology called, appropriately enough, "perceptual organization." This field was founded in large measure by Gestalt psychologists early in this century. Since then it has grown to encompass a vast number of results. Naturally, the hope is that these facts will all fit snugly together once the appropriate theoretical structure is found. At present, however, the psychology of perceptual organization is easily seen as a haphazard potpourri of results that are loosely connected, at best.

My goal in this chapter is to present this field as a coherent topic. It is a challenging task. The major difficulty lies in finding the common thread, the underlying theme, that will unify and organize the disparate pieces of experimental work. If it is correct to assume that the so-called "organizational" phenomena are particularly transparent to the structure of the underlying system, then this common thread should also point us toward the right sort of theories. At least, this is what I believe to be the case and why I have chosen to structure the chapter as I have done.

The first half briefly reviews the field of perceptual organization as psychologists typically study it. I have simplified the task by restricting the phenomena to shape constancy, motion perception, figural goodness, perceptual grouping, and what I call "reference frame" effects. Naturally, restricting the field in this way makes it easier to find a common theme, but I also believe that these are the most basic, powerful, and pervasive phenomena in the domain of perceptual organization. The theme that unifies them is *transformational structure*. When examined in detail, they suggest that perception works by detecting and extracting structure in the form of transformational invariance. They further suggest that the system operates according to principles that maximize the amount of transformational invariance that is extracted. This proposal forms the common thread and points the way to possible theories.

The second half of the chapter presents a computational theory that is compatible with the view presented in the first half. The basic proposal is that a transformational analysis is performed by three closely related subsystems. The first is a system of transformationally related *spatial analyzers*. The second is a system of *invariance analyzers* that compute the extent to which transformational relations exist within the output of the spatial analyzers. The third is a system of *frame analyzers* that relate and organize the output of the spatial and invariance analyzers. A perceptual reference frame is selected by an attentional mechanism that maximizes invariance and stability. This theory is then related back to particular phenomena that initially suggested such a scheme.

Because transformations and group theory will play such a central role in our discussion, it will be useful to review these concepts briefly before diving into the psychological literature. Readers already familiar with these concepts and their role in geometry may skip this section.

2. Transformations, Group Structure, and Invariance

A spatial transformation is a function that maps space onto itself. In other words, it maps each point in space to some uniquely determined point, possibly itself. Some well known examples are the rigid motions (e.g., translations and rotations), projections from three- to two-dimensional spaces, and topological deformations. The mapping can be one-to-one, as in the rigid motions, or many-to-one, as in the projective transformations that map all points along a ray through the projective plane onto the same point on the plane. It cannot be one-to-many, because then it would not be a function.

The structure of spatial transformations has been well studied by mathematicians. Of fundamental importance to us here is the fact that certain sets of spatial transformations have the structure of a *group,* and various subsets of them form *subgroups.* The mathematical entity known as a group is just a set of elements plus a composition operation for putting them together such that a few important properties hold. In the present case, the set consists of the spatial transformations and the operation is just that of performing two transformations in the specified order – e.g., rotating a plane through a 90-degree angle and then reflecting it about a vertical line. Let us denote the set of transformations by T, its elements by $t_1, t_2, \ldots, t_n$, and the composition operation by $*$. There are four basic properties a set must have with respect to its composition operation in order for it to have group structure: closure under composition, associativity under composition, an identity element, and inverses. The set is closed under composition if $t_j * t_k$ is an element of T whenever t_j and t_k are. This insures that when any pair of elements in T are composed, another element of T always results. For example, the set of all rotations about a specified point is closed under composition because when any two of these rotations are performed in order, the result is always a member of the set of rotations about that same point. The second property is that the composition operation is associative: the grouping of composed transformations does not matter in the specific sense that $t_i * (t_j * t_k) = (t_i * t_j) * t_k$. The third requirement is that there must be a unique element of the set, called the *identity* element, that can be composed with any transformation to give that same transformation as a result: formally, $i * t_j = t_j * i = t_j$ for all t_j in T, where i is the identity element. In the set of rotations, rotation through zero degrees is the identity. Finally, each element of the set must have an *inverse* which, when composed with it, gives the identity element as a result: formally, there exists an element, t_j', such that $t_j * t_j' = t_j' * t_j = i$ for all t_j in T. The set of rotations has inverses because any rotation can be "undone" by rotating back in the opposite direction through the same angle. This is important because it allows certain kinds of computations to be carried out using the group, particularly those involving "factoring out" transformations to achieve invariance (identity).

Certain sets of spatial transformations have group structure. As just argued, for example, the set of all rotations of a plane about a given point form a group since it is closed and associative under composition, has an identity element, and has inverses for all its elements. Rotations are actually a subgroup – a subset with group structure – of the group of all rigid motions, including all translations, rotations, and their composites of translations-and-rotations. These, in turn, are a subgroup of the Euclidean congruences, consisting of translations, rotations, and reflections, and their composites. Adding the dilations (radial expansions and contractions) to these gives the still larger group of Euclidean similarities, the primary group underlying Euclidean geometry. This group will turn up repeatedly in the pages to follow. Adding the projective transformations gives the projective group which underlies projective geometry.

As the preceding discussion suggests, group structure turns out to be the fundamental notion in the modern conception of geometry. A geometry is defined as the properties that are *invariant* over the transformations of a particular group. This is the basis of Klein's famous Erlanger Program. It specifies that one figure is equivalent to another figure in a given geometry if and only if there is a transformation in its underlying group that changes the first figure into the second. For example, any triangle with angles of 30, 60, and 90 degrees is equivalent (or "similar") in Euclidean geometry to any other triangle with these same angles because some transformation from the similarity group will map the first triangle into the second triangle such that they are point-for-point identical. When the transformations of the similarity group are studied, certain properties of figures change and others remain the same. For example, line orientation and length vary as figures are subjected to rotations and dilations, whereas angle size remains invariant.

Given the close relation between transformation groups and geometry on the one hand and geometry and perception on the other, it should not be too surprising that transformation groups turn out to be relevant to perception. Cassirer (1944) first brought groups to the attention of psychologists in his illuminating analysis of constancy in perception. This was followed shortly by Pitts and McCullough's (1947) classic paper on a neural networks for computing group invariants. The idea did not "catch on" in psychology despite these efforts until Gibson's (1966, 1979) penetrating discussions of transformations and the importance to perception of invariants over them. More recently, Hoffman (1966, 1978) has presented a formal theory of perception based on Lie transformation groups and Lie algebras. Still, not much attention has been paid to such ideas by mainstream perceptionists, perhaps because group theoretical language is not widely understood in psychology. I suspect, however, that group theoretic approaches to perception will prove invaluable in future theorizing about perceptual organization. At least they must be considered more fully and carefully in the future than they have been in the past.

3. Phenomena of Perceptual Organization

3.1. Shape Constancy

Shape constancy is perhaps the touchstone phenomenon of perceptual organization. It refers to the fact that people see two figures as having the same "shape" or "form" if they differ only in position, orientation, size, sense (mirror-image reflection), or some combination of these. A simple example is shown in Figure 1: people perceive all of these figures as having the same shape despite their many differences. It is easy to see why it is evolutionarily important for an organism to have shape constancy. It is not so easy to see how it is accomplished.

The perceptual intuitions underlying shape constancy are incredibly strong. Euclid built them into his geometry in the form of "similar" figures. Two figures generally have the same perceived shape if they are "similar" by Euclid's definition. It took several thousand years for geometers to break the perceptual shackles that bound them to perceived shape as the basis of geometric equivalence. Psychologists have had even greater difficulty in divorcing themselves from the intuitions of shape constancy. It was not until the Gestalt movement early in this century that the full scope of the shape constancy problem was realized. It is certainly one of the most important problems in perception, and it remains unsolved to this day.

Shape constancy is often presented as primarily a problem of recognizing the shape equivalence of objects "distorted" by projection in depth (e.g.,

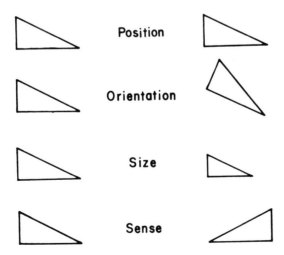

Figure 1. Examples of shape constancy for the same figure at different positions, orientations, sizes, and senses (reflections).

Levine & Shefner, 1981). The classical example is that a circle tilted in depth projects an elliptical image onto the retina, thus distorting its shape by "foreshortening." While this is indeed one aspect of shape constancy, it is but a small part of the problem. Stating it in this way, for example, implies that shape constancy is not a problem for figures that lie in the frontal plane and, therefore, are not "distorted" by projection. But in fact, it is not clear how two figures in the frontal plane are perceived as having the same shape when they differ in position, orientation, size, or reflection. We will see that the problem of shape constancy over depth projections is a straightforward conceptual extension of the more basic problem in the frontal plane.

We said earlier that two figures differing only in position, orientation, size, sense (reflection), or some combination of these properties have the same shape. This is an observation of major importance for theories of perceptual organization. Among other things, it suggests that the perceptual system factors the incoming stimulus information into five fundamentally different but interrelated domains: shape, position, orientation, size, and sense. Whatever the organization of the perceptual system might be, it must reflect this division of information.

Four of these five basic perceptual properties are intimately linked to simple transformation groups. Figures that differ only in position are related by *translations* along a line, those that differ in orientation are related by *rotations* about a point, those that differ in size are related by *dilations* (radial expansion and contraction) about a point, and those that differ in sense are related by *reflections* about a line. This means that, as mentioned earlier, figures having the same shape are related to each other by transformations of the Euclidean similarity group: translation, rotations, dilations, reflections, and their composites. Thus the transformational underpinnings of shape constancy begin to emerge.

Applying this transformational analysis to two-dimensional figures in the frontal plane is straightforward as illustrated in Figure 1. Extending it to shape constancy in depth is only slightly more difficult. First, it is clear that the same transformation group is at work, only in three dimensions rather than two. That is, two objects have the "same shape" in three-dimensional space, if they can be made to coincide by exactly the same set of transformations extended to three-dimensional space: translations, rotations, dilations, reflections, and their composites. However, if one considers the characteristics of the projected *images* that could have arisen from the same three-dimensional shape, the set of image transformations over which shape equivalence extends is considerably larger. In fact, it consists largely of the group of projective transformations that form the basis of projective geometry. Thus the transformational analysis of shape constancy in two- and three-dimensional space is identical at the "object" level and closely related even at the "image" level.

Before leaving the topic of shape constancy, it is worthwhile to consider briefly two classes of theories about how it might be accomplished. Both of them are fundamentally related to the transformational view described above, but in somewhat different ways. We will call them the *invariant features* hypothesis and the *reference frame* hypothesis for reasons that will become obvious once they have been described.

Invariant features. The invariant-features hypothesis assumes that shape perception is mediated by the detection of those geometrical properties of a figure that do not change (are invariant) when the figure is transformed in specific ways. Any given set of transformations partitions the total set of figural properties into two subsets: those that change over the transformation and those that do not. For example, when a figure is rotated through an angle (other than 360 degrees or any integer multiple of it), line orientations change, but angle sizes do not. In fact, angle size is an invariant feature for all figures under the Euclidean similarity group, since it does not change when figures are translated, rotated, dilated, reflected, or any combination of these. As mentioned earlier, each transformation group is associated with its own set of invariant properties. These are just the properties that mathematicians strive to discover when they study a geometry as defined by its transformation group.

The invariant features hypothesis, then, suggests that shape is represented by detecting just those properties of objects that do not change over the relevant set of transformations. As argued above, the set that underlies shape constancy is just the similarity group. Thus, angle-size, relative length of lines, number of lines, number of angles, closedness, connectedness and the like are possible candidates for invariant features of perceived shape because they are all invariant over the similarity group. This means that any pair of figures that have the same shape will be identical with respect to these figures.

If the invariant properties of figures in three dimensions are to be detected solely from properties of their projected two-dimensional *images,* however, the relevant set of transformations is the projective group. Its invariant features are a proper subset of those associated with the similarity group, since the similarity group is a subgroup of the projective group. For example, number of lines, number of angles, connectedness, and closedness are all invariant features of the projective group as well as of the similarity group. These would be good candidates for representing shape in a way that is invariant over projections as well as similarities. Angle size and relative length, however, change over transformations in depth and, therefore, are not invariants of the projective group.

The invariant features hypothesis has long dominated psychological theories of shape perception. Explicitly or implicitly, its assumptions underlie the classical "feature set theories" of pattern recognition proposed by Selfridge and Neissser (1963), Gibson (1969), and many others. Such theories are attractive, at least in part, because of their structural simplicity: shape is represented as a simple set (or list) of attributes. These can be weighted by importance and augmented by transformationally variant features (e.g., line orientation), but the basic proposal is always that a simple list of features is sufficient to explain shape constancy. The question of how these features might get selected – phylogenetically through evolution of the species or ontogenetically through perceptual learning by the individual organism – is an interesting one, but it will not concern us here. We are interested in whether, or to what extent, the invariant-features hypothesis is correct. If it is, we want to know which features are actually used to represent shape.

Reference Frames. The reference frame hypothesis makes use of the underlying transformation group in a somewhat different way. Rather than ignoring properties that vary over the transformations of the group, it assumes that the effects of transformations are neutralized by imposing an "intrinsic frame of reference" that effectively "factors out" the transformation, thereby achieving shape constancy. The "intrinsicness" of the frame simply means that the frame is chosen to correspond optimally with the structure of the figure rather than being imposed arbitrarily.

To see how this scheme can achieve shape constancy, consider an example from analytic geometry. Figure 2a shows three circles, a, b, and c, that have the same "shape," but differ in size, position, or both. Each can be described by an equation within a single reference frame, in this case a standard

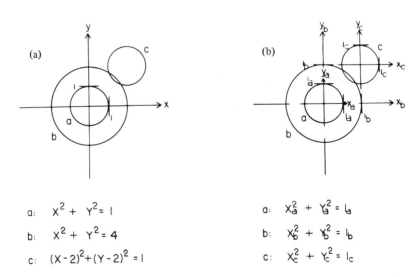

(a)

a: $X^2 + Y^2 = 1$

b: $X^2 + Y^2 = 4$

c: $(X-2)^2 + (Y-2)^2 = 1$

a: $X_a^2 + Y_a^2 = l_a$

b: $X_b^2 + Y_b^2 = l_b$

c: $X_c^2 + Y_c^2 = l_c$

Figure 2. Extrinsic and intrinsic descriptions of circles in different reference frames: equations for circles a, b, and c are given either relative to the same coordinate system (a) or different coordinate systems selected to correspond to the "simplest" description (b).

Cartesian coordinate system. As shown in Figure 2a, the three circles have different descriptions because *a* and *b* differ in size, *a* and *c* in position, and *b* and *c* in both size and position. However, their equations differ in part because they are specified relative to the same arbitrary reference frame. Suppose that *different frames* were used for each circle, ones that were somehow matched to the intrinsic structure of each circle to give the "simplest" description possible. This situation is depicted in Figure 2b. Now the equations are the same for all three circles. The reason for this is that the origin of the reference frame for each circle has been placed at its center and the unit size has been made to correspond to its radius. These are the two fundamental parameters of a circle's structure. By choosing *intrinsic reference frames* that are matched to these aspects of the circle's intrinsic structure, the variations due to size and position are eliminated. They are, in effect, absorbed by the frame itself, leaving the shape description (here represented by the equations) invariant over their differences with respect to these properties.

It should be clear at this point that intrinsic reference frames can compensate for differences in orientation and sense as well as in size and position. These are accomplished by rotating the frame about its origin and reflecting it about its major axis, respectively. The net result is that appropriately chosen intrinsic frames can compensate for structural changes induced by the similarity group: translations, rotations, dilations, reflections, and their composites. This means that if the same frame were always chosen for the same figure, shape constancy would result, since all properties would then be invariant with respect to that frame. The problem, of course, is how the visual system could choose the same frame each time. Part of the answer is that it does not always happen this way, but we will return to this complication in a later section on "reference frame effects."

Before leaving these two theories of shape constancy, we should note their similarities and differences. They are similar in that they both represent shape in terms of invariant properties over the similarity group, assuming that the appropriate frame is always chosen in the reference frame scheme. The differences lie in which properties are available for use and how they are detected. The reference-frame scheme allows virtually any property to be made invariant simply by relating it to the frame. The frame effectively absorbs transformational variation. For example, while line orientation varies over rotation, line orientation relative to an intrinsic frame (that rotates with it) is invariant. The same is true of information about location, size, and sense; each can be made transformationally invariant by selecting the same frame relative to the figure. There is no magic in this. While orientation, position, size, and sense vary over the similarity transformations, *relative* orientation, *relative* position, *relative* size, and *relative* sense are all invariant with respect to an appropriately chosen frame. The problem is how to extract this information. The reference frame hypothesis suggests that it is done by imposing a highly structured set of relationships on the figure in the form of a "frame" that establishes privileged reference standards relative to which information is coded. The invariant features hypothesis is usually applied without using such reference points, thereby reducing the set of invariant features potentially available. Certainly standard theories based on invariant features do not include any such reference system. We will return to these matters in later sections.

It has recently become clear that simple versions of the invariant features hypothesis will not do, at least for certain types of discriminations based on spatial structure. The evidence comes indirectly from an ingenious set of experiments by Shepard, Cooper, and their associates on "mental rotation" (see Shepard & Cooper, 1982). The basic facts that concern us are these: if people are asked to discriminate between pairs of very different shapes at different orientations, it makes little or no difference how these shapes are oriented. The amount of time required to discriminate or even identify them is essentially constant over different orientations. This result is compatible with the invariant-features hypothesis. However, if the task is to discriminate between two figures that differ only by a reflection, performance changes radically. The time required now varies almost linearly with the angle through which one figure must be rotated (see Figure 3).

These results are generally interpreted as indicating that the perceiver must mentally rotate one figure into the same orientation as the other before

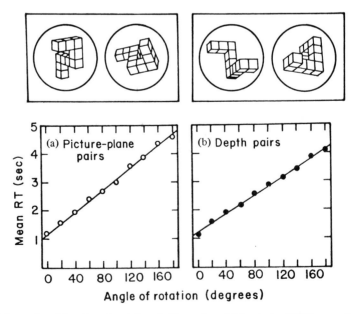

Figure 3. Stimuli and results of Shepard and Metzler's (1971) experiment on mental rotation showing that response time is linearly related to angle of rotation.

making the comparison. Similar results have been reported for figures that differ in size (Bundesen & Larsen, 1975): comparison time varies almost linearly with size disparity. These results are taken as evidence that perceivers perform mental dilations (size scaling operations) to compensate for differences in size. Nobody has yet reported mental translations to compensate for differences in position, but they may yet be found.

It seems, then, that certain kinds of shape comparisons can only be made when figures have the same orientation and size. This is what would be expected from the reference frame hypothesis. The mental transformations sound very much like operations that would be performed to align two frames so that their contents could be compared directly.

3.2. Motion

When parts of the world change their position on a time scale that is neither too fast nor too slow, people perceive motion. Of primary importance for perceptual organization is the structure of this perceived motion. There are countless different ways one could, in principle, experience the same positional change in the physical world, and yet only a few are perceived. Usually there is just one. This must reflect the operation of structural constraints imposed by the system. Once again, we assume that these constraints tell us something about the system's internal structure.

The first observation to be made is that motion *could* be perceived simply as the combined, independent translation of all those points whose locations change. This would be an unstructured motional percept such as one gets while watching random points in random motion. But when certain types of structural regularities exist in the pointwise motion of the objects, the visual system locks into them, and the perceived motion becomes highly organized as a result. In fact, the types of motion people experience are usually the rigid, continuous transformations of the similarity group: translations and rotations. (Reflection is not a continuous transformation and dilation is not a rigid one.) When an object undergoes a rigid translation or rotation, we do not perceive it as simply the simultaneous unstructured movement of many small bits. Rather, we see it as a unified, coherent motion of an object or set of objects. Certainly there are also cases in which we perceive non-rigid motions. A person walking or tree swaying in the wind are obvious examples. But even in these cases, the perceived motion seems to be structured in terms of roughly rigid parts that rotate relative to other roughly rigid parts (Cutting, 1981).

Psychologists typically study the organization of motion perception by investigating very simple stimulus conditions. In such cases the presented information usually underdetermines its interpretation. As a result, the perceptual system shows its "preferences" by achieving one percept rather than others. This often suggests properties that the underlying system might have.

A classical example is the "kinetic depth effect" discovered by Wallach and O'Connell (1953). When a complex, three-dimensional bent-wire figure is projected onto a two-dimensional screen, it is perceived as a flat lin :-drawing figure in two dimensions. However, when that same figure is slowly rotated in depth, the shape of the wire figure is rapidly and vividly perceived in three dimensions as it rigidly rotates about an axis. Note that there is no logical reason why the figure should not be seen as a two-dimensional figure whose shape deforms over time. The visual system thus proves to be highly sensitive to the structure of rigid motion and to prefer this interpretation over nonrigid motion of the two-dimensional image. Note that this implies equally that it prefers interpretations in which shape is invariant (i.e., constancy holds). Thus, preferences for rigid motion and for shape constancy are two different sides of the same coin.

Further evidence supporting the preference for rigid motion and unchanging shape comes from elegant experiments by Johansson (1950) on motion configurations. He showed observers simple and compound harmonic motions like those depicted in Figure 4. When each dot of Figure 4a is moved individually, it is perceived to move on the two-dimensional frontal plane in the direction indicated by the arrow. When pairs of dots move synchronously, however, they are seen as moving together in depth as though fixed to the ends of a rigid bar that alternately recedes and approaches the viewer (Figure 4b). An even more startling effect is obtained when two dots move around a rectangular path (Figure 4c). The dots are again seen as moving in depth as though attached to the ends of a rigid rod, but this time one that pivots around the center of the rectangle as shown in Figure 4d. Once again, there is no logical reason to choose this interpretation rather than one in which the points move independently in different directions on the frontal plane. But the overpowering result is that the visual system prefers to operate as though it were witnessing a rigid object undergoing uniform motion. A similar conclusion arises when the system is presented with image dilations (Figures 4e and 4f). Dilational events can be interpreted either as objects changing in size or translating rigidly in depth. The visual system prefers seeing rigid motion in depth to seeing nonrigid expansion or contraction (Regan, Beverly, & Cynader, 1979).

Yet another line of evidence that points to the same conclusion comes from recent results on apparent rotation (Sheppard & Judd, 1976; Farrell & Shepard, 1981). When two views of the same object at different orientations are shown in alternation (see the views shown in Figure 3), observers perceive smooth continuous rotation of a rigid object back and forth. This occurs, however, only if the rate of alternation is slow enough for the visual system to "construct" the rigid motion interpretation. When the rate is too fast, the object is seen to deform plastically rather than to rotate rigidly. Such deformations are theoretically possible at any rate of alternation, of course, but they appear only when there is insufficient time for the system to complete the con-

Stimulus   Percept

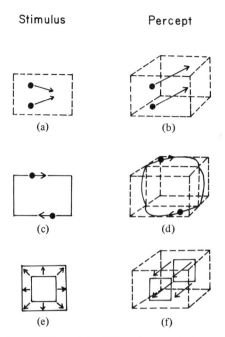

Figure 4. Stimulus conditions (left side) and perceptual responses (right side) to two of Johansson's (1950) moving dot configurations showing that dots are perceived as moving in depth as though attached to the ends of a rigid bar (a-d). Preference for rigid motion is also evident in perceiving an expanding square as moving toward the observer in depth (e and f).

tinuous rigid motion. This conclusion is based on the observed result that the minimum amount of time required to perceive the rigid rotation is a linear function of the angle through which the object is seen to rotate (Shepard & Judd, 1976).

I have observed similar effects in our laboratory using simple three-dot patterns and their 90-degree rotations. At rapid rates of alternation, preferences among possible alternative mappings are strictly consistent with the shortest path solution for each dot as an independent entity. At slower alternation rates, however, the dots tend to move together in rigid two-dimensional rotation. Since the longer paths of rigid rotation are preferred to the shorter ones of non-rigid motion, there must be some structural preference for constant shape and/or rigid motion. We infer that the computational structure of the system is somehow more congruent with such interpretations. Later we will discuss why this might be so.

More recently, Bundesen, Larsen, and Farrell (1981) have extended Shepard's findings in important ways. First, they demonstrated an analogous effect – apparent "looming" – for dilational transformations. Alternating between the same figure at two different sizes produces the perception of smooth, continuous translations in depth, provided the rate of alternation is not too fast. They showed that the fastest alternation rate for which depth motion was seen is a simple function of the distance in depth that the object was perceived to traverse.

More importantly, however, they investigated perceptions that result from simultaneous transformations of rotation and dilation. Phenomenologically, the experience is of an object rotating as it moves back and forth in depth. Quantitative analysis revealed a surprising result: the time required to complete the composite of these two transformations is the sum of the times required for each transformation alone. If they were performed simultaneously, the total time required for the composite transformation should have been much less than the sum of the two individual transformation times. This result is as important as it is unexpected. It suggests that the rotational and translational analyses are performed sequentially rather than simultaneously. The most reasonable interpretation is that the system represents the composite transformation as a rapidly alternating sequence of quantal translations and rotations. The phenomenal impression of simultaneous transformation is difficult to reconcile with a simple sequential model such as a single complete rotation followed by a single complete translation. Such results suggest that other composite transformations might also reveal additive sequential structure. This is a recently uncovered area of perceptual research that promises to provide new information in the near future about the underlying transformational structure of the visual system.

3.3. Figural Goodness

Another important phenomenon of perceptual organization that turns out to be intimately related to transformational structure is what psychologists call "figural goodness." The "goodness" of a figure refers to subjective feelings of order, regularity, and simplicity that arise when the figure is perceived. Certain stimuli, like circles and spheres, are felt to be "good," while others, like random arrays of dots and lines, are felt to be "bad." Figural goodness is not an aesthetic response *per se,* although the two are related (e.g., Berlyne, 1971). Some people prefer highly regular, simple figures, whereas others prefer complex, irregular ones, yet almost everyone agrees on which figures are good (simple, regular, etc.) and which ones are bad, despite their varying aesthetic preferences.

Figural goodness is a subjective state and, as such, is impossible to study directly. What psychologists do is to ask subjects to make ratings of the degree of "goodness" they experience when looking at various figures. People seem to understand the task at an intuitive level once words like "simple," "regular," and "orderly" are mentioned. Psychologists are interested in the stability of these ratings from one person to another (interobserver agreement) and in the systematic ordering they impose on figures. In fact, many years of research have shown goodness ratings to be highly stable over observers, producing powerful, highly replicable results over figures. The reason that these results are believed to be relevant to perceptual organization is that "simplicity," "complexity," and "regularity" seem necessarily to be relative to the internal structure of the system processing the information. What is simple and easy for one system to process may be complex and difficult for another system. As we shall see, figural goodness does in fact relate to the ease with which stimuli are processed, and ease of processing is believed to provide information about the internal structure of the system. I will argue that figural goodness again suggests that the underlying structure of the system is transformational.

Subjective ratings of figural goodness would be of negligible interest were they not related to performance on other perceptual tasks. As it turns out, performance on many tasks depends heavily on how "good" the stimuli are according to such ratings. For example, good figures are better recalled than bad figures, better recognized, more quickly and accurately matched for physical identity, more simply and easily described, and more likely to be generated as completions of partial patterns (see Garner [1974] for review). The general thrust of these results is that the visual system is better designed for handling the kind of structure in good patterns than in bad patterns. If there were an adequate analysis of what that structure is, we would glean some insights into the nature of the system that is "tuned" to it.

There is a long history of attempts to characterize the nature of figural goodness. It began with Gestalt psychologists who first realized its potential importance for understanding visual organization. Unfortunately, they never produced any real theory. For them it was an unanalyzable primitive. They did realize that goodness was related to symmetry in some way, pointing to bilateral symmetry as a prototype, but they never went too far beyond this. The next development was an attempt to ground figural goodness in information theory (e.g., Attneave, 1954; Hochberg & McAlister, 1953). Good figures were shown to have predictable structure and, therefore, to contain little information. Leeuwenberg (1971, 1978) has produced the most complete and sophisticated theory of this type. In his "coding theory," a figure's goodness is equated with the parametric complexity of the code required to generate it.

The analysis of figural goodness that I find most illuminating, however, is based on the construct of transformational invariance. It was first articulated by Garner and Clement (1963) in accounting for goodness ratings made by observers in response to simple five-dot patterns such as those shown in Figure 5. Garner's proposal was that patterns were good to the extent that they were the same as themselves after being subjected to particular transformations. He considered eight transformations: central rotations through angles of 0, 90, 180, and 270 degrees and reflections about vertical, horizontal, left diagonal, and right diagonal lines through the center. When each dot-pattern is transformed in these eight ways, some number of distinguishably different fig-

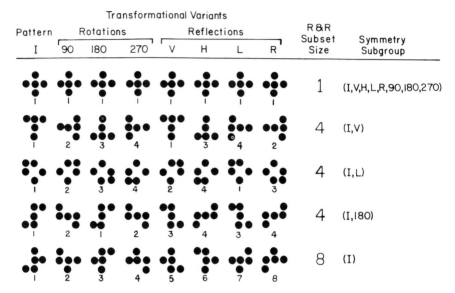

Figure 5. Analyses of figural goodness from the viewpoints of transformational variants (R & R subset size) and invariance (symmetry subgroups).

ures are generated (see Figure 5). This set of figures constitutes what Garner called the "rotation and reflection (or R & R) subset" of the figure. It is the set of *transformational variants* of the figure for these eight transformations. In their experiments Garner and Clement (1963) found that figural goodness varied inversely with the number of figures in the R & R subset: good figures have few transformational variants whereas bad figures have many. Thus, Garner's transformational measure of figural goodness is R & R subset size.

Garner's description of figural goodness is couched in the language of transformational variance. I prefer to describe it in the opposite, but nearly equivalent, language of transformational invariance. Garner's general hypothesis can be restated as follows: good figures have greater transformational invariance (symmetry) than bad figures. In fact, his results can be formulated in group theoretic terms. The eight transformations he used constitute a group of transformations. For each figure there is a set of transformations that leave that figure completely invariant (i.e., the figure is the same as itself after the transformation). Because this set is a subset of the transformation group and also constitutes a group, it is a subgroup of the initial one. It is called the "symmetry subgroup" because it includes the reflectional (or bilateral) symmetries and the rotational symmetries of the figure. I am using the mathematical notion of symmetry here (cf. Weyl, 1952) which is an extension of the normal usage restricted to bilateral symmetry. In mathematics a symmetry is any transformation that leaves its object unchanged. Therefore, the 90-degree rotational symmetry of a square is on the same logical footing as its bilateral symmetry about a vertical line through its center; both leave the square unchanged.

I prefer the language of transformational invariance (symmetry subgroups) to that of variance (R & R subsets) because it provides a more refined analysis, is more easily extended to other sorts of geometrical regularities, and is more transparent to the computational structures that I believe underlie figural goodness effects. I will attempt to justify the first two claims now and the third in the second half of the chapter.

The refinement of Garner's analysis I wish to make concerns differences in goodness due to different types of symmetries. For example, a figure with only one line of reflectional symmetry is given its highest goodness rating when that line is vertical, next highest when it is horizontal, and lowest when diagonal (Chipman, 1977; Palmer & Chase, in preparation). This result is consistent with well known orientational effects in explicit symmetry detection tasks (e.g., Palmer & Hemenway, 1978). The difficulty this poses for Garner's analysis is that the R & R subsets of the same figure at different orientations are identical and, obviously, so are the sizes of these subsets. However, the transformations contained in the symmetry subgroups are distinct (see Figure 5): reflection about a vertical line is not the same as reflection about a horizontal or diagonal line. Therefore, rather than identifying goodness simply with the *number* of transformations in the symmetry subgroup, we identify it with the transformations that comprise the group. This also takes care of another, closely related problem with Garner's measure: namely, that rotational symmetries are distinct from reflectional symmetries. Note that the measure of R & R subset size does not distinguish between a figure with 180-degree rotational symmetry and one with a single reflectional symmetry (see Figure 5). The symmetry subgroup analysis allows for the possibility that different types of symmetries have different effects on perceived goodness. Our results indicate that such differences exist; reflectional symmetries produce more powerful effects on goodness ratings than rotational symmetries (Palmer & Chase, in preparation).

As mentioned above, the general mathematical notion of symmetry is that an object is the same before and after it has undergone a transformation. Rotations and reflections are transformations over which figures can be invariant, but they are not the only ones. Consider for example, an infinite line. If it is translated along itself by any distance, the resulting line is identical to the initial one. Therefore, an infinite line has translational symmetry. A line of equally spaced dots also has translational symmetry, but only for distances that are integer multiples of the distance between adjacent dots. Thus, the notion of symmetry as transformational invariance generalizes naturally to translations.

Symmetry also generalizes to dilations. If an infinite line is dilated outward (or inward) from any point along it, the line does not change. The same is true for an infinite angle dilated from its vertex. Therefore, the notion of symmetry subgroups generalizes naturally to the whole similarity group. For example, the symmetry subgroup for an infinite straight line includes: all rotations through 180 degrees about any point on the line, all reflections about a line normal to the given line, the single reflection about the line itself, all translations by any distance along the line, and all dilations from any point on the line.

The main difficulty in extending the symmetry analysis of figural goodness to translations and dilations is that, strictly speaking, only infinite patterns contain these symmetries. It turns out, however, that a similar analysis can be carried out for *local symmetries* over restricted regions of the figure (Palmer, 1982). Finite patterns contain many such local symmetries of translation and dilation. Although the resulting sets of local symmetries do not have the structure of a group, they still constitute an analysis in terms of transformational invariance. The extraction of local symmetries will turn out to be central in the computational theory to be described later.

I have recently begun to study the influence of both local and global symmetries on figural goodness judgments. In an initial study observers were shown 35 index cards, each of which contained a rectangle and a single small circle at one of the positions shown in Figure 6a. They were asked to rate how good, stable, and natural the position of the circle was within the rectangle on a scale from 1 (bad, unstable, unnatural) to 7 (good, stable, natural). Notice that the rectangle has global symmetries of reflection about its central vertical line

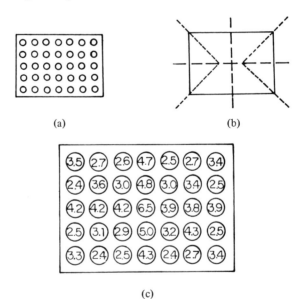

(a)   (b)

(c)

Figure 6. Symmetry structure inside a rectangle as measured by goodness ratings (c) for a small circle placed at any one of 35 positions (a) that lie along global (dashed) or local (dotted) symmetry axes (b).

and about its central horizontal line (dashed lines of Figure 6b). It also has *local* symmetries of reflection about the bisectors of its angles (dotted lines in Figure 6b). The question was whether the goodness ratings would be systematically related to the analysis in terms of global and local symmetries.

The results are displayed in Figure 6c. By far the highest ratings were given to the central position. Here the central symmetries of the circle coincide with both global symmetries of the rectangle. The next highest ratings occurred along the vertical axis of global symmetry. This is consistent with the previous argument that vertical reflection is a more salient transformation for figural goodness than horizontal or diagonal reflection. The next highest ratings occur along the horizontal axis of symmetry, the only other global symmetry present in a rectangle. Next come the dots that lie along the bisectors of the four angles. These are the local symmetries that unify only a limited region of the figure. But they are still perceived to be better, more stable positions for the dot than the remaining ones. Thus, the results conform to the kind of symmetry analysis given above. A rectangular frame organizes the space within it according to at least some of the transformational relationships it contains. Further research is in progress using other shapes as frames and other elements as the targets within the frames. Preliminary results continue to suggest that the controlling factor is alignment of symmetry axes between the frame and the internal figure.

3.4. Perceptual Grouping

The phenomena that are probably most closely associated with the topic of perceptual organization are the so-called "grouping" effects. The most basic observation about grouping is that the perceptual system has an overpowering tendency to put together elements of the visual field in terms of "belongingness." Functionally speaking, this process of grouping is probably the basis for parsing the visual world into discrete surfaces and objects on the basis of relatively simple visual characteristics. Psychologists further suppose that understanding how the system accomplishes this task will provide additional information about its underlying structure (Wertheimer, 1923).

Some standard examples of Wertheimer's grouping effects are shown in Figure 7. In each case the relevant question is, "What goes with what?" The answers are generally provided by simple introspection, but presumably they could be validated and even quantified by appropriate speed and accuracy measurements. In Figure 7a, for example, people almost invariably perceive the dots as organized into vertical columns rather than horizontal or diagonal rows. If the matrix is rotated 90 degrees, however, the same stimulus is spontaneously organized into horizontal rows. Thus, this figure demonstrates the powerful role that spatial proximity plays in perceptual grouping: all else being equal, figures that are closer together tend to be grouped together. This was Wertheimer's first "law" of grouping. The remaining laws identify other fac-

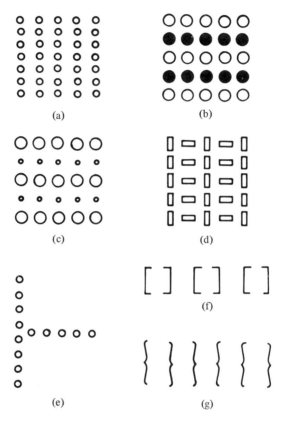

Figure 7. Examples of perceptual grouping by proximity (a), color (b), size (c), orientation (d), continuity (e), closure (f), and symmetry (g).

tors that affect perceptual grouping. Some of the most important ones are demonstrated in Figure 7: similarity in color, size, and orientation, plus continuity, closure, and symmetry.

One of the most potent factors of all is what Gestaltists called "common fate." It cannot be shown in a static display because it concerns the tendency for elements to be grouped together when they move simultaneously in the same direction at the same rate. Even an otherwise homogeneous, unorganized field of random dots will spontaneously organize into separate regions when a subset of the dots begins to move together at the same rate and in the same direction. When the dots stop moving, the grouping rapidly dissolves into a unitary field. Common fate may be viewed as another example of grouping by similarity in which it is the motional properties of the elements that are similar.

I want to claim that grouping phenomena constitute another example of the central role that transformational structure plays in perceptual organization.

The key is to analyze the local symmetries (transformational invariances) that are present in the grouping stimuli. For example, in Figure 7a each dot is related to every other dot by a translation. Various adjacent dot pairs and triples are also transformationally related to other dot pairs and triples. All of these transformational relations mean that the local symmetry set for this figure contains many restricted translational invariances, each of which is defined by a direction, a distance, and a local region for which the symmetry holds. Now, this set can be ordered by translational distance. When it is, the dominant organization into vertical columns emerges as the sets of elements related by the shortest translations. The only other organization that can be perceived without great difficulty is that of horizontal rows, and this is given by the sets of elements related by the next shortest translational invariances. Perhaps sometimes, with great effort, the diagonal organization can be achieved, and this organization is given by the third shortest translational invariances. Thus, the law of proximity can be reformulated by saying that grouping preferences coincide with the ordering imposed by a metric on the set of local translational symmetries in the pattern.

The laws of organization by similarity are straightforward generalizations of the analysis given for proximity, except that the group of transformations involved must be expanded to include (at least) rotations, dilations, and reflections as well as translations: i.e., the whole of the similarity group. Consider Figure 7c, for instance. Each element in the array is related to every other element by either a simple translation or a translation plus a dilation. Even if the minimal translational distances are equal for groupings by rows and by columns, the additional dilations required for the columnar organization make the transformational distance between adjacent horizontal elements smaller than that between adjacent vertical elements.

It is easy to see this in a spatial model. The two-dimensional stimulus array can be thought of as being projected into a higher dimensional space that includes factors other than position. In the current example, the elements of the array project into points in a three-dimensional space that includes two spatial dimensions plus a size dimension. The same-sized elements within a single row project into the same plane of the two-valued size dimension, and alternate rows project into different planes. Thus, the distance between row-adjacent elements in this space is less than that between column-adjacent elements. The organization that is perceived, then, is given by the sets of elements related by the minimal transformations, just as in the case of proximity. The only difference lies in the added dimensions introduced by variations in another factor. In Figure 7b the additional factor is color (a nonspatial transformation of reflectance), in 7c it is size (dilational transformation), and in 7d it is orientation (rotational transformation). Each case can be analyzed in terms of transformational distance, and the result is always that the dominant grouping is given by the sets of elements related by the minimal transformations.

It might sound as though the current transformational account is just restating the Gestalt laws in a new language. In a sense this is true, but there are distinct advantages to the new versions. Perhaps the most important is that it provides a uniform framework within which all of the laws can be interrelated and integrated. The Gestalt formulations treat each factor as a distinct and isolated case. There is no way to formulate what might happen when more than one factor is involved: all the laws are *ceteris paribus* rules. The present account in terms of transformational distance in a multidimensional space suggests a way of formulating the combined effect of several factors. They should combine according to some distance function within the multidimensional transformation space. Although no one, to my knowledge, has yet undertaken to study it, one could certainly determine the nature of the distance function empirically. Would the effects of size and orientation be independent and additive, as was found by Bundesen, Larson, and Farrell (1981) for apparent motion? The present formulation establishes a unified framework within which it is sensible and natural to pose and to answer such questions.

Another advantage of the transformational account is that it is easily extended to the factors of continuity, closedness, and common fate. Consider first the case of continuity. The puzzle is why the dots in Figure 7e are organized as they are when the three dots near the intersection are equidistant from each other. Considering the transformational distance from one dot to another will not do, but consider the transformations by which *pairs* of dots are related to each other. Each *pair* of adjacent dots in Figure 7e is related to each other pair by either a translation or a translation plus a rotation. Extending our previous analysis, we construct a model in which *pairs of dots* project into points in a higher dimensional space that includes both position and orientation. In this transformation space, pairs that are collinear lie in the same orientation plane and those that are not collinear lie in different orientation planes. The minimal dot pairs of Figure 7e thus form two distinct sets when projected into this space, and the sets of pairs that are related by minimal transformations define the grouping that dominates perception. It is the same account as given above for size, orientation, and color except that multi-element units are analyzed. Since closed figures are everywhere continuous, the same general account may be extended for grouping by closedness.

Common fate is another straightforward extension of the analysis in terms of minimal transformations. Motional properties project the elements into a higher dimensional space in which motion separates the two sets into parallel planes. Grouping follows the set of elements that share the common motional parameter just as it does for sizes, orientations, colors, and the like.

There is a more modern literature that bears on perceptual grouping. It is exemplified by the work on texture segregation by Beck (1972) and Julesz (1981) and on feature integration by Treisman (1982). Since these topics are covered in depth by Beck, Prazdny, and Rosenfeld in this volume, I will simply make a few observations on how this work relates to the classical literature and to the present transformational view.

Those properties that allow textures to be segregated "preattentively" are, more or less, what one would expect from the Gestalt literature: orientation, size, density (proximity), and perhaps continuity and closedness. Although Julesz (1975) initially described his results in terms of detecting differences in first-order and second-order global statistics, his more recent theorizing on "textons" (Julesz, 1981) has begun to sound more like Beck's (1972) theory and the classical literature. Textons are texture analyzers that respond to local features of patterns such as the orientation and size of elongated blobs and the presence of line terminals. These local features are closely related to the present view because the underlying processing elements are related by similarity transformations (i.e., all elongated blobs and all line terminals are the same except for their positions, orientations, and sizes).

Quite recently there has been an interesting development in the psychological literature related to perceptual grouping. It comes from Treisman's work on the role of attention in spatial integration of features (e.g., Treisman, Sykes, & Gelade, 1977; Treisman & Gelade, 1980; Treisman & Schmidt, 1982). She has shown that a fundamental difference exists between grouping on the basis of single features (or covarying multiple features) and conjunctions of independent features. Examples are shown in Figure 8. When there is a difference between two half-fields in brightness alone (Figure 8a) or in shape alone (Figure 8b), people are able to segregate the halves automatically, despite the irrelevant and unsystematic variation of the other dimension. However, when the half-fields are defined by the *conjunction* of two feature values (Figure 8c: dark-and-square or light-and-circle vs. light-and-square or dark-and-circle) people do not spontaneously perceive this organization. Treisman argues that these and many other results support her claim that focal attention is required to conjoin features in the same location. The relevance of this to the

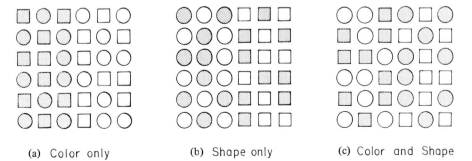

(a) Color only   (b) Shape only   (c) Color and Shape

Figure 8. Examples of spontaneous grouping by color alone (a) and shape alone (b) versus lack of grouping for conjunctions of color and shape (c).

present view is that it suggests that locational dimensions are primary and that other features are organized in terms of them. For example, the "higher-dimensional space" into which the stimulus arrays of Figure 7 get projected may actually consist of several separate spaces, *each* of which contains positional dimensions plus one other, be it size, orientation, or whatever. Then, if a task required conjunctions of such features to be discriminated, it could only be accomplished by an attentional act in which the same position is selected within all higher-order spaces. This possibility is to be contrasted with one in which all dimensions are orthogonally combined so that all featural properties are pre-conjoined, as it were, in the structure of the system itself.

3.5. Frames of Reference

The final category of phenomena I want to discuss is one I believe to be particularly important in understanding perceptual organization. I call them "reference frame effects" because they suggest that the perceptual system constructs descriptions of events in relation to a background structure that resembles a reference frame in analytic geometry. The claim is that properties of objects and events are perceived relative to stable reference standards that unify various parts of the visual field. These reference standards operate like the origin, axes, directions, and unit distances in a standard Cartesian coordinate system, a structured set of assumptions that allow geometrical entities to be described algebraically. Suggestive evidence for this sort of proposal comes from several different domains, particularly shape and motion perception.

3.5.1. Shape

Some of the most compelling reference frame effects are actually counterexamples to shape constancy. Perhaps the simplest and most elegant of these is the fact that when a square is rotated 45 degrees (see Figures 9a and 9b), it is generally perceived as an upright diamond rather than as a tilted square (Mach, 1897). Now, if shape constancy were perfect, as presumably it would be were only invariant features detected, these two figures would be seen as rotational variants of the same shape. Their shapes are often *not* seen as the same, however, at least not in the same sense that, say, upright and 45-degree "A"'s are seen as the same shape in different orientations.

Probably the simplest explanation for this breakdown in shape constancy is that shape is perceived relative to a reference-frame-like structure in which the orientation of the axes is taken as the descriptive standard (e.g., Rock, 1973). This is in contrast to the "invariant features" hypothesis discussed earlier. By the reference frame account, one sees a "square" shape when the sides are oriented parallel and perpendicular to the reference orientation and a "diamond" shape when they are oblique. Thus, perceived shape depends on the orientation of the reference frame relative to the figure. It is not clear what sort of account one could give using the invariant-features approach.

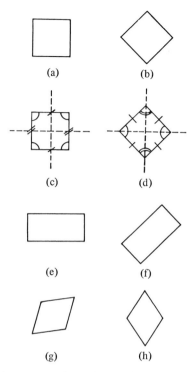

Figure 9. Square and diamond as distinct and ambiguous shapes.

I want to pursue this example in some depth, because I believe that a great deal can be learned from it. What does it *mean* to say that one is square-shaped and the other is diamond-shaped? Certainly the orientation of the sides has something to do with it, but the implications of frames for shape perception run deeper than this. I want to argue that people perceive different geometrical properties of the figure when it is seen as a square rather than as a diamond. Leyton (1982) has presented this argument independently from a group-theoretical viewpoint.

First of all, it is perceptually obvious that the square is bilaterally symmetrical about the bisectors of its opposite sides. It is far less obvious that a square is also symmetrical about the bisectors of its angles, although this can easily be verified by appropriate inspection. For a diamond, the reverse is true; the symmetries about the bisectors of its angles are quite prominent and those about the bisectors of its sides are much less so. These differences are clearly related to the well-known superiority of detecting vertical and horizontal symmetries over detecting diagonal ones (e.g., Goldmeier, 1972; Palmer & Hemenway, 1978; Rock & Leaman, 1963). Such differences in perceived symmetry suggest further differences in what other geometrical relations are perceived. Suppose that only the vertical and horizontal symmetries are detected for both the square and the diamond. What other geometrical relations could be derived from just these two? In the case of a square, these symmetries imply that opposite sides are equal and that adjacent angles are equal (see Figure 9c). Since all adjacent angles are equal, all angles are equal. The vertical and horizontal symmetries of a diamond imply the opposite equivalences: adjacent sides are equal and opposite angles are equal (see Figure 9d). Since adjacent sides are equal, all sides are equal. The net result is that very different properties are salient for squares and diamonds. This further suggests, according to Leyton (1982), that squares should be perceived as most similar to rectangles, in which opposite sides and adjacent angles are equal. Diamonds should be seen as most similar to rhombuses, in which adjacent sides and opposite angles are equal. Although we have not yet verified this conjecture with rigorous experimental methods, the reader can judge for him/herself the phenomenological validity of the claim by comparing Figures 9e-h to 9a and 9b.

Hinton (1979) has made some closely related observations about reference frames and reported some relevant experimental evidence using mental images of three-dimensional cubes rather than perceptions of squares. Hinton asked his subjects to imagine a cube sitting flat on a table (Figure 10a). He then asked them to imagine rotating this cube so that two opposite vertices were vertically aligned (Figure 10b). Once they had accomplished this transformation, he asked them to point to the additional vertices. Nearly everyone pointed to four coplanar vertices of a square on a horizontal plane bisecting the line between the opposite vertices. In fact, this does not define a cube, but quite a different shape (Figure 10c). The correct answer is more complicated: there are actually six vertices that lie alternately on two parallel planes connected by edges that go back and forth between them (see Figure 10b). Hinton (1979) has several other nice demonstrations of this sort. They all suggest that a cube, like a square, has more than one structural organization, depending on the orientation of a frame of reference. Each makes different geometrical relations clear, and people are not at all facile at moving back and forth between them. They are like different descriptions of the same object.

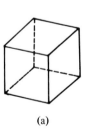 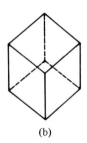 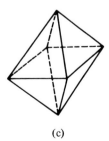

(a)   (b)   (c)

Figure 10. Three dimensional shapes used by Hinton (1979): when subjects imagined rotating a standard cube (a) into an unfamiliar orientation (b), they usually made errors in specifying its shape (c).

What I believe emerges from this type of analysis is that the ultimate basis of a reference frame is the geometrical relations it affords the observer about the structure of the figure. These geometrical relations are just the transformational invariances that define the frame. Since it would be impossible to extract all the possible relations in the input – there being infinitely many – the system must settle for a workable subset, the most stable and useful one it can find. It is this set of transformational relations that I want to identify as the structure of the perceptual reference frame. We will consider its nature more deeply a bit later.

It may seem unreasonable to make such a fuss over these few oddities of shape constancy. However, Rock (1973) has shown them to be a far more pervasive phenomenon, as will be discussed shortly. The importance of these counterexamples lies in their suggestion that reference frames of some sort are implicated in how shape constancy is normally achieved. As is often the case, failures turn out to be at least as illuminating as successes. However, we must not lose sight of the fact that shape constancy is the rule and its failures the exceptions. This means that the frame is usually established in the same orientation relative to the object; only rarely does this fail to occur. We now consider evidence about the conditions under which each result occurs.

Establishing the frame. Rock (1973) was a pioneer in studying how shape and orientation might be perceived relative to a reference frame. He showed that when certain kinds of novel shapes are presented in one orientation and tested for recognition memory in another (see Figure 11a), people are far less likely to recognize them than if they were tested in the same orientation at which they were presented. This performance decrement indicates failure of shape constancy. Rock further showed in many clever studies that the primary factors in determining the reference orientation are environmental and/or gravitational rather than retinal. When observers changed their head orientation by 90 degrees between presentation and test, recognition performance was not disrupted nearly as much as when the orientation of the figures was changed by 90 degrees. Rock took these and related results as evidence that shape is perceived relative to an environmental frame of reference in which gravity defines the reference orientation.

Wiser (1981) has recently refined Rock's analysis in important ways. She has shown that Rock's results do not hold when the stimulus has a well-defined "intrinsic axis." Such figures are recognized as well when they are tested in different orientations as when they are tested in the same orientation. Under exactly the same conditions she (a) replicated Rock's results using figures that, like Rock's, lacked a good intrinsic axis (Figure 11a) and (b) failed to replicate it using figures that possessed a good intrinsic axis (Figure 11b).

In further experiments she showed that when an intrinsically structured figure is presented so that its axis is *not* aligned with vertical, subsequent recognition is fastest when the figure is tested in its *vertical* orientation. Wiser inter-

PRESENT   TEST

Same Orientation   Different Orientation

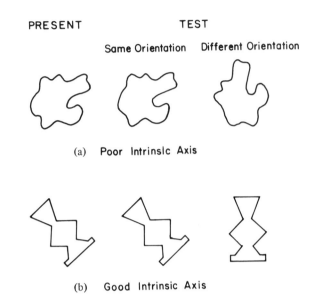

(a)   Poor Intrinsic Axis

(b)   Good Intrinsic Axis

Figure 11. Comparison of figures with poor (a) and good (b) intrinsic axes in a memory task for figures in different orientations.

prets this result to mean that the shape is stored in memory as though it were upright, at least relative to its own reference frame. This result is clearly at odds with Rock's hypothesis that shape recognition is best when figures are presented and tested in the same orientation. It is still consistent with an account in terms of reference frames, however. One merely assumes that the figure's internal frame of reference is established by its own intrinsic structure when that structure is sufficiently strong. Wiser's results then imply that the process of recognition operates most efficiently and effectively when the intrinsic frame of the figure is properly aligned with the extrinsic gravitational frame of its environment. The important difference between this account and Rock's initial one is that the shape of a figure is perceived relative to its own intrinsic frame of reference when it has the type of structure that determines one, and relative to an environmental frame when it does not.

Notice that Wiser's results are consistent with shape constancy. The figure is recognized best in a different orientation because it is perceived as having the same shape. Note also that this only happens when the figure has a good intrinsic axis that "drives" the reference orientation to the same axis both times. When the figure has more than one equally good axis, shape constancy will sometimes fail. Thus, these results support the reference-frame hypothesis of shape constancy.

What is the nature of the stimulus structure that establishes such intrinsic axes? One very potent factor is bilateral symmetry and another is global elongation (Wiser, 1981). As discussed earlier, bilateral symmetry is invariance over reflections about the intrinsic axis, and elongation may be related to approximate translational invariances along an axis. Thus, both sorts of structure are compatible with a transformational account in terms of symmetries.

I have reached similar conclusions about how reference frames are established quite independently by studying a different problem. The basic phenomenon with which I began was the perceived pointing of "ambiguous" triangles. Equilateral triangles are ambiguous in the sense that they can be perceived to point in any of three directions, but only one of them at once (Attneave, 1968). Thus, the triangle in Figure 12a can be seen to point toward either 3, 7 or 11 o'clock and flips back and forth among them. Figure 12b shows that a random field of such triangles all point in the same direction at once and that they all change direction at the same time. I was interested in what happened when several triangles were placed in well-structured configurations. As shown in Figure 12c, triangles aligned along one of their axes of symmetry are strongly biased toward pointing in a direction that coincides with the configural line. When they are aligned along one of their sides (Figure 12d), they point in a direction that is perpendicular to the configural line. Both of these effects have been verified quantitatively using self-report measures (Palmer, 1980) and perceptual performance techniques (Palmer & Bucher, 1981).

In a series of further experiments we have shown that similar bias effects can be induced by placing stripes inside a single triangle (Figures 12e and 12f). Again, stripes parallel to an axis of symmetry bias that direction and stripes parallel to a side bias a perpendicular direction (Palmer & Bucher, 1982). It turns out that stripes on the perceptual ground produce similar, but weaker, results. Another structural factor that produces this type of effect is the presence of a rectangular frame that surrounds a triangle (Figures 12g and 12h). Again, the triangle is biased toward pointing along the long axis of the rectangle when it is aligned with one of the triangle's axes and perpendicular to the long axis when it is parallel to one of the triangle's sides (Palmer, in preparation, a). These are good examples of what might be called "Gestalt" effects: the structure of the whole configuration strongly affects how the parts of that configuration are perceived.

What sort of stimulus structure is responsible for these configural bias effects? At first I was drawn to the elongation of the stimuli that produced biases. Unfortunately, this does not explain why an elongated stimulus would produce a bias along a direction *perpendicular* to its axis of elongation that was at least as strong as the bias *parallel* to its axis. Although one can account for the perpendicular effect by postulating a mediating mechanism such as interaction between perpendicular orientations (e.g., Palmer, 1980; Palmer & Bucher, 1981) it would be preferable to account for it directly. The direct account is in terms of reflectional symmetry (Palmer & Bucher, 1982). It turns out that all of the displays that produce a bias have a single axis of global symmetry along

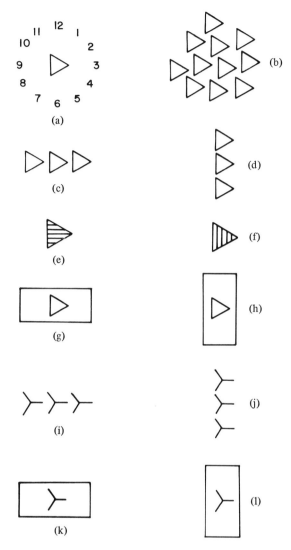

Figure 12. Ambiguous triangles and Ys in conditions that bias perceived direction of pointing.

which the bias occurs. This suggests that the perceptual system first responds to this global symmetry which then affects the perceived orientation of the ambiguous triangle. The bias must be rather strong, because people show its effects even when they try very hard to ignore it (Palmer & Bucher, 1981).

Here is a brief description of my current ideas about why these phenomena arise and how they are related. First, let us assume that people perceive the *shape* of an equilateral triangle relative to an intrinsic, oriented reference frame and that the triangle's *orientation* is given by the relation of this frame to a larger environmental reference frame. Let us further assume that the visual system has a powerful tendency toward establishing the orientation of a reference frame along an axis of reflectional symmetry, if one exists. Equilateral triangles have three such axes in their symmetry subgroup, the dihedral group, D3. Therefore, such triangles are three-ways ambiguous in orientation and direction of pointing, depending on which axis of symmetry is selected for the orientation of the frame (see Figure 13a). The triangle's shape is not correspondingly ambiguous because all of its geometrical properties are invariant over the transformations that relate the alternative frames (i.e., rotations through an angle of 120 degrees and its integer multiples).

When additional elements are added to the display, their symmetries may or may not align with those of the original triangle. In the present cases – configural lines, textural stripes, and rectangular frames – these biasing factors all have a two-fold symmetry subgroup, the dihedral group D2, as illustrated in Figure 13b. The intersection of these two sets of transformations is the symmetry subgroup of the resulting composite display. If it contains more than just the identity transformation, it can contain only one more, and this transformation is reflection about its global axis of symmetry. Now, we have assumed that the reference orientation is established along an axis of symmetry, if one exists. The reference orientation established for the whole display, then, will coincide with its global axis of symmetry (see Figure 13c). The two merely local axes of symmetry of the triangle will be less likely to be selected as a result, and the triangle will tend to be seen pointing along the line of global reflectional symmetry. Note that this analysis holds equally for all three types of bias and for both the axis-aligned and base-aligned versions of each with no further assumptions.

This theory has the virtue of being easily extended to other cases because it appeals only to global reflectional symmetries of stimuli. Therefore, it should apply equally to any other case that has the same symmetry subgroups, D3 and D2. Figures 12i-l show one such extension to Y-shaped figures (symmetry group D3) positioned in configural lines or inside rectangular frames

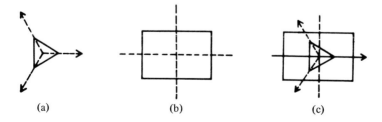

(a)   (b)   (c)

Figure 13. A symmetry analysis of bias effects in perceived pointing of ambiguous triangles.

(symmetry group D2). These Y's, like equilateral triangles, are ambiguous in that they are seen to "point" in one of three directions. They also can be biased toward pointing along axes of global symmetry within linear configurations and rectangular frames.

A second advantage of the present account is that it is easily extended to other symmetry structures. In fact, we have already discussed one important example: the square/diamond ambiguous figure. The ambiguity in shape of the square/diamond is analyzed as follows. This figure has four-fold symmetry (the dihedral group, D4), and so its reference frame can be chosen in any of four ways. However, the descriptions of the figure within these four frames are not all different. The geometrical properties of the figure do not differ between the two axes that bisect the sides. Neither do they differ between the two axes that bisect the angles. But the properties of the figure *do* differ between these two sets of axes, and so the shape of the figure is two-ways ambiguous. As mentioned earlier, the sides are parallel and perpendicular relative to the side-bisector frames, and these produce the "square" shape. The sides are oblique relative to the angle-bisector frames, and these produce the "diamond" shape. The theory further predicts that adding factors with symmetry axes that align with one set or the other should selectively bias these two different shape percepts. Indeed, this is the case, as illustrated in Figures 14a–f. Squares aligned along a 45-degree diagonal (Figure 14c) are generally seen as a tilted column of diamonds, and diamonds aligned along a 45-degree diagonal (Figure 14d) are generally seen as a tilted column of squares (Attneave, 1968; Palmer, in preparation, b). Similar effects due to rectangular frames (Figures 14e and 14f) were demonstrated many years ago by the Gestalt psychologist Kopfermann (1930). The present explanation in terms of symmetry further predicts that analogous effects should result for other figures whose symmetry subgroup is D4. As can be readily observed in Figures 14g and 14h, a "+" and a "×" are ambiguous alternative shapes of the same figure. They too can be biased by being aligned in lines rotated 45 degrees or enclosed in rectangles rotated 45 degrees (Palmer, in preparation, b). Thus, the theory seems to be supported by several sorts of extensions that turn out to be valid, and we are in the process of testing others.

In summary, reference frames seem to be established so that they coincide with intrinsic geometric regularities of figures. The regularities that seem to be important are transformational invariances, particularly symmetries over reflections. It is worth remarking at this point that all the reference frame effects that have been discovered thus far concern orientation. This is an interesting fact that is somewhat perplexing from the reference frame point of view. If perceptual reference frames are truly analogous to those in analytic geometry, they should also contain information about position (origin), dilation (unit distance), and direction (sense). However, no corresponding failures of constancy have been reported due to these factors. There are at least three dif-

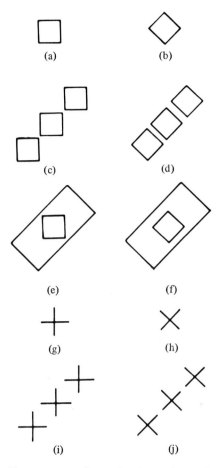

Figure 14. Ambiguous square/diamond and +/× in conditions that bias perceived shape.

ferent explanations that can be offered. One is that all shape properties are truly invariant over position, size, and sense. In this case, reference-frame-like effects will never be found for these variables. A second is that psychologists have not yet looked in the appropriate places with sensitive enough measures. Or perhaps such effects are known, but their connection with reference frames has not been realized. A third explanation is that the stimulus variables that control the establishment of the position and size of the frame may be so unambiguous that the frame is inexorably attracted to the same values for the same shape. Recall that the best reason to believe the reference frame hypothesis at all is that sometimes it fails in the case of orientation. If it were fail-safe for position and size, then one would never observe such effects without detailed theories of precisely how the values were established.

3.5.2. Motion Perception

Motion perception can be thought of as an extension of shape perception into space-time. The naturalness of this extension is reflected in how readily people equate a static shape with the spatial properties of a spatio-temporal event. A single point of light that moves continuously in a circular path bears a close perceptual relationship to the static percept of a circle. It should not be particularly surprising, then, that reference frame effects arise in motion perception as well as shape perception. In fact, they are quite prevalent.

Perhaps the most powerful evidence for the importance of reference frames in motion perception comes from Johansson's (1950) work on configurations of several simultaneous motions. A few of these are shown in Figure 15. Figure 15a schematically depicts an event in which two equal dots

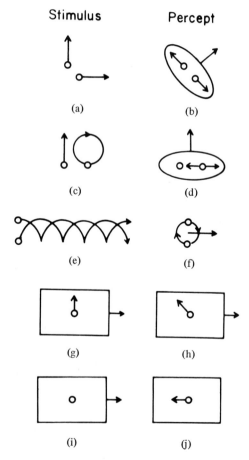

Figure 15. Reference frame effects in the perception of motion configurations studied by Johansson (1950) and others.

move synchronously in harmonic motion, one vertically and the other horizontally. Naturally, each dot alone is seen moving in the way just described. But when both move together, observers usually see the motion depicted in Figure 15b: both dots move together as a single system in harmonic motion along a diagonal path, while the two dots move toward and away from each other along the perpendicular diagonal. The dot motions seem to be decomposed into a vector sum of a *common motion* for the whole system and a *relative motion* for the components within the system (Johansson, 1950). In reference frame terms, two different frames seem to be used simultaneously, one embedded within the other. First, an intrinsic frame is established for the whole system, and the motion of this frame relative to the environment corresponds to the common motion. Second, the dots are perceived to move within the systemic frame rather than within the environmental frame, hence the diagonal motion. Sometimes, however, the independent motions of the two dots are seen relative to the environment without any systemic frame, and so the percept is ambiguous.

Another of Johansson's demonstrations is depicted in Figure 15c. Two dots again move in harmonic motion in phase, and the left dot again moves vertically. The right dot now traverses a circular path whose vertical component is in phase with the left dot's vertical motion. The resulting percept is often quite different from this, however. As depicted in Figure 15d, the two dots form a unitary system moving vertically. Within this frame, the left dot does not move, while the right dot moves horizontally toward and away from the left one. This organization is possible because the circular motion can be decomposed into the vector sum of two orthogonal harmonic motions. One of these coincides with the vertical movement of the left dot, thus forming the common motion of the whole system. The other component of the circular motion then becomes the relative motion of the right dot within the systemic frame.

Yet another interesting case studied by Johansson (and others) is shown in Figure 15e. Two lights move in phase over cycloidal paths. Alone, each light is perceived to move as one would expect from their paths. Moving together, however, they are seen as attached to opposite sides of a circle that "rolls" smoothly as though over a flat surface (Figure 15f). Again, the perceptual system organizes the event so that motions are perceived within embedded systems. The dots are seen as moving over *circular* (not cycloidal) paths with respect to the unseen center of the circle, while the center translates horizontally with respect to the background.

In all of these cases it seems that the perceptual system factors motion into hierarchically structured systems. Motion components shared by the simultaneous movement of several elements are attributed to a multielement system that defines a moving systemic reference frame. The elements of the system are then seen as moving relative to this systemic frame rather than relative to the background. This is what Johansson (1950) calls relative motion. The system as a whole is seen as moving relative to the environmental frame. This is what Johansson calls the common motion. The situation is reminiscent of Einstein's relativity principle in which all motions can be defined relative to different frames of references. However, only certain frames seem to be "natural" to the perceptual system, just as in shape perception. In motion perception too the frame chosen seems to determine which geometrical properties of the event are perceived. For example, the "L" configuration (Figures 15a and 15b) is perceived such that symmetry is evident in the motion of the two dots relative to each other. If one sees two independent motions, this symmetry is not perceived.

The next question concerns how the frame is established. What factors determine what the system is and how it moves? Johansson's (1950) own analysis suggests that common vector components are critical. The visual system seems to organize an event so that the maximum common motion component is factored out as the motion of the whole group of elements, leaving relative motion as the residual. Johansson saw this as an extension of the Gestalt law of common fate. However, it is also possible to view the situation in terms of relative motion being factored out first, leaving common motion as the residual. Cutting and Proffitt (1982) have presented experimental results that they interpret as strong evidence for the primacy of relative motion. However, Pomerantz (1981) has recently reported evidence that seems to contradict their conclusion. He finds that common motion can be selectively attended independently of relative motion, but not vice versa. This is what would be expected if common motion were perceived first. In fact, these two results may not be as antithetical as they seem at first. Relative motion may in fact be extracted before common motion, but common motion may be available to conscious experience before relative motion. Perhaps the perceptual system builds up global percepts (high level frames) from local ones (low level frames), but consciously experiences their contents in the reverse order, from the most global, unified level downward to the most local level. (See Marcel, in press, for a discussion of this general idea.)

Regardless of which level is analyzed first and which is experienced first, one can ask why the systemic frame is established in the way it is. For example, in the "L" configuration, the systemic frame moves with the point that lies midway between the two moving dots (Figures 15a and 15b). However, in the "10" configuration (Figures 15c and 15d) the frame's motion coincides with that of one of the two dots. In fact, these two possibilities seem to be the only stable percepts: the frame is established either at the centroid of the motions of its elements or at one of the elements itself (Cutting & Proffitt, 1982). Logically, of course, there are infinitely many other possibilities.

The few alternatives that are actually perceived correspond to the simplest and most stable ones, as would be expected from the Gestalt law of Prägnanz. The most convincing analysis has been done by Restle (1979). He adapted

Leeuwenberg's (1971, 1978) coding theory to moving configurations and showed that people's preference ordering over the ambiguous alternatives could be predicted by the complexity of the code required to describe them. "Complexity" is defined as a function of the number and type of parameters that must be included. It is presumably this tendency toward simple, parsimonious description that takes advantage of symmetries and regularities in the motion configuration. For example, the reason the frame often gets established at the centroid of the elements is that their motions sum to zero (cf. Cutting & Proffitt, 1982). Similarly, the reason that one element is identified with the systemic frame rather than the other is that the chosen one has the simpler motion (Restle, 1979).

The importance of stability in establishing the systemic reference frame can be shown by changing the relative salience of the two elements. Suppose, for example, that the horizontally moving dot in the "L"configuration were replaced by a large rectangle that enclosed the entire motion of the vertically moving dot (see Figures 15g and 15h). An observer no longer sees a diagonally moving system, but a horizontally moving one tied to the rectangle. The dot is now seen to move diagonally upward and leftward with respect to the rectangle. This change in perceived motion is due to changing the frame of reference from the centroid of the two equally salient dots to the larger, enclosing, and more prominent rectangle. All of these properties make the rectangle the more salient and stable element of the pair. Perhaps part of its stability as a frame is due to the fact that it looks like a background surface on which the dot is moving. Some support for this conjecture comes from the fact that if the dot and the rectangle are separated in depth, the two motions tend to be seen as independent rather than related (Gogel, 1978).

An even more striking demonstration of the influence of stability on motional organization occurs in the phenomenon of "induced" motion. The classical example is that of a small, stationary, luminous dot placed within a larger luminous rectangular frame in a dark room (Figure 15i). When the frame is moved very slowly back and forth relative to the dot, observers usually perceive the dot to be moving back and forth within an unmoving frame (Duncker, 1929) as indicated in Figure 15j. To achieve this illusion, the motion of the frame must be slow enough that it would be perceived as stationary in the absence of the dot. The most reasonable explanation of this result is as follows (Rock, 1975). The threshold for the *relative* motion within the dot and frame system is much lower than that for the *absolute* motion of either the dot alone or the frame alone in an otherwise dark room. Thus, the visual system registers an event in which the dot and frame are definitely moving relative to each other, but neither is moving relative to the environment. The perceptual system resolves this contradiction by attributing motion to either the dot or the frame. The remaining question is why the dot is usually seen as moving and the frame as stationary. The answer seems to be that because the frame is the larger, surrounding, and more stable figure, the system perceives it as stationary and the dot as moving rather than vice versa. Similar induced motion effects can be obtained with a variety of different configurations. The general result is that the system "prefers" to take the larger, more prominent element as the stationary frame.

Let us briefly sum up the nature and importance of reference frame effects for perceptual organization. First, reference frame phenomena provide critical evidence that shape constancy is not mediated solely by invariant features. If it were, then there would be no failures of shape constancy and no ambiguities like those seen for the square/diamond and equilateral triangle. Many of Johansson's (1950) motion configurations show similar ambiguities, suggesting that the same conclusion holds for motion perception: invariant features are not the whole story. Some sort of referential structure is also required, particularly for orientation. The ambiguities and failures of constancy can then be accounted for in terms of selecting different reference frames for the same figure in different contexts.

The reference frame hypothesis requires additional assumptions to account for the exceedingly small number of ambiguous alternatives. This can be handled by proposing that the frame is "data driven": it is largely determined by the structure of the figure rather than being arbitrarily imposed. There are few cases of naturally occurring ambiguity because there are few figures for which the frame can be established in more than one way. In short, the processes that underlie frame selection are powerful heuristics rather than fail-safe algorithms. Their fallibility is revealed in what I have called reference frame effects.

What do these reference frames have to do with transformations and symmetries? We have already mentioned that one of the most powerful rules for selecting the reference orientation is reflectional symmetry. I suspect that other sorts of symmetries are used in selecting other parameters of the reference frame. For example, the origin of the frame might well be established by selecting the point about which there is maximal rotational symmetry.

The other connection between reference frames and the transformational approach lies in the relations among frames. Consider the set of all possible Cartesian coordinate systems. Each is related to all others by some transformation of the Euclidean similarity group. Therefore, transformations are the way of getting from one frame to another and the way of organizing complex data structures consisting of multiple frames. Transformations can thus be seen as the structural glue that holds the perceptual system together.

4. Toward a Theory of Perceptual Organization

The view I have just presented of the field of perceptual organization suggests a number of important design features that the human visual system might have.

(1) The system seems to be constructed to *prefer* certain types of transformational structure to others. This suggests that these are somehow easier to compute. The relevant transformations are just the similarity group of Euclidean geometry: translations, rotations, dilations, reflections, and their composites.

(2) The system seems to be structured to be *transparent* to these same transformations. By this I mean that it is easy for the system to factor out the effects of such transformations, thereby maintaining constancy over them. This feature is accomplished in the present system by establishing a perceptual reference frame. As discussed earlier, the reference frame extracts transformational variance, factoring it out of the total stimulus structure and "absorbing" it into the frame.

(3) The third design feature unifies the first two: the reference frame is selected so that the maximum amount of transformational structure is extracted from the figure or event being perceived. This means, of course, that frame selection is largely a data-driven process. The perceptual system selects the frame whose transformational structure is most compatible with that of the object or event. This will turn out to be a natural by-product of the internal composition of the reference frames.

This last property gives the system a teleological aspect in that the resulting percept will be optimized in terms of its transformational simplicity. This is actually a version of the Gestalt law of Prägnanz: the percept will be as "good" as the prevailing conditions will allow, where "good" is now identified with the amount of transformational invariance that can be extracted by the perceptual system.

These design elements are realized in a computational system consisting of three basic components: spatial analyzers, invariance analyzers and reference frames. They are related to each other hierarchically as depicted in Figure 16. For simplicity the diagram shows the three systems schematically, as though each were a single entity. In fact there may be many layers of each type, resulting in a rich and complex network of interacting elements.

The input to the entire system can be traced back to the sensory surface. We will not concern ourselves here with the properties of the sensory registration process. This is not because they are uninteresting, but only because they do not seem to be particularly relevant to the higher-level organizational structure of the system as a whole. The spatial analyzers operate on the output of the sensory surfaces by computing functions over local regions of space. Their key properties derive from the assumption that they are *transformationally related*. Intuitively, this means that the receptive field of each one "looks like" all the others except for some simple transformation from the group of Euclidean similarities.

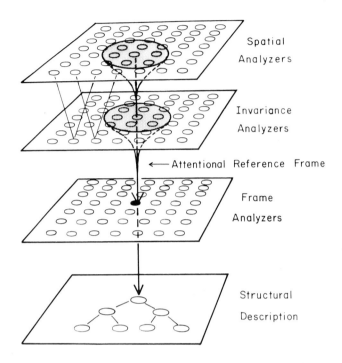

Figure 16. The overall structure of the proposed theory of perceptual organization.

The invariance analyzers make use of these implicit transformational relations by extracting them explicitly. They compute the degree to which transformationally "nearby" spatial analyzers have the same output and, therefore, the extent to which the stimulus contains the corresponding transformational regularity. When these invariances are present simultaneously across different regions of space they reflect local symmetries in the sense discussed earlier. When they are present across both time and space, they reflect local motional properties of stimulation. These motional properties can be further analyzed for regularities by higher-order analyzers. They would extract transformational regularities over extended spatial regions during a motion event.

There remains the crucial problem of putting these piecewise analyzers together into some coherent, organized whole. This is accomplished by the reference frames in the proposed system. They organize lower level analyzers by unifying transformational regularities of the more local analyzers into a coherent whole. The frames actually consist of a relational structure over lower-level analyzers. They make different transformational relations differentially salient, roughly according to their likelihood of occurrence.

Whereas the lower-level analyzers operate in parallel and without interference among each other, the different possible reference frames are in competition. They imply different descriptions or interpretations of the input data in that the same stimulus would be perceived to have different structure and organization within different reference frames. Therefore, one frame must be selected in preference to others. The goal of this selection is to pick the frame that provides the most stable and unified description. The proposed means is to select that frame whose transformational structure accounts for the greatest amount of spatial variation. Since the structure of a frame lies in its invariance analyzers and these analyzers extract symmetries and regularities, this process should select the frame whose structure is most compatible with that of the stimulus. With this schematic description as an intuitive introduction, let us now consider the proposed system more fully.

4.1. First-order Analyzer Space

The processing of sensory information obviously begins at the receptors themselves. Their job is to transduce spatio-temporal patterns of light energy into nearly isomorphic spatio-temporal patterns of neural energy (or electrical energy or whatever). Despite the absolute necessity of this first step, it will be of no further concern in this paper. Let us simply assume that such a transduction process occurs somewhere prior to the level at which we begin and that its resolution in space and time is at least as great as that required by the later processes with which we will concern ourselves. It will be sufficient to think of a transducer array like the retinal mosaic of photosensitive receptors. However, caution should be exercised in any further speculations about the physiological level at which the computational processes discussed here might occur.

The first level of processing to be considered here in detail is accomplished by a systemic structure I called the "first-order analyzer space." It is "first" only with respect to the processes discussed here; several levels of analysis may precede it in the human visual system. The first-order analyzer space is a very large set of elements that process spatial information in parallel. Each element – called a first-order analyzer or "FOA" – computes some real-valued function over the pattern of outputs from the receptor array. Precisely what functions they compute over space is not too important. What *is* important, however, is how these functions are related to each other, for this provides the essential structure for the system as a whole.

The key assumption is that the functions computed by FOAs are *distinct* from each other but *transformationally related* by a limited set of transformations. Not surprisingly, this set is the Euclidean similarity group encountered so frequently in phenomena of perceptual organization: translations, rotations, dilations, reflections, and their composites. The assumption that FOAs are transformationally related means that each analyzer's function is identical to every other analyzer's function except for some transformation of this group. Intuitively, it is probably easiest to conceive of this in terms of the "receptive fields'' for FOAs such as those depicted in Figure 17. If the receptive field of one analyzer is just like that of another analyzer except for being in a different place (*A* versus *B* in Figure 17), then the first is related to the second by a pure translation. Note that the second is related to the first by another translation and that each is the *inverse* of the other. Similarly, if two analyzers differ in size (*A* vs. *C*), orientation (*A* vs. *D*), sense (*A* vs. *E*), or any combination of these properties, then they are related by rotation, dilation, reflection, or one of their composites. Note that in Figure 17 all of these others are also combined with a translation; this is merely for graphical clarity so that the receptive fields do not overlap one another in the diagram.

The assumptions that define the FOA space are actually more abstract and general than the situation depicted in Figure 17 in a number of ways. First, the "receptive fields" shown are merely illustrative. They could be any of a wide variety of functions, as long as they are distinct and are related by transformations of the similarity group. In fact, the few restrictions on the internal structure of analyzers that arise in constructing such a transformationally based system are realized as symmetry conditions; the "receptive field" cannot be symmetrical over a transformation that is part of the first-order space (see Palmer, 1982, for a fuller exposition). Thus, for example, rotationally symmetrical analyzers would not be sufficient to detect rotational structure, at least for analyzers centered on the point of rotation. The only other constraint is that the internal structure of the FOAs be such that a homogeneous field produces zero output. This will be important later when we consider how the reference frames operate.

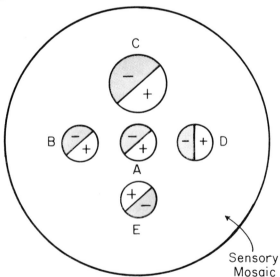

Figure 17. Examples of receptive fields for spatial analyzers in different positions (*A* vs. *B*), sizes (*A* vs. *C*), orientations (*A* vs. *D*), and reflections (*A* vs. *E*).

A second abstraction concerns the way in which transformational relatedness is defined. It does not depend on visual inspection of pictorial representations of receptive fields as described above. That is just a convenient, intuitive way to present the idea. Actually, transformational relatedness is defined over ordered pairs of *functions* that map input patterns from the proximal stimulus (or more accurately, from the output of the sensory surface) into real values. The conditions specifying that one analyzer has a given transformational relation to another analyzer concern the stimulus transformations required to insure that the second analyzer always has the same output as the first analyzer. The important point is that the second analyzer will produce the same output as the first if and only if each pattern is first subjected to the appropriate transformation of the similarity group. For example, suppose analyzer B is related to analyzer A by a rightward translation over one degree of visual angle. Then the output of A for any arbitrary pattern will differ from the output of B for that pattern, but will be identical to the output of B for the same pattern after it has been translated rightward over one degree of visual angle. (See Palmer, 1982, for a more formal exposition of this and related concepts.) In fact, the definition of transformational relatedness in terms of functional equivalence is merely an idealization of how one would determine whether two pattern-processing "black boxes" were related by studying their outputs to patterns and their transformations.

Given that the FOAs are all transformationally related, they constitute a set that can be generated from a single analyzer and the set of transformational relations. The result is a "space" consisting of a set of elements (the analyzers) in many different positions, orientations, resolutions, and both senses (reflections), with a set of structured relations implicit among them. The nature of simple subsets of this space is depicted schematically in Figure 18a for a receptive field and its translations, central dilations, central rotations, and central reflections. If these generating transformations are combined orthogonally, one obtains the set of all first order analyzers in a *functional system*. It is called a functional system because all the analyzers within it compute the *same function* in the sense that they are all related to one another by similarity transformations.

The overall structure of a functional system can be conceptualized as a space in which each analyzer is a point. The dimensions of the space correspond to the position, orientation, resolution (or size), and reflection (or sense) of the analyzers relative to the sensory surface. Since the number of analyzers is certainly finite, the space is only sparsely populated with points. Therefore, it is more appropriate to think of it as something like a discrete lattice structure such as depicted in Figure 18b. The cyclic dimension is orientation (which repeats after 180 degrees of rotation) and the binary dimension is reflection (which repeats after each reflection). The two dimensions of position and the one of resolution are simple orderings. The diagram in Figure 18b is

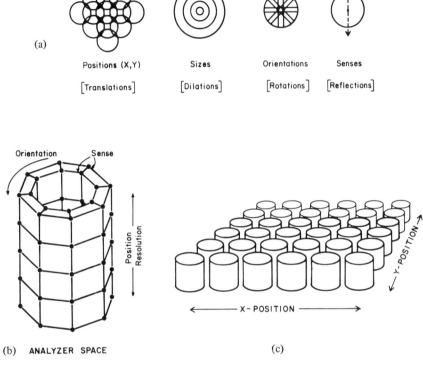

Positions (X,Y)   Sizes   Orientations   Senses

[Translations]   [Dilations]   [Rotations]   [Reflections]

(b)   ANALYZER SPACE   (c)

Figure 18. The structure of the first-order analyzer space in terms of schematic receptive fields (a) and a five-dimensional lattice structure (b,c)

necessarily a simplification of the actual space, since its five dimensions cannot be depicted in real, three-dimensional space. To get a concrete notion of the overall structure of the whole space, one can conceive of the vertical dimension in Figure 18b as resolution and then imagine a whole two-dimensional array of them to represent the two positional dimensions (see Figure 18c).

4.2. Second-order Analyzer Space

The output of the entire first-order system discussed above constitutes a representation of the static, two-dimensional spatial distribution of light intensity over the sensory surface. This representation has many desirable features – as will later be argued in some detail – but many of them are embedded in the *implicit relations* among these outputs rather than in the outputs of individual analyzers themselves. The job of extracting such relations is performed by the *second-order analyzers*. These elements take the outputs of two (or more) first-order analyzers, compare them, and output values determined by the relationship between them. It will also be suggested that there exist third-order analyzers and even higher-order analyzers that perform analogous functions at higher levels. The job of these higher order analyzers is to extract information about *transformational invariance*. This information is crucial to the perception of symmetry (in the extended sense described earlier) and motion.

The fundamental importance of transformational relatedness between two analyzers at the first-order level is that their outputs will necessarily be the same for any two patterns (or local regions of two patterns) that are identical over the particular transformation that relates them. For instance, consider a pattern with reflectional symmetry about a vertical line, such as the letter "A." It is invariant over a reflection about that line. Now consider any first-order analyzer that covers any portion of that pattern. Its output, whatever that might be, *must* be identical to that of any other analyzer related to it by reflection about the same vertical line. This is generally true for all pairs of analyzers related to each other by this particular transformation when presented with this pattern. Moreover, it is true for all such pairs when presented with any pattern having this particular symmetry in its symmetry subgroup.

In general, second-order analyzers (SOAs) are elements that compute transformational relationships among the outputs of first-order analyzers. Thus, they respond to transformational variance and invariance with respect to translations, rotations, dilations, reflections, and perhaps their composites as well. When an SOA compares simultaneous FOA outputs, it computes the extent to which the pattern is the same as itself when subjected to a particular similarity transformation.

This situation can be viewed as a special case of a more general one in which the output of one FOA at one time is compared with that of another FOA at another time. In the special case of symmetry, the two times are simultaneous. If the two times are not simultaneous, then the SOA computes *motion:* the extent to which the pattern at time t_1 is the same as that at t_2 after the action of some transformation from the similarity group. For example, consider an event in which a figure rotates around a fixed point over time. At time t_1 it is in some orientation O_1, and at time t_2 it is in some other orientation O_2. It follows that the analyzer that responds to the pattern at time t_1 will have some output and that this output (at t_1) will be identical to that of some other analyzer at time t_2 that is related to the first by the same rotation that the figure undergoes. In fact, this will be true for all pairs of analyzers related by the same rotation about the same point in this region of this pattern. Indeed, it will also be true for any other pattern that undergoes this same motion.

It is easy to see that this analysis generalizes readily to other transformations in the similarity group such as translations and dilations. If a SOA compares the output of one FOA to another over a non-zero time delay, then its output carries explicit information about motion of the sort associated with the transformation that relates the second FOA to the first. If another SOA compares the same two outputs over a zero time delay, then its output carries explicit information about symmetry of the corresponding sort. Thus, we see that there is a fundamental relationship between the geometrical nature of symmetry and motion when viewed as transformational entities. The fact that the FOA space is transformationally structured allows the SOAs to extract information about symmetry and motion, depending only on whether FOA outputs are compared at the same or different times. Figure 19 shows a scheme that would perform such comparisons: the length of the arrows is proportional to the temporal delay between the FOA (A or B) and the SOA that interrelates them.

If the SOAs compare outputs of transformationally related FOAs to compute invariances, which transformations do they compute? Given the large number of FOAs in the first-order space, they could not possibly compute all possible transformations. There are simply too many of them, since the number of pairs (or n-tuples) of FOAs increases geometrically with the number of analyzers.

As one constraint, it seems reasonable to suppose that the SOAs compare FOAs only over *local regions* of the first-order space. Thus, only relatively "small" transformations would be computed, those related by short distances within the first-order space. This conjecture has some measure of support for motion-sensitive SOAs in that apparent motion breaks down if the distance between the two figures is too great (Kolers, 1972). It is also compatible with what may be related phenomenon in symmetry perception: Julesz (1971) reports that effortless perception of bilateral symmetry in otherwise random dot patterns can be blocked by replacing a thin strip of dots along the symmetry line with a non-symmetrical strip of truly random dots. Apparently the more distant dots – which are still invariant over reflection about the line – are not compared as effectively by the symmetry-sensitive mechanism. This is consistent with the hypothesis that SOAs make primarily local comparisons.

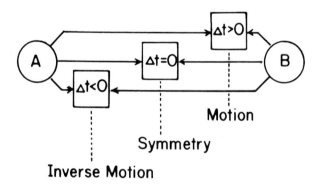

Figure 19. A simple scheme for second-order analyzers that compares the output of two first-order analyzers (A and B) at different temporal disparities to extract information about symmetry or motion.

Another possible constraint on the number of comparisons would be restricting them to those occurring within a single transformational dimension. For example, pure rotations and pure dilations might be computed, but not "spiral" transformations composed of simultaneous rotations and dilations. This possibility is suggested by Bundesen et al.'s (1981) finding that minimum alternation rates are additive for apparent motions consisting of both rotation and dilation. If there were "spiral" motion detectors, they would presumably shorten the time required to perceive these composite transformations. Without them, such composite transformations would have to be represented by some process like vector addition of the component transformations, analogous to representing positions of points by coordinates relative to a set of basis vectors. As mentioned earlier, it is not yet clear – beyond these two cases, at least – which transformations are processed independently and which ones are not. However, it does not seem too unlikely that the system reduces the number of comparisons performed by restricting SOAs to "pure" transformations within a dimension. Note that the limiting case of these two strategies – only local comparisons and only within a dimension – results in the scheme depicted in Figure 18b. The links between nodes correspond to SOAs that compute the minimal transformational relations of the system. It is too simple and local to capture the power and complexity of the visual system, but it gives some idea of how drastically these two assumptions can reduce the number of comparisons that are computed by SOAs. Perhaps a scheme not too much more complicated will suffice.

Now let us consider the overall structure of the second order system. First, notice that it is primarily a system sensitive to image motion. The symmetry-sensitive analyzers are just a subset that compute simultaneous stimulus structure. Therefore, the bulk of SOAs are sensitive to image transformations as they occur in time: translations, rotations, and dilations. Reflections are discrete transformations that seldom occur over time. (In fact, they only occur continuously in the real world in the guise of rotations through a higher-dimensional space, but even then only if the "front" and "back" of the object are indistinguishable.) Therefore, it is unlikely that any SOAs compute reflectional invariance except those that do so simultaneously to extract bilateral symmetry.

It should be clear that the SOAs also form a transformational space, similar in many respects to the first-order space. The primary conceptual difference is that the dimensions of this space are themselves dynamic transformations defined over space-time rather than static spatial properties defined over space alone. Once again, assuming that only local, intradimensional comparisons are made, there would be vertical translations, horizontal translations, rotations, and dilations. Because the SOAs are fundamentally *relational* (i.e., depend on comparing at least two lower level elements), there are two further dimensions for each type of transformation. The first is the bipolar dimension of *direction* that distinguishes the order of the two lower-level analyzers: e.g., up/down, left/right, clockwise/counter-clockwise, and bigger/smaller. Pairs of SOAs that are the same except for this dimension are, in fact, inverses of each other. The second is a dimension of *rate*. It could be computed by varying the temporal disparity (time-lag) between the stimulus conditions that are being compared. It might also be computed by having a single temporal lag, but different transformational distances between the two FOAs. More likely, both of these schemes might be used. The essential thing, however, is that rate of transformation is a dimension of the second order system and that it probably requires two or more analyzers to define it. Thus, even the minimal scheme shown in Figure 18b must be augmented by additional links (SOAs) to represent directionality and rate.

4.3. Third- and Higher-order Analyzers

One can conceive of most of the second-order space as extracting motional properties of the stimulus event much as the first-order space extracts spatial properties at a moment in time. Given this fact and the implicit transformational structure of the SOAs, it is possible to duplicate the same sort of structure at a higher level. Third-order analyzers (TOAs) can be constructed that compare the outputs of SOAs within local regions of the second-order space. Such elements would be sensitive to transformational invariance in motional structure, just as SOAs would be sensitive to transformational invariance in spatial structure.

Again, there are two major classes to consider: those in which the temporal difference between the stimulus conditions is zero (simultaneous) and those in which it is non-zero (sequential). The simultaneous TOAs would process *motional symmetries* and the sequential ones *motional transformations*. Motional symmetries refer to invariances present in events over space at the same time. For example, if the whole visual field is undergoing uniform translational motion, as it would during an eye movement over a static scene, then the optical event has complete translational symmetry (except at the edges) with respect to this type of motion. Third-order analyzers that compared the outputs of SOAs sensitive to translational motion would themselves be sensitive to the sameness of the motions in different parts of the visual field. This sort of process is presumably involved in the spatial grouping of motional events as, for example, in the Gestalt law of "common fate."

Motional transformations refer to systematic changes in events over space and time. Here the possibilities are so numerous that only a hint can be given of the types of visual structure that might rely on such TOAs. One example is the motion gradients studied by Gibson (1966, 1979). For example, the optical flow that occurs during radial expansion as an observer moves toward the fixation point is one in which the rate of translational motion increases systematically for portions of the visual field further from the center. This rich and complex structure might well be extracted by TOAs that compared the rate of translational motions in nearby portions of the visual field. The relations between them are accelerational. That is, if a TOA compares two motional analyzers for *different* rates of the same type of motion and they are found to have the same output, its own output contains explicit information about acceleration/deceleration.

4.4. Property and Symmetry Systems

In broad outline, what the present computational system suggests is something like the following. The first-order analyzers compute some function over space within a transformationally structured system. Its outputs drive the second order system which compares them over transformational and temporal differences. The results are twofold. First, the SOAs compute local spatial symmetries in that they extract information about the degree of simultaneous transformational invariance present in the stimulus. This portion of the SOA system would be sensitive to *transformational relations* of the specified sorts. In addition, the SOAs that include unequal temporal delays in their comparisons are sensitive, in effect, to another type of stimulus property at a higher level: the motions (or displacements) of the similarity group.

The transformational structure implicit in the second-order system is further analyzed by the third-order analyzers. Again, these TOAs compute two kinds of results, depending on whether their comparisons are made between simultaneous or sequential samples of the stimulus. Simultaneous comparisons result in TOAs that carry information about motional invariance (i.e., symmetries in the expanded, group-theoretical sense, but applied to motional properties rather than to spatial ones). Sequential comparisons result in TOAs that carry information about acceleration, because they are sensitive to motional transformations over time. There might be fourth-order analyzers or even higher-order ones that are built on the same design outlined above. They would presumably analyze for further levels of complexity in the transformational structure of stimulus events, but we will not concern ouselves further with these possibilities.

Thus far the system has been described in terms of first-, second-, and third-order levels of processing. It is also possible to see the structure of the system somewhat differently, as divided into just two classes of elements: those that extract a given type of *property* from a local region of the stimulus event in space-time, and those that extract *relations* among two or more such regions simultaneously. As schematically represented in Figure 20, the properties are spatial in the first-order, motional in the second-order, and accelerational in the third-order system. These are related by being temporal derivatives of each other and are, therefore, characterized conceptually by comparing for invariance over temporal disparities. Each of these properties is also compared in purely spatial terms by higher-order analyzers, and the result is a sys-

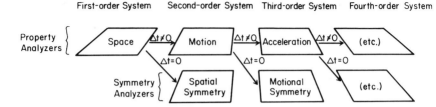

Figure 20. The structure of the analyzer systems classified by hierarchical level and by properties vs. symmetries.

tem that carries information about transformational invariance relations over temporally corresponding regions of the spatio-temporal event. This is the "symmetry system" of invariance analyzers. It will play an important role in explaining how the system's behavior follows the Gestalt law of Prägnanz.

4.5. Reference Frames

All the analyzers within a given system operate in parallel and are driven exclusively by the proximal stimulus event. Thus, their output is a determinate function of the structure of the optical event at the sensory surface. In some sense, then, this output constitutes a representation of the image. At some point this situation must change. The proximal stimulus must be *interpreted* in terms of the distal event that gave rise to it. This representation is not solely determined by the sensory event, but also by the "perceptual set" that the observer takes toward it. Different sets produce different perceptual states as demonstrated by ambiguous figures (e.g., Attneave, 1971). For example, it is difficult to explain why a person sometimes sees a figure as an upright diamond and as other times sees the same figure as a tilted square without invoking some contribution of the perceiver's internal state beyond the sensory image.

At least since Helmholtz (1867) introduced the idea of "unconscious inference," perceptual theorists have appealed to various sorts of theoretical concepts to illuminate the nature of this internal contribution to perceptual interpretation. It doubtless has many different facets at different levels of processing. I wish to concentrate here on an aspect that is particularly relevant to the phenomena of perceptual organization discussed previously.

The general hypothesis is that every perceptual act involves establishing a *reference frame* that selects, organizes, and describes the structure of the event relative to a set of underlying implicit "assumptions." In effect, these assumptions concern the transformational structure of figures and events in general: how they are likely to change from one part of space to another and from one time to another. The underlying transformations I will be concerned with are those of the similarity group: translations, rotations, dilations, reflections, and their composites. In effect, they embody a large portion of a perceiver's tacit knowledge about the geometry of physical space and space-time.

The reference frame schema I will describe is designed to factor out similarity transformations by absorbing them into the perceptual frame itself. In so doing, the effect of these transformations will not be felt on the contents of the frame, and it is these contents that constitute the data of shape perception. This is the basic algorithm for achieving shape constancy, for perceiving motion, and for organizing sensory data. The mechanism that carries out this process takes place within the systems of analyzers already described. Only a few additional assumptions are required to make the selection process choose the "best" frame to suit the structure of the stimulus.

It will be simplest to start with the two-dimensional case and then to generalize from it to the three-dimensional case. Recall from our previous discussion of the "reference frame hypothesis" that a Cartesian coordinate system requires an origin point, an oriented axis, a unit distance, and a sense along the axis to define it fully. These correspond directly to the position, orientation, size, and sense of the analyzers in the first-order space. In effect, each analyzer can be thought of as defining a frame of reference for its local region of the visual field. The fact that each analyzer is related to every other analyzer within its functional system by a similarity transformation is analogous to the fact that each reference frame is related to every other frame by a similarity transformation. Thus, the analyzers themselves can be thought of as implicitly defining a large set of transformationally related reference frames.

The construct of a perceptual reference frame is meant to imply more than mere designation of one FOA among many, however. It is an organizational schema through which other analyzers are related to the defining one in terms of its own internal structure. The defining analyzer serves as the *germ* for a neighborhood of the analyzer space in the sense that the other analyzers can be generated from it by the set of transformations that constitute the frame. Consider an example. Choose any analyzer from the first-order space depicted in Figure 18b. Now consider what happens if that analyzer is "rotated" about its center (i.e., if its receptive field structure is rotated). It generates the set of analyzers related to the defining one by pure rotations, the ones that lie along the same circle in the "horizontal plane" depicted in the diagram. Now consider what happens if the defining analyzer is dilated from its center. This generates the set of analyzers related to the defining one by pure dilations, the ones that lie along vertical lines in the diagram. One can similarly generate a region of the space from any given analyzer by applying such transformations to it.

Thus a reference frame can be thought of as *a collection of transformations that relate an arbitrarily chosen analyzer to its neighbors*. Because these relations are defined relative to the "reference" analyzer, the effect of the frame is to restructure (or interpret) the relations among analyzers in a coherent and self-consistent manner with respect to the *frame* rather than with respect to the sensory surface.

The importance of this restructuring is that it *decouples* processing at or beyond the frame level from the sensory surface. For instance, if the sensory surface were rotated with respect to the world, the output of virtually all the analyzers would change. But if the frame were somehow to *compensate* for the rotation – e.g., by shifting successively from one reference analyzer to its nearest neighbor along the rotation dimension – then the output of the analyzers *as defined by the rotating frame* would not change at all. This is how constancy can be achieved within the present system: the reference frame absorbs the effects of the transformation so that the *contents* of the frame – the outputs of its analyzers defined by its own structure – remain invariant. Another example is shown in Figure 21 for the case of a dilating figure. Three analyzers are shown before and after the transformation both in terms of how they relate to the stimulus (21a vs. e) and in terms of how the stimulus relates to each of them (21b vs. f, c vs. g, and d vs. h). Notice that by shifting from each analyzer to the next larger one (along the arrows shown), the image available to corresponding analyzers is the same before and after the transformation. Thus, constancy is achieved over the dilation by transforming the internal reference frame in a corresponding fashion.

It is worth remarking at this point that the present reference frame system neatly partitions the properties of the percept into position, orientation, size, sense, and shape. The first four are just the defining parameters of the frame

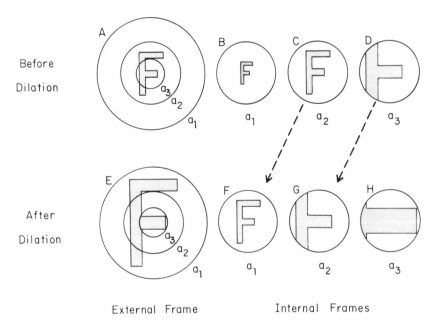

Figure 21. An example of how a transformable internal reference frame achieves constancy over a dilational transformation by compensatory transformations in the analyzer space (arrows).

and the last (shape) is some as-yet-unspecified function of the contents of the frame. We will shortly consider the internal structure of a frame in more detail.

4.6. Visual Attention and the "Mind's Eye"

We have suggested that constancy is achieved in the present system through an internal, transformable reference frame that follows and factors out transformations in the stimulus event. How might this occur? I want to suggest that it is accomplished by a primitive form of *visual attention* that is driven primarily by stimulus structure. I call it "attention" because it is selective. The frame not only restructures the output of the analyzer space, but also makes certain information in it more salient and available and other information less so.

Within the current two-dimensional system an attentional frame can be conceived as a pointer that selects the central defining analyzer. To this pointer is "attached" the transformational structure of the frame. It maps the retinally-based system of analyzers into the frame-based system, effectively decoupling the stimulus from its relationship to the sensory surface. Thus, the perceptual reference frame can be modeled as a mechanism that is positioned within the analyzer space. The output of the frame, then, is just the same as the input to it from some *other* image that is a similarity transformation of the stimulus. The position of the frame in the analyzer space determines which similarity transformation the input image implicitly undergoes as a result of the "positioning" of the frame.

In this way stimulus transformations can be factored out simply by moving the attentional frame around within the analyzer space. Movement along the orientational dimension produces rotations; movement along the resolutional dimension produces dilations; movement along the positional dimensions produces translations; and movement along more than one of these dimensions produces various composite transformations. *The underlying reason that such movements result in output constancy is that they effectively produce the inverse transformation on the image.* For example, when attention moves from a smaller to a larger frame (see Figure 21), the effective size of a constant stimulus relative to the frame is reduced. However, the effective size of a stimulus that enlarges at the same rate the frame does will remain invariant relative to the frame. This is an example of a "visual algebra" in which transformations are counteracted by their inverses to achieve the identity transformation.

The fact that movements of visual attention within the analyzer space produce similarity transformations suggests that the attentional mechanism functions as the "mind's eye." Everyone is familiar with the fact that, phenomenologically speaking, one can "zoom in" on a small region of the visual field for information about fine detail or take a broader view of the whole field without

necessarily moving one's eyes closer to or farther from the region of interest
(Kosslyn, 1981). Similarly, one can take a subjectively different orientation
with respect to a figure by doing something like imagining turning one's head
without actually doing so (Attneave & Reid, 1968), or taking a subjectively different position by attending to a non-central position of the visual field without
actually moving one's eyes (Posner, Nissen, & Ogden, 1978). Such effects of
the "mind's eye" are especially noticeable when inspecting a strong afterimage
(or iconic memory) in which the position, size, and orientation of the image
cannot possibly change. The present suggestion is that these transformations
take place through changes in the attentional reference frame with respect to the
internal analyzer space. They mimic movements of the eye through space
because the effects of such movements are similarity transformations, at least in
two dimensions. It is not quite as simple as this in three dimensions, a problem
we will discuss a bit later.

4.7. Internal Structure of Reference Frames

Thus far we have considered mainly the role of simple FOAs with respect
to the reference frame. But invariance analyzers also play an important part
because they define the transformational structure of the frame explicitly.
Whereas simple property analyzers are merely reinterpreted with respect to the
frame, the invariance analyzers actually *define* the frame, in some sense, and
strongly influence the process of frame selection.

Recall the hypothesis concerning invariance analyzers. Comparisons are
made between fairly local neighbors in a lower-order analyzer space to produce
processing elements that respond to the degree of transformational
variance/invariance in that region of the space. Thus, these elements represent
transformational relations explicitly. Consider now only those invariance
analyzers that relate the reference analyzer to other analyzers within the same
space. These are just the transformations we earlier suggested would generate
the FOAs within the frame relative to its reference analyzer. This subset of the
invariance analyzers would respond to the transformational invariances that are
part of the frame based on that analyzer. This subset would include: translations along the intrinsic orientation of the analyzer, rotations about its center,
dilations with respect to its center, and reflections about its intrinsic orientation.
I have been implicitly assuming that the same reference analyzer also specifies
the orthogonal orientation of the two-dimensional frame. If so, its frame would
also include the set of invariance analyzers for translations along the perpendicular orientation and reflections about the perpendicular axis. Since each of
these analyzers is associated with a kind of local symmetry, the invariance
analyzers within a given frame would make these particular geometrical relations available and others unavailable.

A richer conceptualization of the internal transformational structure of a frame is possible. Rather than the frame consisting of a discrete set of invariance analyzers, one can think of it as defining a metric over the space of invariance analyzers according to how closely related each one is to the reference analyzer. This would provide a continuum of availability for various sorts of geometrical relationships, depending on the distance of its invariance analyzers from the reference analyzer of the frame.

As an example of how invariance analyzers might be involved in frame-dependent perception of structural regularities, consider again the case of the square/diamond. If the reference analyzer is oriented parallel to the sides and positioned at the center, then the symmetries of opposite sides and adjacent angles are central to the frame and highly available within it. This is because they correspond to invariance analyzers about the reference analyzer's intrinsic orientation and its perpendicular. The symmetries of adjacent sides and opposite angles are far less salient because these invariances are not directly related to the reference analyzer. As argued earlier, this pattern of salience is the perceptual organization associated with a square shape. The situation completely reverses when the reference analyzer is one of the two that are oriented at a 45-degree angle to the sides and positioned at its center. Now the equivalence of adjacent sides and opposite angles is associated with invariance analyzers that are central to the frame. The equivalence of opposite sides and adjacent angles is associated with invariance analyzers that are much more distant from the frame. This is the perceptual organization that arises for a diamond shape. Thus, the present hypothesis about how invariance analyzers relate to the attentional reference frame captures at least some of the phenomenal properties of these two different shape interpretations.

Notice that in this example we consider only four frames from the set of all possible frames, a much larger set. These four frames are just the ones that include analyzers for the transformational invariances that apply to the figure being analyzed: its global symmetries of reflection and rotation. This was no accident. A central hypothesis of the present theory is that selection of the attentional frame is largely driven by intrinsic structural properties of the stimulus figure or event. Not just any frame can be chosen, but only those that afford an organization for the figure that is "simple" and "stable." This amounts to requiring that the system operate in accord with some form of the law of Prägnanz. There must be some mechanism that selects the "best" frame, at least locally, from among many alternatives.

4.8. Selecting the Best Frame

As mentioned earlier, shape constancy will hold only if the frame for a given object is chosen so that it bears the same spatial relationship to the object each time. The obvious way to accomplish this is to let the intrinsic properties of the object specify the frame. If the object's position determines the frame's origin, its orientation the frame's axes, its size the frame's resolution, and its reflection the frame's sense, then the frame will be established the same way each time. The difficulty, of course, is managing to do this without first knowing what the object's shape is.

A very general approach to this sort of problem is to generate all possible frames (or at least a very large set of them) and then to pick the "best" from among them. If the schema for evaluating the "goodness" of alternatives is appropriate, the same frame can be chosen time after time. For this to happen, however, the criterion must be stable (invariant) over variations that occur as the result of transformations. The criterion I suggest for choosing among frames is maximizing symmetries. In theory, at least, the frame whose contents are maximally symmetrical for any given figure will always be the same one relative to its contents. Thus, maximizing symmetry should provide constancy for the contents of the selected frame. This hypothesis links the Gestalt principle of Prägnanz to the perceptual constancies.

Perfect constancy would be too good, of course. As we have already observed, sometimes it fails. This will be modeled in the present scheme by biases that operate in frame selection toward salient orientations of vertical and horizontal. These turn out to follow naturally from the structure of the visual world and from uneven distribution of analyzers with respect to orientations.

The mechanism for implementing the symmetry selection scheme obviously rests on the output of invariance analyzers, since they are sensitive to degree of transformational invariance. The basic idea is to have "frame analyzers," one for each frame, that integrate the output of invariance analyzers that constitute the transformations of the frame. One might think that the frame analyzer could simply "sum" its invariance inputs to find the total symmetry value for its frame. In fact, this will not work. What is needed is both invariance as indicated by output at the second-order level and output indicating the presence of variance at the first-order level. Frames should be chosen that contain real differences in the distribution of light (to which the FOAs would be sensitive) as well as structural regularities within them (to which the SOAs would be sensitive).

Consider the following concrete scheme for computing such conditions. Let second-order invariance analyzers compute the absolute value of the difference in output between pairs of FOAs related by a given transformation. Thus, these SOAs actually compute something like "variance" or "irregularity," since they have high output when their FOAs have very different values and zero output when their FOAs have identical values. To compute a value that reflects both high symmetry and high FOA output, the frame analyzer simply adds all its FOA outputs and subtracts all its SOA "variance" outputs. Thus, the frame analyzer will have no output for a homogeneous field, since we assume that the FOAs do not respond to constant input. It will have low output for very asymmetrical contents because the high FOA output will be largely canceled by the high output of the variance units. It will have high output for very symmetrical contents because the FOA output will only be reduced by the low output of the variance units.

All of the frame analyzers make these computations in parallel. Each one computes something like the amount of spatial variability in the pattern that can be accounted for by its own transformational relations. This computation indicates the "goodness" of each frame for capturing the structure of the stimulus in that region. The output of the frame analyzers can be thought of as defining a goodness surface over a "frame space." Peak outputs correspond to those frames that span highly active regions of the first-order space and whose outputs are highly structured by the invariance analyzers of the frame. Valleys correspond to those frames that span inactive regions of FOA space or active regions whose outputs are random with respect to its invariance analyzers. Clearly, these peaks in the output function correspond to frames that ought to be selected for further shape processing. Using frame output to select the "best" frame gives the system Prägnanz with respect to transformational structure.

The system just described is very much like Hinton's (1981) scheme for using frames to map retinotopically defined features onto object-relative features for figures stored in memory. The primary difference between the two systems is that the present scheme uses transformational goodness as the criterion for selection, whereas Hinton uses goodness-of-fit to a memory representation. These criteria are not really all that different if one conceives of the present scheme as being one for matching to a very abstract memory schema for the structure of figures in general, a schema that is embedded in the frame structure itself.

4.9. Representing Patterns

Few patterns are simple enough for their structure to be captured within a single frame. The general scheme for representing patterns in memory is to build a data structure from a sequence of attentional fixations and to store information about the frames together with information about their contents. The frame information defines the global parameters of the object or part being coded: its general position, orientation, size, and sense. The contents of the frame define its shape with respect to that frame: its local symmetries and specific spatial properties. This general proposal is actually quite similar to one I made several years ago (Palmer, 1975) for representing pattern information in semantic networks. Each node of the network represented an object or part and was attached to parametric relations that defined its position, orientation, and size relative to those parameters for its superordinate node. The result was a hierarchy of nodes connected by their transformational frame relations with some information about global shape being attached to each node. This notion of pattern representation is highly compatible with the present system. It is relatively easy to see how such networks could be constructed as the result of a sequence of attentional fixations on salient frames of reference.

One can specify an attentional fixation by just two sorts of entities: one specifying the *frame,* and the other specifying the *contents* of that frame. Each of these can be represented by a structured set of real values. The frame itself can be represented by the coordinates of the frame analyzer within the analyzer space, one parameter for each dimension. These coordinates can be either "absolute" (retinotopic), or relative to some other reference frame, or both. The "contents" of the frame are just the outputs of the spatial analyzers and invariance analyzers within the attentional frame. These can also be specified as a structured set of real values – one for the output of each analyzer – where the structure is defined by the transformational relation of each analyzer relative to the reference analyzer as given by the frame. The contents of an attentional fixation can also be specified in terms of absolute outputs of the FOAs, relative outputs (through the invariance analyzers), or both. We assume that the perceptual code of a whole scene consists of a data structure which stores many such records of attentional fixations.

In order to make a pattern representation useful for later recognition and classification of new patterns, the descriptions of its frames and its contents must be stored relatively rather than absolutely, because it is quite unlikely that the same pattern will be seen again in exactly the same retinal location, orientation, or size. Storing relative information is one obvious way to abstract the representation from its specific retinal parameters. Doing so is particularly easy in the present system because of its transformational and relational structure. The description of one frame relative to another is just the transformational relation between them for each dimension: position, orientation, size, and sense. Then the representation of the frame is very much like a relative address in a computer, where the entire address structure can be relocated simply by changing a single, base address.

The contents of frames can likewise be stored relatively. Doing so simply requires that the relationships between the outputs of the attended analyzers be known. This information is available in the invariance analyzers whose outputs reflect the comparison of pairs of FOAs. Storing relative descriptions of the contents of a frame, then, is easily done by recording the outputs of SOAs within the attentional field. It is here that redundancies in the first-order outputs can be eliminated by storing only those values that are different. In fact, it is possible that the invariance outputs alone are the principal basis of shape information, but this conjecture is beyond the scope of the present paper.

Let us assume, then, that both the frame and the contents are represented by relative descriptions. In other words, the analyzers within the attentional field are specified relative to the reference analyzer and their outputs are specified relative to the output of that analyzer. Now these two references (one for the frame and one for its contents) for recording an attentional fixation can also be described relatively, but they must refer to something outside of the current fixation, namely to the frame and contents of *another* fixation. It makes good sense that each attentional fixation should be related to some more global, superordinate frame, since the more global frame defines the space within which the more local one fits. Consider a person's body, for example. The position, orientation, and size of the frame used to describe the shape of the head would be specified relative to the position, orientation, and size of the frame used to describe the body as a whole. Similarly, the frame for an eye or a nose would be specified relative to that for the head as a whole. In general, the interaction between the structure of the stimulus pattern and the structure of the perceptual system will determine which pieces are natural parts. These natural parts will be coded as local units (consisting of a frame-contents pair) within more global units of a hierarchical structural description (see also Palmer, 1975, 1977).

In general, using other frames as referents for relative descriptions provides a flexible and useful way of storing information. If the analyzers in the attentional field are described as transformations of the reference analyzer, and if the outputs are described relative to some standard (e.g., the output of the centrally attended analyzer), then the whole stored record is coded as a relative description except for the two references. These, in turn, can be specified relative to more global attentional fixations. The net result is that such representations will be accurate descriptions of the pattern and will not really depend on retinal position, orientation, or size in a significant way. These absolute parameters can be factored out of the representation as free variables, because of the transformational structure of the system and the frame-relative coding scheme.

4.10. Emergent Properties and Holism

The doctrine of holism – that the whole is more than (or different from) the sum of its parts – is probably the most central and best remembered tenet of Gestalt psychology. It has various meanings, not all of which will be addressed here. Among them, however, is the assertion that perceptual wholes have "emergent properties" which arise when a set of elements interact in space and time. The essence of this claim is that a whole figure is perceived to have properties that are not shared by or even predictable from its parts in isolation. For example, a single point has only positional attributes, but when a number of points are configured to form a line, the whole line has attributes of length and orientation. The points of which it is composed have neither of these properties, even though they participate in the configural line and determine its holistic properties.

In the present theory, emergent features of wholes arise naturally as the result of analyzing the patterns at different levels of resolution. Each attentional fixation produces a representation for the region of the analyzer space encompassed by the analyzers within the attentional field. This representation is, in a sense, holistic in that it characterizes this region of the perceivable world as a single entity. While it is true that the representation is coded in terms of the output of a number of different analyzers, it is also true that the analyzers themselves are holistic in the sense that they compute a function over a large region rather than just a tiny bit of it. Thus, if one were to attend just to a small region containing a point in a configural line of dots, it would produce the perceptual representation of a filled circle, whatever that might be. But by attending to the line of dots as a whole, a lower level of resolution would be utilized in which the representation would be essentially that corresponding to a line of the appropriate length and orientation, but with lower contrast (cf. Ginsberg, 1971). Hence, there is always a duality inherent in the system: a pattern is simultaneously represented as a whole, unitary figure though an attentional fixation at low resolution and as a collection of parts through a sequence of attentional fixations at higher resolution, each of which represents only a portion of the figure (Palmer, 1977). Each level of analysis reveals aspects of the complete whole that are, in a very real sense, different from and independent of those at other levels. For example, the dots could be replaced by squares of similar size without substantially affecting the line-like representation at low resolution.

This complete separation of local elements from their global configurations will not always hold true, especially when there are few elements that are large with respect to the whole configuration (see Goldmeier, 1972; Kimchi & Palmer, 1982). In such cases, the representation at a low level of resolution will reflect characteristics of the geometrical elements as well as those of the whole configuration. Only when elements are relatively small will the structure of the whole configuration be independent of that of its elements. In such cases, the separation results from the fact that spatial characteristics of the pattern will be analyzed by separate sets of analyzers, the configuration at a low level of resolution and the elements at a high level.

Recent evidence has suggested that low resolution (global) information is processed more rapidly than high resolution (local) information (Navon, 1977). While there clearly must be limiting conditions for this assertion, as Kinchla and Wolfe (1979) and Martin (1979) have shown, it is nevertheless an important finding. It supports the Gestalt claim that the emergent properties of the whole are perceptually more salient than those of its parts, at least for a wide range of stimuli and viewing conditions. This, in turn, strongly suggests that there must be a global, low-resolution analysis of patterns that occurs somewhat independently of that for the smaller parts of which it is composed (see also Broadbent, 1977). The resolution structure in the present theory and the notion of sequences of attentional fixations suggest obvious possibilities for accomplishing this type of processing. Global pattern information could be extracted by an initial attentional fixation at a low level of resolution, whereas local information could be extracted by subsequent attentional fixations at higher levels of resolution.

Another aspect of holism that is captured by the theory is, of course, symmetry and good form in general. These are properties which can only be defined for wholes. They have their representation in the output of the invariance analyzers as structured by the frame analyzers.

4.11. Three-dimensional Frames

Finally, let us consider the important question of how to extend the current system to handle frames in three rather than just two dimensions. First, consider a small piece of what a two-dimensional frame in the frontal place might "look like" in terms of the receptive fields of its FOAs. Figure 22a depicts nine line-like FOAs within a frame defined by the central analyzer. Each of the others can be generated from it by a similarity transformation (in this case, a translation) as indicated by the arrows between them. These arrows correspond to SOAs within the frame. Of course, many other FOAs would also be part of this frame, including those related by dilations, rotations, and reflections, but for now, consider just these nine. Now consider what the corresponding nine analyzers would look like for a frame in a depth plane. As indicated in Figure 22b, they are now elements that differ from the central analyzer in orientation and size as well as in position. However, note that these are just other elements within the same functional system of FOAs. That is, they are all related to each other by similarity transformations in the image plane. The arrows between them represent the same transformational relations (SOAs) as the corresponding analyzers in Figure 22a, within the internal structure of these two frames.

Frames slanted in depth, then, simply map different retinotopic similarity relations into the appropriate frame-relative relations as a function of the angle in depth. In principle, there is no reason why this kind of scheme cannot be carried out to define a whole set of three-dimensional frames. The only structures required are (a) the invariance analyzers that compute the composite transformational relations and (b) the additional frame elements to compute the goodness of the additional frames. The major conceptual difference here is that each frame can no longer be uniquely identified with a single reference analyzer in the first-order space. It is easy to see, for instance, that in Figure 22b the same central reference analyzer could serve as the reference for many different three-dimensional frames, depending on the slant in depth. But the uniqueness of the mapping between frames and FOAs is not really important. The structure of the frame is actually determined by which FOAs and SOAs the frame analyzer includes. Therefore, different frames can be attached to the same reference analyzer because their frame analyzers include different structures of relations over the analyzer space. The conceptual difference is largely that the

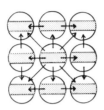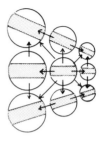

A Reference Frame   B Reference Frame

in Frontal Plane   in Depth Plane

Figure 22. A schematic illustration of how a reference frame in the frontal plane could be generalized to depth planes by restructuring FOAs within the same functional system and the transformational relations between them (arrows).

attentional frame cannot be thought of as a simple "pointer" into the analyzer spaces. Rather, it must be conceived as some more complex or higher-dimensional structure. In principle, however, the extension of the general reference-frame schema to three dimensions is not difficult and should work in exactly the same way as proposed earlier for two dimensions.

5. Organizational Phenomena Revisited

Now that we have covered the basic structure of the present theory, let us briefly consider how it might account for the organizational effects that initially suggested such a scheme.

5.1. Shape Constancy

It should be clear by now how the current system would achieve shape constancy. The "best" frame is selected according to the criterion implicit in the integration of the frame analyzers. Since their outputs depend heavily on relations of transformational invariance, the "best" frame will be quite stable relative to the figure at different retinal positions, orientations, sizes, and senses. As a result, the contents of the frame will not change within it, and perceived shape will be constant over such transformations. This is how the present theory implements the "reference frames" hypothesis for achieving shape constancy.

Shape constancy is not perfect, of course, as Rock's (1973) experiments demonstrate. Figures with poor and ambiguous intrinsic axes are not remembered as well when presented and tested in different orientations as when presented and tested in the same orientation. However, this effect disappears if a figure has a strong and unambiguous intrinsic axis (Wiser, 1981). The most obvious interpretation of these results is that vertical is taken as the reference orientation in the absence of an unambiguous axis. This would also explain the difference in perceived shape for squares and diamonds. Different orientations relative to the figure would be chosen as the basis for the frame when different symmetry axes of the square/diamond are aligned with vertical, assuming that there is a bias toward choosing a vertical frame. Thus, biases toward particular orientations are central to failures of shape constancy.

There are several levels at which such biases could exist within the present system. Biases toward salient orientations relative to the retina could be built directly into the analyzer space. For instance, if there were more "vertical" FOAs, there would be a corresponding tendency to select retinally vertical frames because the frame elements integrate the outputs of their lower level analyzers.

Biases toward environmentally salient orientations would generally occur as a result of the orientational structure of the physical environment. Walls, fields, trees, people, and so forth have strong intrinsic axes that are aligned with the environmental horizontal and vertical. Horizontal and vertical frames would tend to be chosen because high degrees of invariance at these orientations would produce higher outputs in their frame elements. Non-visual biases could also be incorporated by non-visual input to the frame elements. For instance, when the head is turned, there is vestibular information about the change in orientation. This and other information could bias frames that are consistent with gravitation simply by adding to the visual input to frame elements.

5.2. Motion Perception

Motion would be perceived through the converging operations of explicit motion analyzers and "frame tracking" mechanisms. Second-order analyzers that compare for transformational invariance over time are themselves sensitive to image motions of translation, rotation, and dilation. Thus, there would be a bias toward perceiving these kinds of motions. Further, the frame selection mechanism will operate to achieve these kinds of motion rather than plastic deformations. Selecting the same frame relative to the object during a similarity transformation means that the bias toward shape constancy would also operate during motion perception. Thus, shape constancy and preferences for similarity motions are complementary, and their mechanisms converge in phenomena of motion perception.

When two or more components in an event move synchronously, several frames at different levels are needed to capture the structure. The motion of the frame for the whole "system" corresponds to Johansson's "common" motion vector. Within this global frame, local frames are needed to track the motion of the parts in order to maintain their constancy. Such motions are perceived with respect to the more global frame, and these are the "relative" motion vectors.

Frame effects arise in perceiving motion configurations for the same reasons that they do in perceiving static figures. Certain frames are much better than others at capturing intrinsic structure in motion events, and the "best" frame is selected. Some factors that seem to be important in determining which frame is chosen are: the symmetry of the motion event (e.g., Johansson's "L" configuration), the simplicity of the motion (e.g., Johansson's "10" configuration), and the size and complexity of the elements (e.g., induced motion). These seem to have natural correlates in the present system. Motional symmetry is computed by the third-order system: simple motions can be tracked by shorter movements of the attentional frame, and global frame selection should maximize the amount of invariance that can be accounted for. Thus, frame effects in motion perception are natural results of the system's internal structure.

5.3. Figural Goodness

Perceptual goodness effects show that the visual system is sensitive to local and global invariances over the similarity transformations. Following information-theoretic approaches, let us assume that perceived goodness varies inversely with the complexity of the code required to generate it. The questions then become: What is this code and how is it generated within the present system?

While it is not obvious how such coding would happen in detail, it is clear that at least two aspects of the system are well suited to taking advantage of transformational invariances. First, frame selection is strongly biased toward picking frames that maximize the amount of symmetry available in their contents. This allows the system to take maximal advantage of the structure in patterns. Second, the SOAs that compare FOA outputs simultaneously are specifically tuned to symmetries. They can be used to extract transformational invariance such that it need not be explicitly coded: only asymmetries need to be coded in the shape representation process. Thus, the operation of the symmetry analyzers provides an appropriate structure for a system that implicitly "assumes" symmetry and only needs to store asymmetries. A system like the present one may eventually be able to provide a processing mechanism for generating representations like those in Leeuwenberg's (1971, 1978) coding theory.

5.4. Perceptual Grouping

I argued in the initial discussion of grouping phenomena that grouping effects could be conceived as "distance" effects in a higher dimensional space. Distance corresponds to an ordering over transformational relations within the space such that transformationally nearby points get grouped together. This view suggests a frame-based mechanism for grouping effects. Since invariance analyzers extract transformational invariance explicitly, they are implicated in grouping effects. Different frames correspond to different perceptual organizations for the same stimulus, and so they too are implicated in perceptual grouping. The critical link between them is that different frames induce different orderings over invariance analyzers, depending on their transformational distance from the reference analyzer. Thus, different groupings correspond to selecting different frames, and different frames make different relations perceptually salient. Which frame is selected, of course, depends on which transformational relations are "shortest" within the analyzer space.

6. Conclusion

One of the most interesting things about the theory just described is its *systemic nature*. Its crucial properties stem almost entirely from the structure of the system as a whole. Very little seems to depend on the internal nature of the first-order analyzers – i.e., their receptive fields. What does matter are their relations to each other within the system in terms of transformational variance and invariance. These are what allow the higher-order systems to extract information about transformational structure from objects and events, and that is the central purpose of the system. In this sense the current system is consistent with Gestalt theorizing: what matters is the structure of the entire system, not the nature of its atomic components. This further suggests that the properties of the analyzers may actually be determined by optimizing their functional role within the entire system.

The transformational apparatus in the system embodies very basic, low-level *expectations* about the structure of optical events that are encountered by perceivers in the world. As Gibson (1966, 1979) so eloquently pointed out, people are mobile creatures exploring a three-dimensional world. The visual stimulation they receive as a result contains a tremendous amount of optical structure that carries important information about the world. Much of this structure is transformational, and so it is only reasonable that the visual system should be built to extract it.

The second- and higher-order systems can be seen as a dense background of transformational expectations about both the static structure of images of three-dimensional objects and the dynamic structure of how these images are likely to transform over time. Extracting the local spatial symmetries over translations, rotations, reflections, and dilations, then, is seen as a useful heuristic because objects and their images tend to have these sorts of structural regularities. Analyzers for these symmetries amount to expectations that they will generally be present in images. Similarly, higher-order space-time analyzers can be seen as expectational mechanisms for extracting the regularities inherent in the changes that occur in optical events over time. Objects tend to move locally according to rigid transformations in three-dimensional space, so their images change locally according to continuous translations, rotations, dilations, and their composites. These are the sorts of invariances in space-time that the system is structured to extract. As a whole, then, the invariance analyzers embody what the visual system "knows" about the transformational structure of optical events resulting from physical events in space and time. They form a base of "tacit knowledge" about the geometrical fabric of visual events that creatures are likely to encounter in exploring the type of world in which they live.

Using this knowledge of past events to guide expectations about future ones provides the system with a background structure for processing visual information. In fact, it actually defines what constitutes information for the system and what does not. By this I mean that whatever structure in the event is consistent with the background expectations of the information processing system can be coded implicitly by default. For example, symmetries are "assumed" to hold in the absence of information to the contrary. Only asymmetries, then, need to be processed further. They must be analyzed by high-level systems that can extract their structure in the form of higher-level invariances. The process of perceiving that emerges from this view is one which factors out higher and higher level invariances by higher and higher level systemic structures.

The phenomena of perceptual organization with which we began our discussion reveal this transformational structure quite plainly when they are viewed in this way. Object constancy, motion perception, figural goodness, perceptual grouping, and reference frame effects can all be understood as central to the operation of a system that extracts transformational information in the way suggested. These phenomena are, as we initially conjectured, closely related to each other and transparent to the underlying structure of the system that produces them.

Acknowledgment

I would like to thank Michael Leyton for many stimulating discussions about topics and issues presented in this paper. His ideas have certainly influenced me, and no doubt some of them have found their way into this text (and into my own thinking) without proper credit. The discussions of shape structure and shape similarity are particularly close to ideas we talked about together. I would also like to thank Nancy Bucher and Ruth Kimchi for their interesting and stimulating input to the research program that has so strongly influenced my theoretical ideas. Further thanks to Laurie Wagner, Becky Snyder, Dan Morrow, and Nancy Bucher for helping to prepare the manuscript under pressure, and to Susan Van Eyck for her careful editorial comments. The preparation of this chapter was facilitated by Grant 1-R01-MH33103-03 to the author from the National Institute of Mental Health.

References

Attneave, F. Some informational aspects of visual perception. *Psychological Review,* 1954, **61**, 183-193.

Attneave, F. Triangles as ambiguous figures. *American Journal of Psychology,* 1968, **81**, 447-453.

Attneave, F. Multistability in perception. *Scientific American,* 1971, **225** (6), 62-71.

Attneave, F., & Reid, K. Voluntary control of frame of reference and shape equivalence under head rotation. *Journal of Experimental Psychology,* 1968, **78**, 153-159.

Beck, J. Similarity grouping and peripheral discriminability under uncertainty. *American Journal of Psychology,* 1972, **85**, 1-19.

Berlyne, D. E. *Aesthetics and Psychobiology.* New York: Appleton-Century-Crofts, 1971.

Broadbent, D. E. The hidden preattentive processes. *American Psychologist,* 1977, **32**, 109-118.

Bundesen, C., & Larsen, A. Visual transformation of size. *Journal of Experimental Psychology: Human Perception and Performance,* 1975, **1**, 214-220.

Bundesen, C., Larsen, A., & Farrell, J. Mental transformation of size and orientation. In J. Long & A. Baddeley (Eds.), *Attention and Performance IX.* Hillsdale, NJ: Erlbaum, 1981.

Cassirer, E. The concept of group and the theory of perception. *Philosophy and Phenomenological Research,* 1944, **5**, 1-35.

Chipman, S. Complexity and structure in visual patterns. *Journal of Experimental Psychology: General,* 1977, **106**, 269-301.

Cutting, J. E. Coding theory adapted to gait perception. *Journal of Experimental Psychology: Human Perception and Performance,* 1981, **7**, 71-87.

Cutting, J. E., & Proffitt, D. R. The minimum principle and the perception of absolute, common, and relative motions. *Cognitive Psychology,* 1982, **14**, 211-246.

Duncker, K. Ueber induzierte Bewegung. *Psychologische Forschung,* 1929, **12**, 180-259. (Translation in W. Ellis (Ed.), *A Source Book of Gestalt Psychology.* New York: Harcourt, Brace, 1938.)

Farrell, J. E., & Shepard, R. N. Shape, orientation, and apparent rotational motion. *Journal of Experimental Psychology: Human Perception and Performance,* 1981, **7**, 477-486.

Garner, W. R. *The Processing of Information and Structure.* Potomac, MD: Erlbaum, 1974.

Garner, W. R., & Clement, D. E. Goodness of pattern and pattern uncertainty. *Journal of Verbal Learning and Verbal Behavior,* 1963, **2**, 446-452.

Gibson, E. J. *Principles of Perceptual Learning and Development.* New York: Appleton-Century-Crofts, 1969.

Gibson, J. J. *The Senses Considered as Perceptual Systems.* Boston, MA: Houghton-Mifflin, 1966.

Gibson, J. J. *The Ecological Approach to Visual Perception.* Boston, MA: Houghton-Mifflin, 1979.

Ginsburg, A. P. Psychological correlates of a model of the human visual system. Master's thesis, Air Force Institute of Technology, 1971.

Gogel, W. C. The adjacency principle in visual perception. *Scientific American,* 1978, **238** (5), 126-139.

Goldmeier, E. Similarity in visually perceived forms. *Psychological Issues,* 1972, **8** (Whole No. 29).

Helmholtz, H. von. *Treatise on Physiological Optics* (vol. 3). (Translated from the third German edition, *Handbuch der Physiologischen Optik,* Voss, 1867.) New York: Dover, 1962.

Hinton, G. E. Some demonstrations of the effects of structural descriptions in mental imagery. *Cognitive Science,* 1979, **3**, 231-250.

Hinton, G. E. A parallel computation that assigns canonical object-based frames of reference. Proceedings, 7th International Joint Conference on Artificial Intelligence, 1981, 683-685.

Hochberg, J. E., & McAlister, E. A quantitative approach to figural "goodness." *Journal of Experimental Psychology,* 1953, **46**, 361-364.

Hoffman, W. C. The Lie algebra of visual perception. *Journal of Mathematical Psychology,* 1966, **3**, 65-98.

Hoffman, W. C. The Lie transformation group approach to visual neuropsychology. In E. L. J. Leeuwenberg & H. F. J. M. Buffart (Eds.), *Formal Theories of Visual Perception.* Chichester, UK: Wiley, 1978.

Johansson, G. *Configurations in Event Perception.* Stockholm, Sweden: Almqvist and Wiksell, 1950.

Julesz, B. *Foundations of Cyclopean Perception.* Chicago, IL: University of Chicago Press, 1971.

Julesz, B. Experiments in the visual perception of texture. *Scientific American,* 1975, **232** (4), 34-43.

Julesz, B. Textons, the elements of texture perception, and their interaction. *Nature,* 1981, **290**, 91-97.

Kinchi, R., & Palmer, S. E. Form and texture in hierarchically constructed patterns. *Journal of Experimental Psychology: Human Perception and Performance*, 1982, **8**, 521-535.

Kinchla, R. A., & Wolfe, J. The order of visual processing: "top-down," "bottom-up," or "middle-out"? *Perception and Psychophysics*, 1979, **25**, 225-231.

Kolers, P. A. *Aspects of Motion Perception*. New York: Pergamon Press, 1972.

Kopfermann, H. Psychologische Untersuchungen über die Wirkung zweidimensionaler körperlicher Gebilde. *Psychologische Forschung*, 1930, **13**, 293-364.

Kosslyn, S. M. *Image and Mind*. Cambridge, MA: Harvard University Press, 1981.

Leeuwenberg, E. L. J. A perceptual coding language for visual and auditory patterns. *American Journal of Psychology*, 1971, **84**, 307-350.

Leeuwenberg, E. L. J. Quantification of certain visual pattern properties: Salience, transparency, similarity. In E. L. J. Leeuwenberg & H. F. J. M. Buffart (Eds.), *Formal Theories of Visual Perception*. Chichester, UK: Wiley, 1978.

Levine, M. W., & Shefner, J. M. *Fundamentals of Sensation and Perception*. Reading, MA: Addison-Wesley, 1981.

Leyton, M. A unified theory of cognitive reference. Proceedings, 4th Annual Conference of the Cognitive Science Society, 1982, 204-209.

Mach, E. *The Analysis of Sensations*. (Translated from the German edition, 1897.) New York: Dover, 1959.

Marcel, A. J. Conscious and unconscious perception: An approach to consciousness. *Cognitive Psychology*, in press.

Martin, M. Local and global processing: The role of sparsity. *Memory and Cognition*, 1979, **7**, 476-484.

Navon, D. Forest before trees: The precedence of global features in visual perception. *Cognitive Psychology*, 1977, **9**, 353-383.

Palmer, S. E. Visual perception and world knowledge: Notes on a model of sensory-cognitive interaction. In D. A. Norman & D. E. Rumelhart (Eds.), *Explorations in Cognition*. Hillsdale, NJ: Erlbaum, 1975.

Palmer, S. E. Hierarchical structure in perceptual representation. *Cognitive Psychology*, 1977, **9**, 441-474.

Palmer, S. E. What makes triangles point: Local and global effects in configurations of ambiguous triangles. *Cognitive Psychology*, 1980, **12**, 285-305.

Palmer, S. E. Symmetry, transformation, and the structure of perceptual systems. In J. Beck (Ed.), *Representation and Organization in Perception.* Hillsdale, NJ: Erlbaum, 1982.

Palmer, S. E. Frame effects in perceived pointing of ambiguous triangles. In preparation. (a)

Palmer, S. E. Reference frame effects in shape perception. In preparation. (b)

Palmer, S. E., & Bucher, N. M. Configural effects of perceived pointing of ambiguous triangles. *Journal of Experimental Psychology: Human Perception and Performance,* 1981, **7**, 88-114.

Palmer, S. E., & Bucher, N. M. Textural effects in perceived pointing of ambiguous triangles. *Journal of Experimental Psychology: Human Perception and Performance,* 1982, **8**, 693-708.

Palmer, S. E., & Chase, P. A group theoretical approach to Good Gestalt. In preparation.

Palmer, S. E., & Hemenway, K. Orientation and symmetry: Effects of multiple, rotational, and near symmetries. *Journal of Experimental Psychology: Human Perception and Performance,* 1978, **4**, 691-702.

Pitts, W., & McCulloch, W. S. How we know universals: The perception of auditory and visual forms. *Bulletin of Mathematical Biophysics,* 1947, **9**, 127-147.

Pomerantz, J. Asymmetrical integrality of global and local motions. Proceedings, 22nd Annual Meeting, Psychonomic Society, 1981.

Posner, M. I., Nissen, M. J., & Ogden, W. C. Attended and unattended processing modes: The role of set for spatial location. In H. L. Pick & I. J. Saltzman (Eds.), *Modes of Perceiving and Processing Information.* Hillsdale, NJ: Erlbaum, 1978.

Regan, D., Beverley, K., & Cynader, M. The visual perception of motion in depth. *Scientific American,* 1979, **241** (1), 136-151.

Restle, F. Coding theory of perception of motion configurations. *Psychological Review,* 1979, **86**, 1-24.

Rock, I. *Orientation and Form.* New York: Academic Press, 1973.

Rock, I. *An Introduction to Perception.* New York: Macmillan, 1975.

Rock, I., & Leaman, R. An experimental analysis of visual symmetry. *Acta Psychologica,* 1963, **21**, 171-183.

Selfridge, O. G., & Neisser, U. Pattern recognition by machine. In E. A. Feigenbaum & J. Feldman (Eds.), *Computers and Thought.* New York: McGraw-Hill, 1963.

Shepard, R. N., & Cooper, L. A. *Mental Images and their Transformations.* Cambridge, MA: MIT Press, 1982.

Shepard, R. N., & Judd, S. A. Perceptual illusion of rotation of three-dimensional objects. *Science,* 1976, **191**, 952-954.

Shepard, R. N., & Metzler, J. Mental rotation of three-dimensional objects. *Science,* 1971, **171**, 701-703.

Treisman, A. Perceptual grouping and attention in visual search for features and for objects. *Journal of Experimental Psychology: Human Perception and Performance,* 1982, **8**, 194-214.

Treisman, A., & Gelade, G. A feature integration theory of attention. *Cognitive Psychology,* 1980, **12**, 97-136.

Treisman, A., & Schmidt, H. Illusory conjunctions in the perception of objects. *Cognitive Psychology,* 1982, **14**, 107-141.

Treisman, A., Sykes, M., & Gelade, G. Selective attention and stimulus integration. In S. Dornic (Ed.), *Attention and Performance VI.* Hillsdale, NJ: Erlbaum, 1977.

Wallach, H., & O'Connell, D. N. The kinetic depth effect. *Journal of Experimental Psychology,* 1953, **45**, 205-217.

Wertheimer, M. Untersuchungen zur Lehre von der Gestalt. *Psychologische Forschung,* 1923, **4**, 301-350. (Translation in W. D. Ellis (Ed.), *A Source Book of Gestalt Psychology.* New York: Harcourt, Brace, 1938.)

Weyl, H. *Symmetry.* Princeton, NJ: Princeton University Press, 1952.

Wiser, M. The role of intrinsic axes in shape recognition. Proceedings, 3rd Annual Conference, Cognitive Science Society, 1981.

Why the Human Perceiver Is a Bad Machine

D. N. Perkins

Harvard University
Cambridge, Massachusetts

Abstract

The human perceiver exhibits weaknesses that would not be tolerated in a machine perceiving system, for instance gross lapses of metric accuracy and limited abilities for error correction, backup, and top-down control. These weaknesses, taken together with the obvious success of the human perceiver, suggest that the human visual system employs tactics quite different from those of most machine systems: aspects of the scene are processed on a need-to-know basis, errors outside the current need to know are tolerated, only high probability inferences are made, and searches through alternative interpretations are avoided. More attention to the human tactics in machine perception research might yield both a better understanding of our visual system and a fruitful alternative approach to designing machines that see.

The power of nature's own perceiving machine, the human visual system, is routinely acknowledged by contemporary researchers in computer vision. Although only some researchers aim at systems that mimic human perceptual strategies, virtually all strive for performance that matches the human talent for precise and practical perception of real, complex environments. Block worlds or worlds with continuous gradients of texture or intensity are seen simply as temporary test cases along the way to a versatile and robust perceiving system. There is no pretense now that the achievements of machine perception even approach human performance. On the contrary, reports of advances often take care to acknowledge such limitations as a small repertoire of strategies, only a few visual cues used, only a few rather inflexible object models, boundaries between light and dark missed that would be caught by a human perceiver, pattern recognition failures due to occlusion that would not trouble a five-year old nor, indeed, a pigeon (Herrnstein & Loveland, 1964).

It's less often recognized that the tables can be turned. Sometimes the human perceiver behaves in ways and with results that no programmer would ever want to emulate. The human perceiver displays lapses of metric accuracy and top-level control that seem entirely unnecessary. If the human brain were as accessible as the memory of a computer, a novice programmer could debug the system in short order. Why evolution and learning have not debugged the human program long since is a mystery.

Just that mystery is the present concern. Why is the human perceiver a bad machine? Evidently, because for some reason it doesn't matter. The lapses of metric accuracy and control to be discussed are not serious problems for the human perceiver, despite their blatant, arbitrary, and unnecessary character from the standpoint of machine perception. If human perception is to best inform the development of perceiving machines, we need to understand better what the human perceptual strategy is such that it can proceed carelessly in the ways it does.

1. Constraint Handling

Any comparison of human and machine perceptual systems benefits from a common way of talking about them both. A natural starting place is Gibson's (1950, 1966, 1979) thesis that there is abundant information in the light a visual system receives from the environment – much more information than had been recognized by many prior researchers, who stressed the beholder's share of inference and assumption in perceiving. Gibson did perceptual psychology an important service by directing attention away from the inner mechanisms of perception and toward the ways in which light carried information about the layout of surfaces, the independence of objects, and even, in his later concept of affordances, the ways objects could be used (Gibson, 1979). Contemporary efforts to construct machine perceptual systems are striving to take advantage of many of the information sources Gibson emphasized, for instance texture gradients (Witkin, 1981) and motion parallax (Horn & Schunck, 1981).

Although a leader in discovering how much the light had to offer, Gibson also advocated a model of human perception that could hardly be more contrary to the enterprise of machine perception. He held that there was no need for any sort of computation-like perceptual process. The perceiver responded directly to invariants in the light, much as a tuning fork might hum when a tone at its resonant frequency sounded. Gibson seemed to propound not just a black box concept of the human perceptual system, but an *empty* black box.

A convenient way to accommodate both Gibson's extreme position and more moderate ones is to characterize perception in terms of *constraint han-*

dling. The premise of *constraint handling* is that, whatever information there may be in the light, it cannot be gotten out of the light unless the perceiving system of concern incorporates in some manner constraints that determine how the signal should be decoded. For example, the determination of slant from texture necessarily depends on a textural homogeneity constraint; that is, inhomogeneities of texture are to be attributed to the effects of perspective projection rather than to the world imitating such effects. Whatever interpretive process or mechanism relies on texture, it must in some fashion incorporate such a homogeneity constraint. This leaves room for very different implementations, of course. While Witkin (1981) incorporates a textural homogeneity constraint in a complex computational process, Gibson would insist that the human organism somehow responds "directly" to the homogeneity invariant, in the fashion of the tuning fork.

A number of questions can be asked about a perceptual system that refer only to various constraints and their observance, and not at all to any particular computational or other interpretive procedure. For example, what constraints does a system depend upon – textural homogeneity, support of some bodies by others, symmetry, rectangularity? Are unimpeachable assumptions, for instance the rules of projective geometry, in fact strictly observed by the system? Does the system *accurately* determine the output implied by the stimulus and whatever other constraints appear to be operative?

Other questions concern how constraints are realized in whatever sort of processing the system uses. For instance, which constraints are "wired in" and invariable, and which are made dynamically and rescindable in the face of counterevidence? What kind of a procedure computes three-dimensional shapes and orientations from the operative constraints? For example, in certain cases output might be inferred by a table look-up procedure or some psychological analog of it, or by a relaxation process such as occurs in soap films (Attneave, 1982; Perkins, 1982). How are potential rivalries between candidate dynamically made constraints handled? For instance, constraints might compete with one another simultaneously, or occur serially and so compete only indirectly. Biological and machine systems for the perception of shape, orientation, and location of objects in space lend themselves to comparison and contrast in such terms.

2. Some Lapses of Precision

As mentioned in the introduction, the human perceiver displays a number of lapses in inferring three-dimensional shape and slant from various stimuli. These lapses are puzzling because the resolution of the human eye is sufficient to avoid them. Why, then, should the system not function veridically? Let us defer that general question for the moment in favor of some examples.

Shape-slant invariance. As mentioned earlier, projective constraints ought to figure in the operation of any perceiving system, since they derive from the geometry of space and the physics of light. Indeed, the human perceiver does behave very crudely in accordance with perspective constraints, but no more so. Beck and Gibson (1955) investigated a special case of projective constraints called shape-slant invariance. This refers to the fact that a given patch in a projected image, presumably corresponding to some bounded plane surface in space, defines a family of possible surface patches that could project to the image patch. The surface patches more slanted with respect to the line of sight are also more elongated. The experimental question becomes: Whatever interpretation of a shape at a slant perceivers make, will that interpretation fall within the projectively defined family of possible interpretations? Beck and Gibson (1955) report evidence that this is only approximately the case. Under various conditions, subjects' responses to slant and shape judgment tasks were found to lie outside the projective family of options from about 7% to about 25% of the time.

Discriminating rectangular from nonrectangular forms. Perkins (1968) showed that not all box-like shapes, as in Figure 1, can be projections of rectangular solids. A simple rule governs which can and which cannot. For a rectangular interpretation to be geometrically possible, all the angles around the central angle must be greater than 90 degrees, a condition satisfied by Figure 1a, but not 1b. Notice that, in addition, Figure 1b does not *look* rectangular. Experiments have shown that the human perceiver is quite sensitive to this discrimination based on projective geometry, U.S. adults sorting a range of box figures, some meeting the conditions and some not, into rectangular and nonrectangular classes with about 85% accuracy (Perkins, 1972) and youngsters doing nearly as well (Perkins & Cooper, 1980; Perkins & Deregowski, in press). Subjects do this simply by the look of the boxes, no subject ever reporting, upon questioning, the simple mathematically derived rule stated above.

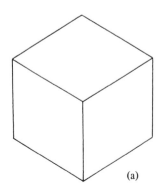 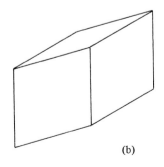

Figure 1. Box shapes satisfying (a) and not satisfying (b) the geometrical conditions for a rectangular interpretation.

To a first approximation, such findings document the practice in human constraint handling of withholding a rectangularity constraint when projective conditions forbid it. But in about 15% of the cases errors do occur. The human perceptual apparatus surely has sufficient sensitivity to do better, since by applying the rule deliberately, if one happens to know it, one can score nearly perfectly on the task.

Furthermore, in certain other cases a much higher error rate occurs. For example, native populations in Zimbabwe have been found to score much lower on the rectangularity discrimination task, although still making the discrimination to a statistically significant degree, a result that fits with many others disclosing difficulties in dealing with pictorial stimuli in such populations (Perkins & Deregowski, in press). The Zimbabwean subjects erred primarily in the direction of identifying as rectangular stimuli which could not be. Perkins (1976) reported a related finding for U.S. subjects as part of another experiment. Working with pictured solids somewhat more complex than rectangular solids, he found that for a condition where no rectangular or symmetric reading was geometrically possible, subjects imposed such readings anyway 30% of the time. (In other conditions, where at least one of three alternative rectangular or symmetric readings was geometrically possible, subjects almost never reported an impossible interpretation.) Finally, in tasks analogous to the rectangularity discrimination task but involving other geometric regularities such as mirror symmetry, higher rates of improperly imposing the regularities appeared (Perkins, 1982). To generalize, occasionally and, for certain perceivers and stimuli, frequently, the human perceiver sets aside geometric constraints that could readily be applied in favor of perceiving stimuli as regular when in fact they could not be.

Slant judgments of rectangular forms. Besides providing mathematically sufficient information for categorical judgments of rectangularity, forms such as those in Figure 1a provide mathematically sufficient information for slant judgments of the edges and faces. Indeed, at any single corner with three visible radiating edges the constraint that the edges are projections of edges at right angles in space suffices to determine the orientation of all three edges in space (Perkins, 1968). The only qualification is that two solutions exist, one concave and one convex, corresponding to the two states of the Necker cube or similar "reversible" figures. In the case of isometric projection, the multiple corners in shapes such as Figure 1b simply provide redundant information. However, in perspective projections, when opposite edges converge, the constraint that these edges correspond to edges parallel in space provides an additional and independent source of information about slants. Yet a third constraint is sometimes relevant: the perceiver may take the edges to be of equal length in space.

How well do human perceptual systems use such constraints to infer slant? Attneave and Frost (1969) reported an experiment testing just this. The subjects looked at shapes like those of Figure 1a monocularly. The experimenters used a half-silvered mirror to superimpose on the subjects' view of the box shapes a stereo view of a wand. The subjects manipulated the wand to report slant. The subjects judged the orientation of an edge by adjusting the wand so that it appeared to be collinear with the edge. Attneave and Frost found that judged slant varied almost linearly with actual geometrically predicted slant, and that the judgments became more accurate as the experimental conditions provided more information: rectangularity alone, rectangularity plus perspective, rectangularity, perspective, and edges in the picture plane compatible with edges of equal length in space. However, even in the last and best condition, the subjects' slant judgments were substantially regressed toward the frontal plane, about 20 degrees regressed at a 60 degree veridical slant from the frontal plane, for instance.

In recent work conducted by Joanne Owens and myself, we have attempted to replicate and extend the experiments of Attneave and Frost to find conditions where slant judgments are better. For pictured rectangular solids without perspective and with edges of equal length in the frontal plane, we have found regressions far worse than that reported by Attneave and Frost. For instance, when a 65 degree slant was predicted, subjects in our early experiments reported only a 38 degree slant on the average. A later experiment presented pictured rectangular corners through irises of various sizes, all of which masked the lack of perspective in the figures, a feature that might induce regression. However, in these and other experiments designed specifically to improve subjects' perceptions of the figures, the average slant when 65 degrees was predicted only rose to the 40-45 degree range. At this point we undertook a comparison between performance on these pictorial figures and an actual solid corner displayed through an iris and viewed monocularly. Even with lighting from one side to create realistic shading on the solid corner's faces, the subjects' judgments showed more regression on this real rectangular form than on the pictured ones!

Such findings do more than demonstrate that human perceivers have difficulty with a certain sort of judgment. They are paradoxical in at least two ways. First of all, human perceivers do rather well on the categorical judgments of rectangularity, despite the occasional lapses discussed previously. Why, then, should their perception of the slants be so extremely regressed? Second, the fact that their responses vary roughly linearly with the correct slants shows that the human perceptual system is picking up the available information; were it not doing so, the slant judgments would be uncorrelated with the correct slants. Since perceivers pick up the information, why don't they take the last simple step of using it correctly? After all, only a scale factor and an adjustment of origin is needed to transform the subjects' characteristically poor performance into a rather accurate one.

Slant judgments of circles. Slant regression effects are not limited to rectangular stimuli. The constraint that an ellipse in the frontal plane corresponds to a tilted circle is a familiar one in perceptual psychology. Eriksson (1967, 1968) reported experiments revealing that subjects' slant judgments of such ellipses were strongly regressed to the frontal plane. Furthermore, in contrast with the slant judgments discussed above, the relationship between correct and perceived slant for the ellipses was strongly curvilinear rather than linear.

Judgments of proportion. Some might question whether judgments of slant relative to line of sight are ecologically natural judgments for the human perceiver to make. If not, then surely judgments of proportion are natural, an ordinary part of seeing the shapes of things in the world. There have been investigations of the perceiver's ability to judge proportion on surfaces seen at a slant. Olson, Pearl, Mayfield, and Millar (1976) reported an experiment where subjects judged the proportions of crosses that appeared on the tops of rectangular boxes viewed at a slant. The results corresponded to a regression of about 15 degrees at 60 degrees veridical slant.

Joanne Owens and myself recently completed an experiment that involved both slant judgments of two edges of pictured rectangular forms, and a proportion judgment of the ratio of the two edges. From each pair of slant judgments, we calculated what the subject's corresponding proportion judgment ought to have been to be consistent with his slant judgments. We compared this to the subject's actual proportion judgments. If proportion were indeed a more "natural" judgment, subjects' actual proportion judgments would prove more accurate than the proportion judgments predicted from their slant judgments. In fact, there was no significant difference. Subjects' slant and proportion judgments often were inconsistent, but not in a systematic way.

An experiment conducted by Mark Miller and myself ventured outside the laboratory altogether. The subjects were asked to judge the proportions of buildings from various angles. They reported their estimates both verbally, and by selecting from a set of pictured rectangles the one closest in proportion to the face of the building in question, the two reporting methods yielding roughly equivalent results. The viewpoints were distant enough from the buildings that stereo vision was no help, but perspective, rectangularity, texture gradients, shading and all other cues normally found in a natural environment were plentiful. Nonetheless, the more foreshortened the face of the building from the perceivers' viewpoint, the more the perceivers underestimated the horizontal relative to the vertical dimensions. For a viewpoint making a 22 degree angle with a frontal viewpoint, estimates ranged from .75 to .97 of frontal estimates, depending on the building; for an 11 degree angle, estimates fell as low as .62 of frontal estimates.

Such results make a comment on our everyday experience of the world. As we stroll down the street glancing about, it would seem that buildings do not appear to have the same proportions when we approach them as when we are opposite them. We live in a rubber world, but do not notice it, perhaps because intellectual assumptions about the stability of object shapes override the messages from our perceptual system per se. I have discovered one circumstance in which the rubber effect is quite plain, once one attends to it. In driving down a road, examine the broken line that signals legal passing. Estimate the proportion of width to length for a bar in the distance. Then estimate the proportion of a bar as close as it still can be seen. You will find the proportions quite different. Now simply observe the bars as the car approaches them. There is a region where the bars seem rather suddenly to stretch in length, the width remaining constant.

Closure. Besides anomalies in the perception of slant, proportion, and categorical shape, there are anomalies in the perception of closure of figures, that is, the geometrically legitimate interconnections of parts within a form. One such anomaly arises with certain figures that involve a cycle of parts. Sometimes the anomaly is detected, as in viewing the well-known Penrose triangle (Penrose & Penrose, 1958). Starting at any corner, perceivers typically scan around the figure, finding that when they have come full circle and should be back to the same location in space, they have ended up substantially behind it. The figure does not close. However, Draper (1978) has presented more complex figures where perceivers apparently lose track of absolute location and take the shape as closed, even though it cannot be (Figure 2).

Another somewhat different problem of closure arises in Figure 3. Huffman (1971) observed that, for this figure to be possible, the edges of the "pyramid," when extended, have to meet in a point. Although they do not do so in Figure 3, nonetheless, the form does not look anomalous.

How can we sum up the human perceiver as a geometer, in light of the above? In past writings (Perkins, 1982; Perkins & Cooper, 1980), I have argued that the human perceiver ought to be called a "sloppy geometer," clearly far from perfect but exhibiting a rather general capacity to infer shape, slant, and other geometrical properties from the stimulus and constraints such as rectangularity, symmetry, and parallelism, which are usually imposed only when they are projectively possible. For some tasks, this still seems to me to

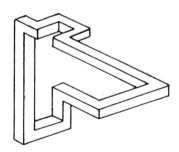

Figure 2. An "impossible figure" with an anomaly much less apparent than that of the Penrose triangle.

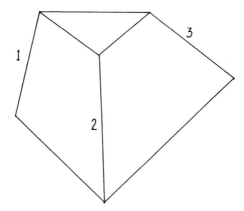

Figure 3. An "impossible figure" that appears possible; the figure violates the condition that the extensions of lines 1, 2 and 3 must meet at a single point.

be apt: for instance, categorical judgments of rectangularity. However, particularly in light of the very weak performance of the participants in recent slant judgment work, it now seems to me that an overall appraisal of "sloppy geometer" may be too optimistic. In many conditions, the human perceiver's performance is not just sloppy, but strikingly poor, and, it would seem, quite unnecessarily poor considering the resolution of the eye and the demonstration that in some cases the perceiver clearly is picking up the necessary information – that is, subjects' judgments vary as a monotonic function of the correct judgment. Perhaps it would be more appropriate to speak of a "qualitative geometer," who extracts a qualitatively tolerably accurate representation with many serious quantitative distortions in orientation and proportion.

3. Some Lapses of Control

Besides discussing the metric accuracy of the human visual system, it's also interesting to consider the control processes that mediate human constraint handling. The points to be made here will not require a finely drawn distinction between control and "lower level" processing, but a simple characterization may be of help. As Gregory (1972; 1981, Chapter 12) has urged, in many ways human perception can be thought of as a process of hypothesis making and testing. For instance, the dynamic constraint handling of a perceiver regarding the Penrose triangle seems to run something like this. The perceiver posits local rectangularity constraints at a corner, leading to an initial interpretation of the corner as a rectangular corner in space. Then the perceiver extends this interpretation around the triangle, accommodating all new evidence adequately until the perceiver comes full circle. At that point, the anomaly emerges, his hypothesis has failed, and the interpretation so far evolved is rescinded in favor of a new start.

Whatever constraint handling processes mediate extending local interpretations to remote parts of the scene, rejecting extensions on the basis of counterevidence, dealing with conflicting information, and so on, will be considered control processes here. In a computer implementation, such processes might well appear as part of the main program, or something close to it. However, there is no implication that the human visual system includes some analog of such a "main program." The control processing might well arise from the interaction of particular local units, for instance through a relaxation process (Davis & Rosenfeld, 1981). Whatever the implementation, the human perceiver exhibits not only metric lapses but what might be called lapses of control, at least relative to what would be expected of a computer implementation. Here are some examples.

Retrying failed constraints. It is a curious feature of the Penrose triangle, the "devil's tuning fork," and similar figures that the human perceiver never seems to come to terms with their anomalies. In the case of the Penrose triangle, the perceiver indeed rescinds the current interpretation on coming full circle and finding the anomaly. But the perceiver immediately initiates a new interpretation with the same sort of local rectangularity constraint that initiated the original one. It may even be the identical interpretation of the same initial corner. No matter how long the perceiver regards the triangle, the futile cycle continues. The constraint handling process seems unable to learn the simple lesson that the rectangularity constraint yields a partial interpretation that, when extended, will be rejected.

Top-down advice of limited influence. A flexible perceiving system ought to accept top-down advice from such sources as the perceiver's conscious conceptual knowledge and assimilate it into the working of the control process as an operative constraint. This is notoriously not the case for the human perceiver. The Penrose triangle again is one example. After some experience with the triangle, one recognizes the perceptual trap it poses, but instructions to one's perceptual system to stop perseverating are ineffective. Another classic example is Ames' room (Ames, 1955; Ittelson, 1952), considerably distorted from a rectangular shape, which appears rectangular when viewed through a peephole as it was designed to be. The room maintains its rectangular appearance even in the face of seeming size distortions of objects within it – size distortions induced by the perceiver's persistent assumption that the room is rectangular – and even in the face of the perceiver's conscious knowledge about what is happening. Only considerable experience in manipulating objects in the room while looking through the peephole gradually leads the viewer to a more veridical perception (Kilpatrick, 1954; Weiner, 1956).

Of course, top-down advice is effective in some circumstances. For instance, it may introduce the perceiver to the reversals of the Necker cube, reversals that are often not seen unless the perceiver is aware of their possibility

(Girgus, Rock, & Egatz, 1977). What sorts of advice the perceptual system will and will not take is an interesting question.

Global interpretations maintained despite local counterevidence. Hochberg (1981) points out that, depending on the fixation point, the perceiver will maintain an orderly global interpretation despite local anomalies. For instance, the viewer focusing on point 1 of Figure 4 can easily maintain a perception in which the figure appears as a backslanting wire cube with an opaque right face, with the catch that the upper front edge appears to parallel the bottom front edge, but to pass behind the right face at 2. There are more reasonable perceptions where the front edge may appear to bend back to pass behind the right face legally, but these are another matter. The real point is that the overall perception sometimes can prevail *while the upper front edge looks anomalous.* In fact, I have no trouble maintaining this odd perception for several seconds while actually fixating on 2. As is the case in more conscious human reasoning, so also in visual processing counterevidence is not always given the decisive role it should have in rejecting an interpretation.

Conflicting evidence resolved by compromise. Any control process may find occasion to deal with conflicting candidate constraints. A fairly sophisticated way to deal with such constraints would be to choose one or the other constraint and try to extend its implications into a consistent interpretation of the scene. If that failed, the other branch could be tried. Or both branches could be tried and the one chosen that led to the best accommodation of other pieces of evidence in the image. Sometimes, the human perceptual system appears to do just this, for instance trying one and then another reading of the "devil's tuning fork" in an attempt to find a suitable interpretation.

However, the visual system often arrives at judgments of slant and shape by a compromise rather than a strategic choice between conflicting cues. The experiment by Attneave and Frost (1969) cited earlier is one example. Regres-

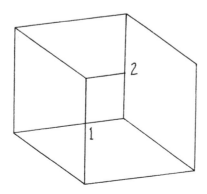

Figure 4. An "impossible figure" for which the perceiver often maintains a global reading at odds with a local discrepancy.

sion toward the frontal plane increased in the conditions where the edges of the pictured rectangular solids were parallel in the picture plane and of equal length in the picture plane, features that Attneave and Frost hypothesized would function as cues promoting a flat interpretation. The result was not, as it might logically be, a bistable response alternating between the cues that said "flat" and those that said "rectangular," but a compromise that does justice to neither aspect. Such compromise effects are very common in research that pits one cue against another.

4. Human Talents

Reviewed here have been aspects of human constraint handling that make the human perceiver seem incompetent. But the human perceiver is highly competent, at least by practical measures. There must be conspicuous abilities in the perceptual system to outweigh the respects in which the perceiver is a bad machine. The following paragraphs explore what some of those talents might be.

High resolution. Binford (1981), writing on machine perception, suggests that much of the power of human visual processing may lie in its high resolution. Of course, mere high resolution in itself accomplishes nothing without higher levels of processing. Nonetheless, Binford maintains that the performance of contemporary machine perception systems is degraded considerably by the coarse grain of current programs for translating light intensity signals into a symbolic representation. In defense of his suggestion, Binford reports experimenting informally with human perceivers and finding that they are sensitive detectors of vertices and boundaries where light intensity changes subtly, more sensitive that the sorts of line finders that are generally available. Some might suggest that this human sensitivity simply reflects the higher level processing the human perceiver can bring to bear, but Binford avers that people make such discriminations well when given just patches of varying illumination, without enough context to make sense of them as arrangements of objects.

Besides examining whether people are better line finders than machine vision systems, one might also investigate whether, given the sorts of data the higher level processing of machine vision systems often have to deal with, people exhibit the sorts of interpretive problems that machine systems do. The occasion of this article prompted me to try a personal experiment along these lines. I will simply relate the results and invite the readers to attempt personal replications. The stimulus was a messy drawer that had not been examined for some time. The viewing conditions involved removal of glasses and a very blurred visual field. The task was to report what sorts of objects in what orientation the drawer contained, moving them out of the way one after another as tentative identifications were made.

Conscious and lengthy hypothesis making and testing proved much more prevalent than in examining less degraded signals. Nonetheless, a number of objects were correctly identified by a combination of rough shape and hue. There were also several errors. A shape correctly identified as a box of candles – a box by appearance and candles by conjectural inference from color and prior knowledge of what was kept in the drawer – was misperceived as to orientation. The end that appeared closest turned out to be furthest away, as the top face of the box reversed in the manner of a face of the Necker cube upon touching it and adding tactile information to the visual. A plastic bag folded in a complicated way was identified as two separate objects due to accidental T-joints, until a touch revealed the connection. A couple of objects turned out to have been missed altogether, even though a recheck revealed that parts of them were easily seen. The parts had not made a coherent enough pattern under low resolution viewing to suggest any sort of object at all. To draw a rather informal conclusion from a rather informal experiment, these lapses of perception under low resolution viewing seem to be very like some of the lapses one often finds in machine vision systems. In summary, Binford's proposal that the high resolution of the human perceiver is critical seems very plausible to me.

Stereo vision. Most human perceivers exercise exceedingly accurate stereo vision at close range. This is one practical answer to the problems of reading slant and shape in space reviewed in the previous section. Although performance may falter when the perceiver regards pictures or scenes at a distance, when hands-on activities are the concern, stereo does away with those limitations.

Superior recognition ability. One of the most striking characteristics of human perceivers is their pattern recognition ability. People often recognize things that are only partly visible or considerably distorted. For instance, research on the recognition of faces has shown that photographs of familiar individuals are recognized in a high percentage of cases, even by youngsters, when only the head from the eyebrows up is visible (Goldstein & Mackenberg, 1966), or when the faces are extremely blurred (Harmon, 1973). Concerning distortion, people recognize caricatures, including caricatures of individuals they have never seen caricatured before (Perkins & Hagen, 1980), although recognition is not as good as with photographs of those individuals. But one hardly needs to turn to the psychological laboratory for examples of recognition under suboptimal conditions. Consider how easy it is to recognize the form of a person in a hydrant, the face of a person in a house, the shapes of trees or landscapes in mottled walls, an effect da Vinci was familiar with and recommended as a means of artistic invention. Such recognitions are not, of course, mistakes. The perceiver is well aware that the human shape in the hydrant or the landscape in the wall is not a real instance. But the recognitions occur nonetheless, and demonstrate the remarkable stretch the human recognition apparatus has (cf. Perkins, 1978; Perkins & Hagen, 1980).

Indeed, the flexibility of pattern recognition may extend through much of the animal kingdom. Herrnstein and Loveland (1964) used a reinforcement paradigm to train pigeons to differentiate photographs that did and did not include human figures, many of the humans being partially obscured, in different parts of the pictures, and in different states of dress or undress. After extended training, the pigeons exhibited the discrimination to an impressive degree. Thus highly abstract and variable classes such as "pictures that display human beings" seem likely to be within the reach of many biological perceiving systems – and no extant machine systems.

Of course, recognition in the sense of perceptual identification of objects as fitting generic categories like "person" or "chair" cannot provide an adequate account of the general flexibility of perception. Not only do we occasionally encounter objects of unfamiliar kinds, but, more to the point, nearly any object has a somewhat individual form and surface details which, in time of need, are perceived. Not only stereo vision, but perceptual strategies dependent on other cues, such as texture gradients, shading, and motion parallax, must provide such details. However, presumably recognition provides important input to such visual strategies by guiding expectations as to what sorts of details might be encountered.

Sketchy representation. The previous paragraphs have stressed the human talents for high resolution input, precise stereo vision, and quick, flexible recognition. By these measures, the human perceiver might seem to be a strikingly precise performer, the reservations discussed earlier notwithstanding. However, it has to be recognized that one of the great resources of the human perceiver is sketchiness. Broadly speaking, we can take the aim of the perceiver's constraint handling to be the construction of a representation of the stimulus that recovers information about object categories and spatial shape, orientation, and position. Clearly, the human perceiver constructs a representation that is, for the most part, very sketchy. The hedge off to the right is likely to be seen only as a mass of such and such a shape, color, and texture, rather than the details of its dendritic structure being registered. The picket fence will be coded as a row of pickets of uncertain count. And so on. Only in the immediate focus of foveal vision is there likely to be a very detailed encoding, and anyone who looks and looks again at such stimuli as works of art or aerial photographs of military installations knows that, even there, a good deal gets missed.

Pattern recognition is a resource that serves well the purposes of sketchy representation. A place in the perceiver's representation can be occupied by an identity label of some sort, rather than by a coding of the details of the object. Over in the rear left is an "elm tree," let us say. If the perceiver needs to access visual properties of the elm tree without actually looking again at the scene, this can be done by retrieving standard visual attributes of elm trees stored under the perceiver's generic representation of the tree, as, for instance, in Kosslyn's (1980) model of mental imagery.

However, pattern recognition is not the perceiver's only recourse. For example, a perceiver may encode a hedge as a tangle of branches with a certain envelope. If the hedge is of a familiar type, the three-dimensional texture of its branches might be encoded simply as a category label. If not, however, the perceiver would have to construct some kind of a textural predicate to represent the texture of the individual hedge. The perceiver's representation would also indicate that this predicate applied over the entire envelope of the hedge, an envelope that also might be novel and hence not representable by a stock category. So there must be a combinatorial capacity for constructing sketchy representations of objects, besides the capacity to pigeonhole objects into familiar object categories.

Need to know. Hand in hand with the use of sketchy representations goes the "need to know" control principle of human perception. We do not bother to develop a detailed representation of portions of a scene unless there is a motive for such a representation. In complex physical manipulative tasks, we scan from place to place, establishing detailed representations in whatever local region is germane at that instant.

Flexible inferencing. It's natural to ask "what leads to what" in a constraint handling system. Which clues does the system start with, which constraints get applied first, and what happens next along the way to constructing a representation? The human perceiver seems very flexible in this respect. Color may provide the crucial clue to the identity of a blurred object, while, on the other hand, the object if sharply seen can be identified perfectly well without color. A dot may indicate an eye in the context of other dots disposed in a face-like configuration in a circle. On the other hand, an eye shape rendered in detail will be recognized as an eye even if anomalously located in a surrealist painting. The known shape of an object can serve as a clue to its slant, but, on the other hand, Wallach and Moore (1962) demonstrated that stimulus information about slant independent of shape can facilitate the perception of shape. It seems that the perceiver can start any place where there is good information and proceed from there. In metaphorical terms, one could think of an adequate representation flowing out from regions of high information density to regions of low information density.

Biederman (1981) reports a series of experiments that can be interpreted as supporting this contention. Biederman notes that a reasonable and often proposed order of processing begins with generic cues such as support and interposition, along with object identification, and moves on to incorporate semantic constraints reflecting schemata incorporating semantic information, schemata that viewers have for various sorts of scenes. For instance, hydrants are expected to occur on the edges of sidewalks, not up on the walls of buildings.

The reported studies argue that semantic relations come into play at least as quickly as generic cues of interposition and support. Biederman accordingly suggests that scene schemata figure importantly even in early phases of processing. But one can add that since not all scenes have scene schemata – it is easy to construct an utterly unfamiliar layout of objects that is nonetheless interpreted adequately – the perceiver can proceed by multiple routes. Biederman's experiments also document impressive autonomy and rapidity in reading scenes. A single fixation suffices to extract considerable information and prior exposure of the same scene up to four times appears not to alter the processing, which apparently runs off so rapidly and perfectly that short-term learning has nothing to add.

The flexibility of inferencing identified in such cases appears to be a general characteristic of human perception. Another case in point is reading. Kolers (1977), comparing reading text and reading pictures, emphasizes that the orderly layout of text does not necessarily lead to an equally orderly series of fixations. Like readers of pictures, who scan here and there to pick up whatever information they need, the reader of text often departs from the sequence and syntax of the text to interrogate the page for desired information. This may involve anything from glancing back to check a point, to the case of a highly skilled reader who routinely reads down the left hand page and up the right hand page.

Maximizing with minimal search. It seems clear that the human perceiver deals with some stimuli by integrating diverse cues, each insufficient in itself, into an overall interpretation that is sound. This, of course, happens in recognition of familiar object types such as tables and chairs. However, it also happens when the stimulus is not of a familiar type. Figure 5 displays a picture I have frequently used to argue this point. We have no trouble making out the three dimensional shape of the form, even though we probably never have encountered just this shape before. Apparently, the perceiver works on a Gestalt criterion, making an interpretation that exhibits many geometric regularities such as rectangularities, symmetries, and parallelisms (Attneave, 1982; Barrow & Tenenbaum, 1981; Leeuwenberg, 1971, 1982; Perkins, 1982; Perkins & Cooper, 1980).

By what sort of a constraint handling process might perceivers arrive at such interpretations? The options range widely, including spontaneous processes of a system relaxing to a minimum energy state (Attneave, 1982) and more computation-like hypothesis making, extension, and testing processes (Perkins, 1982). The latter reference also contains a comparative discussion of several possibilities. The aim here is simply to suggest that, whatever the mechanism, very little trial and error is involved. That is, the process rarely constructs interpretations of much of the stimulus that then have to be abandoned in favor of even better interpretations that accommodate even more of the stimulus.

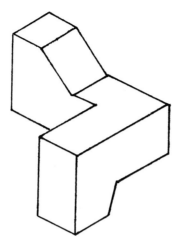

Figure 5. An unfamiliar shape perceived readily because it incorporates geometric regularities like rectangularity and parallelism.

The argument for this point depends on noting when rescinding occurs and what the experience of it is like. We find it in the viewing of anomalous figures such as the Penrose triangle or the devil's tuning fork. When rescinding happens, it surprises the perceiver. The very fact that such surprises seem limited to a few cleverly constructed stimuli and the quite rare natural occurrences of like oddities suggests that rescinding of a major chunk of a perception rarely occurs. To put it another way, the human perceptual system rarely makes that big a mistake in the first place.

Further assurance on this point comes from recognizing that maximizing procedures that rarely make big mistakes are quite feasible. One possible model, for example, is a production system (Newell & Simon, 1972) where the patterns that trigger the productions are restrictive enough so that rarely does a production have to be rescinded on the basis of later evidence. For instance, a system hypothesizing that a certain corner was rectangular might gather considerable local evidence before so venturing. Then again, the criteria for extending a hypothesis to an interpretation of other parts of the stimulus could be severe enough so that bad hypotheses damp out very quickly. Another possible model is some physiological analog of a physical relaxation process, such as occurs in soap films, where interlocking forces ensure that the system will converge rather directly to a minimal energy state (cf. Attncave, 1982; Perkins, 1982).

Taken with the previous point about flexible inferencing, the present one about minimal search suggests that the human perceiver proceeds along paths of highly reliable recognitions and inferences in order to avoid backup or construction of parallel alternatives. This might seem too straightforward to argue for, were it not for the fact that processes involving considerable search are real options in machine perception. A case in point that involves symmetry constraints is Kanade's (1981) procedure for reconstructing the three-dimensional shape of an object from a single view. In a first pass, Kanade uses an extension of the line labeling system evolved by Huffman (1977), Clowes (1971), and Waltz (1975), to generate all legal labellings of the segments as convex, concave, or occluding on one side or the other. In a second pass, Kanade's procedure applies a skewed symmetry principle, which assumes that skewed symmetries in the figure correspond to true bilateral symmetries in its three-dimensional interpretation. For instance, the parallelograms that make up the faces of a projected rectangular form are skewed symmetric around axes that bisect them parallel to their edges, and hence are interpreted as rectangles.

Kanade's procedure appears to contrast with the human in the amount of search it does. While Kanade's procedure involves two passes, one for line labelling and another for applying the skewed symmetry heuristic, the human perceiver appears to take both steps at once, generating an interpretation of part of the figure – for instance one corner of a Penrose triangle – which includes both the occluding and convexity-concavity status of the edges and assignment of orientation in three-space to the edges. Then the perceiver extends this interpretation to try to accommodate other parts of the object (Perkins, 1982; Perkins & Cooper, 1980). Also, while Kanade's procedure keeps track of multiple interpretations, the human procedure seems to pursue only one at a time, attempting another only if the first fails, as in the case of viewing the Penrose triangle. (Although not relevant to manner of search, another difference is that the human perceiver often uses a rectangularity constraint even when it is not associated with a skew symmetry constraint, finding interpretations that Kanade's procedure will not – cf. Perkins, 1976.)

5. The Human Style and the Machine Style

Some deficits of the human perceiver have been reviewed, along with some assets. How do these strengths and weaknesses fit together? Can they both be seen as reflections of a unified style of constraint handling in the human visual system? Perhaps they can, as follows. The human perceiver depends upon high resolution input to a very rapid and flexible pattern recognition and inference making system. One might characterize the constraint handling style with three principles. *Need to know:* proceed beyond a sketchy representation only on a need-to-know basis. *Reliable steps:* construct a representation of the stimulus through a series of reliable recognitions and inferences, following whatever course promises high reliability. *Avoid search:* rarely explore deeply alternative representations.

The three principles work well together. Settling for sketchy representations, which the need-to-know principle usually allows, leaves the system free to avoid search and take reliable steps, since the constraint handling need not wring the last drop of information from the input. Taking mostly reliable steps naturally reduces the need for search and favors sketchy representations, since there will be a level of detail to which the process could not proceed and maintain reliability. Likewise, avoiding search recommends taking reliable steps as a way of doing so and a need-to-know principle as a way of eliminating occasions where search might be required.

Of course, the harmony of the three depends on an input of high resolution, rich with information. For instance, a highlight, a shadow line, a corner allowing a rectangular interpretation, an intensity gradient, a texture gradient, or simple combinations of such features, might be the information that, in a particular case, reliably discourages or encourages a particular extension of the interpretation-so-far. With less resolution and fewer sorts of features, there would be a need to know beyond the reach of reliable steps that avoided search.

With this in mind, how is the sometimes extreme metric inaccuracy of the human perceiver to be understood? We recognized at the outset that somehow, the frequent metric errors of the perceiver must be benign. We can relate this to the three principles by saying that reliability must be read as reliability relative to the need to know. With this in mind, benign mistakes make good sense in terms of the three principles. Fixing a parameter in a somewhat ad hoc way rather than maintaining it as an unknown helps to avoid search, and consequent steps can still be reliable relative to the need to know. For many situations, exact metric information is not critical, so why waste the time to compute it? In hands-on situations the human perceiver can call on stereo vision and feedback from touch for an accurate reading of the world.

It was argued earlier that the control process by which representations of the stimulus are developed was not very sophisticated, often opting for compromise rather than exploration of alternatives, accepting top-down advice only sporadically, re-trying failed interpretations, and so on. This, too, is understandable in light of the constraint handling strategy sketched above. With high-reliability recognition and inferencing, there is no particular need for a sophisticated control process. Any path to a representation will do, so long as it is a reliable one, because it will end up in the same place as any other reliable path. Compromise between conflicting cues avoids searching to determine independently the fitness of the two alternatives, and the result is a benign mistake.

A corollary of this perspective on control is that, although the human perceiver can be considered a hypothesis maker and tester of sorts, the perceiver is not a very good one. That is, the hypothetico-deductive method with its constant proposing and rejecting of hypotheses is not his bread-and-butter strategy. It is a fallback strategy that operates rather inefficiently when the principal one – being right the first time, or at least right enough for the purpose – fails.

We began with the observation that a human is a bad machine. The reason seems to be that a human perceiver has quite a different style of visual processing, a style that explains the human lapses in terms of some of the human strengths. One might say that, just as humans are bad machines, so are machine perceiving systems bad humans – not merely in the sense that they do not see as well, but in the sense that they employ a different style of visual processing.

What is the current "machine style?" As Brady (1981) has recently emphasized, most extant research on machine perception is concentrating on particular classes of visual cues – shape from shading (Ikeuchi & Horn, 1981), shape from texture (Witkin, 1981), shape from models (Brooks, 1981), and so on – rather than on building integrated systems. There is a commitment to the most general possible analysis of how the impingement of light on surfaces generates a given sort of information in the reflected light, and to the construction of general algorithms that suffice to recover all of that information with as few additional ad hoc assumptions as possible. The algorithms tend to involve considerable search and construction of parallel alternative interpretations, as with Kanade's (1981) procedure discussed earlier. The result is what has been called a "physicist's system." In contrast, the human style yields what might be called a "pragmatist's system," casual about benign mistakes, opportunistic about whatever varied cues provide reliable paths of inference in the situation, wary of searches, sketchy unless there is a need to know, and so on.

Does any advice for the development of machine systems come from this contrast between the human and the machine styles? Certainly not just that machine systems that integrate different sorts of cues will need to be built, since this has been the intent of such research all along. Rather, the human example suggests that such physicist's care with the contributing components – shading, texture, Gestalt regularities, models, and so on – may not be the only, or even the best, course. A pragmatist's perceiving machine might be designed with a more slapdash approach to some hypothetico-deductive aspects of control, but with constraint handling that incorporates the three principles outlined earlier.

There is no implication here that a pragmatist's system would be easy to construct. Neither is there any suggestion that the work on physicists' systems has been a waste of time. Although evolution may have left the human system a poor physicist and a good pragmatist, the ultimate machine perceiver would best be good both ways. The point is simply that the human style deserves attention in machine perception research, both because this might lead to better machines, and because it might illuminate just what about the human style, only broadly characterized here, makes it work so well.

Acknowledgment

I thank Joanne Owens for her able assistance with certain of the investigations reported here and in preparing this paper. The research was performed at

Project Zero, Harvard Graduate School of Education, Cambridge, Massachusetts, with support from National Science Foundation Grant No. BNS-7924746. The opinions expressed here do not necessarily reflect the positions or policies of the supporting agency.

References

Ames, A. *An Interpretative Manual for the Demonstrations in the Psychology Research Center, Princeton University,* Princeton, NJ: Princeton University Press, 1955.

Attneave, F. Prägnanz and soap bubble systems: A theoretical explanation. In J. Beck (Ed.), *Representation and Organization in Perception.* Hillsdale, NJ: Erlbaum, 1982.

Attneave, F., & Frost, R. The determination of perceived tridimensional orientation by minimum criteria. *Perception and Psychophysics,* 1969, **6**, 391-396.

Barrow, H. G., & Tenenbaum, J. M. Interpreting line drawings as three-dimensional surfaces. *Artificial Intelligence,* 1981, **17**, 75-116.

Beck, J., & Gibson, J. J. The relation of apparent shape to apparent slant in the perception of objects. *Journal of Experimental Psychology,* 1955, **50**, 125-133.

Biederman, I. On the semantics of a glance at a scene. In M. Kubovy & J. R. Pomerantz (Eds.), *Perceptual Organization.* Hillsdale, NJ: Erlbaum, 1981.

Binford, T. O. Inferring surfaces from images. *Artificial Intelligence,* 1981, **17**, 205-244.

Brady, J. M. Preface – the changing shape of computer vision. *Artificial Intelligence,* 1981, **17**, 1-15.

Brooks, R. A. Symbolic reasoning among 3-D models and 2-D images. *Artificial Intelligence,* 1981, **17**, 285-348.

Clowes, M. B. On seeing things. *Artificial Intelligence,* 1971, **2**, 79-116.

Davis, L. S., & Rosenfeld, A. Cooperating processes for low-level vision: A survey. *Artificial Intelligence,* 1981, **17**, 245-263.

Draper, S. The Penrose triangle and a family of related figures. *Perception,* 1978, **7**, 283-296.

Eriksson, E. S. The shape-slant invariance hypothesis in static perception. *Scandinavian Journal of Psychology,* 1967, **8**, 193-208.

Eriksson, E. S. Two-dimensional field effects and static slant perception. *Scandinavian Journal of Psychology,* 1968, **9**, 19-32.

Gibson, J. J. *The Perception of the Visual World.* Boston, MA: Houghton-Mifflin, 1950.

Gibson, J. J. *The Senses Considered as Perceptual Systems.* Boston, MA: Houghton-Mifflin, 1966.

Gibson, J. J. *The Ecological Approach to Visual Perception.* Boston, MA: Houghton-Mifflin, 1979.

Girgus, J. J., Rock, I., & Egatz, R. The effect of knowledge on reversibility of ambiguous figures. *Perception and Psychophysics,* 1977, **22**, 550-556.

Goldstein, A. G., & Mackenberg, E. J. Recognition of human faces from isolated facial features: A developmental study. *Psychonomic Science,* 1966, **6**, 149-150.

Gregory, R. L. A look at biological and machine perception. In B. Meltzer & D. Michie (Eds.), *Machine Intelligence 7.* Edinburgh, Scotland: Edinburgh University Press, 1972.

Gregory, R. L. *Mind in Science: A History of Explanations in Psychology and Physics.* Cambridge, UK: Cambridge University Press, 1981.

Harmon, L. D. The recognition of faces. *Scientific American,* 1973, **229** (5), 70-84.

Herrnstein, R. J., & Loveland, D. H. Complex visual concept in the pigeon. *Science,* 1964, **146**, 549-551.

Hochberg, J. Levels of perceptual organization. In M. Kubovy & J. R. Pomerantz (Eds.), *Perceptual Organization.* Hillsdale, NJ: Erlbaum, 1981.

Horn, B. K. P., & Schunck, B. G. Determining optical flow. *Artificial Intelligence,* 1981, **17**, 185-203.

Huffman, D. A. Impossible objects as nonsense sentences. In B. Meltzer & D. Michie (Eds.), *Machine Intelligence 6.* Edinburgh, Scotland: Edinburgh University Press, 1971.

Huffman, D. A. Realizable configurations of lines in pictures of polyhedra. In E. W. Elcock & D. Michie (Eds.), *Machine Intelligence 8.* Edinburgh, Scotland: Edinburgh University Press, 1977.

Ikeuchi, K., & Horn, B. K. P. Numerical shape from shading and occluding boundaries. *Artificial Intelligence,* 1981, **17**, 141-184.

Ittelson, W. H. *The Ames Demonstrations in Perception: A Guide to their Construction and Use.* Princeton, NJ: Princeton University Press, 1952.

Kanade, T. Recovery of the three-dimensional shape of an object from a single view. *Artificial Intelligence, 1981,* **17**, 409-460.

Kilpatrick, F. P. Two processes in perceptual learning. *Journal of Experimental Psychology,* 1954, **47**, 362-370.

Kolers, P. A. Reading pictures and reading text. In D. N. Perkins & B. Leondar (Eds.), *The Arts and Cognition*. Baltimore, MD: Johns Hopkins University Press, 1977.

Kosslyn, S. M. *Image and mind*. Cambridge, MA: Harvard University Press, 1980.

Leeuwenberg, E. L. A perceptual coding language for visual and auditory patterns. *American Journal of Psychology*, 1971, **84**, 307-349.

Leeuwenberg, E. L. Metrical aspects of patterns and structural information theory. In J. Beck (Ed.), *Representation and Organization in Perception*. Hillsdale, NJ: Erlbaum, 1982.

Newell, A., & Simon, H. *Human Problem Solving*. Englewood Cliffs, NJ: Prentice-Hall, 1972.

Olson, R. K., Pearl, M., Mayfield, N., & Millar, D. Sensitivity to pictorial shape perspective in 5-year old children and adults. *Perception and Psychophysics*, 1976, **20**, 173-178.

Penrose, L. S., & Penrose, R. Impossible objects: A special type of visual illusion. *British Journal of Psychology*, 1958, **49**, 31-33.

Perkins, D. N. Cubic corners. MIT Research Laboratory of Electronics, Quarterly Progress Report 89, 1968, 207-214.

Perkins, D. N. Visual discrimination between rectangular and nonrectangular parallelopipeds. *Perception and Psychophysics*, 1972, **12**, 396-400.

Perkins, D. N. How good a bet is good form? *Perception*, 1976, **5**, 393-406.

Perkins, D. N. Metaphorical perception. In E. Eisner (Ed.), *Reading, the Arts and the Creation of Meaning*. Reston, VA: National Art Education Association, 1978.

Perkins, D. N. The perceiver as organizer and geometer. In J. Beck (Ed.), *Representation and Organization in Perception*. Hillsdale, NJ: Erlbaum, 1982.

Perkins, D. N., & Cooper, R. How the eye makes up what the light leaves out. In M. A. Hagen (Ed.), *The Perception of Pictures II: Dürer's Devices: Beyond the Projective Model*. New York: Academic Press, 1980.

Perkins, D. N., & Deregowski, J. A cross-cultural comparison of the use of a Gestalt perceptual strategy. *Perception*, in press.

Perkins, D. N., & Hagen, M. A. Convention, context and caricature. In M. A. Hagen (Ed.), *The Perception of Pictures I: Alberti's Window: The Projective Model of Pictorial Information*. New York: Academic Press, 1980.

Wallach, H., & Moore, M.E. The role of slant in the perception of shape. *American Journal of Psychology*, 1962, **75**, 289-293.

Waltz, D. Understanding line drawings of scenes with shadows. In P. H.

Winston (Ed.), *The Psychology of Computer Vision.* New York: McGraw-Hill, 1975.

Weiner, M. Perceptual development in a distorted room: A phenomenological study. *Psychological Monographs,* 1956, **70** (16).

Witkin, A.P. Recovering surface shape and orientation from texture. *Artificial Intelligence,* 1981, **17**, 17-45.

Spatiotemporal Interpolation in Vision[1]

T. Poggio

Massachusetts Institute of Technology
Cambridge, Massachusetts

M. Fahle

Universität Tübingen
Tübingen, Federal Republic of Germany

Abstract

Stroboscopic presentation of a moving object can be interpolated by our visual system into the perception of continuous motion. The precision of this interpolation process has been explored by measuring the vernier discrimination threshold for targets displayed stroboscopically at a sequence of stations. The vernier targets, moving at constant velocity, were presented either with a spatial offset or with a temporal offset or with both. The main results are:

1) Vernier acuity for spatial offset is rather invariant over a wide range of velocities and separations between the stations (see Westheimer & McKee, 1975).

2) Vernier acuity for temporal offset depends on separation and velocity. Optimal acuity decreases with increasing separation.

3) Blur of the vernier pattern decreases acuity for spatial offsets, but improves acuity for temporal offsets.

4) A temporal offset exactly compensates the equivalent spatial offset only for a small separation and optimal velocity. Otherwise the spatial offset dominates.

A theoretical analysis of the interpolation problem suggests a computational scheme based on the assumption of constant velocity motion. This assumption reflects a constraint satisfied in normal vision over the short times and small

---

[1]Further details can be found in (Fahle & Poggio, 1981).

distances normally relevant for the interpolation process. A reasonable implementation of this scheme only requires a set of independent spatiotemporal channels, that is, receptive fields, with the different sizes and temporal properties revealed by psychophysical experiments, but it requires direction selective properties.

Finally, this computational scheme may be relevant for machine vision in order to provide high resolution data for objects in motion and avoid motion smear.

Part I

1. Introduction

Since the first measurements of vernier acuity in 1892 by Wülfing in Tübingen, the extraordinary accuracy with which the human eye can estimate the relative positions of lines or other features in the visual field has represented a long-standing puzzle in vision research. Acuity of this type, also called hyperacuity, can be measured in a variety of situations (Westheimer & McKee, 1977b). A typical example is the acuity found in reading a vernier (see inset of Figure 7a). This can be as high as 5″ of arc (Westheimer & McKee, 1975), that is, 0.02 mm at 1 meter distance.

Most remarkably of all, vernier acuity is not affected by movement of the target in a velocity range from 0°/sec to at least 4°/sec (Westheimer & McKee, 1975). This means that a subject can detect the relative position of two lines to within a fraction of a receptor diameter (and spacing) while the whole pattern is moving across 70 receptors in 150 msec. Recently, evidence has been accumulating which suggests that the visual system is able to perform a very precise temporal interpolation as well, by reconstructing the spatial pattern of activity at moments intermediate between discrete temporal presentations (Barlow, 1979). The most telling demonstration, apart from cinematography, was introduced by Burr (1979a; see also Morgan, 1980) and is shown in the inset of Figure 7c. Vernier line segments are displayed stroboscopically at a series of stations to portray a moving vernier; an illusory displacement occurs if the line segments are accurately aligned in space but are displayed with a few milliseconds' delay in one sequence relative to the other. Not only do the segments appear to move smoothly from one station to the next but also, between the strobes, they are seen to occupy positions between those where they are actually exposed. The accuracy of detecting the displacement is again in the vernier acuity range, provided that the target moves at constant speed and elicits a clear sensation of motion. One is forced to conclude that not only spatial but also temporal interpolation is performed in the visual system to preserve acuity (and resolution) for objects in motion (see Barlow, 1979).

It is clear that the attainment of such spatiotemporal accuracy does not break any physical law (see Westheimer, 1976). As pointed out by Barlow (1979) and by Crick et al. (1981), the classical sampling theorem allows a correct reconstruction of the visual input from a set of discrete samples in space and time since the LGN signal is bandlimited in temporal and spatial frequency by the photoreceptor kinetics and the eye's optics respectively. In particular, Crick et al. have suggested, similarly to Barlow, that the fine grid of granule cells in layer IVc of the striate cortex performs an interpolation on the output of the LGN fibers, with the goal of representing the position of zero-crossings (the boundaries between activity in an ON and OFF ganglion cell layer) with a very high accuracy (see also Marr & Hildreth, 1980; Marr et al., 1979).

Although spatiotemporal interpolation can be well understood in terms of information theory, the astonishing performance of the visual system seems to require an algorithm and corresponding mechanisms of great ingenuity and precision. An understanding of visual interpolation may also be quite interesting from a purely information processing point of view. High resolution, smear-free real time imagery could benefit significantly from this study of human vision. Here we investigate some properties of this spatiotemporal interpolation. In particular, we examine its performance for a range of "sampling intervals" in space and time.

2. Methods

The vernier target used in these experiments consisted of a thin vertical bar made up of two segments. The stimuli were generated on a Tektronix 604 display equipped with P31 phosphor (fading to 1% within 250 μsec) under the control of analog electronics. Each bar was intensified for 0.1 msec at Δt msec intervals at n successive stations horizontally displaced by a separation Δx. Each of the two segments making up the bar was 24′ high and 1.5′ wide intensified to a luminance of about 50 times detection threshold on a background of 10 cd/m². During an experimental run, a target was presented every 3 seconds. Brief displays of $n \Delta t = 150$ msec, with randomized direction of motion (terminating at the central fixation point), were used to prevent effective pursuit eye movements (Westheimer, 1954). The experiments measured:

a)   the acuity for detection of real vernier offsets of the two segments by δx seconds of arc;

b)   the acuity for detection of apparent vernier offsets produced by delaying the presentation of the lower or upper segment, displayed at the same sequence of stations, by δt msec;

c)   the acuity for detection of mixed vernier offsets produced by a real spatial offset δx together with a temporal delay δt of opposite sign.

In a forced-choice task the subject was required to signal whether the bottom segment was displaced to the right or to the left of the top segment by setting a binary switch. In some experiments a buzzer signaled to the subject his incorrect decision. Acuity was determined by the standard criterion of 75% correct identification. Responses (at least 150 for each data point) were collected from three observers who performed all the experiments reported here. The three subjects (TV, AK, and HW) were male students aged between 19 and 25. One of them (HW) has normal sight; the two others attain normal vision with correcting spectacles. They were naive concerning the goals of this study. The main effects have been verified with several other observers, including the two authors.

Since we wanted to study how vernier acuity depends on separation between the stations (Δx) and on the velocity (v) we had to choose whether to keep constant the number of stations (n) or the total presentation time ($T = n\Delta t$). Furthermore, T had to be kept below 150 msec to avoid pursuit eye movements (Westheimer, 1954). In all experiments reported here T is constant ($T = 150$ msec) and, as a consequence, the number of stations n is variable ($n = 2$ to 95).

Notice that the strobe interval Δt, the distance between stations Δx and the velocity v are not independent variables. Any two of them suffice to determine the third since $\Delta x = v \Delta t$. Since the total presentation time was constant, the total distance over which the pattern travelled changed as a function of velocity v. Although in most cases the stimulus remains restricted to the fovea, at high velocities the stimulus is to a large extent parafoveal (at $v = 80°/\text{sec}$, the total distance is about 12°).

2.1. The Spatial Type of Acuity: Dependence on Velocity (v) and Separation (Δx)

The results for spatial offsets (with simultaneous presentation of the two segments at each station) are shown in Figures 1a and 1b. The main result is that spatial acuity is relatively independent of the separation between the stations and of the velocity of the target up to rather large velocities. These data confirm and extend Westheimer and McKee's results (1975), which showed that vernier acuity is unaffected by rate of movement from 0°/sec up to 4°/sec. Our results imply that this type of vernier acuity is relatively independent of Δt, the strobe interval.

Figure 1. Vernier resolution threshold of spatial offset for different separations Δx between the stations as a function of velocity. (a) shows the data from subject AK; (b) from subject TV. The standard deviation of the data is about 25% of the threshold value for (a) and 20% for (b). In (a) the point for $\Delta x = 1'$ and $v = 10°/\text{sec}$ was measured masking the beginning and the ending of the trajectory; the same procedure did not change the threshold for the point at $v = 2.6°/\text{sec}$, $\Delta x = 1'$. Of the two points at $\Delta x = 2.5'$ and $v = 25°/\text{sec}$ in (a), the worse value was measured under the "masking" condition whereas the better one was measured in the standard way. In (b) the point at $\Delta x = 2.5'$ and $v = 25°/\text{sec}$ was also measured with zero offset at the first and last station. (From Fahle & Poggio [1981].)

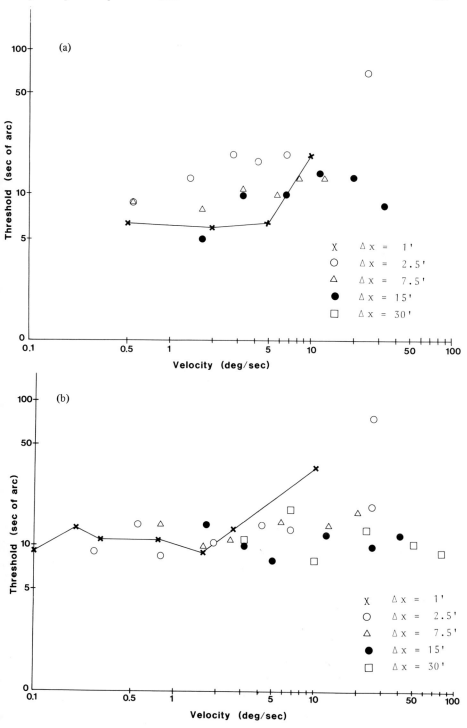

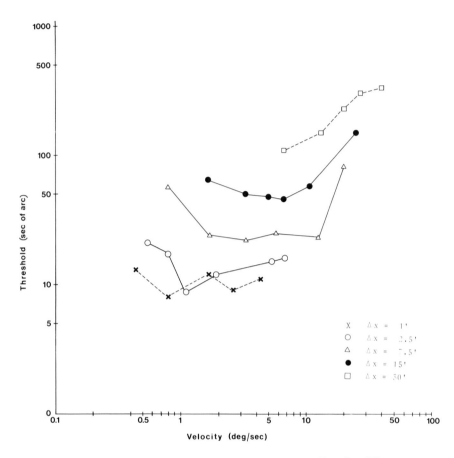

Figure 2. Vernier resolution thresholds of temporal offset for different separations between the stations as a function of velocity. (a) shows the data from subject *AK*, (b) from subject TV. The standard deviation is about 20% of the threshold values for subject *AK* and 18% for subject TV. (From Fahle & Poggio [1981].)

Acuity for continuous motion – not studied in this paper – is expected to deteriorate at high velocities. We were able to support this expectation with several subjects at the limit of the velocity range available to our set-up. Figure 1a, for instance, shows that at separation $\Delta x = 1'$ and velocity $v = 10°/sec$, acuity is worse. This value thus represents an upper bound to the limit velocity expected for continuous motion. Our data and informal experiments suggest that at large separations (certainly for $\Delta x > 7.5'$) acuity is invariant over a very wide range of velocities.

In this experiment we measured pure vernier acuity. The appearance of the spatiotemporal pattern changed considerably for too low or too high strobe rates. Whereas a sharp bar smoothly moving was perceived at intermediate velocities, at very low strobe rates (10 Hz) the display began to flicker and a

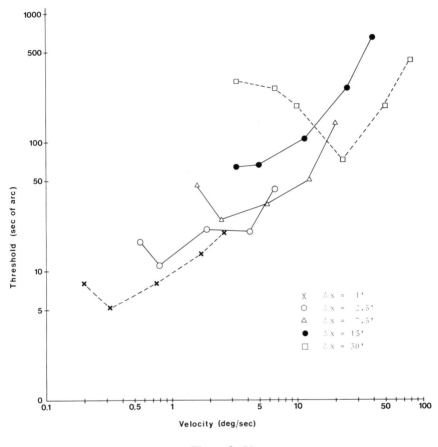

Figure 2. (b)

sequence of flashes of an isolated vernier was perceived. For high (60 Hz) strobe rates and large separations, many of the stations seemed to be simultaneously illuminated, giving the impression of a stationary grating of parallel verniers. Under these conditions, vernier resolution remained essentially unchanged (see Figure 1), although the spatiotemporal pattern was perceived in a distorted form. At high strobe rates and a separation of $\Delta x = 1'$, motion perception was strong but the vernier pattern was considerably smeared: acuity was then lower.

2.2. The Temporal Type of Acuity: Dependence on v and Δx

Figures 2a and 2b show the results for temporal offsets. The accuracy of detecting the equivalent displacement is in the classical vernier acuity range (compare Burr, 1979a,b): the best value for observer AK was 8'' for spatial and 5'' for temporal offset at comparable separations and velocities. Our main

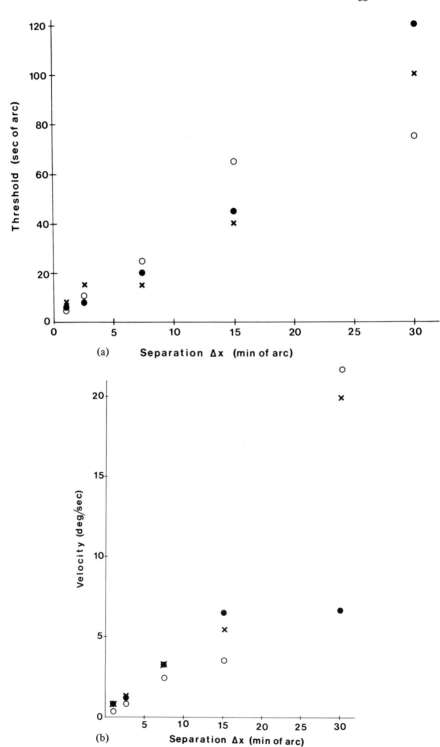

(a)   Separation Δx (min of arc)

(b)   Separation Δx (min of arc)

new result is that although acuity does not break down for large separations between the stations, at least up to half a degree, it deteriorates significantly almost in proportion to Δx (see Figure 3a).

Vernier acuity of this temporal type is bad at low and high speed. As already clearly demonstrated by Burr (1979a,b) apparent motion is necessary for temporal offsets to be seen as spatial offsets. In our experiments, deterioration of acuity at low velocities could be due to the speed per se as well as to the lower number of stations (because our total presentation time is constrained to $T = 150$ msec the stimulus consisted, at the lowest velocities, of two stations). In any case, deterioration of acuity at low velocities can be linked with a decreased sensation of motion (see discussion).

A second important result is that the range of velocities for which temporal interpolation is good shifts upwards for larger separations between the stations. The fact that at higher separations higher velocities are required for good resolution suggests that a more revealing parameter is the time interval Δt between the strobes. In fact, at any separation Δx, temporal interpolation is optimal for a temporal interval Δt between 50 msec and 20 msec.

2.3. The Effect of Blur on Spatial and Temporal Acuity

Standard vernier acuity is known to be affected, as one would expect, by attenuation of the high spatial frequencies of the vernier pattern (see for instance Stigmar, 1971). Is temporal interpolation also degraded in the same way?

We have performed some experiments to answer this question by placing a ground glass screen at 1 cm in front of the display. When a sharp line is viewed through such a ground glass screen the resulting light distribution has an approximately Gaussian line spread function with a width at half-height of at least 15′, corresponding to a cutoff frequency of around 3-4 cycles/deg. Our data show that in the experimental situation of Figure 2, blur of the pattern *improves* acuity at large separations and velocities. Figure 4 compares directly for the same observer and for the same separation the effect of blur on spatial and temporal interpolation. Westheimer's type of acuity is degraded by blur, whereas Burr's type of acuity improves dramatically with blur (at high velocities). Out of five observers only in one case did blur of the pattern cause a reduction in temporal vernier acuity at high separations and velocities.

These data again show that temporal hyperacuity has different characteristics from spatial hyperacuity.

Figure 3. (a) shows the best vernier resolution threshold (with temporal offset) for each separation Δx as a function of separation. The data are from three subjects (partly from Figures 2a and 2b). ○ *AK*; ● TV; × HW. In (b) the velocity v for which optimal vernier resolution is found is plotted against the separation Δx. Same data as in (a). (From Fahle & Poggio [1981].)

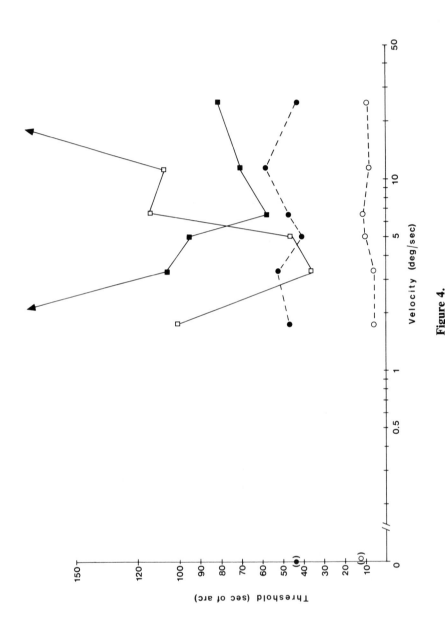

Figure 4.

2.4. Spatial vs. Temporal Offset

The apparent offset produced by a temporal delay δt should follow the ideal relationship $\delta x^t = v \delta t$. As shown by our data the sign of the offset is indeed correctly detected. Does its size also satisfy this relation? How faithful, in other words, is temporal interpolation? To answer this question we measured the temporal delay δt needed to compensate for a given real spatial offset δx for different conditions.

Figure 5 shows that for a separation $\Delta x = 2.5'$ and a velocity $v = 1.1°/\text{sec}$ the temporal offset $\delta x^t = v \delta t$ matches rather closely the real spatial offset δx. Under these conditions spatiotemporal interpolation is indeed rather precise (compare Burr & Ross, 1979). It is not so for higher velocities and/or larger separations (Figure 5). The temporal offset needed to compensate for a real spatial offset is then much larger.

Part II

1. Spatiotemporal Interpolation: How Is It Done?

The results described in Part I constrain the problem of hyperacuity tightly enough to justify a theoretical analysis of how spatiotemporal interpolation may be done in the visual system. The precise meaning of interpolation in terms of our visual stimuli is a well-defined question, and this is the main point to discuss.

1.1. A Simple Illustration

Figure 6 illustrates a very simple scheme for achieving spatiotemporal interpolation of a visual pattern. The elements of this scheme could be – but do not need to be – interpreted as cells with associated receptive fields and temporal impulse responses. Visual input is sampled in space by an array of cells with a sampling density high enough to preserve the whole of the spatial information (in accordance with the sampling theorem). The input is then reconstituted in more detail on a finer grid of cells by convolving the sampled values with the function sinc x. In effect each cell of the interpolation layer weights its inputs according to a center-surround receptive field. A variety of filters (i.e., "receptive fields") are capable of performing a correct interpolation, especially in two spatial dimensions (see Crick et al., 1981).

Figure 4. The effect of blur on spatial and temporal interpolation as a function of velocity for a separation between the stations $\Delta x = 15'$. Vernier resolution of a spatial offset is measured with (●) and without (○) blur. Vernier resolution of a temporal offset is also shown with (■) and without (□) blur. The screen was blurred as described in the text. Notice that the first point for spatial offset is for $v = 0°/\text{sec}$. The observer is TV. The standard deviation is about 20% of the threshold values. (From Fahle & Poggio [1981].)

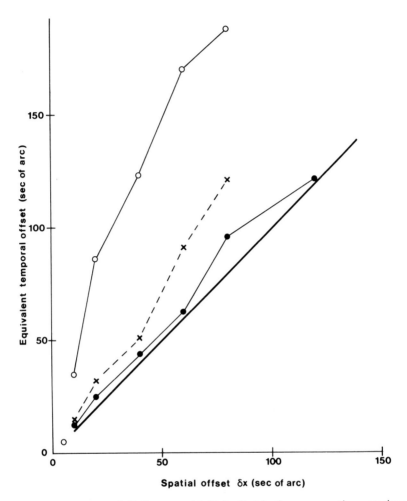

Figure 5. Temporal ($\delta x'$) vs. spatial (δx) offset in the compensation experiment. The ordinate shows the temporal offset in equivalent spatial units $\delta x' = v \cdot \delta v$ needed to compensate the spatial offset shown in the abscissa. ● is for a separation between the stations $\Delta x = 2.5'$ and a velocity $v = 1.11°/\text{sec}$ ($\Delta t = 37$ msec); × is for $\Delta x = 2.5'$ and $v = 5.28°/\text{sec}$ ($\Delta t = 7.9$ msec); ○ is for $\Delta x = 7.5'$ and $v = 4.11°/\text{sec}$ ($\Delta t = 30$ msec). Larger separations yield an even greater mismatch. The continuous diagonal indicates the locus of perfect compensation. Subject TV. (From Fahle & Poggio [1981].)

If the input intensity distribution is presented at discrete instants in time, temporal interpolation can be achieved by suitable temporal low-pass properties of each individual pathway. If the temporal interval between presentations is small enough the effect of the filter is to reconstruct the original continuous temporal input. Spatial interpolation can then operate at each instant of time.

Figure 6b shows the Fourier interpretation of the spatial interpolation process (interpolation in time can be interpreted in a similar way). The effect of sampling is to replicate the original spectrum in an infinite number of side lobes. Spatial interpolation – i.e. reconstruction of the original function from its samples – is accomplished by filtering out all side lobes but the central one, which is the original spectrum.

This model is probably the simplest conceivable scheme. In it, interpolation in space and time are performed independently, since the temporal dependence of the input is not constrained in any way. We now consider the conditions under which this scheme can be effective.

1.2. Remarks on Interpolation

Before embarking on an analysis of various interpolation schemes, it is appropriate to make a few general points which arise from the discussion so far.

First, the process of computing intermediate values from samples does not depend on the existence of a finer retinotopic grid of "cells," where the results are represented. All filtering transformations indicated in Figure 6 could be carried out at a rather symbolic level for only a few distinguished points. Thus, it is important to keep separate the problem of a process from the problem of representing its output. This paper is directly concerned only with the first issue.

Second, the goal of the interpolation process may be far more modest than a full reconstruction of the input distribution. As suggested by Crick et al. (1981), the aim of interpolating the ganglion cells' activity is to provide the position of the zero-crossings (where activity switches from the on-center to the off-center cells) with high accuracy. This can be achieved by using very simple interpolation functions such as a normal center-surround receptive field (Marr et al., 1980).

1 3. More Complex Interpolation Schemes Are Required

The scheme of Figure 6 can provide a correct reconstruction of a spatiotemporal input sampled at intervals $\Delta\zeta$ (in space) and $\Delta\tau$ (in time) only when the input function is bandlimited in spatial (by f_x^τ) and temporal (by f_t^τ) frequencies in such a way that $\Delta\zeta \leqslant 1/2f_x^c$ and $\Delta\tau \leqslant 1/2f_t^c$ (Theorem 1 in the Appendix). The image which reaches the retina is indeed bandlimited in spatial frequencies to less than about 60 cycles per degree by the diffraction limited optics of the eye. Furthermore, a temporal cutoff is imposed at the level of the photoreceptors by their limited temporal resolution. The scheme of Figure 6 can therefore correctly reconstruct an image sampled at intervals of less than $30''$ in space (for the 2-D case see Crick et al., 1981). Temporal samples of

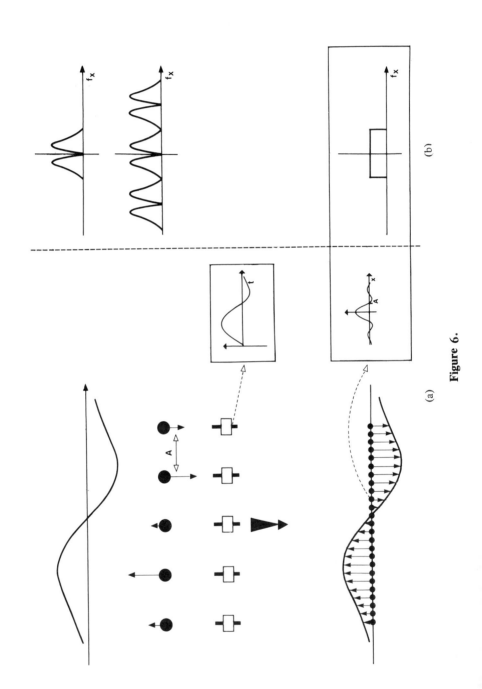

Figure 6.

the photoreceptor activity could be interpolated under similar conditions (though *regular* temporal sampling in our visual system is highly implausible).

Since the spacing of the photoreceptors is almost exactly matched to the eye's optics, interpolation in normal vision – when the image is a continuous function of time and space – can be accounted for by simple schemes like that of Figure 6. In particular, such models could account for the vernier acuity measured with real continuous motion of the retinal image. When, however, motion of an object is simulated by presenting the image at discrete positions at separate instants, the conditions of Theorem 1 are in general no longer satisfied. In our experiments we present to the eye an image which is already sampled either in time (Westheimer type of stimulus) or space (Burr type of stimulus) or both. We enforce on the system arbitrary sampling intervals Δx and Δt, and this is done *before* the bandlimiting operations of the eye's optics and of the receptor kinetics come into play. Under these conditions the input function $g(x,t)$ is not ensured to be appropriately bandlimited before spatial or temporal sampling occurs. The scheme of Figure 6 should for instance perform poorly when the input function is sampled in space at intervals Δx significantly coarser than the photoreceptor array. Burr's and our data, however, show that our visual system performs significantly better then. We are clearly forced therefore to consider other types of interpolation schemes.

2. Interpolation Schemes

2.1. The Spatiotemporal Spectrum of a Moving Vernier

Our analysis of alternative interpolation schemes begins with the description in frequency space of the physical stimuli corresponding to Westheimer's and Burr's experimental situations. When a spatial pattern $g(x)$ moves continuously at constant speed, the resulting spatiotemporal distribution of excitation on the retina has a simple representation in the Fourier space of temporal (f_t) and spatial (f_x) frequencies. Its Fourier transform takes values only on the diagonal line shown in Figure 7a with a slope equal to the velocity (see Fahle & Poggio, 1981). For each spatial frequency contained in the pattern, there is

Figure 6. (a) A simple scheme for spatiotemporal interpolation. The input pattern is sampled by an array of "cells." Spatial interpolation is accomplished on a finer interpolation grid of cells, each one weighting the sampled values with a sinc-shaped receptive field (shown in the lower inset). Temporal interpolation is obtained by filtering with an appropriate low-pass or band-pass filter each of the input channels (its impulse response is shown in the upper inset). Thus a series of discrete frames of a moving pattern can be interpolated (see Theorem 1 in Appendix) into a continuous temporal function in each of the channels. The spatial input distribution outlined here represents an intensity edge as seen by center-surround ganglion cells. (b) The spatial interpolation process in Fourier space. Interpolation is equivalent to filtering out the side lobes originated by the sampling process. Temporal interpolation can be interpreted in a similar way. (From Fahle & Poggio [1981].)

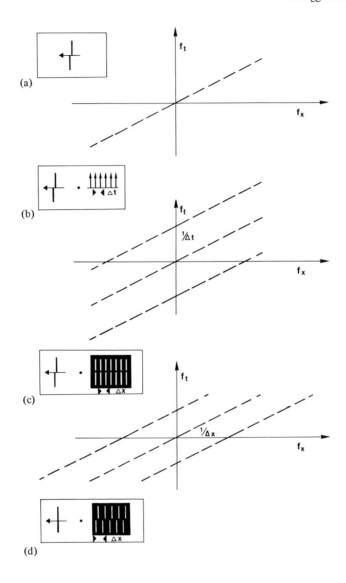

Figure 7. (a) The support on the f_x-f_t plane of the Fourier spectrum associated with continuous motion of a vernier (see inset) at constant velocity $-v$. The slope of the line is v, and $g(f_x,f_t) = g(f_x)$ on that line. Curtailing the duration of motion to T = 150 msec spreads the line into a bar-like support, corresponding to a sinc function. Under our experimental conditions all line supports shown in Figure 7 are spread in this way along the f_t axis. (b) The support of the Fourier spectrum associated with Westheimer's type of experiment. The inset indicates that displaying the vernier stroboscopically at a sequence of times with an interval Δt is equivalent to "looking" at the continuous motion of a vernier through a series of temporal "slits." This has the effect of replicating the spectrum of (a) along the f_t axis in an infinite number of side lobes. The distance of the lobes on f_t is $1/\Delta t$. The line encounters the f_x axis at

a unique temporal frequency corresponding to it. Curtailing the duration of motion (in our case to $T = 150$ msec) spreads the Fourier transform over a large area of temporal and spatial frequencies, changing the narrow line into a wider area. The spread (along the f_t axis) is the same for all our data. Thus the line supports shown in Figure 7 must be interpreted as being spread along f_t as a sinc function. For $T = 150$ msec the width of the spread is about 14 Hz for the central lobe of the sinc function and 28 Hz for the central lobe plus the first negative side lobe on both sides. The retinal stimulus elicited by continuous motion of a vernier at constant velocity can be described in this way (see Fahle & Poggio, 1981). The upper and the lower segments have the same line support on the f_x-f_t plane. Their Fourier transforms differ at all frequencies only by a phase factor which mirrors the spatial offset. The correct detection of this information underlies positional acuity.

Figure 7 summarizes the description of the two basic stimulus configurations used in this paper according to the derivation outlined by Fahle and Poggio (1981). Westheimer's experimental situation is equivalent to looking at the continuous motion of a vernier through a series of equidistant narrow *temporal* slits within which the pattern is briefly visible (see Figure 7b). Burr's experimental situation ideally corresponds to a vernier moving behind a spatial window with a series of equidistant narrow slits (see Figure 7c). The spatial or temporal windows affect differently the spectrum of the retinal input. As indicated in Figure 7, in the Westheimer situation the complex spatial spectrum of the pattern, which contains amplitude and phase information, is replicated an infinite number of times along the temporal frequency axis. The distance between successive replicas is $1/\Delta t$ ($1/v \Delta t = 1/\Delta x$ on f_x) for the case of Figure 7b and $1/\Delta x$ on f_x in Figure 7c. An important observation is that in Figure 7b (Westheimer stimulus) all lobes at any given f_x support exactly the same complex spectrum g. This is not so in Figure 7c (Burr stimulus), where, instead, all lobes have the same g at any given f_t. We re-emphasize that Figure 7 describes the physical properties of the different stimuli without any reference to the human visual system.

$1/v \cdot \Delta t = 1/\Delta x$ (if $\Delta x = 1'$, the distance of the side lobes on f_x is 60 cycles/deg). Notice that for any f_x, each lobe supports the same complex Fourier spectrum $g(f_x)$. (c) The support of the Fourier spectrum associated with Burr's type of experiment. Displaying the line segments of a vernier at the same position but with a slight delay is equivalent to looking at the continuous motion of a vernier through the spatial window depicted in the inset (transparent slits in an otherwise opaque screen). This corresponds to replicating the spectrum of Figure 8a along the f_x axis. The distance of the lobes is $1/\Delta x$, where Δx is the interval between successive slits in the spatial window. At a given f_x, the Fourier spectrum $g(f_x)$ of different lobes is in general different. (d) The support of the Fourier spectrum associated with the compensation experiment is the same as in Figure 8c. The different window corresponding to this stimulus (see inset) corresponds, however, to a different complex Fourier spectrum. (From Fahle & Poggio [1981].)

2.2. Computational Aspects of Interpolation: The Constant Velocity Assumption

More effective interpolation schemes are feasible if general constraints about the nature of the visual input are incorporated directly in the computation. The key observation here is that the temporal dependence of the visual input is usually due to movement of rigid objects, and that in everyday life motion has a nearly constant velocity over the times and distances which are relevant to the interpolation process (T < 100 msec and x < 1°). The constant velocity assumption leads to a more specific form of the sampling theorem, given in the Appendix (see also Crick et al., 1981), which states formally what is intuitively clear: the spatiotemporal sampling rate can become very low without losing information. Interpolation schemes based on the constant velocity assumption exploit the equivalence of the time and space variables ($x + vt$). From the point of view of filtering this means that spatial and temporal interpolation cannot be performed independently as in the simple scheme of Figure 6. In the Fourier domain the constant velocity assumption constrains the spectrum of the visual input to lie on the line of support shown in Figure 7a. In the ideal case of infinitely long motion the side lobes generated by sampling either in time (Figure 7b) or space (Figure 7c) can always be excluded by means of appropriate filters, if the precise value of v is known (e.g. by measurements). The recovery of the original spectrum (Figure 7a) corresponds to an ideal interpolation for arbitrarily large sampling intervals (if v is known and different from zero). In the realistic case of finite duration of motion finite sampling intervals are enforced by the spread of the Fourier spectrum into a larger area, but the same basic arguments still apply.

An interpolation scheme of this type could be implemented simply by measuring the exact velocity of movement and then reconstructing the spatiotemporal trajectory of the pattern for either temporal or spatial information. Another, more attractive possibility is suggested by the idea, supported by much psychophysical evidence, that in the human visual system there exist several channels at each eccentricity, i.e. several sets of receptive fields tuned to different spatial sizes and with different temporal properties. We imagine, following Burr (1979b), that these channels have somewhat overlapping supports covering the region of the f_x-f_t Fourier plane which corresponds to the sensitive range of the visual system. "Stasis" channels are tuned to high spatial frequencies (small receptive fields) and low temporal frequencies (sustained properties); "motion" channels are tuned to low spatial frequencies (large receptive fields) and high temporal frequencies (transient properties). Thus, each channel is tuned to a different range of velocities, centered on the ratio between the optimal temporal and spatial frequencies characteristic for the channel; stasis channels, for instance, are tuned to low velocities whereas motion channels are tuned to high velocities. Figure 8b shows a set of idealized "velocity channels" of this type. Since each channel has its own cutoff in temporal

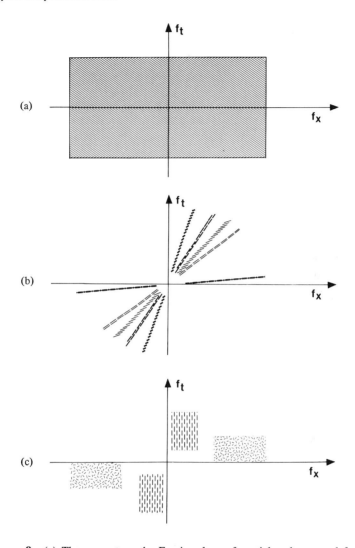

Figure 8. (a) The support on the Fourier plane of spatial and temporal frequencies of an interpolation filter corresponding to a scheme such as Figure 6. (b) The support on the Fourier plane of a set of spatiotemporal filters ideally tuned to different velocities. A large number is needed to cover all velocities of interest. The filters are assumed to be direction-selective, since they only operate in the Fourier quadrants corresponding to positive $v = f_t/f_x$ in $g(x + vt)$. A spatial pattern moving at constant velocity ands sampled at spatial intervals Δx has on this plane the support shown by Figure 7c. To avoid aliasing, the low velocity filters can be "switched off" by information about the velocity of the motion. (c) A more realistic set of filters, broadly tuned to different velocites. The stasis channel is tuned to low temporal and high spatial frequencies and thus to low velocites. The motion channel is tuned to high temporal and low spatial frequencies and thus to high velocities. Intermediate channels (not shown here) may be also present. The hatched areas represent the support of such directional filters. Nondirectional filters would also have a symmetric support in the other two quadrants. (From Fahle & Poggio [1981].)

and spatial frequency, interpolation may be performed independently and with different characteristics within each channel. In the Burr type of experiment stasis channels could correctly interpolate only patterns displayed at small separations and low velocities, whereas motion channels could be effective (but not so accurate) at large separations and high velocities by filtering out the side lobes arising from the coarse spatial sampling. The complementary argument applies for coarse time sampling. As indicated in Figure 8b the stasis channels may suffer from aliasing at values of Δx for which the motion channels interpolate correctly. We assume, then, that in this scheme the wrong channels are switched off by use of velocity information.

Figure 8c shows a more realistic interpolation scheme of the same basic type. Instead of many channels, each one sharply tuned to velocity and inactivated when the pattern does not move at its characteristic velocity, there are a few channels coarsely tuned to velocity and without any specific velocity sensitive inactivation, apart from possible directional selective properties. In this specific algorithm spatial and temporal interpolation may even be performed independently *within* each channel (directional selective properties aside); with appropriate filter functions, then, Figure 6 may be a rough description of each of the channels of the scheme of Figure 8c.

In the light of this analysis we turn now to a detailed discussion of our experiments. Our main question concerns, of course, which type of interpolation scheme is actually used by our visual system.

3. Discussion

3.1. Westheimer's Acuity: Recovery of Spatial Offset

a) In Fourier terms, the aim of the interpolation process is to filter out the side lobes, preserving only the central lobe, as the latter represents the Fourier spectrum of a continuously moving bar.

When both the time interval Δt between presentations and the velocity v are small, interlacing of the side lobes in the Fourier spectrum is negligible. Temporal low-pass properties of the visual pathway, as in the model of Figure 8a, suffice for eliminating the side lobes and thus achieving a correct interpolation. When Δt is large, however, interlacing is considerable in the sense that, even for the scheme of Figure 8c, there are one or more channels which mix the main lobe with at least one of the side lobes. Because of the spread associated with the short duration of the motion sequence, actual overlap between the lobes can be significant. It turns out, however, that this does not represent a problem from the point of view of the spatial acuity measured in our experiments. At each f_x the complex Fourier spectrum of all side lobes is exactly the *same*. Thus, the spatial spectrum is correct irrespectively of the temporal frequency and independently of the number of side lobes contained in the support of the interpolation filters. At large Δx and high v, the presence of the side lobes turns out to be even beneficial for vernier acuity (consider Figures 7 and 8c); under these conditions high frequency channels, which would not be stimulated by continuous motion, can obtain the correct spatial information from the side lobes, which are an artifact of the discrete time presentations (see Figure 7b). On the whole, and in the absence of a sophisticated interpolation process that always excludes all side lobes (such as the scheme of Figure 8b), one expects vernier acuity to be rather invariant for a wide range of separations and velocities. For very small separations ($\Delta x = 1'$) or for continuous motion, the upper velocity limit is linked to motion smear (the effect of smear on acuity is similar to blur of the pattern, via the temporal cutoff of vision). Our data conform well to these expectations. Notice that the presence of side lobes at high velocities and large separations corresponds to the perception not of a moving bar but of a briefly illuminated stationary grating – which carries, however, the correct spatial information. In this sense at large Δx and high v interpolation fails to retrieve the "correct" spatiotemporal pattern, but still preserves spatial acuity (even at extremely high speeds). An irony of the failure of interpolation is that, as a consequence, hyperacuity of the Westheimer type is better at large Δx and high v than in the case of successful interpolation (and of continuous motion).

b) The qualitative interpretation of our data in usual space-time variables is straightforward. Spatial interpolation, for instance by appropriate receptive fields, takes place correctly for each frame (i.e. for each station) even when temporal interpolation fails. Since our forced-choice task measures only spatial acuity, performance is in this case independent of the interpolation of the temporal dependence of the visual input.

c) These results suggest that spatiotemporal interpolation is not performed by the "ideal" interpolation scheme of Figure 8b. For temporal aspects should then be retrieved correctly at all Δt, while acuity for high velocities should be exactly as bad for continuous motion. The one-channel scheme of Figure 8a could explain these data on positional acuity; but as pointed out by Burr (1979b, 1980) the image should then be inevitably smeared at all but very low velocities.

3.2. Burr's Acuity: Interpolation of Temporal Offset

a) In Burr's experiment the situation is quite different. For any given f_x the side lobes contain different parts of the original spectrum. Thus when more side lobes lie in the support of the same channel (in Figure 8a or Figure 8c) there is a mixture of spatial frequencies, detrimental to acuity. One understands, therefore, that acuity deteriorates considerably (see Figure 2) with increasing overlap among the side lobes (large separations between stations). At any given (large) separation, low velocities bring about a considerable overlap between the side lobes. Higher velocities reduce the degree of overlap at the expense of high spatial frequency information, which is filtered out by the temporal cutoff(s) of the visual pathway (between 20 and 50 Hz; see for instance Kelly, 1979). Thus one expects to find, for each separation Δx, an optimal velocity at which the side lobes just avoid overlap. Assuming a spread of 15 Hz (see Section 2.1), the optimal velocity (in degrees/sec) should be $v = 30\Delta x$ (Δx in degrees), which is in rough agreement with the data of Figure 3b. When the velocity approaches zero the line supports in Figure 8c all tend to lie on the f_x axis (notice that, because of the finite presentation time T, the supports effectively overlap). In this situation information about the offset cannot be retrieved. In the limit of very high velocity the set of lobes approaches the line spectrum of a stationary grating with no offset. Notice that we assume for the scheme of Figure 8c that the vernier threshold is higher when some of the channels signal zero to offset while the others still "see" the correct offset.

b) When the temporal component of the filters fails to interpolate between temporal frames motion is perceived as discontinuous. As a consequence the spatial interpolation process correctly signals zero spatial offset for each frame. The critical strobe interval which yields optimal temporal interpolation is not very different between the channels (see Figure 3a).

c) The scheme of Figure 8b should clearly interpolate much better than our visual system does. Though its performance may worsen at high velocities, as for the continuous motion, it should be rather invariant with respect to Δx, the separation between the stations. Figure 3a shows that this does not happen. The opposite conclusion holds for the scheme of Figure 8a. Its performance should deteriorate rapidly for separations Δx between the stations larger than the distance between photoreceptors, which is in conflict with Burr's and our data. An interpolation scheme of the type of Figure 8c seems consistent with these results: while small, slow "receptive fields" would be unable to interpolate correctly at large separations Δx, large, fast receptive fields could perform a correct interpolation, if the velocity is appropriate.

The fact that spatial acuity is extremely good at separations up to 2.5' suggests that the interpolation channels are direction-selective. As Figure 8c shows, this is equivalent to switching off the mechanisms responding to the wrong direction of motion. Otherwise these mechanisms may be excited by side lobes which partly lie in the wrong quadrant. For instance at $\Delta x = 2.5'$ and low velocities the side lobes, centered at 24 Hz, substantially invade the wrong quadrants. Small nondirectional receptive fields would pick up this wrong information (see Figure 8c).

3.3. Effect of Blur

a) The interpolation scheme outlined in Figure 8c makes a rather strong prediction about the effect of blur. In the Westheimer case blur can only *degrade* vernier acuity, since it eliminates the high frequency channels. Blur of the Burr stimulus, however, should *improve* acuity at least at large separations and high velocities, since it eliminates side lobes which signal the absence of an offset. Our data are fully consistent with this expectation. A more perceptual but equivalent description of the effect of blur is this. At high velocities and large separations there is a strong sensation of a grating of thin, unbroken lines, corresponding to the side lobes seen by visual mechanisms tuned to low temporal and high spatial frequencies, and a weak impression of a single moving target with a clear offset, corresponding to the main lobe seen by mechanisms tuned to lower spatial and higher temporal frequencies. This ambiguity is removed, as already noticed by Burr (1979a), by blur of the screen, which suppresses the high frequency grating.

b) In other words, blur eliminates the contribution of the small receptive fields which are unable to interpolate correctly at large separations and therefore signal zero offset. The large receptive fields, however, remain largely unaffected by blur.

c) The effectiveness of blur in improving vernier acuity at large Δx shows that our visual system does not normally have the intrinsic possibility of switching off the wrong channels as assumed in the scheme of Figure 8b.

3.4. Spatial vs. Temporal Compensation

a) This stimulus situation corresponds to looking at the continuous motion of a vernier through the spatial window shown in the inset of Figure 7d. The resulting Fourier support is again as in Figure 7c: here, however, the main lobe signals no offset corresponding to precise spatiotemporal compensation, whereas the other lobes all signal the spatial offset between the upper and lower grating of the window. In other words, exact compensation between space and time is realized only in the main, correct lobe. Thus, the spatial offset should dominate as soon as the side lobes are ''seen'' by some of the channels of Figure 8c. This is increasingly so for larger separations Δx between the stations. Correspondingly, the perception of the stationary grating carrying spatial offset information (the broken slits in the window of Figure 7d) is expected to dominate at large separations and velocities. Again our data are consistent with these expectations. Even at relatively small separations between the stations (see Figure 5) the system does not achieve a perfect interpolation – that is, removal of all side lobes. Only in this case would the temporal offset exactly cancel the spatial offset. As expected, blur improves compensation, since it helps to remove the ''wrong'' side lobes, which carry information only about the spatial offset.

b) This experiment combines Burr and Westheimer stimuli. Since spatial interpolation always retrieves the spatial offset, this dominates for all cases in which the temporal component of interpolation is not fully correct.

4. Conclusions

To summarize, the psychophysical experiments reported here suggest that spatiotemporal interpolation in the visual system, remarkable though it is, is far from being perfect and flawless. Ideal interpolation is equivalent to filtering out the side lobes in the Fourier spectrum arising from the discrete presentations. As our data suggest, our visual systems do not seem to use a very sophisticated spatiotemporal interpolation process. The side lobes are not effectively filtered out under all conditions. Spatiotemporal interpolation, then, can be considered as a direct consequence of the spatial and temporal properties of early vision, in terms of an interpolation scheme of the type of Figure 8c. The existence of independent channels tuned to different spatial and temporal frequencies seems sufficient to account for the spatiotemporal interpolation revealed by our experiments. A detailed theoretical analysis with the help of appropriate computer experiments is necessary for a quantitative evaluation of interpolation models of this type.

4.1. Neural Mechanisms

Interpolation can be regarded as a spatiotemporal filtering of the input transmitted from the retina. This is the point of view taken in this paper. We cannot advance any hypothesis as to where this filtering stage may be localized in the brain on the basis of our psychophysical data alone. Throughout this paper we have used the term "interpolation" without necessarily implying a direct reconstruction of the pattern of visual activity, say its zero-crossing profile in the various channels, somewhere in the visual pathway. As we mentioned earlier, the interpolation scheme suggested by our data may be implemented as an "implicit interpolation," that is, as a computational process involving manipulation of symbolic quantities: or it may depend on an "explicit reconstruction" of a coded version of the array of photoreceptor activity on a fine retinotopic grid of neurons. These extreme possibilities – and all those in between – can be implemented in a variety of ways.

For instance, activity may be reconstructed on the fine grid of layer IVc β by spatiotemporal properties of intracortical connections; all or part of the channels may thus be independently realized by such connections. To achieve the required spatiotemporal filtering, connectivity in this layer should implement an appropriate spatiotemporal receptive field, including the required directional selective properties. For instance, a "motion" channel which is bandpass in f_t and lowpass in f_x requires a broad center-surround receptive field and a temporal impulse response with a positive peak followed by a negative phase (compare Burr, 1979b; see Barlow, 1979; Crick et al., 1981; von Seelen, 1973). Spatial interpolation, with the goal of providing the precise location of zero-crossings, may rely on rather sharply localized inhibition between on and off layers (Crick et al., 1981). Interestingly, data about vernier acuity for strongly blurred patterns (see for instance Figure 5 and Stigmar,

1971) would imply that the coarse "motion" channel still requires a rather fine interpolation grid, only about six times sparser than the "stasis" channel (on a linear scale).

On the other hand, a specific, more symbolic process could read the outputs of retinal ganglion cells and perform the correct interpolation for any desired position and time. In this case interpolation would be implicit and mixed with the decision process itself.

In the first case, the decision routine (is the upper segment to the right or to the left?) would operate on an interpolated version of the image. Thus, "reprogramming" of the vernier routine may not be expected to affect the interpolation process, contrary to the second case, in which different detection strategies may directly influence interpolation.

In any case our data are not inconsistent with the idea that the spatial frequency-tuned channels present in early human vision are the interpolation filters themselves. If this is so, at least some of the corresponding neurons might be located not earlier in the visual pathway than where interpolation takes place. Although directional selectivity must be an important property of the interpolation filters, neurons in the interpolation grid do not need to be direction-selective themselves. Finally, it is interesting that several of the properties of classical vernier acuity as characterized mainly by Westheimer and coworkers can be explained by neural schemes of this type (Westheimer & Hauske, 1975; Westheimer & McKee, 1977a,b, 1979).

4.2. Interpolation: Why?

All of our analysis has stressed that the interpolation needed for high positional acuity can be regarded as a specific filtering of the visual image. A filter that correctly interpolates the visual input automatically avoids any defect in the neural representation of the visual image, since it reconstructs the original input. It avoids in particular motion smear (see Burr, 1980); also it "fills in" the eventual gaps, either in space or time, where or when the visual input is missing. Quite interestingly, uniform velocity – assumed in the filtering scheme we favor – is a necessary requirement for avoidance of motion smear (Burr, 1979b). It is thus possible, extending a suggestion of Barlow (1979), that the goal of interpolation is broader than simple precision and involves some "correction" in the visual representation of the visual input by enforcing what we may call "continuity of the visual field" in space and time. Whether such ideas may be developed in detail remains of course to be seen. A different possibility, already mentioned, is that hyperacuity does not involve the "automatic," parallel interpolation of the whole image, but only the operation of a specialized routine on a small region of the image to answer specific questions, like the right-left choice in a vernier task.

4.3. Significance for Information Processing and Machine Vision

Any visual processor with human-level performance must be capable of analyzing time-varying imagery. The analysis starts with the spatio-temporal interpolation of the raw visual input. The spatial resolution of the photosensitive image available for processing is limited by the sampling density of the photosensitive elements in the sensor and by their noise. Image motion introduces the additional problem of temporal resolution. The limiting factors are the frame rate and the integration time determined by the sensitivity of the photosensitive elements. This is of little consequence for a stationary scene, but for moving targets it poses the problem of motion smear.

The problem of high spatiotemporal resolution can be partially overcome by using better sensors with larger arrays and higher frame rate. There are, however, technological and physical limits to the spatiotemporal resolution that can be achieved in this manner, since increasing the spatial and temporal sampling rate reduces the number of photons per sensor element per cycle. Consider that since the number x of photons is Poisson distributed, their standard deviation is $\sigma = \sqrt{x}/2$. The number of distinguishable levels in thus roughly $n = 2\sqrt{x}$. Thus 8 bits of resolution ($n = 256$) needs about $2^{14} = 10^5$ photons. Note that the light intensity of a *bright* surface is 10^4 cd/m^2, and this means 10^4 photons per 50 msec per sensor, assuming a sensor efficiency similar to the human cones (see Barlow, 1981)!

Fortunately, the performance of a given sensor can be improved by appropriate spatiotemporal interpolation schemes. As we have seen, using such processes the human visual system achieves an extremely high spatiotemporal resolution (compared to the sampling density of the photoreceptors and their integration time), which is, in addition, smear-free.

There are various methods for reconstructing the original signal at high resolution by interpolating values measured at widely spaced intervals. The best known approach to this problem is based on the Shannon sampling theorem and on its various extensions. For static images interpolation of this type can provide a resolution much higher than the original sampling grid. Since in our framework the positions of zero-crossings (and not the grey level values) are important, Hildreth and Poggio have examined the problem of interpolating the values of the $\nabla^2 G$ convolution in order to obtain precisely the location of zero-crossings. Analytical arguments, supported by computer experiments, have shown that the position of a zero-crossing can be interpolated precisely in terms of very simple interpolation functions, even just by linear interpolation. For time-varying images the situation is more complicated. In the classical sampling theorem, interpolations in space and time are performed independently, since the temporal dependence of the input is not constrained in any way. Interpolation algorithms based on the constant-velocity assumption discussed earlier could achieve higher spatio-temporal resolution for objects in motion, as long as the constant-velocity assumption is not grossly incorrect, despite low spatial and temporal sampling rates. Computer experiments are presently planned to explore the performance of specific implementations of the interpolation scheme suggested in this paper.

Appendix

We consider a one dimensional pattern $g(x)$. Arbitrary, non-rigid movement of this pattern produces a spatiotemporal image $g(x,t)$. Rigid movement of the same pattern at constant speed gives an image $g(x,t) = g(x-v/t)$. We state here the classical sampling theorem for the first case and an appropriate modification of it for the second case.

Theorem 1 (classical sampling theorem)

If a signal $g(x,t)$ is bandlimited in spatial and temporal frequencies it can be recovered exactly by independent interpolation in space and time of its sampled values, provided that the sampling separations $\Delta\zeta$ and $\Delta\tau$ are such that $\Delta\zeta \leqslant 1/2f_x^c$, where f_x^c are the spatial and temporal bandwidths.

Theorem 2 (Crick et al., 1981)

Assume that the spatiotemporal signal $g(x,t) = g(x-v/t)$. The function g can then be reconstructed at the desired resolution from its spatial (temporal) samples. The required sampling density can be decreased arbitrarily by knowledge of the velocity v. If only the sign of the velocity is available the maximum sampling distance can be twice the classical limit for stationary patterns.

Comments

a) The proof of these results can be easily obtained from diagrams in the f_x-f_t Fourier plane (see Figure 7; Crick et al., 1981).

b) Theorem 1 requires the function $g(x,t)$ to be bandlimited before sampling takes place, since overlap of the frequency lobes as an effect of sampling usually leads to an irretrievable loss of information. This condition is not needed in Theorem 2. Overlap never occurs (for infinitely long motion) even when the pattern $f(x)$ is not bandlimited in spatial frequency. Any desired part of the original spectrum can be recovered exactly (without aliasing) by an appropriate interpolation filter.

c) The spatiotemporal filter implementing the interpolation depends on v. Assume, for instance, that we wish to endow an interpolation scheme with direction selective properties (i.e. to use information about the sign of v); it can be shown that the new spatiotemporal filter is obtained by adding to the spatiotemporal impulse response its Hilbert transform with a sign controlled by the sign of v (in the case of Figure 7 the Hilbert transform of the spatial point spread function is an odd function).

References

Barlow, H. B. Reconstructing the visual image in space and time. *Nature,* 1979, **279**, 189-190.

Barlow, H. B. Critical limiting factors in the design of the eye and visual cortex. *Proceedings of the Royal Society, London,* 1981, **B212**, 1-34.

Burr, D. C. Acuity for apparent vernier offset. *Vision Research,* 1979, **19**, 835-837. (a)

Burr, D. C. On the visibility and appearance of objects in motion. Doctoral dissertation, Cambridge University, 1979. (b)

Burr, D. Motion smear. *Nature,* 1980, **284**, 164-165.

Burr, D. C., & Ross, J. How does binocular delay give information about depth? *Vision Research,* 1979, **19**, 523-532.

Crick, F. H. C., Marr, D. C., & Poggio, T. An information-processing approach to understanding the visual cortex. In F. Schmitt (Ed.), *The Organization of the Cerebral Cortex.* Cambridge, MA: MIT Press, 1981. (Also MIT AI Memo 557, 1980.)

Fahle, M., & Poggio, T. Visual hyperacuity: Spatiotemporal interpolation in human stereo vision. *Proceedings of the Royal Society, London,* 1981, **B213**, 451-477.

Kelly, D. H. Motion and vision: II. Stabilized spatio-temporal threshold surface. *Journal of the Optical Society of America,* 1979, **69**, 1340-1349.

Marr, D., & Hildreth, E. Theory of edge detection. *Proceedings of the Royal Society, London,* 1980, **B207**, 187-217.

Marr, D., Poggio, T., & Hildreth, E. Smallest channel in early human vision. *Journal of the Optical Society of America,* 1980, **70**, 868-870.

Marr, D., Ullman, S., & Poggio, T. Bandpass channels, zero-crossings, and early visual information processing. *Journal of the Optical Society of America,* 1979, **69**, 914-916.

Morgan, M. J. Analogue models of motion perception. *Philosophical Transactions of the Royal Society, London,* 1980, **B290**, 117-135.

Seelen, W. von. On the interpretation of optical illusions. *Kybernetik,* 1973, **12**, 111-115.

Stigmar, G. Blurred visual stimuli. II. The effect of blurred visual stimuli on vernier and stereoacuity. *Acta Ophthalmalogica,* 1971, **18**, 364-379.

Westheimer, G. Eye movement responses to a horizontally moving visual stimulus. *Archives of Ophthalmology,* 1954, **52**, 932-941.

Westheimer, G. Diffraction theory and visual hyperacuity. *American Journal of Optometry and Physiological Optics,* 1973, **53**, 362-364.

Westheimer, G., & Hauske, G. Temporal and spatial interference with vernier acuity. *Vision Research,* 1975, **15**, 1137-1141.

Westheimer, G., & McKee, S. P. Visual acuity in the presence of retinal-image motion. *Journal of the Optical Society of America,* 1975, **65**, 847-850.

Westheimer, G., & McKee, S. P. Integration regions for visual hyperacuity. *Vision Research,* 1977, **17**, 89-93. (a)

Westheimer, G., & McKee, S. P. Spatial configurations for visual hyperacuity. *Vision Research,* 1977, **17**, 941-947. (b)

Westheimer, G., & McKee, S. P. What prior uniocular processing is necessary for stereopsis? *Investigative Ophthalmology and Visual Science,* 1979, **18**, 614-621.

Wülfing, E. A. Ueber den kleinsten Gesichtswinkel. *Zeitschrift für Biologie,* 1982, **29**, 199-202.

Isolating Representational Systems

Michael I. Posner
Avishai Henik[1]

University of Oregon
Eugene, Oregon

Abstract

The representation of even a simple stimulus event involves parallel activation of a number of separable internal codes. Some of these codes are available only to neurophysiological studies, some only to cognitive methods and some to both forms of investigation. We review the parallel activation of simple features of visual objects such as orientation, motion and color and their combination into the conjunctions of features that form objects. Rather surprisingly, output of feature codes appear to be available to some forms of cognitive study. Feature activation carries with it implicit information on location but dissociations between features argue that the precise conjoining of features into an integrated object appears to be a separable act. Studies of visual imagery reveal the operations of higher levels of representation that seem to play a role in pattern recognition and provide evidence favoring the ability to choose a level of detail appropriate to goals. Language stimuli follow similar coding principles but provide a special opportunity to apply cognitive methods to the study of cross-modality representation. The interplay between many forms of representation rather than a single unifying principle may underlie the richness of human cognition. Nonetheless the ability to isolate coding systems by experimental methods or brain injury supports efforts to simulate aspects of pattern

---

[1]Avishai Henik is regularly at the Department of Behavioral Science, Ben Gurion University of the Negev, Beersheva, Israel.

HUMAN AND MACHINE VISION

recognition and also suggests why the simulation of any one aspect may not fully duplicate human performance.

The problem of how the nervous system represents simple visual patterns has been with psychology since the inception of the discipline. The difficulty of the issues involved is caught very well in a recent chapter by Palmer (1978). He says:

> . . . This chapter was born of an ill-defined feeling that we, as cognitive psychologists, do not really understand our concepts of representation. The field is obtuse, poorly defined and embarrassingly disorganized. Among the most popular terms, one finds: visual codes, verbal codes, spatial codes, physical codes, name codes, image codes, analog representations, first-order isomorphisms, second-order isomorphisms, multidimensional spaces, templates, features, structural descriptions, relational networks, multicomponent vectors and even holograms . . . (p. 59).

We have learned from a century of research in color vision that what may be opposed theories at one level of analysis can be complementary from a view of the nervous system as entailing many levels of information integration. That a level view is also appropriate for the analysis of form can be seen clearly from an experiment reported a number of years ago by one of the editors of this volume (Beck, 1966a,b). He asked people to rate the perceptual similarity of an upright and a tilted T and of an upright T and backward L. Phenomenologically, the T and tilted T appeared to be more similar in the subjects' ratings. This may in part be because we have learned that orientation is unimportant in dealing with visual form since we see them tilted in many ways. On the other hand, when systematic studies of texture grouping were done by Beck it was clear that the T and tilted T were more differentiated by the visual system than the T and backward L. What appears as similar at one level of analysis appears as quite distinct at a different level of analysis. In the study of form as in the study of color, what appear to be opposing or fundamentally different views when the system is seen as a single processing system appears quite different when differential levels of analysis are involved.

A feature of both the neurophysiological and the information processing approaches to visual perception has been an attempt to isolate different levels of integration. Substantial progress in isolating a number of processing levels has already taken place. Unfortunately, the complexity of the picture emerging undercuts the idea of any simple model for their combination. There is a serious problem with the postulation of different integrative levels. Levels may proliferate so that every conceivable attribute is assigned to a different level. In that case, the level concept would be identical to a logical analysis of all the attributes of visual forms that could be processed by the nervous system and would provide no constraints to visual information processing. To avoid this problem, it is necessary to have some type of empirical analysis to validate whether a particular logical attribute is represented by a separate system. Both neuroscientists and cognitive scientists seek such methods.

In the neurosciences the method used is to show that there is a particular cell or cell group that is maximally responsive to changes of the type postulated. For example, if one can find cells in the visual system that are sensitive to changes in spatial frequency but not to other known changes of visual form one postulates a level at which a spatial frequency analysis is made. Fortunately, it appears often to be the case that cells sensitive to particular attributes are located in a spatially accurate map of the retina (Cowey, 1979).

For the cognitive scientist, the issue is more difficult. In cognitive studies, we only have available the output of the system as a whole. In order to know if that output comes from a particular isolable representation, it is necessary to show that subjects can make responses based upon a particular aspect of visual form that are unaffected by manipulations of other aspects. According to the experiment described earlier, one can show for example that texture discrimination is affected by line orientation but not by line arrangement. This suggests that the level giving rise to texture discrimination takes into account orientation but does not conjoin lines. A similar logic is used to separate the visual and phonological coding of individual words and letters. Manipulations such as changes in color or brightness tend to affect visual system processes but to be ineffective in manipulating the time course of processes based on the phonological code. Similarly, the perception of motion is, under some conditions, independent of the form of the object which is seen to move. The subject seems to be able to base his output or percept on the motion system independent of form. These arguments for the isolability among different attributes or codes of stimuli are at best somewhat indirect. When there is a convergence between psychological and physiological methods one has more assurance of isolability.

The complexity of the problem of representation is usually avoided by considering one aspect. Sometimes we talk about the recognition of a pattern as being evidenced by our ability to sense a change, by our ability to name, or to determine whether two patterns are the same or different. Many programs that deal with pattern recognition consider only a part of the complex representation that is coded by the human nervous system. No doubt this simplified approach can be satisfactory for certain purposes. However, we should bear in mind that the human representation of a visual experience consists of many levels, some isomorphic to the form itself, some symbolically related to the stimulus. No one of these representations should necessarily be taken as the full measure of the recognition of the pattern.

Two issues have dominated the discussion of the process of recognition. The first of these is whether the levels involved in recognition should be seen as serial or parallel. There seems very little question now from a variety of physiological and psychological evidence (Posner, 1978; Stone, Dreher, & Leventhal, 1979) that a combination of serial and parallel processing is required to simulate human performance.

The second major issue has been the degree to which the system involved in recognition of patterns should be seen as driven by the evidence presented in the stimulus as opposed to being driven by expectancies. Marr and Nishihara (1978) argue that one should confine the effects of expectancy to relatively high levels of the system leaving the lower levels to gain from the stimulus evidence the strongest representation possible. Physiological mechanisms undoubtedly exist to allow even the earliest stages of processing to be affected by expectancies. A major finding of cognitive psychology in this regard is the difficulty people have in separating bottom-up and top-down evidence accumulated in the process of perception. According to signal detection theory, the contributions of external information and internal expectancies are inseparable in the activation of the individual recognition units. Introspectively, expectancy and evidence are inextricably combined so that the subject himself cannot by conscious effort separate the two.

Evidence for parallel processing and the difficulty of separating top- down from bottom-up processing make it necessary to examine several separate questions in order to explore principles underlying representations. To this end we select three major topics for analysis. The first topic, "Object Recognition", examines the representations involved in representing a visually presented object with stress on bottom- up processes. Evidence from psychological and physiological experiments is examined in an effort to sort out the important features involved in the representation of the object. We argue that both physiological and cognitive experiments can provide evidence even about relatively early stages of the pattern recognition process. The ability of cognitive experiments to deal with the early stages of processing seems to require that such stages can be coupled directly to output under certain circumstances.

A very different approach to representational systems is to begin with the pure conscious content. Representation in consciousness without the presentation of the stimulus allows the cognitive scientist to examine the characteristics of conscious manipulations. Studies of visual imagery provide constraints on the manipulation of such conscious images.

In the first two sections, visual pattern recognition and visual imagery are contrasted as methods for understanding representation. Of course, representations are not necessarily visual but involve a number of modalities. As a model for this process, it is useful to have systems in which information from more than one modality has contact with highly overlearned common representations. Two systems in which this is so for the skilled reader are language and spatial location. Psychological methods allow for the exploration of cognitive systems that combine auditory and visual information in the recognition of words and spatial position.

Thus in this paper we examine object recognition, conscious imagery and cross-modality processing, in order to explore arguments about the fundamental nature of the recognition process that could be of interest to those working in artificial pattern recognition and simulations of human functions.

1. Visual Pattern Recognition

A major development in our understanding of object recognition has been the idea that attributes of objects such as orientation, color, motion, are mapped into separate neural systems or channels (see Cowey [1979] for a review). This view received great impetus from the anatomical findings of Zeki (1976) that the monkey visual system employs separate maps in the prestriate cortex that deal with color, orientation and motion information. These findings suggest that every visual stimulus is disassembled at relatively high levels of the nervous system into separable or isolable codes which represent its attributes.

What is most important in this doctrine is the possibility that codes at this level of the nervous system may be available for output. If so, it should be possible to design psychological as well as physiological tests that allow study of the characteristics of these codes. There have been several approaches to the question of psychological analysis of individual attributes of objects. One useful method has been to flood the visual field with information and study those features that allow easy figure segregation.. Using this method, Beck (1966a,b) has shown that line orientation will lead to texture segregation while line arrangement will not. Moreover, in cluttered visual fields the distinction between line orientations is much easier to make than distinctions between forms.

Similarly, Treisman and Gelade (1980) show that simple attributes such as color or shape will lead to figural segregation whereas conjunctions will not. Another approach based on this same general logic has also been investigated by Treisman and Gelade (1980). Their assumption is that individual feature levels will show parallel processing. They found, for example, that searching for a single feature, e.g., a blue or tilted letter, can occur at the same rate irrespective of the number of items in the visual display; the target simply appears to pop into view. However, when the search requires an explicit conjunction of simple features, for example, looking for a green T in a background which consists of other green and T figures, it appears that each item must be approached individually and searched serially. The reason for this slow search, they argue, is that features may combine in an illusory way, that is the greenness of one figure might combine with the T-ness of another, and to guard against such illusory conjunctions a serial search process is used. Whether or not this is the correct explanation, it does appear that for simple features very rapid processing of large visual fields is possible whereas conjunctions of features do not provide as convincing evidence for parallel processing.

Another psychological test for isolating attributes is whether each attribute can be examined without showing influence of context. For example, it has been known that under certain conditions of real or apparent motion it does not matter whether the objects at the ends of the motion are identical in form or not. There is also evidence that the thresholds for appearance of motion are considerably below those at which form information is being processed. Insofar as form and motion information may be apprehended separately without influencing one another, one can argue that they are separable features of visual experience. Psychophysical evidence can be supplemented by chronometric evidence indicating that, for example, matching or sorting of form information is uninfluenced by the colored background or the color of the information, etc.

Insofar as the psychological tests give independence between attributes one argues that the attributes are registered in separate systems. In one sense these behavioral tests are much stronger than the physiological experiments. They argue that not only are there independent representations of the form but that they have access to either subjective experience on the one hand or output mechanisms on the other. Thus, according to these studies the object is not processed as a whole even though the whole object may be the basis for conscious awareness.

Evidence of the psychological reality of attributes arises in reports of striking dissociations. For example, a subject shown a large array of letters and asked to report the red one will, on a reasonable proportion of trials, report with great conviction the identity of an adjacent letter as the red one (Snyder, 1972). Similarly, in a rapidly presented serial list subjects will frequently report a temporally adjacent item as the red target (Broadbent, 1977; Lawrence, 1971). The color seems to dissociate from its form and join a form adjacent in time or space. These illusory conjunctions can occur with considerable frequency and seem to be as real percepts as the correct target detections. In one such report Treisman and Schmidt (1982) showed colored letters flanked by black digits. Subjects complained that on some trials the digits were colored, not black, as the experimenter had promised, even though no colored digits were shown. These dissociations argue for the separation among attributes as well as for some temporal and spatial error in their assembly.

It is usually the case that our perceptual system ties various attributes in an orderly way which corresponds to their structure in the world. In this process temporal and spatial aspects play a major role. However, both physiological and behavioral evidence on independent attributes argue for some sense of separability of simple features of visual objects. This sense of separability may require that at least under some circumstances the individual attributes be available for later processing parallel with or instead of the assembled object.

A rather different way of breaking down the attributes of visual objects for entry into a pattern recognition system is based on the idea of frequency analysis. It has been known for one hundred years that the auditory system analyzes complex tones into their individual frequencies. In the last twenty years this approach has been taken to the study of visual objects as well. Kabrisky (1966) originally proposed a whole-field Fourier analysis as a candidate model of how the visual system might begin the process of object recognition. Much of the work in recent years has focused on frequency analysis that might be performed over relatively small parts of the visual field by cells in the striate cortex. Albrecht, De Valois, and Thrall (1980) have presented evidence which favors an early frequency analysis by striate cortex cells. It seems difficult to resolve the question of whether these cells can best be seen as loosely tuned to appropriate features or poorly tuned to spatial frequencies (McKay, 1981). The discussion often boils down to whether the cells are more sensitive to a single line slope or more sensitive to a line appearing among others that make up a grating.

The distinction may not seem too important, but the idea introduced by a spatial frequency approach, that is to view the visual system in terms of separate channels tuned to local or global features, provides a number of useful insights. Low spatial frequencies reveal the general locations and outlines of objects. Thus, Ginsburg (1979) has demonstrations in which a highpass filtering of the visual image gives a better representation of the commonalities of the visual scene than are present when much more detail is included.

A number of experimental studies have begun to provide some useful boundary conditions on the balance between local and global information that might be present in brief glimpses of stimuli. Some of these studies have argued that the global information sometimes thought to be carried by Y-cells is available for information processing before the more detailed local information. This idea led Navon (1977, 1981) to suggest that stimuli in which local and global conflict will show more interference of the global on the local than vice-versa. Whether global or local features are processed first depends upon their relative size, location with respect to the fovea, and other factors (McLean, 1978; Miller, 1981; Navon, 1981). Thus no behavioral evidence has been able to demonstrate a universal advantage to the global information. The distinction between a feature approach and a frequency analysis approach does emphasize a basic problem in the area of recognition. This issue is the way in which spatial location is employed in the recognition of objects. Clearly, we are able to recognize an object irrespective of where it falls in the retinal mosaic. Thus there is generalization of information over large areas of visual space. On the other hand, the visual system does seem to be organized to maintain point to point representations. Treisman and Gelade (1980) argue that the reason that one must guard against illusory conjunctions is that they are not conjoined automatically by virtue of their common location. They argue that attributes such as color and form are spatially free floating. More specifically they propose that "identifying a conjunction necessarily requires us to locate it spatially, whereas identifying a feature may not."

The idea that features are free floating and only tied to location for example by a later attentive process appears to conflict somewhat with the relevant physiology. Consider, for example, the attributes of color and form. Areas of the visual cortex that respond selectively to these visual features have small receptive fields; while cells in the inferotemporal cortex which correspond to configurations of features (Gross, Rocha-Miranda, & Bender, 1972) have dramatically large receptive fields. Accordingly, inferotemporal cortex has been thought to involve pattern learning in which it is important to be able to generalize responses to the same configuration irrespective of spatial position (Gross & Mishkin, 1977). Thus, the physiological data suggest that each feature is tied implicitly to a position within a spatial map (Cowey, 1979) while configurations can only be tied to space when they are actively related back to the individual feature positions that provide their input or receive location input from systems that serve orienting.

A major question is how this location information feeds into systems subserving conjunctions. A resolution of this issue might allow us to understand the way in which spatially separated material either in the frequency domain or in separate maps can be conjoined in more complex visual perceptions. One solution to this question may be that the information in separate maps is conjoined by the implicit spatial location present in the maps.

Nissen (1980) tested this prediction in a set of perceptual tasks which examined the relationship among the accuracy of reporting by location, color or form. On each trial, subjects viewed a stimulus display consisting of a colored form at each of four locations. In separate blocks, they were either asked to identify the object present at a particular location by giving its color and form, the location and form of a particular color, or the color and location of a particular form. Nissen found that the color and form information were accessed independently but that they were both tied to location. Thus in the location cue condition the accuracy of form and color responses were independent of each other.

These results conform to findings that orienting to a spatial location enhances even the detection of a visual event in a dark field. Thus when a subject knows something about where an event will occur his simple reaction time is speeded and his threshold for detection is also reduced (Posner, Nissen, & Ogden, 1978). Information about the identity of a stimulus does not produce similar effects on detection (Posner, Snyder, & Davidson, 1980).

These experiments suggest that whatever type of representation is used, location is the most powerful single organizing principle. Location can be used to organize individual attributes into a single overall percept. Perhaps subjects can select information for conscious processing either by spatial location or by frequency. It is quite possible that the two are related. Concentration on a small area of space may be related to selection of high spatial frequency information. There is good evidence that selection of information by spatial location enhances both relevant and irrelevant information that occurs at that position. In this sense the spatial position serves as a channel or conduit for the entry of all types of information into consciousness. No similar demonstrations have yet been made for spatial frequency. We do not know whether a subject who tunes in a particular spatial frequency enhances all information that arrives at that frequency. If spatial frequency and location both serve as input channels the percept at any given moment will clearly depend upon the selections made. We turn now to the study of such percepts.

2. Imagery

In dealing with object recognition we have been guided by a view of information as accruing passively over time in separate systems. The neuroscience research based on single cell recording from different synaptic way stations represents a central method of gaining information about such representations. As we have seen, while there are cognitive methods of getting at some of these early levels of representation, the neuroscience approach has been dominant. On the other hand, the neuroscience approach has not given us any way of dealing with the problem of how information is represented in consciousness. In cognition the problem of consciousness has often been discussed under the topic of imagery. Imagery is the content currently in consciousness. In recent years the term imagery has usually been applied to visual representations. It seems useful to distinguish between the conscious visual representations we call visual images and those that may be visual in character but which are not currently in front of the mind which we call visual codes.

In the first section of this paper we dealt with coding that is visual in character but derived from visual input. In order to deal with visual images in their purest form it is useful to separate them from the presentation of visual objects. This is usually done by allowing the subject to assemble the visual image based upon some verbal instruction or upon some visual information which is different from the image. What is most impressive about these studies is the fact that systematic measurement is possible in the absence of sensory input.

Early evidence on the construction of visual representations in consciousness employs the same type of mental chronometry that we have referred to in the study of object recognition. For example we (Posner, 1978) studied matching of two successive visual letters with matching a visual letter following auditory input. The result indicated that the auditory input could be converted to a representation that was as efficient as visual matching in one to two seconds.

More impressive was the work of Shepard and his associates (Cooper & Shepard, 1973; Shepard & Metzler, 1971) showing that subjects can develop a visual representation, so when presented with a rotated letter they do not need to rotate the letter to the upright before deciding whether it is a correct or mirror image letter. Shepard's results not only gave clear evidence of visual imagery but suggested its concrete nature by showing that there was very little advantage in rotation if the subject did not know the identity of the specific letter that was being rotated. Cooper (1976) showed that the rate of rotation in imagery can be predicted from the rate of mental rotations of physically presented letters. Later Shepard and Judd (1976) presented evidence that the laws governing mental rotation taking place in imagined representation were similar to those occurring when subjects receive two visual impressions of the same object and perceive the rotation in apparent motion. These experiments provide evidence that experimental techniques can deal in detail with the characteristics of visual images and that visual imagery employs some of the same machinery as involved in the perception of objects.

This last contention has led to a great deal of dispute about the nature of representation of visual images (Kosslyn, 1980; Pylyshyn, 1981). Presumably no one disputes that the visual perceptions represent aspects of our visual world in a form that is unlike representation by language. The real question seems to boil down to the level at which this information is represented in absence of input.

Two recent suggestions with respect to representation of spatial relations come from the works of Marr (Marr, 1978; Marr & Nishihara, 1978), and Attneave (1982). Marr suggested a framework which consists of three levels of representation. The first level, called the primal sketch, makes explicit the intensity changes and the local two-dimensional geometry of an image. The following 2-1/2 D level of representation deals with properties of surfaces. That is, it gives orientations and discontinuities of visible surfaces, relative to the viewer's vantage point. The highest level is the 3-D sketch, which is an object-centered representation. This level includes "volumetric primitives" that make explicit the space occupied by the object and not only the visible surfaces. Perceptual processing is equivalent to transforming one format into the next, from the first which is the more peripheral to the third which is more central. Attneave suggests a soap bubble system that can account for monocular perception of tridimensional space. In this sytem the spatial representation uses a 3-D field ("some sand-box in the head": Attneave, 1972). The soap-bubble model moves towards stability (simplicity) by an interplay of forces within the analog model. This system is a viewer-centered system. It generates a 2-1/2 D representation.

Pinker (1980) reports a set of experiments dealing with imagery of three dimensional scenes. He asks his subjects to generate an image and to scan an imagined point from one object to another. Scanning time increases linearly with distance in the third dimension. Thus, it seems that people have the ability to preserve 3-D information. In another experiment he finds that 2-D information is also preserved. Moreover, the subjects are capable of adopting various viewpoints and of performing the task on objects seen from a new perspective. He thus suggests that neither a 2-, 2-1/2 D nor 3-D representation is sufficient in itself for the performance evidenced. A more promising description would incorporate several types of interacting representations. The system suggested is one in which the surface array, in which the image is represented, resembles a 2-1/2 D viewer centered representation, like the one suggested by Attneave. The information stored in long term memory, the deep representation, contains a 3-D object centered representation. Images are constructed by applying a certain algorithm to the deep representation.

There is striking similarity between operations performed upon mental images and those that occur in visual perception. Recent studies have somewhat expanded the number of similarities between visual perception and images. For example, Finke (1980) reported that acuity functions for images fall off with distance from the fovea in the same way as was found for sensory stimuli and that visual images can be used to produce adaptation in prism experiments. The methods used required subjective judgments, that is, they are psychophysical rather than chronometric and may be subject to demand characteristics based upon the person's expectation of what data should be found (Pylyshyn, 1981).

In some more recent studies subjects have been asked to image more complex visual displays in an effort to understand how they can scan through such representations. The studies suggest the importance of physical distance in determining the time to move between objects (Kosslyn, 1980; Pinker, 1980) and the close relations between internal scans and eye movements (Weber & Malenstrom, 1979). If we assume that scanning of visual images provides information on the nature of central mechanisms that select information in the visual world it would be possible to study such scanning under simpler conditions. It has been shown that subjects can move attention in accordance with cues in order to enhance detection of increments in luminance (Bashinski & Bachrach, 1980; Posner, 1980; Remington, 1980). These shifts of covert attention can be studied by examining changes in the efficiency of taking in information from different places in visual space with eyes fixed. By the use of probes between fixation and target events, it is possible to show that as attention moves from one position in the visual field to another it passes through intermediate positions (Shulman, Remington, & McLean, 1979; Tsal, 1981). These results suggest the analog nature of shifts of covert attention in the visual field and are similar to the results obtained from imaged scenes. Moreover, in this simple situation it is proven possible to link purely behavioral studies to those studying cognition deficits following brain injury and recording electrical activity from alert animals (see Posner, Cohen, & Rafal, 1982).

Cohen (1981) asked whether attention moves in retinal coordinates or in environmental coordinates. In his experiments a visual cue is presented in the periphery. The subject is asked to move his eyes as soon as possible after the onset of the cue. Now a target is presented. If the target is presented at the same environmental location as the cue it does not occupy the same location (as the cue) on the retina. If the target is displaced in the same way as the eyes (same distance and direction) the cue and the target occupy different locations environmentally but the same location retinally. The results show facilitation in retinal coordinates -- that is, when the target appears at a location which is different from the cued location environmentally but the same retinally. Why is attention coupled with retinal rather than environmental coordinates? A possible answer is that retinal location is preserved in the retinotopic maps in the prestriate cortex. These maps provide a means to conjoin attributes by aligning the separate spatial maps as Nissen (1980) suggested.

These results must be considered suggestive rather than definitive. They do tend to integrate studies of visual imagery, visual attention and what is known about the physiology of cortical visual maps. They thus provide a starting point for empirical explorations and for computer simulations.

In particular, the study of imagery provides a means of observing characteristics of higher levels of the visual system. There are a number of techniques for observing the characteristics of visual representations constructed by a person. The evidence suggests that these same levels of processing are involved in object recognition in the sense of our ability to relate visual input to information stored in memory. The combined study of object recognition and imagery gives some promise of allowing the development of a more complete theory of how bottom-up and top-down procesing shape our conscious experience of the visual world that surrounds us.

3. Cross-modality Processing

So far all the information presented in this paper deals only with the problem of visual representation. A vast majority of work in the area of pattern recognition has involved the processing of visual stimuli. The general methods for study of top-down and bottom-up processing that are reflected in this paper apply to the study of auditory or tactile information as they do to visual information. However, there are important differences between modalities that might need to be reflected in our understanding of issues of representation.

In studying objects that cross modalities it is particularly useful to explore situations in which information from more than one modality can in principle be mapped into a single representation. Two areas that seem to be best suited for this goal are the study of the processing of language information in which, at least for the skilled reader, auditory and visual information maps into a common semantic system, and the area of spatial orienting, where any object in the environment may be seen as arising from some location and giving rise to attributes that might extend over more than one modality. For example, a person occupies a position in space that gives rise to both visual and auditory stimulation. It might be thought that calling a person's attention to some position in space would amplify all modalities that come from that spatial position. If this were the case, one could argue that the object is the source of selection and the representation in the nervous system reflects the environment by treating as a unit all stimulation from a given spatial position. This view would be very compatible with the Gestalt idea that objects are primary psychological representations. To explore this idea we (Posner, 1978, Chapter 7) studied reaction time to targets that could be either tactile or visual and that could arise either from the left or right side of space. We compared cues that informed subjects as to the spatial position of the target but not the modality of input with those that informed them about the modality of input but not the spatial position. It was found that modality cues were effective but spatial position cues were not. It was as though it is more important to know the neural analyzing system (vision or touch) than to know the environmental location.

These results agree with models of separable attributes in vision. Stimuli arising from a common position in space are disassembled into their modality-specific attributes rather than represented in a common multi-modal spatial position. When we first found this result with visual and tactile stimulation we were quite surprised because most information about the recognition of objects was based on the Gestalt rather than the attribute view. The subsequent work of Treisman and Gelade makes it less surprising that the subjects should be able to output responses from a system in which the modalities are separate rather than conjoined by spatial location.

Language is another area in which cross-modal input has access to highly over-learned common responses. Both the written and spoken words can be interpreted in a common semantic framework. We have been interested in how separable the representations of visual and auditory words are. In one experiment that seems analytic to this question (Hanson, 1981), subjects were presented with simultaneous visual and auditory words. For the visual word, the task was either classification into upper or lower case (physical code), detection of a particular phoneme (phonological code), or classification into a particular category (semantic code). For the auditory word, the physical operation was classification into male or female voice while the phonological and semantic operations were the same as for vision. An individual subject worked either with the visual or auditory task. The results showed that the unattended modality had clear effects on the phonological and semantic levels but no clear effects on the physical level. For example, classifying an auditory stimulus as to whether it occurred from a male or female voice was neither enhanced nor inhibited by the fact that the same word occurred on the visual modality. A subject making a case classification was not affected by identity of the auditory word. On the other hand, for phoneme monitoring and for semantic classification, there were cross-modality facilitation and interference effects. This suggests that the phonological and semantic representations of words are not independently mapped for the different modalities but correspond to a common internal representation.

Recently we have been taking advantage of the well-known Stroop effect to survey the organization of spatial position and language with stimuli in visual and auditory modalities. The paradigm that we chose for this investigation was to have subjects respond with a manual left or right response to the words "left" or "right" that occurred either visually to the left or right on the display or over earphones to the left or right ear. There were four different tasks: respond to the visual word, auditory word, visual position or auditory position. For each of these tasks there could be either one, two or three sources of incompatibility. For example, in the visual word task, the sources of incompatability could be auditory word, visual position or auditory position.

Our preliminary results suggest that responses based on language processing (e.g., reading or speech perception) tend to produce stronger effects on one another regardless of input modality. Similarly, spatial information (ear of entry or side of screen) seem to affect one another more than either affects the language input of the same modality. These studies fit with many findings that the higher levels of the brain tend to be organized in accord with cognitive systems, e.g., language, spatial information, emotion, etc., rather than by modality of input. In a sense they also conflict with findings cited earlier that argue for selection based on the input modality for detecting visual-tactile targets. In fact there need be no conflict since it appears that people are capable of selection of output based upon many different levels of analysis. New kinds of experiments will be necessary to disentangle the role of sensory modality, motor system and central cognitive systems in determining the interference between tasks.

4. Conclusions

A major goal in the study of representations is to analyze the problem into manageable constituents. For the research scientist this depends not upon logic alone but also upon available methods. We have attempted to review some of the methods currently in vogue for studying component representation from the early attributes through construction of the phenomenal object. The analysis of these representations constitutes a view of the raw materials out of which our perceptual world is constructed.

This survey serves to reinforce the complexity of the problem of representation. Moreover, the parallel activation of different forms of representation and their access to output systems argues that phenomenal experience or behavior may be based on a number of simultaneously acting representations of the input. This suggests that the problem of human pattern recognition is not unitary but consists of a number of subproblems based upon the goal of the person at the time he receives the input information. The degree to which the phenomenal world is a construction from isolated attributes seems to be a surprisingly powerful aspect of current ideas of representation.

Acknowledgment

The writing of this paper was supported in part by the Alfred P. Sloan Foundation and in part by IBM grants to the University of Oregon for the support of work in Cognitive Science.

References

Albrecht, D. G., DeValois, R. L., & Thrall, L. G. Visual cortical neurons: Are bars or gratings optimal stimuli? *Science,* 1980, **207**, 88-90.

Attneave, F. Representation of physical space. In A. W. Melton & E. J. Martin (Eds.), *Coding Processes in Human Memory.* Washington, DC: Winston, 1972.

Attneave, F. Prägnanz and soap bubble systems: A theoretical exploration. In J. Beck (Ed.), *Organization and Representation in Perception.* Hillsdale, NJ: Erlbaum, 1982.

Bashinski, H. S., & Bacharach, V. R. Enhancement of perceptual sensitivity as the result of selectively attending to spatial locations. *Perception and Psychophysics,* 1980, **28**, 241-248.

Beck, J. Effect of orientation and of shape similarity on perceptual grouping. *Perception and Psychophysics,* 1966, **1**, 300-302. (a)

Beck, J. Perceptual grouping produced by changes in orientation and shape. *Science,* 1966, **154**, 538-540. (b)

Broadbent, D. E. The hidden preattentive processes. *American Psychologist,* 1977, **32**, 109-118.

Cohen, Y. Internal and external control of orienting. Doctoral dissertation, University of Oregon, 1981.

Cooper, L. A. Demonstration of mental analog of an external rotation. *Perception and Psychophysics,* 1976, **19**, 296-302.

Cooper, L. A., & Shepard, R. N. Chronometric studies of the rotation of mental images. In W. G. Chase (Ed.), *Visual Information Processing.* New York: Academic Press, 1973.

Cowey, A. Cortical maps and visual perception. *Quarterly Journal of Experimental Psychology,* 1979, **31**, 1-17.

Finke, R. A. Levels of equivalence in imagery and perception. *Psychological Review,* 1980, **87**, 113-132.

Ginsburg, A. P. Visual perception based on spatial filtering constrained by biological data. Proceedings, International Conference on Cybernetics and Society, 1979.

Gross, C. G., & Mishkin, M. The neural basis of stimulus equivalence across retinal translation. In S. Harrad, R. Doty, L. Goldstein, J. Jaynes, & G. Krauthamer (Eds.), *Lateralization in the Nervous System.* New York: Academic Press, 1977.

Gross, C. G., Rocha-Miranda, C. E., & Bender, D. B. Visual properties of neurons in inferotemporal cortex of the macaque. *Journal of Neurophysiology,* 1972, **35**, 96-111.

Hanson, V. L. Processing of written and spoken words: Evidence for common coding. *Memory and Cognition,* 1981, **9**, 93-100.

Kabrisky, M. *A Proposed Model of Visual Information Processing in the Human Brain.* Urbana, IL: University of Illinois Press, 1966.

Kosslyn, S. M. *Image and Mind.* Cambridge, MA: Harvard University Press, 1980.

Lawrence, D. H. Two studies of visual search for word targets with controlled rates of presentation. *Perception and Psychophysics,* 1971, **10**, 85-89.

Marr, D. Representing visual information. In E. M. Riseman & A. R. Hanson (Eds.), *Computer Vision Systems.* New York: Academic Press, 1978.

Marr, D., & Nishihara, H. K. Visual information processing: Artificial intelligence and the sensorium of sight. *Technology Review,* 1978, **81**, 2-22.

MacKay, D. M. Strife over visual cortical function. *Nature,* 1981, **289**, 117-118.

McLean, J. P. Perspectives on forest and trees: The precedence of parts and wholes in visual information processing. Doctoral dissertation, University of Oregon, 1978.

Miller, J. Global precedence in attention and decision. *Journal of Experimental Psychology: Human Perception and Performance,* 1981, **7**, 1161-1173.

Navon, D. Forest before trees: The precedence of global features in visual perception. *Cognitive Psychology,* 1977, **9**, 353-383.

Navon, D. The forest revisited: More on global precedence. *Psychological Research,* 1981, **43**, 1-32.

Nissen, M. J. The role of localization in conjoining color and form. Procedings, American Psychological Association, 1980.

Palmer, S. E. Fundamental aspects of cognitive representation. In E. Rosch & B. D. Lloyd, (Eds.), *Cognition and Categorization.* Hillsdale, NJ: Erlbaum, 1978.

Pinker, S. Mental imagery and the third dimension. *Journal of Experimental Psychology: General,* 1980, **109**, 354-371.

Posner, M. I. *Chronometric Explorations of Mind.* Hillsdale, NJ: Erlbaum, 1978.

Posner, M. I. Orienting of attention. *Quarterly Journal of Experimental Psychology,* 1980, **32**, 3-25.

Posner, M. I., Cohen, Y., & Rafal, R. D. Neural systems control of spatial orienting. *Philosophhical Transactions of the Royal Society, London,* 1982, **B298**, 187-198.

Posner, M. I., Nissen, M. J., & Ogden, W. C. Attended and unattended processing modes: The role of set for spatial location. In H. L. Pick & I. J. Saltzman (Eds.), *Modes of Perceiving and Processing Information.* Hillsdale, NJ: Erlbaum, 1978.

Posner, M. I., Snyder, C. R. R., & Davidson, B. J. Attention and the detection of signals. *Journal of Experimental Psychology: General,* 1980, **109**, 160-174.

Pylyshyn, Z. W. The imagery debate: Analogue media versus tacit knowledge. *Psychological Review,* 1981, **88**, 16-45.

Remington, R. Visual attention, detection and the control of saccadic eye movements. *Journal of Experimental Psychology: Human Perception and Performance,* 1980, **6**, 724-744.

Shepard, R. N., & Judd, S. A. Perceptual illusion of rotation of three-dimensional objects. *Science,* 1976, **191**, 952-959.

Shepard, R. N., & Metzler, J. Mental rotation of three-dimensional objects. *Science,* 1971, **171**, 701-703.

Shulman, G. L., Remington, R. W., & McLean, J. P. Moving attention through visual space. *Journal of Experimental Psychology: Human Perception and Performance,* 1979, **5**, 522-526.

Snyder, C. R. R. Selection, inspection and naming in visual search. *Journal of Experimental Psychology,* 1972, **92**, 408-431.

Stone, J., Dreher, B., & Leventhal, A. Hierarchical and parallel mechanisms in the organization of visual cortex. *Brain Research Reviews,* 1979, **1**, 345-394.

Treisman, A. M., & Gelade, G.   A feature integration theory of attention. *Cognitive Psychology*, 1980, **12**, 97-136.

Treisman, A. M., & Schmidt, H.   Illusory conjunctions in the perception of objects. *Cognitive Psychology*, 1982, **14**, 107-141.

Tsal, Y.   Movements of attention across the visual field.   Proceedings, American Psychological Association, 1981.

Weber, R. J., & Malenstrom, F. V.   Measuring the size of mental images. *Journal of Experimental Psychology: Human Perception and Performance*, 1979, **5**, 1-12.

Zeki, S. M.   The functional organization of projections from striate to prestriate cortex in the rhesus monkey.   *Cold Spring Harbor Symposia on Quantitative Biology*, 1976, **15**, 591-600.

A Sketch of a (Computational) Theory of Visual Kinesthesis

K. Prazdny

Fairchild Camera and Instrument Corporation
Palo Alto, California

Abstract

Visual factors play an overwhelming role in the perception of egomotion. While it has been shown experimentally many times that vision nearly always dominates the kinesthetic sense, the mechanism by which the visual system computes egomotion from retinal information remains unknown. This paper outlines, using a computational approach to visual perception, a mechanism possibly underlying the perception of egomotion. The major components are:

1) The assumption that the relevant inputs are the retinal velocities.

2) The algorithm to decompose a set of retinal velocities into the constituent translational and rotational components (describing relative motion).

3) A decision procedure which (using the decomposition algorithm) operates in the far periphery to find the largest consistent subset of relative motion parameters computed locally. Such a subset is then taken to constitute a stationary surface, and the relative motion parameters computed in its region form a description of egomotion.

Some supporting psychological evidence is briefly mentioned.

When an observer is passively transported along a rectilinear path in a uniform motion the only sensory information available about egomotion is supplied by vision. Even under these circumstances, however, egomotion is clearly perceived (Rock, 1968). In curvilinear motion (where kinesthetic information about changing acceleration is available), vision seems to be a dominant source of information for egomotion (Dighans et al., 1972; Lishman & Lee, 1973).

Egomotion is distinct from locomotion, as it is possible to locomote without any displacement with respect to the surrounding, or to be displaced without locomoting. The fact that locomotion is neither a sufficient nor a necessary condition for the perception of egomotion is intimately linked to the phenomenon of induced motion of the self. It has been shown experimentally (e.g., Duncker, 1929), and is confirmed by everyday experience, that a stationary observer will erroneously perceive himself as moving and his surrounding as stationary if the surrounding moves as a whole in a smooth way. Familiar examples are the "magic swing" or the "train illusion." Rock (1968) in his investigation of the basis of position constancy during passive movement of the observer concluded that the induced movement of the self is a very important and pervasive phenomenon playing a dominant role even in situations when the observer is actually moving (i.e., when there is kinesthetic information available to compute the egomotion). These phenomena reflect (in the view advanced here) an important characteristic of the operation of the visual system: to perceive the self as moving one has to perceive (a part of) the environment as stationary. Once such a reference surface is established, the observer regards all motions in the retinal regions corresponding to this surface as egomotion.

In all the cases mentioned above, it was a large stimulus array (encompassing the majority of the visual field) which was actually moving with respect to the stationary observer. However, it was recently shown (Johansson, 1977) that stimulation by retinal motion over only a few percent of the far periphery is sufficient to elicit the perception of egomotion even when contradictory information for a static state is available over the foveal regions of the retina. When the foveal regions were similarly stimulated while the far periphery signaled no motion the observers typically did not perceive any egomotion; rather, the retinal motions were interpreted as object motion.

While the experimental evidence about the role of vision in the perception of egomotion abounds, the mechanism by which these computations are accomplished is unknown. In this paper we outline a mechanism possibly underlying the perception of egomotion. (Egomotion is defined here as the motion of the center of polar projection which we identify with the [instantaneous] position of the observer.) The mechanism has two parts.

1)   The first part is a decomposition algorithm. This is a mechanism for computing the parameters of relative motion of the observer with respect to a surface projected onto a retinal region from retinal velocities in that region.

2)   The second part consists of a procedure for finding a reference surface which will be perceived as "stationary" and with respect to which the observer will judge his own motion.

The two parts are not necessarily sequentially ordered. The starting point of our analysis can be found in Gibson's hypothesis that

> . . . there is literally no such thing as a perception of space without
> the perception of a continuous background surface (Gibson, 1950,
> p. 6).

It is suggested that the perception of egomotion is brought about, in part, by an (explicit or implicit) decision on the part of the visual system as to what part of the visual field corresponds to such a continuous background surface. This surface is then taken as the frame of reference and perceived as stationary, and the motions on the retina in the regions corresponding to this surface (or surfaces; the region does not have to be contiguous) are interpreted as caused by egomotion (the motion of the observer with respect to this surface). The exact form of this decision procedure is unclear, at present, except for the fact that the retinal loci at the far periphery are apparently much more important in this respect.

Gibson (1979) suggested that the induced movement of the self is due to relations between the optic array and the edges of the visual field:

> . . . the upper and the lower edges of the field of view sweep over
> the ambient array in swinging; the field of view wheels over the
> array in tilting; and the lateral edges of the field of view sweep
> across the array in turning (Gibson, 1979, p. 186).

His explanation of why is it that such retinal motions are attributed to the observer has to be found in what can perhaps be called the "principal hypothesis of ecological optics" which relates events, egomotion and surface layout:

> . . . an event is specified by a local change of the ambient array
> while locomotion is specified by a global change of the ambient
> array. The surface layout of the environment is specified by invari-
> ants of structure (Gibson, 1977, p. 161).

While the hypothesis has its attraction the available experimental evidence does not seem to support this notion; the problem of deciding on what is a local or global change in the array is apparently rather complex. In fact, Johansson's finding that

> . . . a continuous optical flow in very narrow bands in the far peri-
> phery of the visual field (some thousandths of the retinal area) is
> regularly interpreted as locomotion (Johansson, 1977, p. 373).

makes problematical the distinction between local and global transformations of the optical array as underlying the perception of object motion and egomotion respectively.

The fact that a very limited and unstructured retinal motion in the far periphery apparently supersedes cognition, experience and the available conflicting kinesthetic information suggests that the far periphery is predisposed (or hard-wired) to interpret retinal motions as due to egomotion; one of its main functions may well be the control of the organism's attitude (i.e., position and motion) with respect to the environment. On the other hand, as noted by Johansson (1977, p. 374), the visual system's reaction to onset of motion and to ongoing motion differs. When "isolated" motion stimulation in the periphery suddenly occurs, the visual system reacts with a reflexive change of fixation and attention, and interprets the retinal motion as object motion. A continuous motion stimulation, however (possibly constrained by retinal symmetry), causes the perception of egomotion.

This distinction between sudden motion and ongoing motion stimulation may perhaps explain the phenomenon typically found in laboratory investigations of induced motion of the self: it takes some time (on the order of 10 s) before egomotion is experienced (Rock, 1975). (When "sudden" becomes "ongoing" [i.e., when motion occurs over extended spatio-temporal regions], retinal motions tend to be interpreted as egomotion.) The area over which the stimulation has to occur is apparently also important, although exactly in what way is not yet clear. Rock (1968) found that a single visible point did not induce the experience of egomotion; rather, it was interpreted as moving with respect to the observer (situated in dark room). It seems that some conditions on the elements comprising the potential reference frame must be met before it can function as such. These conditions seem to be the position of the motion stimulation in the visual field, perhaps their extent and some notion of numerosity. (In Rock's [1968] experiment, when there were many points visible, the perceivers reported a sensation of egomotion instead of seeing the dots as moving.) More research is needed to pinpoint the exact nature of this hypothesized decision procedure (what happens, for example, if the far periphery is presented with two conflicting stimuli indicating different kinds of egomotion).

The idea that the far periphery is specialized (or predisposed) to interpret retinal motions as resulting from observer motion relative to a (stationary) reference frame accords with some recent neuropsychological ideas that the visual system consists of two separated and rather specialized sub-systems: one ambient, processing space at large around the organism, and the other focal, specialized in processing of detail. Apparently, the mechanism mediating the ambient vision is much older:

> . . . vision of space at large, detected by the whole retina, is mediated through sub-hemispherical mechanisms; but foveal vision of detail, essential to full conscious perception of the substance of surroundings and of identities of objects, is mediated by a hemispheric or cerebral cortical mechanism (Trevarthen, 1978, p. 114).

It is hypothesized here (see above) that the mechanism underlying the perception of egomotion consists of a decision procedure which labels parts of the visual field as a stationary reference frame, and that this occurs mainly (though probably not exclusively) in the far periphery. Speculatively, the relative motion parameters may be computed locally at many different parts of the far periphery simultaneously and then tested for consistency. (Each retinal region can be taken as a reference and the corresponding retinal motions interpreted as caused by observer motion relative to the underlying surface.) The largest consistent subset of such loci would then be chosen to be the (stationary) frame of reference with respect to which the corresponding relative motion parameters would constitute a description of egomotion.

Assuming this is indeed the case, the next step is thus to show how to use the retinal motions to arrive at a description of observer motion causing them. We suppose here that the retinal motions can be sensed instantaneously as (an approximation to) retinal velocities. Because any curvilinear motion of the observer relative to a reference surface can be (instantaneously) uniquely decomposed into a translation and a rotation about an axis through the vantage point (Whittaker, 1944), our description of egomotion will be to specify, at each instant, the instantaneous direction of motion of the observer (the direction of the translational component), and possibly the amount and plane of instantaneous rotation.[1] One feature of such a decomposition is the inability to compute the magnitude of the translational component (see, e.g., Nakayama & Loomis, 1974; Longuet-Higgins & Prazdny, 1980); only the ratio of translational velocity magnitude to the distance can be obtained. While this in itself may be seen as a drawback, it is a feature which is also found to hold for the visual system. As reported by Johansson (1977, p. 375)

. . . The perceived speed of locomotion seems to be a function (probably not linear) of perceptual distance to the centrally fixated point.

In other words, the speed of the translatory motion component is a function of phenomenal distance.

In the following we briefly outline a method for obtaining the instantaneous direction of (relative) motion from image velocities on the planar projection surface (for a more detailed discussion see Prazdny, 1981). While a planar projection surface will be used for convenience, it should be noted that the results are independent of any particular projection surface; a simple transformation converts one representation into another (see, e.g., Lee [1974] for an example of the use of yet another [cylindrical] projection surface). We assume that the image velocities are generated by observer motion relative to a rigid and opaque surface.

The distribution of velocities on the projection surface due to relative motion contains information not only about the three-dimensional relationships between the texture points projecting onto the projection surface, but also about

---

[1] It is doubtful that the rotational component provides useful information.

the relative motion itself. As stated above this relative motion of the observer with respect to a surface is (instantaneously) decomposable into a translation and a rotation. However, only the translational component contains information about surface orientation or relative depth (see, e.g., Longuet-Higgins & Prazdny, 1980). The translational "retinal"[2] field ("retinal" motions caused by observer translation) consist of motions along straight lines all intersecting at one common point, the focus of expansion (FOE). This point corresponds to the point where the (three-dimensional) vector specifying the instantaneous direction of observer motion (the vector tangent to the motion path described by the observer at a given instant) pierces the projection surface (Figure 1). The rotational velocity field ("retinal" velocities due to rotation) can be (instantaneously) decomposed into two orthogonal components: one which we will call hyperbolic (the component vectors are tangent to hyperbolas through given image plane loci) due to rotation about an axis (through the center of polar projection) parallel to the projection plane, the other a circular field (the component vectors are tangent to circles [centered at the "fovea"] through given

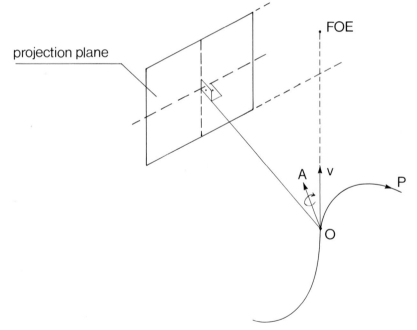

Figure 1. Observer O moves along a 3-D path P. The instantaneous direction of motion (**v**) pierces the projection plane at the focus of expansion (FOE). The motion of O along the curvilinear path P is (instantaneously) equivalent to a simultaneous translation with velocity vector **v**, and a rotation with angular velocity vector **A**.

---

[2]The word "retina" will be used to denote the projection surface (plane, here). The "fovea" refers to the center of this surface.

image plane loci) due to rotation about an axis (through the center of polar projection) perpendicular to the projection plane (Figure 2).

We see thus that each "retinal" velocity vector can be decomposed into three vectors, one due to translation, and two due to rotation. The decomposition is not arbitrary; it is constrained by the fact that the translational vector components all have to meet at one common FOE. Suppose now that arbitrary values for the two rotational component vectors are chosen. Because the three vectors simply add at a "retinal" locus to produce the observed "retinal" velocity at that locus, subtracting the two rotational components from the given "retinal" velocity produces the translational component. If all such obtained "remainders" intersect at one point on the projection surface we know that the values chosen for the rotational component are correct. Conceptually, the whole decomposition process is thus simply a constrained minimalization of a function of three variables (the direction and magnitude of the hyperbolic field, and the magnitude of the circular field [see Prazdny, 1981]). Observe that only a few retinal velocities are needed, i.e., the mechanism does not require a dense optical flow field. The constrained minimalization procedure is biologically feasible. Ullman (1979) has shown that it can be implemented in a locally connected network of elements performing the same computations.

For an illustration of the operation of the decomposition algorithm, refer to Figure 3.[3] In Figure 3a, an optical flow field is shown produced by simulat-

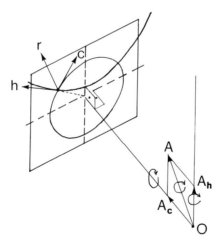

Figure 2. Any rotation **A** can be decomposed into two simultaneous rotations, one parallel to the projection plane (A_h) which causes a "retinal" point Q to move along a hyperbola, the other perpendicular to the projection plane(A_c) which causes Q to move along a circle centered at the "fovea." If the observer rotates with angular velocity **A**, the point Q moves (instantaneously) simultaneously along the circle (with velocity **c**) and along the hyperbola (with velocity **h**), i.e. it moves with velocity **r**.

[3] A simple Nelder-Mead minimalization procedure (Nash, 1979) was used in our computer implementation. Much faster and more efficient algorithms are available.

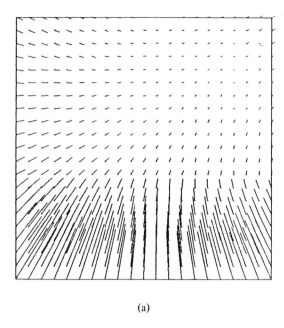

(a)

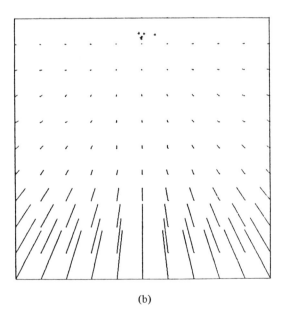

(b)

Figure 3. (a) An optical flow field. (b) Its translational component and FOE. See text for further explanation. The "+" symbols denote the FOEs computed from neighborhoods at different loci on the projection plane.

ing an observer moving curvilinearly relatively to two planar surfaces. This field consists of somewhat noisy velocity vectors (the errors are on the order of about 10 to 20 minutes of arc). Figure 3b shows the translational velocity field obtained from the optical flow field in Figure 3a by the process outlined above. As can be seen, the instantaneous direction of observer motion was located rather precisely. (Each cross symbol in Figure 3b corresponds to an FOE computed using four "retinal" velocities at loci separated by a visual angle of approximately 5 degrees of arc.) Note the simplicity of the method. It does not involve projective or geometrical relations, and is based only on relationships and data defined and measurable on the projection surface. The instantaneous direction of egomotion (the egoheading) is defined directly as a locus on the retina. The visual direction along which a particular image element lies is also not required (as was the case in, e.g., Prazdny, 1980). The method can, but does not have to be, implemented as a (spatially) local computation.

Conclusion. This paper has presented an outline of a mechanism possibly underlying the perception of egomotion as determined by purely visual factors in monocular stimulation. The hypothesized mechanism relies on an algorithm by which retinal velocities can be used to compute the parameters of relative motion (of the observer with respect to a surface) which caused the velocities. Assuming that the far periphery is geared to interpret retinal motions as mainly due to egomotion, this algorithm can then be used at many loci in the periphery simultaneously to obtain a set of independent estimates of relative motion parameters. It is hypothesized that the largest consistent subset (i.e., the largest set of retinal points with velocities producing the same [or roughly the same] parameters of relative motion) is chosen to represent a reference surface which is then taken to be stationary. The relative motion parameters computed in its region constitute a description of egomotion; the observer perceives himself as moving in a given way (determined by the computed parameters of motion) relative to this surface. The data on which the scheme operates are retinal velocities. Because the method is essentially a constrained minimization procedure, the data can contain relatively large amounts of noise. This is important, for it is known that the information in the periphery is of low resolution, so that the velocity estimates might be inaccurate. A noteworthy feature of the method is the fact that the location of the focus of expansion may be computed using only a few retinal velocities. A dense optical flow field is not a prerequisite and can, in fact, be computed as a consequence of locating the FOE.

While the visual kinesthetic information is regarded here as an important source of egomotion information, it should not be inferred that mechanical kinesthesis does not play a part in the perception of egomotion. It is clear that it indeed does (for example, the extra-retinal signals do co-determine perceptions; see, e.g., Prazdny & Brady, 1981). Information about ocular rotations is not required in the proposed mechanism if one is interested only in the instantaneous direction of observer motion. If information about the environmental rotation of the observer is also required, ocular rotation information would be needed. It would be sufficient, for this purpose, to supply information only about the status of the eye (i.e., moving/stationary). The nature of the interaction between the two sources of kinesthetic information is, however, at present still only vaguely understood.

While the method for decomposing the retinal velocity field can in principle be applied to "local" events to characterize the motion of individual objects with respect to the observer, there are some reasons to believe that a different mechanism may underlie the perception of object motion. For one thing, the method delivers an unambiguous characterization of relative motion of the object while it is known that such perceptions, especially with respect to the rotations, are often ambiguous (e.g., the rotating trapezoid illusion). Egomotion, on the other hand, is never (or only very rarely) an ambiguous phenomenon. It is thus plausible that egomotion, as opposed to perception of object motion (mediated by more focal mechanisms), is processed by peripheral mechanisms along the principles outlined above. The link between the retinal velocities and the perception of egomotion is, unfortunately, not so simple. For example, it has been found (Oman, Block, & Huang, 1980) that the visually induced self-motion sensations adapt rapidly (after 1 to 3 hours) to left-right visual reversals while wearing reversing goggles. The precise implications of this finding for the proposed mechanism of visual kinesthesis are, at present, not clear.

References

Dighans, J., Held, R., Young, R. L., & Brandt, T. Moving visual scenes influence the apparent direction of gravity. *Science,* 1972, **178**, 1217-1218.

Duncker, K. Ueber induzierte Bewegung. *Psychologische Forschung,* 1929, **12**, 180-250.

Gibson, J. J. *The Perception of the Visual World.* Boston, MA: Houghton-Mifflin, 1950.

Gibson, J. J. On the analysis of change in the optic array. *Scandinavian Journal of Psychology,* 1977, **18**, 161-163.

Gibson, J. J. *The Ecological Approach to Visual Perception.* Boston, MA: Houghton-Mifflin, 1979.

Johansson, G. Studies on the visual perception of locomotion. *Perception,* 1977, **6**, 365-376.

Lee, D. N. Visual information during locomotion. In R. B. MacLeod & H. L. Pick (Eds.), *Perception: Essays in Honor of James J. Gibson.* Ithaca, NY: Cornell University Press, 1974.

Lishman, J. R., & Lee, D. N. The autonomy of visual kinesthesis. *Perception*, 1973, **2**, 287-294.

Longuet-Higgins, H. C., & Prazdny, K. The interpretation of a moving retinal image. *Proceedings of the Royal Society, London*, 1980, **B208**, 385-397.

Nakayama, K., & Loomis, J.M. Optical velocity patterns, velocity sensitive neurons, and space perception. *Perception*, 1974, **3**, 63-80.

Nash, J. C. *Compact Numerical Methods for Computers*. New York: Wiley, 1975.

Oman, C. M., Block, O. L., & Huang, J. Visually induced self-motion sensation adapts rapidly to left-right visual reversal. *Science*, 1980, **209**, 706-708.

Prazdny, K. Egomotion and relative depth map from optical flow. *Biological Cybernetics*, 1980, **36**, 87-102.

Prazdny, K. Determining the instantaneous direction of motion from optical flow generated by a curvilinearly moving observer. *Computer Graphics and Image Processing*, 1981, **17**, 238-248.

Prazdny, K., & Brady, M. Extraretinal signals combine with induced motion. *Perception and Psychophysics*, 1981, **29**, 403-406.

Rock, I. The basis of position constancy during passive movement of the observer. *American Journal of Psychology*, 1968, **81**, 262-265.

Rock, I. *An Introduction to Perception*. New York: Macmillan, 1975.

Trevarthen, C. Modes of perceiving and modes of acting. In H. L. Pick & I. J. Saltzman (Eds.), *Modes of Perceiving and Processing Information*. Hillsdale, NJ: Erlbaum, 1978.

Ullman, S. Relaxation and constrained optimization by local processes. *Computer Graphics and Image Processing*, 1979, **10**, 115-125.

Whittaker, E. T. *A Treatise on the Analytical Dynamics of Particles and Rigid Bodies*. New York: Dover, 1944.

Environment-Centered Representation of Spatial Layout: Available Visual Information from Texture and Perspective

H. A. Sedgwick

State University of New York
New York, New York

Abstract

This paper reviews the information for spatial layout that is made available to a stationary, monocular observer by texture and perspective, and it discusses the constraints on the validity of such information. The concept of an environment-centered representation of spatial layout is introduced, and it is shown that several forms of available information can lead directly to an environment-centered representation without having to pass first through a viewer-centered representation.

1. Introduction

1.1. The Available Information for Spatial Layout

The topic of space perception is currently, and has been for some time, the subject of considerable theoretical controversy, the crux of the issue being how to conceptualize the perceptual processes underlying space perception. Among the theoretical formulations currently receiving attention are the ecological approach of Gibson (1961, 1979), the Gestalt, or Prägnanz, theory currently supported by Attneave (Attneave & Frost, 1969), the Helmholtzian theory of unconscious inference (Rock, 1977; Hochberg, 1981), and the computational approach described by Marr (1978). Recent attempts to confront the differences between these approaches are fairly numerous (see, for example, Attneave, 1972; Hochberg, 1974; Ullman, 1980; Fodor & Pylyshyn, 1981; Turvey, Shaw, Reed, & Mace, 1981).

Copyright © 1983 by Academic Press, Inc.
All rights of reproduction in any form reserved.
ISBN 0-12-084320-X

In spite of the diversity of these theoretical approaches, there is, it seems to me, a good deal of common ground between the investigators following them, particularly between much of the work that Gibson and his colleagues have done and some of the work done in the last few years in the area of computer vision. One important commonality lies in the attention given to the problem of perceiving the spatial layout of visible surfaces in the environment. This emphasis is central to Gibson's concept of the *visual world* (1950a), to Marr's concept of the 2-1/2 D sketch (1978), and to Barrow and Tenenbaum's concept of *intrinsic images* (1978). A second, equally important commonality lies in the analytical approach taken to this problem; the investigation of the relations between structures in the light from a complex environment reaching a visual system and structures in the environment itself is treated as an important prerequisite for the study of how the visual system perceives the three-dimensional layout of environmental surfaces. Such an investigation involves analyzing these relations and the conditions under which they apply; the emphasis is on finding relations of sufficient generality that they can be applied prior to the recognition of objects in a scene and of sufficient specificity that they allow an unambiguous determination of three-dimensional structures. Such an investigation does not concentrate exclusively on any particular visual system but rather is concerned with the conditions within which all visual systems, whether biological or machine, must function. Different individuals or groups give more or less prominence to this task, talk about it in different terms, and make somewhat different uses of the results obtained, but this common underlying aspect of their research is still clearly present and provides one basis for fruitful exchanges. In this paper we consider some results of this work, coming both from experimental psychology and from computer science.

Gibson (1961) used the term *optic array* to refer to the structured array of light reaching a *point of observation* (in humans the nodal point of the eye is taken to be the point of observation). For our present purposes we can think of the optic array at a point of observation either as the collection of nested solid visual angles subtended at that point by the visible surfaces and surface elements of the environment or as a spherical surface of unit radius surrounding the point of observation and onto which the surfaces and edges of the environment are projected. These two descriptions of the optic array are clearly interchangeable. It is also sometimes convenient to think of a local portion of the environment as being projected onto a plane tangent to the unit sphere (and thus perpendicular to the line of regard); projections on this *projection plane* or "picture plane" are easier to illustrate and, for projections close to the line of regard, are nearly identical to projections on the spherical optic array.

Following Gibson (1979, p. 307) we shall refer to the relations that exist between the structure of the environment and the structure of the optic array at a point of observation as *available information* because these relations are available to any organism or machine capable of making use of them and because it is the existence of these relations that makes it possible to obtain information about the environment through vision. Clearly, however, any particular structure in the optic array is only specific to a particular structure in the environment within certain *constraints*. Thus in any discussion of available information we must be explicit about the constraints under which it applies.

Visual information whose availability does not depend on the observer's limited depth of field or on the observer's taking up multiple points of observation, either through motion or through binocular vision, is often referred to as *pictorial information* because much of it can in principle be simulated with pictures. Discussions of such information tend to consist of disconnected lists of pictorial "cues" such as aerial perspective, interposition, texture gradients, and so forth. In this paper we shall consider an important subset of these cues – those arising from texture and from perspective – and shall try while discussing them to consider some of the interrelations among them. We shall restrict our considerations to spatial layouts composed of planar surfaces and straight edges. Thus, for the remainder of the paper surfaces are assumed to be planar and edges to be straight when other descriptions are not explicitly given.

1.2. Representations of Spatial Layout

It has become common in recent years to describe the visual system as forming one or more internal *representations* of the scenes it views. It is not the purpose of this paper to examine the appropriateness of this conceptualization, but we shall consider some of the ways in which the specific representations that are posited interact with the search for available information; for this purpose it is sufficient to observe that the concept of representation appears at least to be useful heuristically at the present stage of some investigations.

Marr (1978), characterizing representations by the information they make *explicit,* has made the point that our choice of which representations to use in building visual systems is crucial both in shaping our investigations and in determining how successfully and how efficiently the systems we build will function. Clearly, the information that we wish to make explicit will influence what available information we look for and, on the other hand, our knowledge of what information is available, as well as of the computational costs of obtaining and manipulating it, will help determine our choice of representation.

Marr (1978) postulates that the explicit description of the three-dimensional spatial layout of surfaces is first represented in what he calls a 2-1/2 D sketch; fully three-dimensional representations are then derived from the 2-1/2 D sketch. Marr refers to this 2-1/2 D sketch as a *viewer-centered* representation, which means that the spatial information about each surface location is represented in terms of its spatial relation to the viewer; specifically, what are represented are the *range,* which is the distance from the point of observation to a point on the surface, and the *orientation,* which is the orientation at the surface relative to the line of sight, of each visible surface location (additionally, surface discontinuities are represented). Thus, with a flat, extended surface, such as the ground plane, both range and orientation vary smoothly along the surface. Also, whenever the point of observation moves, the range and orientation of every surface location change and need to be updated in the representation. The *intrinsic image* conceptualization of Barrow and Tenenbaum (1978) incorporates a similar viewer-centered representation of spatial layout (see Figure 1).

Gibson (1950a), following a tradition of interest in the perceptual constancies that goes back to Descartes, stressed that our naive visual perception is of the *visual world,* that is, a stable spatial layout of extended surfaces having fixed spatial relations to one another that remain unchanged as we move about (see Gibson, 1950a, pp. 26-43, for a much richer description of the visual world). Gibson further characterized our normal visual environment as *terrestrial,* meaning that he assumed that there is generally a continuous, more or less horizontal, ground surface on which the objects of the environment rest and by means of which they are given definite locations relative to one another (1950a, pp. 6-7; 1966, pp. 154-163; 1979, pp. 8-16). What this approach suggests to me is that we might want to try thinking in terms of what I shall call an *environment-centered* representation of spatial layout. An environment-centered representation would specify spatial information about surface locations in terms of their relations to the fixed environment rather than in terms of their relations to the viewer. In such a representation, for instance, the orientation assigned to a planar surface would be constant across its whole extent and the assigned orientation would not change when the point of observation moved.

Clearly, given a complete viewer-centered representation of spatial layout, it is possible to derive an environment-centered representation of its visible surfaces – or vice versa (given that the location of the point of observation is included in the environment-centered representation). But, going back to

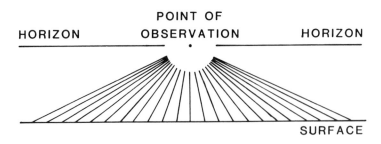

Figure 1. Viewer-centered representation of a surface. This diagram shows a 2-D slice through the point of observation and some surface in space. The slant angle and range at which a line of regard meets the surface change continuously along the surface. (After Sedgwick [1973] p. 4.)

Marr's dictum, our choice of representation will influence what we look for in the optical structure reaching the eye, and it will determine the kinds, and perhaps the amount, of computation or processing that a visual system has to do. In this paper we shall explore some of the implications of an environment-centered representation for the potential usefulness of the available visual information that we are considering.

2. Texture

2.1. Definition of Texture

Gibson (1950a) characterized the surfaces of the environment as being textured, and he initiated the analysis of *texture* as a source of information for spatial layout. This analysis has been carried further by a number of people and has also given rise to many empirical studies, but here we shall consider only some main points in the analysis of available texture information.

Characterizations of texture in this context have generally been highly schematized because the concern is not with how the texture of any particular surface is perceived but with how texture in general might serve the perception of spatial layout. The essential characteristic of surface texture in this respect is its spatial uniformity; texture elements do not generally change systematically in size, shape, or density of distribution across the extent of a material surface. Surface texture is sometimes entirely regular, like a tile floor, and such regularity is often convenient to assume in analyses. Gibson pointed out, however, that to be useful in perceiving spatial layout, the elements of surface texture need only have a statistical tendency toward even scatter (1950a, pp. 77-78); Purdy (1960) has given a brief analysis of the mathematical relation between variability in distribution of texture elements and the resulting variability in the available information that depends on texture. On the other hand, systematic variation in textural characteristics across a surface, although uncommon, does occur, and Gibson speculated that such occurrences might give rise to predictable misperceptions (Gibson & Flock, 1962). A useful discussion of the relation between a surface texture's regularity and the accuracy and precision of the spatial information it makes available has recently been provided by Stevens (1981).

Gibson also pointed out that the level of surface organization that is treated as texture can be variable depending on the needs and point of observation of the visual system. For instance, what are treated as objects at one time may be treated as elements of surface texture at another (see, for example, Figure 2). Purdy (1960) suggested that the visual system may make smooth transitions from one level of the hierarchy of textures on a surface to another as the distance of the surface increases and so may always be able to find a level of texture that lies within the processing constraints of the system.

Figure 2. Objects creating a surface texture. The barrels in this line drawing may be treated as individual objects by a perceptual system or their tops may be treated as circular texture elements defining a surface. (Photograph by Gendreau and used in Gibson [1950a] p. 83.)

2.2. Gradients

How might texture be used in layout perception? One potential source of spatial information arises from the systematic gradients of projected texture that slanted surfaces produce in the optic array. Gibson (1950a) first called attention to the existence of texture gradients and gave qualitative descriptions of their characteristics; more detailed and mathematical analyses of their properties have been carried out by others.

Gibson (1950b) introduced the term *optical slant* to refer to the angular relation with which a surface, at a given point, is intersected by the line of sight. The essential characteristic of texture gradients is that for each point on a surface the texture gradient of the surface's projected image is specific to the optical slant of the surface. In principle this allows the three-dimensional orientation of the surface at every point to be recovered from the texture gradients in the optic array.

Two angles are required to specify optical slant, one which indicates the *direction* of slant and another which indicates the *amount* of slant (Stevens,

1980, discusses a variety of ways in which the two degrees of freedom that underlie slant can be expressed).

Direction of slant is very simply specified in the optic array. As Phillips (1970) and Stevens (1980) have pointed out, the direction of slant is specified by the direction in which the texture gradient is steepest. Another criterion, which, however, produces the same result, was used by Flock (1964), who took direction of slant as the direction perpendicular to the normal meridian – where the normal meridian is the optic array meridian along which optic array changes in projected texture are symmetrical in either direction. Of course in the immediate vicinity of the point these changes are nearly zero and in a projection plane, rather than an optic array, they are precisely zero; thus we could say that the direction of slant is perpendicular to the zero-gradient direction.

The amount of slant is then specified in several ways by the texture gradient along the direction of slant. The first mathematical analysis of how texture gradients specify amount of optical slant was carried out by Purdy (1960). Purdy developed his analysis using a regular texture grid such as that shown in Figure 3. He defined several different angular measures of this texture: *texture size* ($W = ML$), the solid angular area of a unit of texture; *texture density* ($N = 1/W$), the number of texture elements per unit solid angle; *texture compression* ($C = M/L$), the ratio of the angular size of the radial dimension to the angular size of the tangential dimension; and *linear perspective* (L), which is based simply on the angular width (L) of the texture element. For each texture measure, Purdy then defined a corresponding gradient as the *relative rate of change* of the texture measure as the line of regard, whose direction (F) is measured relative to the horizontal, is swept along in the direction of the slant:

$$G_W = (dW/dF)/W,$$
$$G_N = (dN/dF)/N,$$
$$G_C = (dC/dF)/C, \text{ and}$$
$$G_L = (dL/dF)/L.$$

Making some reasonable approximations, Purdy then showed that each of these texture gradients specifies the amount of optical slant (R) of the surface at that location:

$$G_W = 3/\tan R,$$
$$G_N = -3/\tan R,$$
$$G_C = 1/\tan R, \text{ and}$$
$$G_L = 1/\tan R.$$

Slant relative to the line of regard is what would be needed for a viewer-centered representation such as that proposed by Marr (1978). Gibson (Gibson

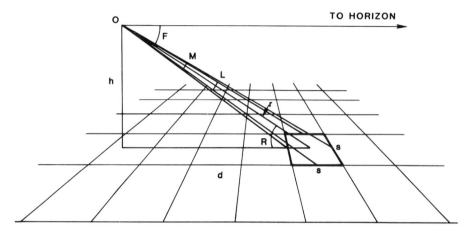

Figure 3. A regular texture grid. O is a point of observation at a perpendicular height h from the surface and at a distance d along the surface from an s by s square texture element. A line of regard from O meets the texture element at an angle R, the optical slant, and makes an angle F with the horizontal; the distance (range) from O to the texture element along the line of regard is r. The angular width that the texture element subtends at O is L; its angular depth is M. (After Purdy [1960] p. 6.)

& Cornsweet, 1952) distinguished between *optical slant,* which is relative to the line of regard and so changes continually as the line of regard is swept along the surface, and what he called *geographical slant,* which is the orientation of the surface relative to some fixed environmental reference such as the ground plane and consequently is constant throughout the extent of the surface. Geographical slant would be needed for an environment-centered representation. Purdy (1960) pointed out, however, that there is a simple relation that transforms optical slants into an environment-centered form: the geographical slant (S) of a surface with respect to the horizontal is given by the difference between the optical slant (R) of the surface at any point and the angular distance (F) of this point from the horizontal (see Figure 4), i.e.,

$$S = R - F.$$

To make this transformation, the direction of the reference plane, in this case the horizontal, must be available.

As was mentioned above, Gibson asserted that the appropriate basic model for the underlying structure of a terrestrial environment is an extended, more or less horizontal ground plane on which rest the various objects in the environment. The distances to these objects from an observer standing on the ground are then measured along the ground plane. As Purdy pointed out, any such distance (d), relative to the height (h) of the observer's eye, is very simply specified by any of the texture gradients that he analyzed. This is so because of a simple trigonometric relation involving optical slant (R), namely,

$\tan R = h/d$ (see Figure 3). If we consider, for example, the gradient of texture size (G_W), then because $G_W = 3/\tan R$ we can see immediately that

$$d/h = G_W/3.$$

2.3. Scale

If we consider the surface of the ground to be regularly textured, then that texture provides a constant scale across the ground that is invariant under projective transformation. The *relative sizes* of objects at different distances along the ground are then given, as Gibson (1950a, pp. 174, 180-181) pointed out, by their relation to the texture scale of the ground, e.g., an object that covers three by two units of texture at one distance is the same size as an object that covers three by two units of texture at another distance (see Figure 5). Likewise, *relative distances* between objects are specified by the number of texture elements between them along the ground. Size and distance here are specified relative to a *scale factor,* which is the size of the texture element.

What is critical here, as Gibson noted, is that the scale be taken at the point, or along the edge, where the object makes contact with the ground (see Gillam [1981] for a discussion of the persistent failures to appreciate this point). As Gibson also noted, we must assume here that *optical contact* with the ground specifies *physical contact* with the ground unless there is information to indicate otherwise. It is not hard to contrive situations that would deceive a monocular stationary observer (see Figure 6; also see Gibson, 1950a, Figure 72), but we would not expect such situations to occur commonly under ordinary circumstances.

2.4. Range and Scale

The *range* of a point on a surface is its distance from the point of observation. Proposed viewer-centered representations typically contain the range of every visible point in the scene (Marr, 1978; Barrow & Tenenbaum, 1978). Texture provides information about range because texture element size (s), visual angle (L), and distance (r) are linked together by the simple trigonometric relation $s/r = 2\tan L/2$ (see Figure 3), which for fairly small visual angles, such as texture elements would generally subtend, is closely approximated by $s/r \approx L$ (measured in radians). To obtain absolute distance from this relation, the absolute size of the texture element must be available. Even without this information, however, all distances are specified relative to each other if texture element size is assumed to be uniform:

$$r_1/r_2 \approx L_2/L_1,$$

i.e., the relative distance of texture elements is inversely related to their relative angular size. Both of these potential sources of information are well-known,

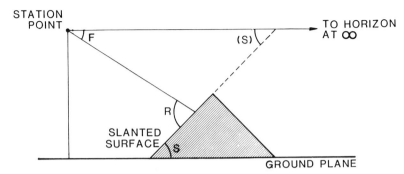

STATION
POINT

F

(S)

TO HORIZON
AT ∞

R

SLANTED
SURFACE

S

GROUND PLANE

Figure 4. Geographical slant, optical slant, and the horizon. A surface slanted up from the ground plane is here seen from the side. S is the geographical slant of the surface. The dotted line shows that the surface would make the same angle (S) if extended to meet the visual ray to the horizon. R is the optical slant of an arbitrarily chosen location on the surface, and F is the projective angle between that location and the horizon. (After Sedgwick [1980] p. 81.)

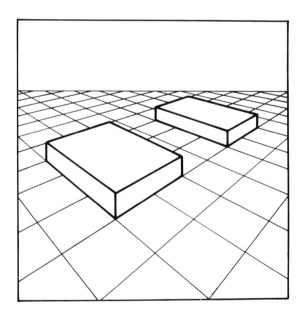

Figure 5. Size and distance specified by the texture of the ground plane. In this perspective view the relation between each of the two objects and the grid covering the ground plane specifies that the objects both have the same ground plane dimensions (three by two squares) and are separated by a distance of two squares. (From Sedgwick [1980] p. 75.)

the former "cue" to distance commonly being referred to as *familiar size* (Hochberg & Hochberg, 1952) and the latter being referred to as *relative size* (Hochberg & McAlister, 1955). Stevens (1981) has recently emphasized their application to textured surfaces. As Stevens points out, these relations between size, distance, and visual angle cannot be applied indiscriminately to visual angle measures of texture size. They are only valid for texture measures taken perpendicular to the line of sight; in any other direction a projective compression occurs in the visual angle subtended by the texture measure so that the simple inverse relation between texture size and distance is no longer valid. This constraint can generally be satisfied by choosing the texture measure so that it lies along the zero-gradient direction; Stevens refers to this dimension of a projected texture element as its characteristic dimension.

We have seen above that texture provides a uniform scale over any extended regularly textured surface. Scale, however, is not a value that enters directly into those viewer-centered representations that we have discussed. Instead, it seems to be assumed that scale or size would be derived when needed from range and visual angle information by an operation that takes appropriate account of viewer-centered orientation; this is essentially the inverse operation to obtaining range from texture size. On the other hand, scale or size does enter naturally into an environment-centered representation, whereas range does not. That is, the physical size of a given surface bears a fixed relation to the texture scale of the surface; this value is invariant under projective transformation yet is available in the optic array. An environment-centered representation thus might explicitly contain scale rather than range information, with range being derived, as discussed above, when needed.

3. Perspective

3.1. Introduction

Let us now turn to the perspective structures in the optic array and consider some of the ways that they can specify spatial layout. We have already seen, in discussing Purdy's work (see Section 2.2), that linear perspective can be analyzed as just another form of texture gradient specifying the optical slant of each surface location. Such an analysis, however, because it is based on a viewer-centered representation of direction, i.e., optical slant, is more complicated than it need be and obscures the most distinctive characteristics of perspective structures. Another way to think of perspective is as a system whose fundamental grouping principle is parallelism. For a given line in space, consider the set of all lines that are parallel to it. When lines from this set are projected onto a plane, their projections all converge toward a single point, the vanishing point (see Figure 7); thus they form a set that can be characterized by this single point. In an environment-centered representation the concepts of

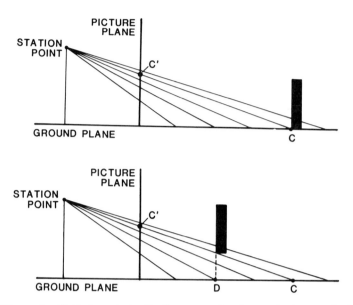

Figure 6. Optical contact. In the upper drawing projective contiguity at C' corresponds to physical contact between the object and the ground at C. In the lower drawing projective contiguity at C' is deceptive because the object is floating in the air above D. (From Sedgwick [1980] p. 73.)

direction and *parallelism* are closely related; a straight line has the same direction all along its length, and sets of straight lines are parallel when they all have the same direction. Thus, in an environment-centered representation the perspective transformation groups every set of lines that have the same direction in three-dimensional space into a set of converging lines characterized by a single vanishing point. In a viewer-centered representation, however, the direction assigned to a single straight line changes smoothly along its length so that there is no simple relation between the directions of two parallel line segments. This is not to say that parallelism cannot be *computed* from such a representation but only that it is not the most natural representation for forming sets of parallel lines. In the remainder of this discussion we shall consider perspective structures within an environment-centered representation and in doing so we shall see that they can specify aspects of spatial layout in a very simple yet quite powerful way. In our discussion we shall be applying in a somewhat different context a well-known body of knowledge concerning perspective structures in representational displays (see, for example, Ware, 1900; Sedgwick, 1980).

3.2. Vanishing Points

Consider the optic array at a given point of observation. Any straight line in three-dimensional space that *does not* pass through the point of observation determines a plane containing both the line and the point of observation.

PICTURE PLANE

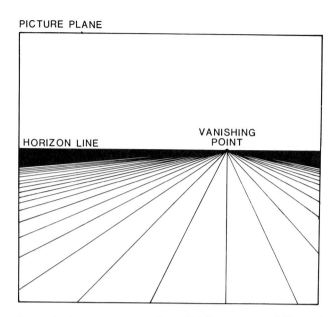

HORIZON LINE

VANISHING POINT

Figure 7. Projective convergence of parallel lines to a vanishing point. (From Sedgwick [1980] p. 53.)

The projection of that line in the optic array is then simply the intersection of that plane with the spherical optic array. Because the plane is chosen to pass through the point of observation, the line's projection is half of a great circle in the optic array. If a straight line in three-dimensional space *does* pass through the point of observation, then its projection will clearly be the two diametrically opposed points at which the line pierces the sphere of the optic array. Now consider the set of all straight lines parallel to a given line, i.e., the set of all straight lines having a given direction. Exactly one of these lines passes through the point of observation and so has two diametrically opposed points as its projection; the projections of all the rest are thus great semi-circles in the optic array. A little reflection will show that these great semi-circles must all intersect in the same two diametrically opposed points that are the projections of the single line passing through the point of observation. Furthermore, these two points are the vanishing points of the given set of parallel lines. Every distinct set of parallel lines has a distinct pair of diametrically opposed vanishing points in the optic array, and conversely, every distinct pair of diametrically opposed points in the optic array is a pair of vanishing points corresponding to a set of parallel lines in three-dimensional space. A similar relation between parallel lines and vanishing points holds concerning the projection plane with the difference that there is only one vanishing point for each set of parallel lines (with the exception that sets of lines that are parallel to the projection plane have no vanishing point).

From the basic relation that exists between vanishing points and sets of parallel lines it follows immediately that a vanishing point of a set of parallel lines *specifies* their direction. Their direction simply equals the direction of a line of regard from the point of observation to the vanishing point. This is necessarily true because such a line of regard is identical to the single line in the set of parallel lines that passes through the point of observation and has the vanishing point as its projection (see Figure 8).

It also follows that for any two non-parallel straight lines in three-dimensional space the angle between them, if they intersect, is equal to the optic array angle between their vanishing points. Thus, for instance, the three-dimensional angles between edges meeting at a vertex of an object would be given in the optic array by the angles between the vanishing points of the edges.

Because the vanishing point of any line can be thought of as the projection of a point that is infinitely far from the point of observation, any finite motion of the point of observation will be negligible in comparison, leaving the location of the vanishing point in the array completely unchanged. This means that all the vanishing points in an optic array, and the angular relations between them, remain invariant across the transformations in the optic array produced by movement of the observer (and also across the binocular transformations, i.e., the changes in optic array structure produced in going from the point of observation of the right eye to that of the left eye, or vice versa).

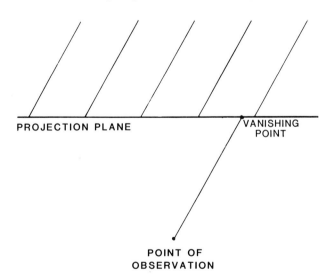

Figure 8. Environment-centered line orientation determined by the lines' vanishing point. The line of regard to the vanishing point of a set of parallel lines is the member of that set whose projection is a single point, and so the line of regard's orientation is the same as the environment-centered orientation of all the other lines in the set. (After Sedgwick [1980] p. 54.)

Hay (1974) pointed out these geometrical facts and suggested that the set of vanishing points belonging to the outlines and "inlines" of surfaces constitute a powerful way of analyzing the perspective relations in the optic array, for instance, in analyzing perspective distortions in pictures seen from the wrong viewpoint (an analysis of the sort suggested by Hay has recently been carried out in detail by Rosinski and Farber [1980]). Hay also suggested that this vanishing point structure *might* be a form of information useful to a visual system, perhaps even the human visual system. Hay gave "the picture plane pattern made by a set of vanishing points" the picturesque name the *ghost image* because, as Hay said, "the ghost image has no physical embodiment, yet it controls the perspective features of the optical image" (Hay, 1974, p. 270). We might also refer to this pattern as the *vanishing point structure* of the optic array.

What Hay proposed could be thought of as an environment-centered representation of the scene in terms of its vanishing points. Figure 9, based on a figure of Hay's, shows the vanishing points (ghost image) in the projection plane of three sets of parallel lines lying in three different surfaces; Figure 10, also based on Hay, shows how the angles between the surfaces can be reproduced in the optic array angles between the vanishing points.

3.3. Horizons

In spite of its considerable interest, Hay's analysis is flawed, I think, by a failure to distinguish clearly between the perspective properties of lines and surfaces. In the previous figure from Hay, the optic array angle between the vanishing points only equals the angle between the surfaces because the direction of the outlines and "inlines" of each surface is identical to the direction of slant of the surface. That this is not a necessary condition is obvious if we reflect that any flat surface contains an infinite number of sets of parallel lines, all lying in the surface but each having a different direction. In Figure 11, Hay's figure has been redrawn so that the angles between the vanishing points no longer equal the angles between the surfaces; in particular, the surfaces S_1 and S_3 are still parallel even though their outlines have different vanishing points.

What does characterize the projection of a surface is that all the vanishing points of the projections of lines lying in the surface lie along a straight line – the horizon of the surface. I would suggest, then, that although vanishing points are useful in considering the directions of line segments, in considering the orientation of a surface in the environment the appropriate perspective structure to consider is its horizon. (Note that "horizon," in the sense used here, does not imply "horizontal"; the horizon of a surface may take on any orientation, depending on the orientation of the surface.)

For all the relations that we noted above between lines and vanishing points, analogous relations hold between surfaces and horizons. Projections of

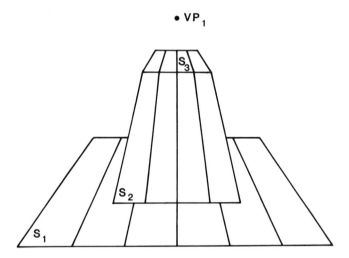

Figure 9. Hay's "ghost image": perspective view. This diagram, redrawn in somewhat modified form from Hay (1974, p. 269), is a perspective view of three surfaces, S_1, S_2, and S_3. Both the outlines and "inlines" of surfaces S_1 and S_3 converge toward the vanishing point VP_1; the outlines and "inlines" of S_2 converge toward VP_2. The horizontal lines of the perspective view can be considered to converge toward a third vanishing point, VP_3, at right infinity (left infinity) on the projection plane. (After Hay, "The Ghost Image: A Tool for Analysis of the Visual Stimulus," in *Perception: Essays in Honor of James J. Gibson,* edited by R. B. MacLeod and H. L. Pick, Jr., pp. 268-275. Copyright © 1974 by Cornell University. Used by permission of the publisher, Cornell University Press.)

parallel surfaces converge to a single horizon on the projection plane or in the optic array. Every straight line in the projection plane or great circle in the optic array is the horizon of a different set of parallel surfaces in space.

All surfaces parallel to a given surface in space have the same environment-centered orientation. The orientation of any surface in space is thus specified by the orientation of its horizon; this is necessarily true because a plane passing though the point of observation and intersecting the projection plane or the optic array along the surface's horizon will necessarily be parallel to the surface. It follows from this that the angular relation between any two surfaces is given by the angular relations between their horizons in the optic array (see Figure 12). This allows for surface orientations to be specified either relative to some reference plane, such as the horizontal, or simply relative to

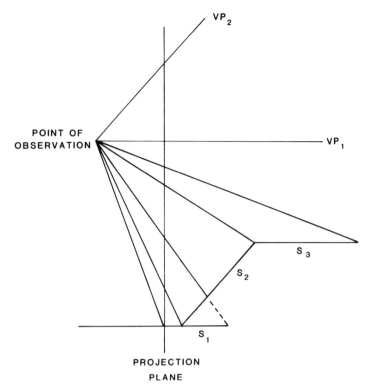

Figure l0. Hay's "ghost image": side view. This diagram, redrawn in somewhat modified form from Hay (1974, p. 269), is a side view of the three surfaces in Figure 9. The line of regard to VP_1 can be seen to be parallel to S_1 and S_3, whereas the
line of regard to VP_2 is parallel to S_2. (After Hay, "The Ghost Image: A Tool for the
Analysis of the Visual Stimulus," in *Perception: Essays in Honor of James J. Gibson,*
edited by R. B. MacLeod and H. L. Pick, Jr., pp. 268-275. Copyright © 1974 by Cornell University. Used by permission of the publisher, Cornell University Press.)

each other (e.g., the size of every dihedral angle between two surfaces is specified in this way).

Just as we referred to the pattern of vanishing points belonging to the projections of line segments in the environment as the vanishing point structure of
the optic array (what Hay called the "ghost image"), we might refer to the pattern of horizons of surfaces as the *horizon structure* of the optic array. The
horizon structure and the vanishing point structure taken together would constitute the *perspective structure* of the optic array arising from a given environment. *The perspective structure then directly specifies the environment-
centered orientation in space of every line segment and every surface in the
environment.* Like the vanishing point structure, the entire perspective structure
is invariant across changes in the optic array produced by movement of the

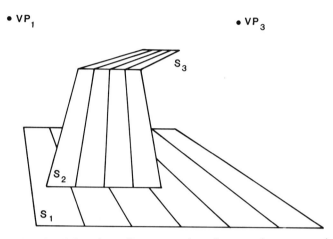

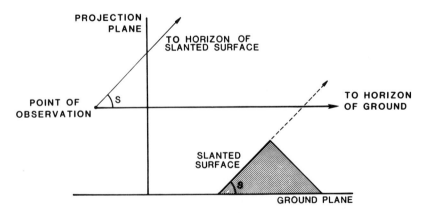

Figure 11. The "ghost image" representation of perspective structure is incomplete. In this further modified version of Hay's diagram, the surfaces S_1 and S_3 have the same geographical slant but the vanishing points VP_1 and VP_3 of their outlines and "inlines" are different.

point of observation (or by the binocular transformation). The specification of orientations by the perspective structure is so simple that it might itself *be used as* an environment-centered representation of surface and edge orientation.

Figure 12. Horizon relation specifying slant. This side view shows that the visual angle between the horizon of a slanted surface and the horizon of the ground is equal to the geographical slant of the surface. (After Sedgwick [1980] p. 86.)

4. Finding the Perspective Structure

4.1. Introduction

Except in special cases (see Section 5), the perspective structure of a scene is not explicitly present in the optic array. That is, because the edges and surfaces of a scene are only finite in extent, their actual projections do not include their vanishing points or horizons. Nevertheless, given certain constraints, much or all of the perspective structure of a scene may be implied by structures that are present in the optic array. When this occurs the perspective structure may be said to be *implicit* in the optic array. The ways in which this may occur are numerous and highly interrelated, so it is possible to indicate only a few of the principal ones here.

4.2. Finding Vanishing Points

The projection of a single line segment does not generally specify the location of its vanishing point, but its vanishing point is constrained to lie somewhere along the extended line that contains the projection. The vanishing point of two or more parallel line segments, however, is simply the location at which their projections would intersect if extended. A question that remains, however, is how, given an optic array, to determine which sets of lines in the array are the projections of parallel lines in space. Two general answers may be given to this question. One is that non-perspective, contextual information may indicate, more or less strongly, that some projections arise from parallel lines in space (e.g., in what is recognized to be a room, the edges of the floor boards might be taken as being at least roughly parallel). The second answer is that the perspective relations between some projections may suggest parallelism. In particular, three or more arbitrarily chosen projective lines generally do not intersect, if extended, in a single point. If they do, this is generally an indication either that they arise from parallel lines in space or that they arise from lines that are converging toward an actual intersection in space. Context or the specific details of their configuration may help to distinguish between these two possibilities.

Although we are not for the most part considering information that arises from motion of the point of observation, we might note here that the invariance under motion of the perspective structure in an optic array makes motion a powerful method of discovering the perspective structure. For instance, consider the case discussed above of projective lines converging toward a point of intersection. If the implicit point of intersection is due to a fortuitous configuration of projections of actually unrelated lines, then it will disappear or shift into a new configuration when the point of observation moves; if the implicit point of intersection arises from lines that are converging toward an actual intersection in space then its location in the optic array will change when the point of observation moves (unless the point of observation happens to be moving directly toward or away from it); but if the implicit point of intersection is the vanishing point of parallel lines then its location in the optic array will remain unchanged when the point of observation moves. Even the vanishing point of a single line segment is specified when the point of observation moves; it is the optic array location toward which the line segment's changing projection continues to point.

Returning to consideration of the static optic array, we may observe that the vanishing point of a line segment must lie on the horizon of the surface containing the line segment. Thus, if that horizon is known, the vanishing point of a single line segment is specified as the optic array intersection of that horizon line with the line containing the projection of the line segment. This intersection may be more accurately determinable in some situations than the intersection specified by the projections of two parallel line segments because in the latter case the lines containing the projections may be nearly parallel themselves so that their point of intersection may be quite difficult to determine accurately. The extreme of this situation is *parallel projection,* which never occurs in optic arrays from real environments but is often closely approximated in local regions of the optic array. The more closely parallel projection is approximated the more difficult it becomes to determine vanishing points from the intersections of the projections of parallel lines, but the vanishing point of each line segment can remain clearly specified by the intersection of the line containing the line segment's projection with the horizon of the surface containing the line segment. Thus, local approximations to parallel perspective do not in principle eliminate perspective information, as is sometimes asserted.

4.3. Finding Horizons

Let us now consider how the horizons of surfaces might be implicit in the optic array. In general, the ways of specifying a surface's horizon that we shall consider here all grow out of projective relations between line segments or texture elements lying on the surface. Thus a prior question is how, given a particular optic array, it can be determined which line segments or texture elements belong to which surfaces. This is a difficult question, and we shall not attempt to answer it here beyond noting that, as in the question of identifying projections of parallel lines (see Section 4.2), both non-perspective contextual information and constraining perspective relations may be pertinent.

Given a set of parallel lines lying on a surface, their vanishing point must lie on the horizon of the surface. As we saw in discussing Hay's work, however, one vanishing point does not specify the horizon of a surface, just as one point does not specify a line. An infinite number of differently oriented surfaces can all contain sets of parallel lines having the same vanishing point; they are only constrained in that the edges of the dihedral angles formed by their intersections will all be parallel and will have their orientation specified by the vanishing point (this is true, for instance, of the floor, ceiling, and side walls of a straight corridor).

Two vanishing points, however, do form a horizon; thus any two sets of parallel lines lying on a surface specify the horizon of the surface (see Figure 13). Note that it is not necessary for the two sets of parallel lines to be at right angles to each other or for the lines in either set to be regularly spaced (thus any parallelogram lying on a surface specifies the horizon, and hence the slant, of the surface – if its sides are taken to be parallel).

A single set of parallel lines lying on a surface can specify the horizon of the surface if the lines are regularly spaced. It is easiest to see how this works with a planar projection. Any particular surface in space, one of a family of parallel surfaces, is completely defined by two lines in the projection plane – one line is its horizon line, which is common to the family, and the other line is the line formed by the surface's intersection with the projection plane, a line referred to as its *trace line* (see Figure 14). The trace line and the horizon line are necessarily parallel because when two parallel planes (the surface plane and a parallel plane through the point of observation) are cut by a third plane (the projection plane) the lines of intersection are necessarily parallel. If we now consider any set of parallel lines (except those parallel to the projection plane) lying on the surface, the projections of these lines will converge to a vanishing point on the horizon and will intersect the trace line with regular spacing. This is necessarily so because the trace line actually lies on the surface plane and any line cutting across a series of regularly spaced parallel lines on a surface will be cut into equal divisions (see Figure 14 again). Finally, it follows from simple geometry that any other line in the projection plane that cuts the converging lines at equal intervals will be parallel to the trace line and hence to the

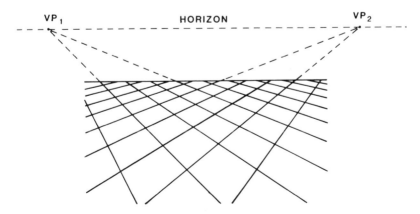

Figure 13. Two vanishing points specify the horizon of a surface. Each vanishing point is located at the point of convergence of the projections of a set of two or more parallel lines lying on the surface.

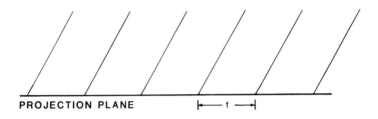

PROJECTION PLANE |←— t —→|

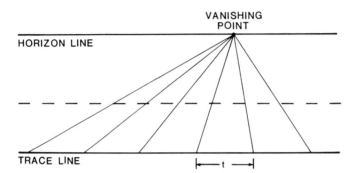

VANISHING
POINT

HORIZON LINE

TRACE LINE |←— t —→|

Figure 14. The horizon line and the trace line. The upper drawing is a view looking edge-on to the projection plane of a set of parallel lines lying on a surface. The lines intersect the projection plane at regular intervals (t). The lower drawing is a perspective view of the same parallel lines. The spacing (t) is the same as in the upper drawing because the trace line is the actual intersection of the surface with the projection plane. Any line, such as the dotted line in the lower drawing, that cuts the converging lines at equal intervals will necessarily be parallel to the trace line and the horizon line. (After Sedgwick [1980] p. 55.)

horizon line. Thus, if we can find such a line it will specify the orientation of the horizon line in the picture plane; this orientation, taken together with the vanishing point of the single set of parallel lines, completely specifies the location of the horizon line. (Stevens, 1981, notes this method of determining direction of slant but does not note its application to determining amount of slant.) We may observe here that if parallel lines lying on the surface are not regularly spaced then it will not, in general, be possible to find any line in the projection plane that is cut into regular intervals by their projections.

Let us now look briefly at the more general case of finding the horizon of a surface that has a covering of texture elements rather than parallel lines. The horizon-specifying optic array structures that we have already discussed are still relevant here because, even though a textured surface's features may include no explicit parallel lines, they still include many sets of implicit parallel lines. Consider, for example, the textured surface of circular barrel tops drawn in Figure 2. Assuming the circles all to have equal diameters, then any two circles define a pair of parallel lines, one that is tangent to the left sides of the two circles and another that is tangent to their right sides (see Figure 15). Because the tangency relation is projectively invariant, we can find the projections of these parallel lines directly in the optic array. Each such pair of parallel lines specifies a vanishing point, and any two vanishing points specify the horizon, as we have already seen; thus the horizon of the surface of circular barrel tops is specified many times over (see Figure 16). Small random fluctuations from the circularity or equal size of the texture elements will cause the vanishing points determined by the element pairs to be scattered around the horizon line. If these fluctuations are large, however, the vanishing points become so scattered that it seems unlikely that they could be of much use in finding the horizon. Thus the example of the barrel tops is a rather special case.

A more general but somewhat more complicated set of optic array structures specifies the horizon of any surface that has a statistically regular texture. We have already noted that the orientation of the horizon line is the same as that of the trace line (see Figure 14) and that the orientation of the trace is in

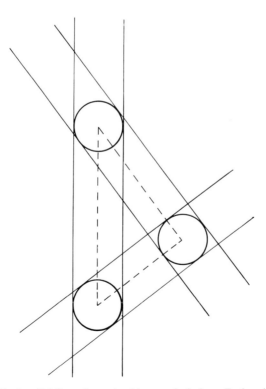

Figure 15. Parallel lines determined by equal circles. Each pair of circles determines a pair of parallel lines whose orientation equals the orientation of the (dashed) lines connecting the circles' centers.

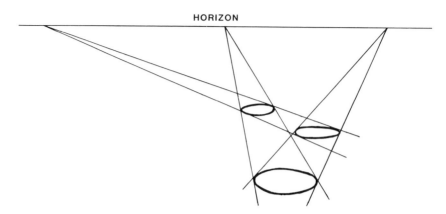

Figure 16. The horizon determined by circular texture elements. The projections of equal circles determine the projections of pairs of implicit parallel lines. The vanishing points at these lines specify the horizon of the surface.

turn the same as that of any projective line that is cut into equal intervals by the projection of a set of equally spaced parallel lines lying on the surface. Similarly, the orientation of the trace and of the horizon is the same as the zero-gradient direction (see Section 2.2) of any projected regular texture.

With the orientation of the horizon thus determined, all that is needed is one vanishing point to specify the horizon's location. In fact, however, an indefinitely large number of sets of parallel lines (and, of course, their vanishing points) are implicitly specified by the relations that exist between the projections of equal, non-collinear sets of texture elements lying along zero-gradient lines. Consider an implicit line lying in the projection plane and oriented along the zero gradient of texture density. Mark off some arbitrary, convenient number of texture elements anywhere along that line. Now consider another, distinct line in the projection plane that is also oriented along the zero gradient. Mark off the same number of texture elements anywhere along that line. Because they are both oriented along the zero gradient, these two lines must be the projections of two parallel lines lying on the surface. Thus, the implicit line that connects the left endpoints of these two sets of texture elements and the implicit line that connects their right endpoints are necessarily also the projections of two parallel lines lying on the surface. The intersection of these two implicit lines is a vanishing point and so specifies the location of the horizon (see Figure 17). Any two equal sets of texture elements chosen in this way specify another vanishing point along the horizon.

Stated algebraically rather than geometrically, the visual angles, e.g., $L(N)_1$ and $L(N)_2$, subtended by equal amounts of texture (i.e., an equal number, N, of texture elements) measured along the zero-gradient direction are proportional to their angular distances, F_1 and F_2, from the horizon:

$$L(N)_1/L(N)_2 = F_1/F_2$$

(this relation is exact in the projection plane and is closely approximated in the optic array). If there is local fluctuation in the size and distribution of texture elements, then the larger N is taken to be, the smaller is the variation that these fluctuations introduce into the specified location of the horizon. If the local fluctuation is small or non-existent, then the relation between the angular widths, L_1 and L_2, of pairs of individual texture elements ($N = 1$) accurately specifies the location of the horizon.

Gibson (1950a) and Purdy (1960) unified the discussion of perspective information and texture information by showing that perspective could be treated as a special case of texture gradient. We have seen an indication in this section that such a unification can also take place in the opposite direction with texture information being treated as one of the many ways of specifying the perspective structure of the optic array.

5. The Terrestrial Horizon

5.1. Introduction

I would like now to turn to consideration of a special case, the horizon of the ground plane, which I shall refer to as the *terrestrial horizon*. Gibson, as

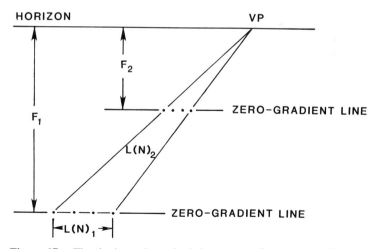

Figure 17. The horizon determined by zero-gradient texture elements. Any implicit projective line along which the texture gradient is zero, i.e., along which the projected texture elements are regularly spaced, is parallel to the horizon. Equal numbers of texture elements ($N = 4$ in this diagram) on any two such lines determine the projections of two implicit parallel lines whose vanishing point (VP) specifies the location of the horizon.

indicated above (see Section 1), argued repeatedly for the special status of the ground plane for terrestrial animals like ourselves, both because it is our own surface of support and because it is also the continuous surface of support underlying most other objects in our environment. He emphasized that the horizon of the ground creates in the optic array a fundamental division between the structured terrain below it and the relatively unstructured sky above (Gibson, 1950a, pp. 60-61; 1966, pp. 156-159; 1979, pp. 162-164). Parma, Hanson, and Riseman (1980) have recently made use of this fundamental optic array structure as a filter in scene analysis; constraints can be placed on the labeling of regions in a scene depending on their relation to the terrestrial horizon; assuming a flat ground plane, bits of sky cannot be found below the horizon, for instance, nor can bits of road be found above it. My own interest (Sedgwick, 1973, 1980) has been in the ways that the sizes and distances of objects resting on the ground can be specified by their projective relations to the terrestrial horizon. Although of most interest in this special context, most of these relations can be generalized to other surfaces and their horizons.

Let us first note, at the risk of repetition, that the terrestrial horizon is (unless it is occluded) an invariant feature of the optic array. No matter how an observer moves or rotates, the terrestrial horizon remains the horizontal great circle of the optic array (see Figure 18). Furthermore, any fixed point on the horizon maintains an invariant direction in the optic array. The terrestrial horizon thus provides a possible explicit environment-centered framework for the optic array.

5.2. The Horizon-ratio Relation

The sizes of all objects resting on the ground are specified, to within a scale factor, by their projective relations with the terrestrial horizon. Any line of regard to the terrestrial horizon of course is parallel to the ground plane. Such a line of regard thus intercepts any object resting on the ground at a height equal to the height of the point of observation. This provides a constant scale for size, regardless of distance, across the entire ground plane (see Figure 19). On this basis we can derive the following ratio (see Figure 20):

$$v/h = (\tan E + \tan F)/\tan F,$$

which I refer to as the *horizon-ratio relation*. It specifies the height, v, of every object resting on the ground plane relative to the height, h, of the point of observation in terms of the optic array angles E and F subtended between the terrestrial horizon and the top and bottom, respectively, of the object. For objects that are not too big relative to their distance this ratio can be approximated by

$$v/h \approx V/F, \text{ where } V = E + F.$$

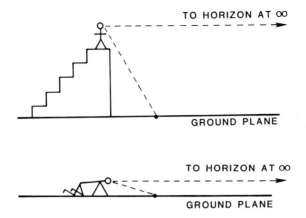

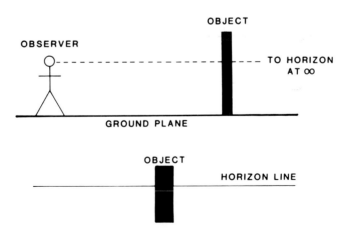

Figure 18. Optic array invariance of the terrestrial horizon. No matter how far or in what direction the observer moves, the horizon remains fixed at eye level in the optic array. Every ground plane location that is at a finite distance from the observer changes its projective location in the optic array as the observer moves. (After Sedgwick [1980] p. 52.)

In other words, the height of any object resting on the ground plane, relative to the height of the point of observation, is given, approximately, by the ratio of the visual angle subtended by the object to the visual angle subtended between the base of the object and the horizon line. Figure 21 illustrates several exam-

Figure 19. Scale specified by the terrestrial horizon. The upper figure is a side view, showing an observer and an object standing on the ground plane. The lower figure is a front view, showing the object from the point of view of the observer. The observer's line of sight to the horizon maintains a constant height above the ground so that in the optic array the horizon necessarily intersects the object at the height that is equal to the height of the point of observation.

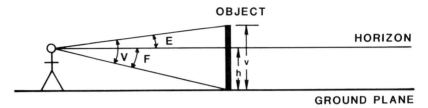

Figure 20. The horizon-ratio relation. The height (v) of every object standing on the ground plane relative to the height (h) of the point of observation is specified by the optic array angles E and F: $v/h = (\tan E + \tan F)/\tan F$. When E and F are fairly small, this relation is closely approximated by the simple horizon-ratio relation: $v/h \approx V/F$. (After Sedgwick [1973] p. 15.)

ples of the horizon-ratio relation. If we assume the height of the point of observation to be 5 feet, then the tree in the drawing is about 15 feet high, i.e., its total height in the projection plane is about three times the portion of it that lies below the horizon. Note that it is not necessary for an object to actually intercept the horizon – the bush is specified as being about half the height of the point of observation, or 2-1/2 feet high. Note also that the horizon-ratio relation specifies size independently of distance; the telephone poles are all at different distances but have the same horizon-ratio relation and consequently the same height – about four times the height of the point of observation, or 20 feet. Finally, it is implicit here that the horizon-ratio relation specifies *relative* height of different objects on the ground independently of the height of the point of observation; whatever is the height of the point of observation, the telephone poles are all the same height, are 1-1/3 times taller than the tree, and are 8 times taller than the bush.

There is also a simple horizon relation that specifies distance, d, along the ground relative to the height of the point of observation. Referring to Figure 22, we can see that

$$d/h = 1/\tan F.$$

This simply means that the visual angle subtended by the constant scale factor is inversely related to its distance. Likewise, the relative distances, d_1 and d_2, of any two locations along the ground are specified by

$$d_1/d_2 = \tan F_2/\tan F_1 \approx F_2/F_1.$$

5.3. Finding the Terrestrial Horizon

Let us now briefly consider the question of locating the terrestrial horizon in the optic array. Clearly, the structures discussed earlier by which the horizon of any plane might be located are applicable in the special case of the terrestrial horizon. What is more specific to the terrestrial horizon, however, is that it may effectively be *explicitly* present in the optic array when the visible portion of the ground plane is very extensive. Figure 23, for example, is a

Figure 21. An illustration of the horizon-ratio relation. The heights of the tree, the bush, and the telephone poles are all specified, relative both to each other and to the height of the point of observation, by their projective relations to the terrestrial horizon. This relation holds independently of distance. (From Sedgwick [1973] p. 11.)

stimulus photograph used in an early size matching study of Gibson's (1950a, pp. 183-186). The boundary of the field where it meets the distant hills is not the true horizon of the ground plane, but this visible horizon can be estimated to lie less than 5 minutes of visual angle away from the true horizon and so to permit good approximations to the horizon-ratio relations of all of the stakes standing on the field (in comparison, the ground texture is scarcely visible at the location of the more distant stake).

The error that results when some visible boundary along the ground is used to approximate the true terrestrial horizon can be calculated in some simple cases. When the ground plane is level out to a boundary, as is very nearly the case in Figure 23, then that boundary forms a visible horizon whose closeness, in the optic array, to the true horizon of the plane depends directly on the boundary's distance from the point of observation. A line of sight to that visible horizon comes gradually closer to the ground rather than running parallel to it (see Figure 24), so that the horizon scale, rather than remaining constant,

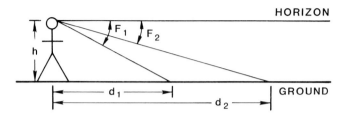

Figure 22. Horizon distance relation. The point of observation is at a perpendicular height h above the ground surface and at distances d_1 and d_2 from two locations on the ground; the optic array angles between these locations and the horizon are F_1 and F_2, respectively. (After Sedgwick [1973] p. 52.)

decreases with distance. Thus, the farther away an object is, the larger it will be according to the horizon-ratio relation. The error will be small, however, as long as the distance of the approximate horizon is large relative to the distance of the object.

Another possibility for specifying the terrestrial horizon is simply that it is always at eye level, and eye level may be specified by various kinds of non-

Figure 23. An approximate terrestrial horizon. The visible boundary of the plowed field in this size-at-a-distance experiment forms a very close approximation in the optic array to the true terrestrial horizon, which is hidden by hills in the distance. (From J. J. Gibson, *The Perception of the Visual World*, pp. 184-185. Copyright © 1950, renewed 1977 by the Houghton-Mifflin Company. Used by permission. This book is now available from Greenwood Press, Westport, CT).

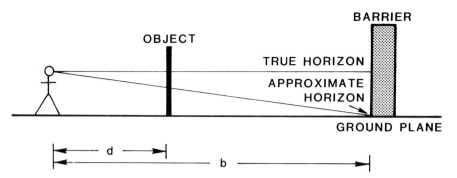

Figure 24. The horizon-ratio relation with an approximate horizon. A line of sight to the visible approximate horizon in this illustration comes closer to the ground plane with increasing distance from the observer. Thus, the horizon-ratio relation specifies sizes that are too large (overconstancy). The farther away an object is, the greater the overconstancy. If v' is the height specified by the horizon-ratio relation and v is the true height, then $v'/v = b/(b-d)$. (After Sedgwick [1973] p. 67.)

visual proprioceptive information. Establishing the terrestrial horizon in this way may be quite useful in some situations, but in humans the error in estimating eye level without visual information can be fairly large – around a couple of degrees (MacDougal, 1903; Matin et al., 1982) – so that in many situations the visible horizon, even if it is only approximate, may give much more precise information about the location of the true horizon than either eye level or the linear perspective of texture gradients.

6. Conclusion

This paper has reviewed some of the information for spatial layout that is made available by texture and perspective. Attention has been restricted to structures in the optic array that are invariant over, and hence do not depend on, motions of the point of observation; thus all the information discussed here would be available to a stationary, monocular observer. Attention has been further restricted to spatial layouts composed of planar surfaces and straight lines. An effort has been made, within this rather narrow domain, to indicate the constraints within which each form of information is valid and to examine some of the relations that exist among the different forms. Nevertheless, this treatment is far from exhaustive, and much more could be said about how these forms of available information and their constraints interact.

The term "available information" is used to refer to optic array/environment relations that exist independently of any particular visual system. Thus the analysis of available information only provides the starting point for investigations of a number of other important questions such as what available information human observers (or other organisms) normally make use of, what available information human observers (or other organisms) can learn to

make use of, and what available information can be made use of most effectively in various types of machine vision systems.

As a step toward considering such questions about utilization, the interplay between types of available information and types of representation of spatial layout was discussed. The concept of an environment-centered representation of spatial layout was introduced and was contrasted with the viewer-
centered representation of spatial layout that has been a guiding concept in
some recent computer vision investigations. Several forms of available information were shown to lead naturally to an environment-centered representation
without having to pass first through a viewer-centered representation. This
leads to the conclusion that we should not assume that a viewer-centered
representation is more easily or more directly attainable by a visual system than
an environment-centered representation nor should we assume that any
environment-centered representation is necessarily derived from a prior viewer-
centered representation. On the contrary it is possible either that viewer-
centered and environment-centered representations are derived in parallel from
the available information or that various types of viewer-centered representations are derived as needed from an environment-centered representation (this
last possibility is close in spirit to the position put forth by Gibson [1950a] in
his discussion of the visual world and the visual field).

Acknowledgment

I am grateful to Barbara Gillam for her helpful comments on this chapter.
The Houghton-Mifflin Company has been unable to locate the photographer
Gendreau concerning permission to use the photograph from Gibson (1950a)
that is reproduced in Figure 2; nevertheless, I am grateful to the Houghton-
Mifflin Company for their efforts in this regard.

References

Attneave, F. Representation of physical space. In A. W. Melton & F. Martin
(Eds.), *Coding Processes in Human Memory.* New York: Wiley, 1972.

Attneave, F., & Frost, R. The determination of perceived tridimensional orientation by minimum criteria. *Perception and Psychophysics,* 1969, **6,**
391-396.

Barrow, H. G., & Tenenbaum, J. M. Recovering intrinsic scene characteristics
from images. In A. R. Hanson & E. M. Riseman (Eds.), *Computer
Vision Systems.* New York: Academic Press, 1978.

Flock, H. R. Some conditions sufficient for accurate monocular perception of
moving surface slants. *Journal of Experimental Psychology,* 1964, **67,**
560-572.

Fodor, J. A., & Pylyshyn, Z. W. How direct is visual perception?: Some reflections on Gibson's "ecological approach." *Cognition,* 1981, **9,** 139-196.

Gibson, J. J. *The Perception of the Visual World.* Boston, MA: Houghton-Mifflin, 1950. (a)

Gibson, J. J. The perception of visual surfaces. *American Journal of Psychology,* 1950, **63,** 367-384. (b)

Gibson, J. J. Ecological optics. *Vision Research,* 1961, **1,** 253-262.

Gibson, J. J. *The Senses Considered as Perceptual Systems.* Boston, MA: Houghton-Mifflin, 1966.

Gibson, J. J. *The Ecological Approach to Visual Perception.* Boston, MA: Houghton-Mifflin, 1979.

Gibson, J. J., & Cornsweet, J. The perceived slant of visual surfaces – optical and geographical. *Journal of Experimental Psychology,* 1952, **44,** 11-15.

Gibson, J. J., & Flock, H. The apparent distance of mountains. *American Journal of Psychology,* 1962, **75,** 501-503.

Gillam, B. False perspectives. *Perception,* 1981, **10,** 313-318.

Hay, J. C. The ghost image: A tool for the analysis of the visual stimulus. In R. B. MacLeod & H. L. Pick, Jr. (Eds.), *Perception: Essays in Honor of James J. Gibson.* Ithaca, NY: Cornell University Press, 1974.

Hochberg, C. B., & Hochberg, J. E. Familiar size and the perception of depth. *Journal of Psychology,* 1952, **34,** 107-114.

Hochberg, J. E. Higher-order stimuli and inter-response coupling in the perception of the visual world. In R. B. MacLeod & H. L. Pick, Jr. (Eds.), *Perception: Essays in Honor of James J. Gibson.* Ithaca, NY: Cornell University Press, 1974.

Hochberg, J. E. On cognition in perception: Perceptual coding and unconscious inference. *Cognition,* 1981, **10,** 127-134.

Hochberg, J. E., & McAlister, E. Relative size vs. familiar size in the perception of represented depth. *American Journal of Psychology,* 1955, **68,** 294-296.

MacDougall, R. The subjective horizon. *Psychological Review Monograph,* Supplement 4, 1903.

Marr, D. Representing visual information. In A. R. Hanson & E. M. Riseman (Eds.), *Computer Vision Systems.* New York: Academic Press, 1978.

Matin, L., Picoult, E., Stevens, J. K., Edwards, M. W., Jr., Young, D., & MacArthur, R. Oculoparalytic illusion: Visual-field dependent spatial mislocalizations by humans partially paralyzed with curare. *Science,* 1982, **216,** 198-201.

Parma, C. C., Hanson, A. R., & Riseman, E. M. Experiments in schema-driven interpretation of a natural scene. In J. C. Simon & R. M. Haralick (Eds.), *Digital Image Processing*. Dordrecht, Holland: Reidel, 1981.

Phillips, R. J. Stationary visual texture and the estimation of slant angle. *Quarterly Journal of Experimental Psychology*, 1970, **22**, 389-397.

Purdy, W. C. The hypothesis of psychophysical correspondence in space perception. General Electric Technical Information Series, R60ELC56, 1960.

Rock, I. In defense of unconscious inference. In W. Epstein (Ed.), *Stability and Constancy in Visual Perception: Mechanisms and Processes*. New York: Wiley, 1977.

Rosinski, R. R., & Farber, J. Compensation for viewing point in the perception of pictured space. In M. A. Hagen (Ed.), *The Perception of Pictures I: Alberti's Window: The Projective Model of Pictorial Information*. New York: Academic Press, 1980.

Sedgwick, H. A. The visible horizon: A potential source of visual information for the perception of size and distance. Doctoral disssertation, Cornell University, 1973.

Sedgwick, H. A. The geometry of spatial layout in pictorial representation. In M. A. Hagen (Ed.), *The Perception of Pictures I: Alberti's Window: The Projective Model of Pictorial Information*. New York: Academic Press, 1980.

Stevens, K. A. Surface perception from local analysis of texture and contour. MIT AI TR-512, 1980.

Stevens, K. A. The information content of texture gradients. *Biological Cybernetics*, 1981, **42**, 95-105.

Turvey, M. T., Shaw, R. E., Reed, E. S., & Mace, W. M. Ecological laws of perceiving and acting: In reply to Fodor and Pylyshyn. *Cognition*, 1981, **9**, 237-304.

Ullman, S. Against direct perception. *Behavioral and Brain Sciences*, 1980, **3**, 373-415.

Ware, W. R. *Modern Perspective* (revised edition). New York: Macmillan, 1900.

Recent Computational Studies in the Interpretation of Structure from Motion

Shimon Ullman

Massachusetts Institute of Technology
Cambridge, Massachusetts

Abstract

The interpretation of structure from motion is the recovery of three-dimensional shape from the transformation induced in the image by objects in motion.

This paper reviews, compares, and extends the results obtained to date in the computational study of the interpretation of structure from motion. Comparisons are also made between perspective versus orthographic, velocity-based versus position-based, and discrete versus continuous interpretation schemes.

In comparing the different schemes, two main issues are discussed. The first is the uniqueness problem: under what conditions a two-dimensional transformation has a unique three-dimensional interpretation. The second is the robustness problem: whether the interpretation be stable under small changes in the input.

With respect to the uniqueness problem, the conditions that are known to guarantee the uniqueness of the solution are summarized. With respect to the robustness problem, it is argued that schemes based on the instantaneous velocity field are unstable when applied locally.

1. Introduction

When objects move in the environment, the images they cast upon our retinas undergo complex transformations. The human visual system can interpret these transformations to recover the three-dimensional (3-D) structure of the viewed objects and their motion in space.

This capacity to interpret structure from motion has been demonstrated in a number of studies (see Braunstein [1976] and Ullman [1979a] for reviews). Its earliest systematic investigation was carried out by Wallach and O'Connell (1953) in the study of what they have termed "the kinetic depth effect." In their experiments, an unfamiliar object was rotated behind a translucent screen, and the shadow cast on the screen by a distant light source was observed from the other side of the screen. In most cases, the viewers were able to describe correctly the hidden object and its motion, even when each static shadow projection of the object was unrecognizable, and contained no three-dimensional information.

The original kinetic depth experiments employed primarily wireframe objects which projected as sets of connected lines. Later studies (e.g. Braunstein, 1962; Green, 1961; Ullman, 1979a) established that 3-D structure can be perceived from displays consisting of unconnected elements in motion, and under both continuous and apparent motion conditions.

Additional demonstrations of motion-based interpretation were provided by the remarkable experiments of Johansson (1973, 1975). These demonstrations were created by filming human actors moving in the dark with small light sources attached to their main joints. Each actor was thus represented by up to 13 moving light dots. The resulting dynamic dot patterns created a vivid three-dimensional impression of the actors and their motion.

In this paper, I shall review the main results obtained to date in the computational study of the interpretation of structure from motion. These studies examine the problem from a theoretical standpoint in an attempt to attain two main goals. The first is what may be called the underlying computational theory of the task (Marr & Poggio, 1977; Marr, 1982). This theory tries to explain how the interpretation can be achieved in principle by any biological visual system or by a man-made device. The second computational goal is to develop and compare different schemes that can actually recover 3-D structure from motion. The study of these schemes, their relative merits and shortcomings, and the comparison of their performance with that of humans, can lead to a better understanding of the interpretation scheme embodied in the human visual system.

In the next section, I shall analyze the computational problem, and describe a scheme for recovering structure from motion. Only a brief outline will be given, since a detailed description can be found in (Ullman, 1979b). Section 3 will review recent alternative schemes and additional computational results obtained to date. Finally, Section 4 will make some comparisons among the main different schemes. In particular, the recovery of structure from continuous velocity fields under orthographic and perspective projections will be examined in some detail.

2. The Rigidity-based Interpretation of Structure from Motion

In this section we shall assume that the motion is given as a sequence of discrete frames, each one depicting a collection of unconnected elements. Figure 1 (taken from Ullman, 1979b) shows an example in which the elements are lying on the surfaces of two coaxial cylinders. The 3-D coordinates of all the dots are stored in a computer's memory, and their projection on the frontal plane is computed and presented on a CRT screen. The imaginary cylinders are then rotated (up to about 10 degrees between frames), and their new projection is computed and displayed on the screen. Each single static view of the cylinders appears as a random collection of dots. However, when the changing projection is viewed in a movie-like fashion, the elements in motion across the screen are perceived as two cylinders whose shapes and angles of rotation are easily determined.

How can this interpretation be achieved? The fundamental underlying problem is the ambiguity of the interpretation: there are many different motion patterns in space that could produce the same two-dimensional motion of the elements on the screen. To resolve this ambiguity, the interpretation scheme must incorporate some additional constraints that would rule out most of the possible interpretations and force a unique solution, which in most cases is also the correct one.

A possible constraint, suggested originally by Wallach and O'Connell (1953), is a rigidity constraint. That is, the preferred interpretation is the one in which the elements move together as a rigid object rather than a collection of elements moving independently in space. The suggested rigidity constraint raises, however, a number of problems. The first is the question of uniqueness: if the same 2-D transformation of the elements can be produced by different 3-D objects, participating in different 3-D movements, then rigidity must be

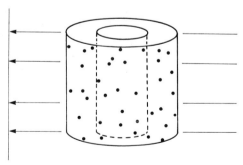

Figure 1. The interpretation of structure from motion. The dots comprising the two cylinders are projected on the screen (the outline of the cylinders is not shown in the actual presentation). The 3-D structure of the cylinders can be recovered from the motion of the dots across the screen. (After Ullman, 1979.)

rendered an insufficient constraint for the 3-D interpretation task. A second problem is the possibility of false targets: if the elements participate in fact in a non-rigid motion in space, and if their 2-D projection happens to have a rigid interpretation, then this false rigid solution will be forced upon the elements. Finally, there is the multiple object problem: if the observed elements belong not to a single rigid object, but to distinct objects moving independently, then the collection as a whole will fail to have a rigid interpretation. A rigidity-based interpretation must, therefore, be applied somehow to relevant sub-collections of the elements, and not to the entire scene at once.

A detailed analysis of these problems can be found in (Ullman, 1979a,b), together with a discussion of possible implications for human motion perception. Here I shall only sketch briefly the main computational results. It has been shown by Ullman and Fremlin (in the "structure from motion" theorem, Ullman, 1979a), that given three distinct views of a moving object, it can be determined unambiguously whether they represent a single rigid object, and if they do, then the 3-D structure can be recovered uniquely. The object is defined here as a collection of identifiable points, and to obtain uniqueness, the object must contain at least four non-coplanar points.

Rigid objects in motion are thus determined uniquely on the basis of information that is local in both space and time. This result provides answers to the problems raised above. Uniqueness of the solution is guaranteed, provided that the moving object contains at least four non-coplanar points. The possibility of false targets is eliminated: it can be shown that the probability that four points that in fact do not belong to a single rigid object will happen to have a rigid interpretation is negligible. Finally, the locality of the interpretation suggests a solution to the multiple object problem. If the interpretation is restricted to local groups of about four elements each, then, due to the contiguity of objects, many of these groups will lie within a single object, and therefore their interpretation will not be affected by the additional objects. Given a scene containing several rigid objects in motion, a correct interpretation can therefore be obtained using the following scheme: divide the image into local groups of about four elements each, test each group for a unique rigid interpretation, and combine the results obtained for the different groups.

3. Alternative Schemes and Additional Results

The analysis outlined in the previous section has been formulated in terms of distinct views, distinct identifiable points, and a parallel projection of the moving objects. In this section I shall examine similar results obtained under somewhat different formulations.

3.1. Perspective vs. Parallel Projection

The structure-from-motion theorem mentioned above assumed parallel, or orthographic, projection. Unlike perspective projection, orthographic projection is formed by parallel light rays that are perpendicular to the image plane. Under such projection the interpretation is unique up to a possible reflection about the image plane. This is an inherent ambiguity, since the projections of a rotating object, and its mirror image rotating about the same axis in the opposite direction, coincide under parallel projection. It is not surprising, therefore, that under orthographic projection, and in the absence of any additional source of 3-D information, human observers experience spontaneous depth-reversals of the objects, accompanied by reversal in the observed direction of rotation.

The natural projection of 3-D objects onto the retina is a perspective rather than parallel projection (i.e., the projection rays are not parallel but meet at a common point). Although it has been demonstrated (e.g., in the original kinetic depth experiments) that structure can be perceptually recovered from the parallel projection of moving objects, it is of interest to analyze the rigidity-based interpretation under perspective projection. Experiments with computer algorithms (some of which are described in Ullman, 1979a) have suggested that seven or eight points in two perspective views are sufficient for a unique interpretation. Six points are sometimes sufficient for a unique interpretation, but not always. Longuet-Higgins (1981) has proposed an elegant algorithm for recovering 3-D structure from two perspective views of eight points. The computation required is particularly simple, involving primarily the solution of eight linear equations. This analysis did not provide, however, a uniqueness proof.

Recently, Tsai and Huang (1981) provided a comprehensive analysis of the uniqueness problem under two perspective views. Their results established that seven points guarantee a unique interpretation, provided that (i) they do not lie on a pair of planes, one of which passes through the origin, and (ii) they do not lie on a single cone containing the origin. This means that except for a few cases where the points happen to form some special configurations in space, the interpretation will be unique. Tsai and Huang provided in addition simple algorithms for the recovery of the motion and structure parameters.

On The Meaning of "Computational Experiments." Those who associate "experiments" primarily with the testing of human subjects may wonder what is meant here by experiments with computer algorithms. The answer lies in the fact that it is often easier to find a solution to a problem (at least an approximate one) than to prove its existence and uniqueness. Suppose, for example, that it is conjectured that the three-dimensional structure of an object can be recovered from its changing projection by solving a certain system of linear equations. The coefficients in these equations will be variables that assume different values for different objects in motion. It may be difficult to show that the system has in general exactly one solution. It is a straightforward procedure, however, to solve the equations for particular examples and verify the solution and its uniqueness. If we test the solution for a variety of examples and consistently recover the correct solution, we have some reason to believe that the interpretation scheme is in general correct. It is still possible, however, that under certain conditions the interpretation scheme will fail. Such conditions may be difficult to discover, and this is one reason why a comprehensive analytic analysis is more satisfactory than computational experiments.

3.2. The Use of Velocity Information

So far, the changing projection of the moving objects has been described in terms of discrete movie-like frames. It should be noted, however, that the formulation of the structure-from-motion theorem in terms of discrete views does not imply that the input image must be discrete rather than continuous. If a continuous motion extends long enough to contain three distinct views (and the qualification for "distinct" will depend on the accuracy of the imaging system), then it contains sufficient information for a unique interpretation. The theorem states this fact without excluding the possibility of implementing the computation using a continuous scheme.

A distinctive property of the scheme outlined in Section 2 is that the information used was expressed entirely in terms of the positions of the elements at different times. An alternative formulation (explored by Clocksin, 1980; Prazdny, 1980) uses the velocities of the points as well as their positions.

The input to the computation consists then of a single perspective view in which the positions of the moving elements and their velocities are specified. This velocity-based formulation can be viewed as the limiting case of two frames, as the time interval between them approaches zero. The uniqueness problem then takes the following form: given the position and velocity of N points in the image, determine whether or not they belong to a single moving object, and find the 3-D structure of the object and its motion in space.

A preliminary theoretical problem is to determine the number N for which this recovery problem has a unique solution. Mathematically, this problem is still unresolved. A counting argument of equations and unknowns shows that at least five points would be necessary. A computer algorithm implemented by Prazdny (1980) suggests that five points might also be sufficient. Since the computer algorithm proved sensitive to errors in the input, especially when the viewed object was small, it seems that a robust recovery algorithm would require more than five points. This problem of robustness is examined again in more detail in Section 4.3.

3.3. The Use of a Continuous Velocity Field

The scheme outlined in Section 2 relied on a small number of discrete elements. An alternative mathematical approach is to assume that the velocity of points in the image is known everywhere within a given region. The information is sometimes assumed to be known only locally, for example, the velocity field and its spatial derivatives are known at a single point. This formulation can be thought of as a limiting case of the discrete formulation, as the distance between points approaches zero.

The most complete analysis to date of this problem has been provided by Longuet-Higgins and Prazdny (1980). Their analysis has established that the velocity field at a point has at most three different interpretations. More precisely, it showed that for nonplanar surfaces, given the velocity field and its first and second spatial derivatives at a point, there are at most three solutions to the surface orientation at that point. The analysis also provided a scheme for computing the solutions. The possible improvement of this result, in particular a determination of whether the solution is in fact unique, poses an open question for future research.

In an analysis of the orthographic continuous velocity field, Hoffman (1980) has shown that 3-D structure can be recovered uniquely (up to the unavoidable reflection about the image plane) if both the velocity and the acceleration fields are known within a region.

3.4. Restricted Motion and the Interpretation of Biological Motion

The discussion so far has examined the general case of unrestricted motion. Additional results have been obtained for situations where certain limitations are imposed upon the motion of the viewed objects. In this section, I shall review the main results, together with their implications for problems in visual perception.

Results of considerable interest were obtained recently by Hoffman and Flinchbaugh (1981, with a contribution by Horn). They have shown that for a rigid rod moving in a plane, its length and orientation in space are determined uniquely by three views (under parallel projection). A similar result was obtained for two rigid rods hinged together end-to-tail. When such a pairwise-rigid configuration moves in a plane, its 3-D structure and motion are uniquely determined by only two parallel projections.

These results offer a powerful tool for the interpretation of biological motion (the kind of interpretation demonstrated by Johansson's experiments). Using 3-D measurements of motion in space, Hoffman has established that the arm and leg motion during locomotion often conforms to the planar motion constraint. Hoffman and Flinchbaugh found that when the planarity-based scheme is applied to data obtained from Johansson-like experiments, a correct interpretation of the 3-D structure and motion of the moving light dots is often obtained. Their results suggest that the human visual system may incorporate processes that are capable of detecting pairs or triplets of feature points engaged in planar motion, and apply to them the planarity-based interpretation scheme. Figure 2 illustrates the application of the planarity-based interpretation scheme

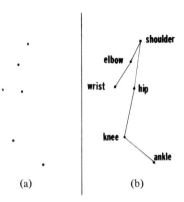

(a)   (b)

Figure 2. The interpretation of biological motion. (a) Six unconnected dots representing parts of the human body in motion. The motion of these dots can give rise to a vivid perception of a moving person (after Johansson, 1973). (b) Links made along the rigid connections. These links and their 3-D structure can be recovered by the planarity-based interpretation scheme. (After Hoffman, 1982.)

to six dots representing the human body in motion. The unconnected dots are shown in Figure 2a and the rigid connection in Figure 2b. These connections and their 3-D structure can be established by the planarity-based scheme for all the rigid links that obey the planarity constraint.

A similar scheme that can cope with somewhat less restricted motion was developed by Webb and Aggarwal (1981). They have considered the interpretation problem for an object assumed to rotate continuously about a fixed axis. (In general, the axis of rotation may change with time.) They have applied this scheme successfully to instances of biological motion, but no mathematical results regarding the information required for a unique interpretation have yet been obtained.

3.5. Vertical Rotation and Horizontal Translation: The Computation of Depth from Stereoscopic Disparity

A recent result by Longuet-Higgins (1982) established uniqueness for the case of an object that rotates about the vertical axis and translates in the horizontal plane. When the motion is restricted in this manner, three points are in general sufficient for a unique interpretation. The significance of this result lies in its possible applicability to the computation of depth from stereoscopic disparity. In the stereoscopic case the object is in fact fixed, and the different views are obtained by the different viewing positions of the two eyes. This problem is formally equivalent to the interpretation of structure from motion, given only two views. The computation of depth from stereoscopic disparities requires, in principle, knowledge of the direction of gaze of the two eyes. Longuet-Higgins' result suggests that this information can be obtained without reliance on non-visual information. Assuming that the horizontal meridians of the two eyes coincide accurately, three points (non-meridional, and not all lying in the vertical plane) are sufficient for the recovery of the two directions of gaze.

The main results regarding the unique interpretation of structure-from-motion discussed so far are summarized in Table 1 for general (upper half) and restricted (lower half) motion.

4. Remarks Concerning the Different Schemes and their Relevance to Perception

The previous section examined the interpretation problem in a number of different formulations. Since a primary goal of these studies is to provide a computational basis for the study of perceptual phenomena, it is worthwhile to compare the relevance of the different schemes to human perception. Three questions will be examined briefly in this section. The first has to do with the applicability of mathematical results and algorithms to biological visual systems, the second has to do with the use of positions vs. velocity fields, and the third with parallel and perspective projections. On the first two of these questions I shall limit myself to a brief discussion of a few selected points.

4.1. Mathematical Algorithms and Biological Visual Systems

The results outlined in the preceding two sections were formulated in terms of mathematical propositions and algorithms. Two difficulties are sometimes raised regarding the applicability of such results to biological visual systems. The first is that, unlike an electronic computer, a biological system cannot be expected to solve the equations used in deriving the mathematical results. The second is that a biological system does not have access to the perfectly accurate data used in the mathematical abstraction.

A comprehensive examination of the first objection would go beyond the immediate goals of this review. The main answer lies, however, in the distinction between different levels of analysis: competence vs. performance (Chomsky, 1965), or computational vs. algorithmic (Marr & Poggio, 1977).

The computational studies aim primarily at establishing principles such as rigidity or planarity that apply to any visual system facing the problem of interpreting structure from motion. Certain equations may be used in the derivation of such principles, but it does not follow, of course, that a system utilizing these principles would have to solve these equations in the process of interpreting structure from motion.

The problem of accuracy in the measurements and computation is an important one. To be of practical value, the interpretation scheme must be robust: small errors in the measurement of position and velocity, for example, should not lead to a complete breakdown of the interpretation scheme. This

Table 1

Uniqueness of the interpretation of structure from motion.
The main results obtained to date are summarized for general
motion (upper half) and restricted motion (lower half).

Unrestricted Motion
Discrete Points & Views
4 points in 3 orthographic views (Ullman, 1979)
7 points in 2 perspective views (Tsai & Huang, 1981)
Discrete Points & Velocities
5 points and their velocities in a single perspective view (Prazdny, 1980)
Velocity Field & its Derivatives
Up to 3 solutions for general motion (Longuet-Higgins & Prazdny, 1980)
Unique solution from velocity and acceleration under orthographic projection (Hoffman, 1980)
Application: The recovery of 3-D structure from unrestricted motion

Restricted Motion
3 orthographic views of 2 points in planar motion 2 orthographic views of 3 points in a "hinged" configuration, planar motion (Hoffman & Flinchbaugh, 1982)
Application: Biological motion
3 non-meridional points, vertical axis and horizontal translation (Longuet-Higgins, 1982)
Application: The recovery of depth from stereo disparities

means that computational studies should not only explore what is possible under idealized conditions, but also examine the effects of small perturbations and errors. An example of such an analysis will be given in Section 4.3 below.

4.2. The Use of Positions vs. Velocity Fields

All of the structure-from-motion schemes examined above used certain measurements as their "inputs," and recovered the 3-D structure and motion space as "output." Different schemes used different inputs; some used the positions (in the image) of the moving elements at different times, while others used their retinal velocities.

As noted in Section 2, the difference between the two formulations is not that velocity-based schemes are continuous and position-based discrete. A position-based computation, for instance, can use the continually changing positions, without necessarily requiring discrete "snapshots," but also without using velocity measurements.

It is not clear at present which formulation, the position-based or the velocity-based, is more directly relevant to human motion perception, since the measurements employed by the human visual system are not fully known. It may prove valuable, therefore, to explore in more detail different schemes that are based on different types of inputs. The comparison of such schemes with the performance of the human visual system could provide some clues regarding the types of measurements employed by the visual system in the interpretation of structure from motion.

4.3. Parallel and Perspective Velocity Fields

The projection of the external environment available to our eyes is perspective rather than parallel. What, then, is the relevance of the parallel projection studies? There are two answers to this question. The first is that humans can recover structure from motion under orthographic projection. The second answer lies in the fact that under local analysis, perspective and parallel projections are nearly identical. (Local analysis means here that the surface patch under analysis is restricted to a small part of the visual field, so that the dimensions of the patch are small compared to its overall distance from the viewer.) If the interpretation scheme is to be robust and insensitive to small errors, it must also be capable of coping with the minor differences between the two projections, and either projection can therefore be assumed. One can, in fact, use the two kinds of projection as a test for stability. If a given interpretation scheme can operate under perspective projection but fails under orthographic projection, it cannot be a stable local interpretation method.

In the remainder of this section, I shall argue that (i) the recovery of 3-D structure from the instantaneous velocity field is impossible under orthographic projection, and (ii) for perspective projection, the recovery is unstable under local interpretation.

4.3.1. The Orthographic Velocity Field

In this section we shall derive a complete characterization of the 3-D information that can be recovered from the instantaneous velocity field under orthographic projection.

Consider two surface patches such as S_1 and S_2 in Figure 3. The figure shows a cross-section of the surfaces from a side view. They are assumed to be rotationally symmetric with respect to rotations around the observer's line of sight, so that S_1 for example is a part of the surface of a sphere. The observer is assumed to view the objects along the y-axis, which is his line of sight, or depth axis; x is the observer's horizontal axis, z the vertical, and the x-z plane is called the image plane. The objects are assumed to be fixed at one point which is taken as the origin of the coordinate system (Q in Figure 3).

Let O_1 and O_2 denote the parallel projections of S_1 and S_2, respectively (on the x-z plane); P_1 and P_2 will denote the perspective projections of S_1 and S_2; $O_i{}'$ and $P_i{}'$ (for $i = 1,2$) will denote the parallel and perspective velocity fields, respectively; $u(x,z)$ will denote the projected velocity in the horizontal direction, and $v(x,z)$ in the vertical.

The surface S_2 in Figure 3 was obtained from S_1 by the following transformation: if (x,y,z) is a point on S_1, then (x,ky,z) is a point of S_2, for a fixed constant k ($k = 5$ in Figure 3). S_1 is assumed to rotate by some angular velocity ω_1 about the vertical axis, and S_2 rotates about the same axis with angular velocity $\omega_2 = \omega_1/k$. This transformation (including the rotations) will be called "depth scaling," since the depth coordinate y is scaled by a constant k.

Since the rotation is about the vertical z-axis, under orthographic projection all the points will travel in the image plane parallel to the x-axis. The

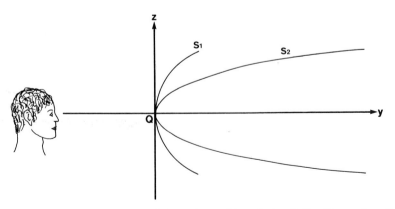

Figure 3. The ambiguity of the orthographic velocity field. The surface S_2 is obtained from S_1 via depth-scaling. Any motion of S_1 can be "mimicked" by S_2 in such a way that their velocity fields will coincide.

projected velocity of a point on S_1 having spatial coordinates (x_1, y_1, z_1) will be $u_1 = -y_1\omega_1$. The corresponding point on S_2 (i.e., the point on S_2 whose projection coincides with that of (x_1, y_1, z_1)) has a projected velocity $u_2 = -y_2\omega_2$. Since by definition, $y_2 = ky_1$, $\omega_2 = \omega_1/k$, it follows that $u_1 = u_2$ for every image point, and therefore the velocity fields of S_1 and S_2 coincide.

In this example, the objects were assumed to rotate about the vertical z-axis, but S_2 can in fact "mimic" S_1 under arbitrary rotation. This claim can be established by noting that any rotation in space can be decomposed into two components: one is a rotation about an axis lying in the frontal x-z plane, the other is a rotation about the line of sight which is perpendicular to the x-z plane. We shall call the first component xz-rotation and the second y-rotation. A general rotation of S_1 can be decomposed therefore into an xz-rotation with angular velocity ω_1 and a y-rotation with angular velocity θ_1. Let S_2 rotate by $\omega_2 = \omega_1/k$ about the same axis in the x-z plane, and assume $\theta_2 = \theta_1$. It is not difficult to see that with this choice of ω_2 and θ_2 the orthographic velocity fields of S_1 and S_2 will coincide.

Unlike the previous case where the two objects had a common rotation axis, in this case (assuming $\omega_1 \neq 0$, $\theta_1 \neq 0$) the two objects would rotate about different spatial axes. We conclude that under orthographic projection, two objects can differ drastically in their shapes (e.g., in terms of their surface orientation and curvatures), axes of rotation, and rotation speeds, and yet induce identical velocity fields.

Since the constant k used in the definition of S_2 can be chosen arbitrarily, the velocity fields admit not only two distinct interpretations, but an infinite family of surfaces, for depth scaling by different factors k. (The definition of depth scaling includes the appropriate relation between the rotation components, $\omega_2 = \omega_1/k$, $\theta_2 = \theta_1$. It is also assumed that $\omega_1 \neq 0$.) It can further be shown that this family completely characterizes the set of confusable objects. We can summarize these claims in the following proposition:

The depth scaling proposition

> If a non-planar surface S_1 is a possible rigid interpretation for a given orthographic velocity field, then S_2 is another possible interpretation if and only if it is obtained from S_1 via depth scaling.

The proof of this proposition is given in the Appendix. It serves to give a complete characterization of all the possible interpretations of the orthographic velocity field. Its first implication is that this interpretation is always non-unique. The second implication is that although the interpretation is non-unique, properties that are invariant under depth scaling can be recovered from the orthographic velocity field. For example, the depth ratio y_i/y_j for two points with image coordinates (x_i, z_i) and (x_j, z_j) can be recovered uniquely from the velocity field; extremal points in y and inflection points can also be recovered. As a special case, the Appendix establishes that if the object is planar (and assuming $\omega \neq 0$), then it would be possible to determine its planarity from the velocity field. The orientation of the plane, however, would be impossible to recover.

As it turns out, the depth-scaling ambiguity of the solution can be resolved with the addition of a single view. More specifically, if the projected positions and velocities of five points in a general configuration (i.e., no four of which are coplanar) are given at time t_1, and the positions of the same five points are given at a later time t_2, then the 3-D structure can be recovered uniquely (up to the unavoidable reflection ambiguity about the image plane). Since the emphasis in this section is on the instantaneous velocity field, the proof of this claim will be omitted.

In summary, under orthographic projection, the interpretation of the velocity field is non-unique. For non-planar objects, if S_1 is a possible interpretation, then S_2 is another possible interpretation if and only if it is related to S_1 via depth scaling. Only properties that are invariant under depth scaling can therefore be recovered from the orthographic velocity field. For planar objects, there are exactly two distinct interpretations up to depth scaling. The structure can be recovered uniquely for as few as five points in a general configuration if a single view is given in addition to the velocity field. A more detailed analysis, including the planar case, is given in the appendix.

4.3.2. The Perspective Velocity Field

The perspective velocity field is in a sense richer in information than the orthographic field. While the orthographic field admits infinitely many interpretations, in the perspective case the velocity field, even in an arbitrarily small neighborhood, can have no more than three interpretations (Longuet-Higgins & Prazdny, 1980). For a sufficiently small patch of surface, perspective and orthographic projections are not very different, however, and one may suspect therefore that the local interpretation of perspective velocity fields is unstable. The argument is roughly as follows. Under orthographic projection, two widely different objects S_1 and S_2 can have similar, and even identical velocity fields, O_1' and O_2', respectively. For a sufficiently large viewing distance, the perspective velocity fields P_1' and P_2' become similar to O_1' and O_2' respectively. Consequently, P_1' and P_2' are also closely similar, and therefore slight errors in the measured velocity fields can have large effects on the interpreted 3-D shape.

This argument can be made more precise. Let S_1 and S_2 in Figure 4a be two surfaces rotating about the z-axis. As before, the direction of gaze is along the y-axis. S_2 is derived from S_1 in this case by extending the ray from the viewing point to each point on S_1 so that $y_2/y_1 = k$ for a given constant k. The surface S_2 depends in this case on the viewing point: when the viewing distance increases, the shape of S_2 in the perspective case approaches the ortho-

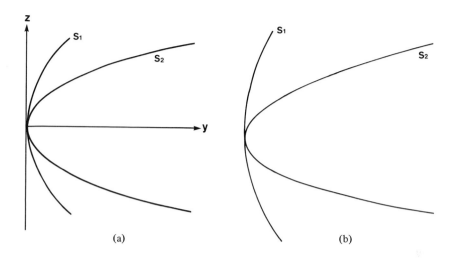

Figure 4. The instability of the local interpretation of perceptive velocity fields. Under local analysis, small errors in the measured velocity field can induce large errors in the interpreted 3-D structure. The difference in the velocity fields induced by S_1 and S_2 in (a) will be less than 3.2% when the surface patches occupy one degree of visual angle, and about 6% at twice this size. The maximum difference in (b) is about 5% when the objects occupy one degree of visual angle and 10% at twice this size.

graphic depth scaling of S_1 by the constant k. (In Figure 4 it was assumed that the ratio of viewing distance to object size is such that the object occupies one degree of visual angle.) As before, we will choose $\omega_2 = \omega_1/k$. Finally, let H denote the distance of the origin Q from the observer.

The projected velocity of p_1 (a point on S_1 with $y \neq 0$) is (u_1, v_1). This velocity will be measured as an angular velocity, i.e., the number of deg/sec a point travels in the observer's field of view. The angular velocity of the corresponding point on S_2 is (u_2, v_2). The claim is that when H is sufficiently large, the vectors (u_1, v_1) and (u_2, v_2) become arbitrarily close. That is, as H grows, the ratio between their magnitudes $\sqrt{u_1^2 + v_1^2}/\sqrt{u_2^2 + v_2^2}$ approaches 1, and their difference in direction, measured by $(v_1/u_1) - (v_2/u_2)$, approaches 0. (Assuming $y \neq 0$ in the region, except at the origin.)

The proof of this claim is straightforward and will therefore not be detailed. It can be derived from the expressions given below for the speed ratio and the direction difference.

The speed ratio (squared) is:

$$\frac{u_2^2 + v_2^2}{u_1^2 + v_1^2} = \frac{[(H + ky_1)y_1 + x_2^2/k]^2 + z_2^2 x_2^2/k^2}{[(H + y_1)y_1 + x_1^2]^2 + z_1^2 x_1^2} \frac{(H + y_1)^4}{(H + ky_1)^4}$$

where

$$x_2 = x_1 \frac{H+ky}{H+y} \qquad z_2 = z_1 \frac{H+ky}{H+y}$$

The difference in direction is:

$$\frac{v_2}{u_2} - \frac{v_1}{u_1} = \frac{z_2 x_2/k}{(H+ky_1)y_1 + x_2^2/k} - \frac{z_1 x_1}{(H+y_1)y_1 + x_1^2}$$

It can be seen from these expressions that for any point with $y_1 \neq 0$ the speed ratio approaches 1 and the difference in directions approaches 0 as H grows. Under certain conditions, which will not be elaborated, this will also happen uniformly within a region.

As a result, it is possible to construct drastically different objects whose velocity fields within a region are almost identical. That is, the differences in speed and direction at any given point within the region can be made arbitrarily small. As explained in the analysis of the orthographic case, this ambiguity is not restricted to rotation about the vertical axis, but can arise for arbitrary rotations as well.

The surfaces in Figure 4 illustrate that this problem can be quite severe. When the viewing distance is such that the surfaces in Figure 4a occupy one degree of visual angle, the differences in their perspective velocity fields within the entire one degree patch will not exceed 3.2%. At half the viewing distance, the maximum error is about 6%. For the surfaces in Figure 4b, the maximum difference is about 5% when the object occupies one degree of visual angle, and about 10% at twice this size.

5. Concluding Remarks

The comparisons discussed in the last sections can be combined with psychological experiments to gain further insight into the possible use of instantaneous velocity fields in the recovery of structure from motion by the human visual system. One can test, for example, whether shape parameters, such as surface orientation and curvature, can be recovered by the visual system for objects subtending one or two degrees of visual angle. Even a moderate success in this task would suggest that either our visual system measures velocities with high precision, or that the interpretation under these conditions does not rely on the instantaneous velocity field. The first of these alternatives does not seem attractive. Computational studies have indicated that the measurement of the velocity field is in general a difficult task (see e.g., Marr & Ullman 1981; Ullman, 1981), and it is probably unrealistic to expect these measurements to reach the level of precision required for the interpretation task. The ability to interpret correctly the 3-D structure of small objects can therefore provide evidence against the use of the instantaneous velocity field in this task. One may expect instead to find in the visual system the capacity to integrate information over time periods that allow sufficient excursion of the moving object, rather than to base the interpretation on instantaneous velocity measurements.

Acknowledgment

I thank E. Hildreth for her help and comments. This report describes research done at the Artificial Intelligence Laboratory of the Massachusetts Institute of Technology. Support for the laboratory's artificial intelligence research is provided in part by the Advanced Research Project Agency of the Department of Defense under Office of Naval Research contract N00014-75-C-0643 and in part by National Science Foundation Grant MCS77-07569.

Appendix

This appendix uses the same notation as Section 4 above. As before, S_1 and S_2 are two rotating rigid surfaces. The rotation of S_1 can be decomposed into a component with angular velocity ω_1 (assumed to be non-zero) about an axis in the frontal x-z plane (xz-rotation), and a second component (y-rotation) with angular velocity θ_1 about the line of sight y. The corresponding components for S_2 are ω_2 and θ_2. S_2 will be called a *depth scaling* of S_1 if:

1. For every point (x, y, z) on S_1, (x, ky, z) is a point on S_2 for some constant $k \neq 0$.

2. The rotations ω_1 and ω_2 are around the same axis, and $\omega_2 = \omega_1/k$.

3. $\theta_1 = \theta_2$.

The depth scaling proposition

If a non-planar surface S_1 is a possible rigid interpretation for a given orthographic velocity field, then S_2 is another possible interpretation if and only if it is obtained from S_1 via depth scaling.

Proof: We have seen in Section 4.3 that if S_2 is obtained from S_1 via depth scaling then their orthographic velocity fields coincide. The converse statement that remains to be shown is that if S_1 and S_2 have identical velocity fields, then they are related via depth scaling.

This property clearly holds in the vertical rotation case where S_1 and S_2 both rotate about the z-axis. If the angular velocities of S_1 and S_2 about the vertical axis are ω_1 and ω_2 respectively, then the projected velocities at point (x, z) in the image are $-\omega_1 y_1$ and $-\omega_2 y_2$ respectively. This implies that for the velocity fields to coincide the depth ratio y_2/y_1 at any given point in the image must equal ω_1/ω_2.

To show that the proposition holds under general rotation we shall establish the following claim: if two rotating (non-planar) objects S_1 and S_2 have the same orthographic velocity field then $\theta_1 = \theta_2$ (where θ denotes, as before, the rotation component about the line of sight).

Let θ_1 be the y-rotation of the first object. By rotating the velocity field induced by S_1 by $-\theta_1$ about the line of sight, the y-rotation component of S_1 is cancelled, and the only component that remains is the xz-rotation about some axis in the x-z plane. The resulting velocity field will have the property that all the velocity vectors will now be parallel to each other. We shall next show that $-\theta_1$ is the only y-rotation which, when added to the velocity fields, results in a parallel velocity field. The implication will be that given a velocity field, the y-rotation component is uniquely determined and hence $\theta_1 = \theta_2$.

Without loss of generality we can assume that in the resulting parallel velocity field all the projected velocity vectors are in the direction of the x-axis. This means that the image velocity component in the z direction $v(x,z) = 0$ at every image point (x,z), and the velocity field $(u(x,z), v(x,z))$ can therefore be described as $(u(x,z), 0)$. Let us now add to this field a y-rotation with some angular speed θ. The combined velocity field will now be $(u(x,z)-z\theta, x\theta)$. This will again be a parallel velocity field only if either

$$u(x,z)-z\theta = 0$$

or

$$\frac{x\theta}{u(x,z)-z\theta} = c$$

for some constant c. In either case it can be readily verified that the velocity field $u(x,z)$ must have been induced by a planar surface. In the second case, for example, $u(x,z) = (\theta x/c)+\theta z$ and the original rotation of S_1 was assumed to be about the z-axis. For rotation about the z-axis with angular velocity ω the velocity field is given by

$$u(x,z) = -\omega y(x,z)$$

It follows that

$$y = -\frac{\theta}{\omega c}x - \frac{\theta}{\omega}z$$

which is the equation of a plane.

We conclude that for a non-planar object, starting from a parallel velocity field, a y-rotation by any amount will destroy the parallelism of the field. It follows that given the velocity field of S_1 (which is also the velocity field of S_2), there is one and only one y-rotation, by $-\theta_1$, that will create a parallel velocity field. This rotation is defined by the velocity field itself, independent of the inducing object, and hence $\theta_1 = \theta_2$.

We can now add to both S_1 and S_2 a rotation component $-\theta_1$ that will cancel rotation about the line of sight. Their velocity fields will still coincide, but now both objects rotate about a common axis in the image plane. Without loss of generality, this axis can be labeled the z-axis, and the general rotation can thereby be reduced to the case of vertical rotation.

The velocity fields of planar surfaces

For a planar surface, when the velocity field is parallel to the x-axis, it will have the following form:

$$u(x,z) = \alpha x + \beta z$$

That is, the field depends linearly on x and z. This velocity field is illustrated in Figure 5. Two special lines are marked in this figure: line a (coinciding in this case with the z-axis), which is perpendicular to the field direction, and line b (the dotted line in Figure 5) along which the field is nullified.

In describing the 3-D interpretation of this field we shall make use of the following definition. A plane that is not parallel to x-z must intersect it along a straight line, which will be called the "tilt line" of the plane. We then have

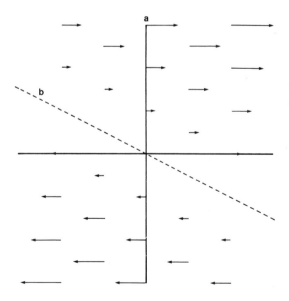

Figure 5. The velocity field of a planar surface. Line a is perpendicular to the field direction, and the field is nullified along b. This field has two different interpretations. In one a is the rotation axis and b is the tilt line. In the second interpretation the roles are switched: a is the tilt line, and b is the projection of the rotation axis.

the following proposition:

> There are exactly two planar interpretations of the velocity field in Figure 5. In one interpretation a is the axis of rotation and b is the tilt line. In the second interpretation the roles are switched: a is the tilt line, and b is the projection of the rotation axis (the axis itself lies on the planar surface). Each of these interpretations is determined as before up to depth scaling.

The proof of this proposition is rather straightforward and will therefore be sketched briefly. Let $(\Omega_x, \Omega_y, \Omega_z)$ be the angular velocity vector of the rotating object. The velocity of a point (x, y, z) is given by the vector product $(\Omega_x, \Omega_y, \Omega_z) \times (x, y, z)$. The velocity field is required to have the form $u(x, z) = \alpha x + \beta z$, $v(x, z) = 0$. Therefore

$$\Omega_y z - \Omega_z y = \alpha x + \beta z$$

$$\Omega_x y - \Omega_y x = 0$$

One solution to these equations arises when $\Omega_y = 0$. This implies that $\Omega_x = 0$ (if y is not identically 0), and $y = -(\alpha x + \beta z)/\Omega_z$. This solution corresponds to a plane whose tilt line is b, rotating about the vertical axis.

If $\Omega_y \neq 0$ then $\Omega_x \neq 0$ and $y = (\Omega_z/\Omega_x)x$. The surface must therefore be a plane with its tilt line coinciding with the z-axis.

In conclusion, the following statements summarize the analysis of the information content of the orthographic velocity field, for both planar and non-planar objects. Given the orthographic velocity field:

1. It is possible to determine whether the inducing object is planar.

2. If it is non-planar, then the interpretation is determined exactly up to depth scaling.

3. If it is planar, then there are two distinct solutions up to depth scaling.

References

Braunstein, M. L. Depth perception in rotation dot patterns: Effects of numerosity and perspective. *Journal of Experimental Psychology,* 1962, **64**, 415-420.

Braunstein, M. L. *Depth Perception Through Motion.* New York: Academic Press, 1976.

Chomsky, N. *Aspects of the Theory of Syntax.* Cambridge, MA: MIT Press, 1965.

Clocksin, W. F. Perception of surface slant and edge labels from optical flow: A computational approach. *Perception*, 1980, **9**, 253-269.

Green, B. F. Figure coherence in the kinetic depth effect. *Journal of Experimental Psychology*, 1968, **62**, 272-282.

Hoffman, D. D. Inferring shape from motion fields. MIT AI Memo 592, 1980.

Hoffman, D. D., & Flinchbaugh, B. E. The interpretation of biological motion. *Biological Cybernetics*, 1982, **42**, 195-204.

Johansson, G. Visual perception of biological motion and a model for its analysis. *Perception and Psychophysics*, 1973, **14**, 201-211.

Johansson, G. Visual motion perception. *Scientific American*, 1975, **232** (6), 76-88.

Longuet-Higgins, H. C. A computer algorithm for reconstructing a scene from two projections. *Nature*, 1981, **293**, 133-135.

Longuet-Higgins, H. C. The role of the vertical dimension in stereoscopic vision. *Perception*, 1982, in press.

Longuet-Higgins, H. C., & Prazdny, K. The interpretation of a moving retinal image. *Proceedings of the Royal Society, London*, 1980, **B208**, 385-397.

Marr, D. *Vision*. San Francisco, CA: Freeman, 1982.

Marr, D., & Poggio, T. From understanding computation to understanding neural circuitry. In E. Poppel et al. (Eds.), *Neural Mechanisms in Visual Perception. Neuroscience Research Program Bulletin*, 1977, **15**, 470-488.

Marr, D., & Ullman, S. Directional selectivity and its use in early visual processing. *Proceedings of the Royal Society, London*, 1981, **B211**, 151-180.

Prazdny, K. Egomotion and relative depth map from optical flow. *Biological Cybernetics*, 1980, **36**, 87-102.

Tsai, R. Y., & Huang, T. S. Uniqueness and estimation of three-dimensional motion parameters of rigid objects with curved surfaces. University of Illinois at Urbana-Champaign, Coordinated Science Laboratory Report R-921, 1981.

Ullman, S. *The Interpretation of Visual Motion*. Cambridge, MA: MIT Press, 1979. (a)

Ullman, S. The interpretation of structure from motion. *Proceedings of the Royal Society, London*, 1979, **B203**, 405-426. (b)

Ullman, S. Analysis of visual motion by biological and computer systems. *Computer*, 1981, **14** (8), 57-69.

Wallach, H., & O'Connell, D. N. The kinetic depth effect. *Journal of Experimental Psychology,* 1953, **45**, 205-217.

Webb, J. A., & Aggarwal, J. K. Visually interpreting the motion of objects in space. *Computer,* 1981, **14** (8), 40-46.

On the Role of Structure in Vision

Andrew P. Witkin
Jay M. Tenenbaum

Fairchild Camera and Instrument Corporation
Palo Alto, California

Abstract

People are able to perceive structure in images, apart from the perception of tri-dimensionality, and apart from the recognition of familiar objects. We impose organization on data (noticing flow fields, regularity, repetition, etc.) even when we have no idea what it is we are organizing. What is remarkable is the degree to which such naively perceived structure survives more or less intact once a semantic context is established: the naive observer often sees essentially the same things an expert does, the difference between naive and informed perception often amounting to little more than labeling the perceptual primitives. It is almost as if the visual system has some basis for guessing *what* is important without knowing *why*. Our objective in this paper is to understand the role of primitive structure in visual perception. What does this level of organization *mean?* What, if anything, is it good for? If it is useful, how do we get it and, once gotten, use it? We will argue that perceptual organization is a primitive level of inference, the basis for which lies in the relation between structural and causal unity: the appearance of spatiotemporal coherence or regularity is so unlikely to arise by the chance interaction of independent entities that such regular structure, when observed, almost certainly denotes some underlying unified cause or process. This view will be shown to have broad implications for computational theories of vision, providing a unifying framework for many current techniques of early and intermediate vision, and enabling a style of interpretation more in keeping with the qualitative and holistic character of human vision.

1. Introduction

In recent years, research on human and machine vision has been brought together by an emerging view of vision as a computational process whose underlying principles apply equally to natural and artificial systems. This view has already contributed significantly both to our understanding of human vision and to our ability to build better machine vision systems.

A great deal of research has been predicated on a hierarchic computational framework, proceeding through a series of stages from the raw image to high-level symbolic descriptions. A keystone of this model is an intermediate description of local surface characteristics in the form of numeric arrays known as intrinsic images (Barrow & Tenenbaum, 1978) or as the 2-1/2 D sketch (Marr, 1978). It has been widely believed that such a description is the proper goal for early and intermediate vision, that it can be recovered using arrays of local processes, and that it greatly facilitates higher level description into surfaces, volumes and objects.

It is too early to know how far this approach will take us, either as a successful design or as a model for biological perception. We do have questions about the feasibility of recovering local quantitative surface descriptions reliably, and about the utility of those descriptions once obtained, More fundamentally, though, we are concerned that the current emphasis on recovering local quantitative surface descriptions is too narrow.

Almost entirely neglected has been an area that lies at the very heart of human perception: the ability to impose organization on sensory data – to discover regularity, coherence, continuity, etc., on many levels. We feel that this capability is more important in its own right than recovering numeric depth arrays, and moreover we doubt that three-dimensional structure can be recovered without it.

People's ability to perceive structure in images exists apart from both the perception of tri-dimensionality and from the recognition of familiar objects. That is, we organize the data even when we have no idea what it is we are organizing. What is remarkable is the degree to which such naively perceived structure survives more or less intact once a semantic context is established: the naive observer often sees essentially the same things an expert does, the difference between naive and informed perception amounting to little more than labeling the perceptual primitives. It is almost as if the visual system has some basis for guessing *what* is important without knowing *why*.

Regardless of the process by which structure is perceived, and of the principles underlying that process, the close relation between naive and informed perception makes the utility of this capability quite clear: by anticipating semantically important relationships, it provides a basis for interpretation in terms of stored models, and for learning new models. For this reason alone, we regard understanding perceptual organization as a prime objective for computational vision.

Our aim in this paper is to explore the role of structure in perception. We will begin by reviewing the prevailing computational model of vision: a hierarchy of representations built around a quantitative description of local surface characteristics. The strengths and weaknesses of this model will be identified. We will then pursue a tentative and frankly speculative investigation into the functional role of perceptual organization, into some of the principles that appear to underlie its ability to serve that role, and into some of the means by which those principles are carried into practice.

We will argue that the aim of perceptual organization is the discovery and description of spatio-temporal coherence and regularity. Because regular structural relationships are extremely unlikely to arise by the chance configuration of independent elements, such structure, when observed, almost certainly denotes some underlying unified cause or process. A description that decomposes the image into constituents that capture regularity or coherence therefore provides descriptive chunks that act as "semantic precursors," in the sense that they deserve or demand explanation. The implications of this argument for computational vision will then be explored. Perceptual organization will be shown to contribute centrally to recognition, to three-dimensional interpretation, to early description, and to induction and inference.

2. Background

2.1. A Computational Model

Recent research in computational vision has been pursued largely within the general framework of models like that sketched in Figure 1 (Marr, 1982; Barrow & Tenenbaum, 1981b). In this prevailing view, a vision system consists of a succession of levels of representations. The initial levels are constrained by what is possible to compute directly from an image, while higher levels are dictated by the information required to support goals (such as navigation or object recognition). In between, the order of representations is constrained by what information is available from preceding levels and what is required by succeeding ones. Processing is primarily data-driven (i.e., bottom-up), though it can be responsive to goals and expectations at the higher levels.

The initial level is an explicit description, in local terms, of significant spatial and temporal intensity changes in the image (corresponding to edge and line elements, blobs, flow vectors, and so forth), giving their type, position, orientation, and other relevant parameters, such as width and contrast. Such information is typically obtained by convolving the image with local operators at a variety of scales and suitably combining the results obtained at each point (Marr & Hildreth, 1980).

The information about intensity changes is conveniently represented in the form of an array of feature descriptors that preserves the local two-dimensional geometry of the image. Marr called this the "raw primal sketch." He also envisaged a full primal sketch that would include information about more global structure, such as curvilinear organization and groupings, but this idea has not been seriously pursued.

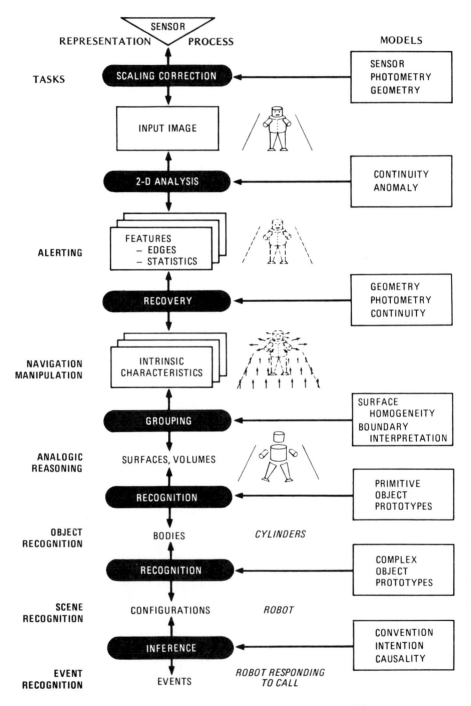

Figure 1. Computational architecture for a general-purpose vision system.

Figure 2. Distance and reflectance images obtained using a laser range finder.

The next level describes local properties of visible surfaces in a viewer-centered coordinate frame. Included are properties directly related to geometric structure, such as distance and orientation (relative to the viewer), as well as photometric properties such as reflectance (i.e., albedo), specularity, luminosity, and transparency. The properties are recovered by numerous local processes operating on information in the original image and the primal sketch, for example stereopsis, shading, texture, contour and flow (for a survey, see Barrow & Tenenbaum, 1981b; Marr, 1982).

The recovered local surface properties are placed in a set of numeric arrays in registration with the image array. Each array contains values for a particular characteristic of the surface element visible at corresponding points in the sensed image. Each array also contains explicit indication of local boundary elements due to discontinuities in value or gradient. Such arrays have been called ''intrinsic images.'' The distance and orientation images together correspond to Marr's concept of a 2-1/2 D sketch.

Figure 2 shows a pair of simulated intrinsic images from a simple office scene (obtained by direct physical measurement using a scanning laser range finder.) The left image is a range map, with distance encoded by brightness (brighter points being nearer). The right image is a reflectance (albedo) map. Note that aside from a small amount of cross-talk, the range data are uncorrupted by reflectance variations and the reflectance data are unaffected by ambient lighting and shadows. This decomposition, in principle, greatly facilitates higher levels of processing.

The third level represents elementary three-dimensional structures in the scene, such as surface patches and compact volumes. Each is described in terms of natural axes of symmetry to which surface and volume primitives are attached. This description captures global properties (e.g, shape), in a viewpoint-independent coordinate frame, as required for tasks such as navigation and object recognition (Marr, 1978).

Surface primitives are patches over which range and orientation vary continuously. These patches can be recovered by simple iconic grouping processes, operating on the intrinsic images (Duda et al., 1979; Fischler & Barrett, 1980). Volume primitives are structural prototypes such as ellipsoids or generalized cylinders (a cylinder whose axis can be an arbitrary space curve, and whose cross-sectional shape can vary continuously along the axis [Binford, 1971]). Clustering and parameter estimation techniques have been developed for fitting generalized cylinders and ellipsoids to data in a distance array (Agin & Binford, 1976; Nevatia & Binford, 1977; Gennery, 1977).

Generalized cylinders, illustrated in Figure 3, have considerable intuitive appeal for describing objects with well-defined axes, including polyhedra, prisms, solids of revolution, and many natural objects (Figure 4). Marr and Nishihara (1977) extended the representation to include the important concept of description at multiple scales. Thus, generalized cylinders can be used to represent both a detailed description of part of an object, and a gross description of the whole object (Figure 5).

Given a description of a scene in terms of surface and volume primitives, the next step is to recognize instances of objects similarly described. Precisely what constitutes an object depends largely upon experience and convention, since boundaries between objects are often visually indistinguishable from boundaries between surfaces of a single object. (Contrast a cup and saucer with a cup and its handle, for example.) Thus while processing up through extraction of elementary surfaces and volumes is primarily data-driven, iconic, and domain-independent, higher-level processing tends to be goal-driven, symbolic, and governed by semantic models and the expectations of the perceiver.

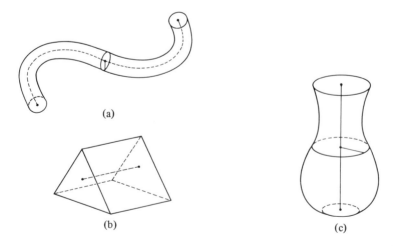

(a)

(b)   (c)

Figure 3. Solids representable by generalized cylinders (axis and cross-section.)

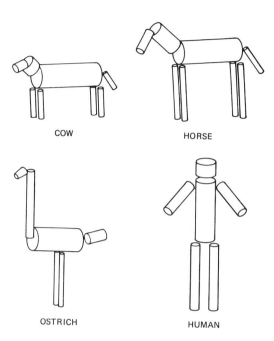

Figure 4. Simplified cylinder models of animals. (From Marr & Nishihara, 1977.)

Objects are modeled by three-dimensional configurations of surface and volume primitives having specified attributes (e.g. color, size) and relationships (e.g. attached to, above). Recognition involves using attributes of the scene primitives to select a small set of possible object models. Each such hypothesis is then verified by establishing a detailed correspondence between model and scene primitives, and testing whether the attributes and relationships are consistent. In the course of verification, it becomes clear which scene fragments should be grouped into a larger whole (Marr & Nishihara, 1977; Brooks, 1981).

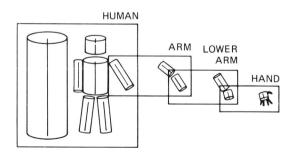

Figure 5. A hierarchical cylinder model.

At this point, the boundary between vision and cognition becomes blurred, and our ideas on architecture become less certain. Presumably, objects are grouped into yet higher-level configurations, comprising scenes and events, and arbitrarily complex reasoning processes may be required.

2.2. Commentary

Within the framework provided by the above model, the major focus of research during the past five years has been on the recovery of local quantitative surface properties, such as depth and surface orientation. This level of representation marks a critical transition from pictorial information to information about the scene itself. It is seen as the natural goal for early and intermediate vision: a description that can be obtained by local, data-driven processes and that greatly facilitates subsequent, higher-level organization into surfaces, volumes and objects.

Historically, this emphasis arose from widespread frustration with the segment-and-label paradigm of early machine vision research. Direct segmentation into namable objects was seen as hopeless because of the enormous gulf that separated primitive image features from models of objects. An early expression of this realization appeared in a paper by Zucker et al. (1975), but they were somewhat vague as to the nature of the intermediate representations and models that were needed to bridge the gap. It was later argued that the segmentation-and-label problem was not only hopeless, but ill-defined, because what constitutes an object depends heavily on task and context (Marr, 1982).

Against this backdrop, a line of research initiated by Horn's work on shape from shading (Horn, 1975) was to many people a revelation. First of all, Horn set forth a concrete and tangible intermediate objective: the recovery of surface orientation at each point in the image. Second, he expressed the relation between this desired quantity and the measured intensity in the image using a precise model for the imaging process. It was then clear that recovering shape from shading quantitatively required that sufficient additional constraints be applied to solve the resulting differential equations. What was refreshing about this approach was above all the crisply defined objective, and the ability to express the problem in terms of precise physical models. It was gradually realized that Horn's framework could be generalized to the recovery of a variety of scene parameters under a variety of physical models and constraints (e.g. shape and depth from stereo, motion, texture, contour, etc.)

A number of specific modules of the above type have now been implemented, each recovering a particular surface characteristic using a particular information source (e.g., orientation from shading). These modules have three key properties in common.

First, they use quantitative models of image photometry and geometry to express the relation between measurements at a point in the image and the properties of the corresponding surface point. These relations are underdetermined: photometrically, the light intensity at each point in an image can result from an infinitude of combinations of illumination, reflectance, and orientation at the corresponding scene point. Geometrically, the distance at each point in the image is lost in the projection from the three-dimensional world, resulting in the kinds of ambiguities depicted in Figure 6.

Second, since these point-wise relations are inherently underdetermined by the image data, additional constraining assumptions are required. Two types of assumptions have been widely used: artificial domain restrictions (e.g. assumptions of planarity, point illumination, constant albedo, etc.); and more generally, applicable idealizations of the natural world (e.g. rigidity, continuity, isotropy, etc.), the latter typically justified by appeal to the basic cohesiveness of matter over space and time. These assumptions are used to tie together surface properties at neighboring points, in effect resolving the ambiguities at each point by solving the local imaging equations over a neighborhood simultaneously. Thus, while the imaging process constrains the three-dimensional scene location of each image point to lie on a ray in space, the continuity of surfaces in three-dimensional scenes further constrains distance and orientation values to be continuous over most of the image (except at edges, corresponding to surface boundaries). Similarly, the rigidity of most objects constrains values to vary smoothly over time.

In addition to coupling assumptions like continuity and rigidity, boundary conditions are also required. Edges are an important source of such boundary conditions. The pattern of brightness variation on both sides of an edge pro-

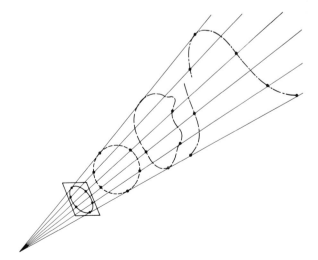

Figure 6. The ambiguity of projection: an infinity of space curves project into the same image curve.

vides strong clues as to the type of scene event responsible and thus to which scene characteristics are actually discontinuous at that edge. For example, across an edge resulting from a shadow boundary, hue is approximately constant, texture elements are continuous, and there is usually high contrast. When these conditions are observed, it can be assumed that only illumination is discontinuous, and the ratio of incident illumination across the edge can then be determined. Sometimes, the interpretation of brightness edges as scene events allows explicit values to be determined for some characteristics. For example, surface orientation is uniquely determined everywhere along one side of an extremal occluding boundary (where a surface curves smoothly out of view) (Barrow & Tenenbaum, 1978). Given suitable boundary conditions and integrating constraints, the value at each point can, in principle, be uniquely determined.

Third, the techniques were implemented (or at least conceptualized) in terms of locally interacting retinotopic networks, in part reflecting the local character of the models and constraints, and in part reflecting a desire to produce implementations consistent with the architecture of the visual cortex. These networks solved the coupled systems of local image equations by iteration, propagating values in from the boundary conditions established along edges. The solution process is analogous to numerical relaxation techniques long used in the solution of classic boundary value problems (e.g, determining the distribution of temperature or potential over a region).

On the whole, the performance and generality of such recovery techniques has been unimpressive. Those techniques that rely on weak, general assumptions such as isotropy have proved fragile and error-prone; while such assumptions may be frequently valid, they also tend to be violated fairly often. Those that rely on artificial domain restrictions (e.g., smooth Lambertian surfaces, uniform albedo, point source illumination) clearly do not apply in complex natural scenes such as Figure 7. In such scenes, surfaces have complex reflectance functions, often with strong directionalities (e.g., the surface froth of cascading water). Illumination patterns are equally complex, encompassing such phenomena as shading gradients from nearby or extended sources, and secondary illumination by light reflected from nearby surfaces (e.g., sunlight filtering through trees). The combined effect is that parameters frequently change in fairly arbitrary ways from point to point, making any local continuity-type assumptions quite tenuous.

Thus, we currently have a number of modules each of which imparts a grain of truth, but none of which is particularly reliable by itself. It is conceivable that the outputs of the various modules can be integrated in such a way that their strengths and weaknesses complement each other. Here, the main problems are deciding which method(s) to believe at each point, how to combine their results into a consistent overall interpretation, and how to fill in information in areas where no information is locally available. These are recognized

Figure 7. A complex natural scene.

to be very difficult problems (Barrow & Tenenbaum, 1981b) and no satisfactory solutions have yet been put forward (although they are currently receiving attention; e.g., Terzopoulos, 1982).

It may turn out, however, that the problems with current recovery techniques are inherent in the local, quantitative nature of the approaches they use. It is easily demonstrated (e.g., by looking through a reduction tube) that, in general, very little information about surface or boundary characteristics can be gleaned from small image neighborhoods that are viewed out of context. Thus, a local surface patch often appears much more perpendicular to the viewer than it is, a phenomenon well-known to psychologists as "regression to the frontal plane"; occlusion, shadow and reflectance boundaries will often appear locally indistinguishable. Since the information available from a local neighborhood is underdetermined, we know that some form of spatial constraint is therefore required. However, the simple context-independent assumptions (e.g., local continuity) that have been almost universally employed seem both too weak and of limited applicability. Reliance on local continuity, in particular, seems quite questionable in natural scenes, where, as was previously observed, parameters can change arbitrarily from point to point. At the least, what seems needed are constraints that are more sensitive to spatial structure in the image data.

It is also clear that detailed analytic models of surface photometry and the image-forming process are not essential for surface perception; humans can routinely perceive surface structure in images for which no such models are available. For example, they can see three-dimensional surfaces in range finder images such as Figure 2 (in which range is encoded fairly arbitrarily by brightness), and in electron micrographs such as Figure 8 (in which the shading corresponds roughly to that which would be produced by a negative light source at the viewpoint), not to mention oil paintings and line drawings. Indeed, even with ordinary photographs, one almost never has access to the parameters that would be needed to exploit imaging models in a quantitative way: the position and orientation of the camera, the focal length and aperture of the lens, the exposure time and speed of the film, the photographic and printing processes through which the image passed, etc.

Virtually the only thing that imaging transformations as diverse as laser range finding and electron microscopy have in common is that they all tend to preserve the basic spatial structure of a scene – continuity, discontinuity, proximity, and so forth. Somehow, human surface perception must be exploiting this structure in a qualitative way. Some evidence to support this conjecture is

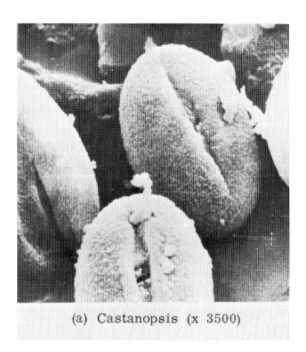

(a) Castanopsis (x 3500)

Figure 8. An electron micrograph.

provided by an experimental investigation of the importance of quantitative photometry in recovering shape from shading (Barrow & Tenenbaum, 1981a). In this study, a variety of artificial shadings were imposed on several surface contours (see Figure 9). It was found that perceived shape was essentially unaffected by wide variations in shading (eg. ranging from the black silhouette in Figure 9a to the linear fall-off of intensity from the highlight in Figure 9b, to the quadratic falloff in Figure 9c.) The only caveat was that the intensity transformation had to be structure-preserving; it could not introduce any discontinuities or alter the direction of the brightness gradient. Thus, people were confused by the shape of the surface depicted in Figure 9f, where the contour of Figure 9a has been rotated to imply a direction of curvature orthogonal to that implied by the shading.

The above experiment suggests that human surface perception does, indeed, depend more on qualitative, structural features, such as the relation between the shape of a bounding contour and the direction of the brightness gradient over the enclosed region, than it does on quantitative photometry. Exactly how structural features are used in recovering surface characteristics is unknown, but one intriguing possibility is that they fill the need, identified above, for spatial constraints that are stronger and more reliable than local continuity assumptions. This and other possibilities will be discussed at length in Section 4.2.[1]

So far, we have focused solely on the question of how arrays of surface properties might be recovered. There is, however, a more fundamental issue: Even if a depth map could be reliably obtained, how would it be used? As is apparent from Figure 2, a depth map is still fundamentally an image, with distance replacing brightness as the dependent variable. Being just an array of numbers, it is difficult to think of tasks that a depth map directly supports. For example, while raw depth values may suffice for elementary obstacle avoidance, grasping or recognizing objects requires that the depth data first be organized into larger structures corresponding, e.g., to continuous visible surfaces. Although perhaps better defined, this process is a great deal like that of segmenting a brightness image. One might therefore expect the traditional difficulties encountered in segmentation to arise. This expectation has indeed been borne out by a number of experiments on segmentation of range and brightness arrays (Duda et al., 1979). To be sure, range facilitates segmentation by enhancing depth discontinuities and attenuating shadow and reflectance edges. However, all the fundamental difficulties due to noise and low contrast

---

[1]Given the ease with which many exotic, structure-preserving forms of imagery can be interpreted (e.g., range maps, electron micrographs), it seems surprising that people have such difficulty interpreting ordinary photographic negatives. One conjecture, due to Pentland (private communication, 1982a), is that the inversion of light and dark confuses primitive mechanisms for dealing with shadows, resulting in figure-ground reversals.

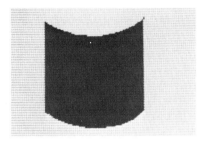

(a) A CYLINDRICAL PATCH IN SILHOUETTE

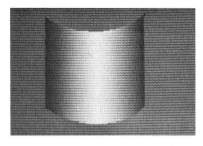

(b) A CYLINDRICAL PATCH WITH SHADING
     FALLING OFF LINEARLY ON BOTH SIDES
     OF THE HIGHLIGHT

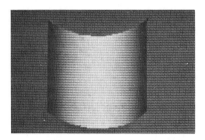

(c) A CYLINDRICAL PATCH WITH SHADING
     FALLING OFF QUADRATICALLY FROM
     THE HIGHLIGHT

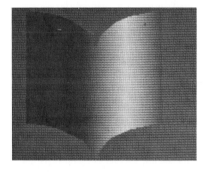

(d) A CUSPED SURFACE WITH SHADING
     FALLING OFF LINEARLY

(e) A CUSPED SURFACE WITH SHADING
     FALLING OFF QUADRATICALLY

(f) A CYLINDRICAL PATCH WITH
     INCONSISTANT SHADING

Figure 9. An informal experiment demonstrating the relative importance of shading and boundary curves as determinants of surface shape. Different one-dimensional shading gradients appear to have little effect on the perceived shape of a surface defined by a given outline. The direction of the shading gradient does seem important, however, as a qualitative cue for line sorting. Case (f), where the shading gradient runs orthogonal to the cylindrical curvature implied by the contour, is difficult to interpret. (a) A cylindrical patch in silhouette; (b) a cylindrical patch with shading falling off linearly on both sides of the highlight; (c) a cylindrical patch with shading falling off quadratically from the highlight; (d) a cuspate surface with shading falling off linearly; (e) a cuspate surface with shading falling off quadratically; (f) a cylindrical patch with inconsistent shading.

edges remain (see, for example, the gradient edge at the intersection of the top and side of the desk). Range contributes incrementally, in much the same way as color; it adds a dimension that simplifies some decisions, but it certainly does not make segmentation trivial.

In summary, the keystone of current computational approaches to vision is an intermediate description of local surface characteristics in the form of numeric arrays (known as intrinsic images or the 2-1/2 D sketch). Such arrays, it turns out, are hard to recover on the basis of physical models and local constraints. More fundamentally, although depth maps can be helpful, it is not obvious that they are helpful enough to deserve their present stature as a central goal of vision. To the extent that their utility hinges on imposing further organization and that some of the problems entailed in recovering depth stem from insufficient prior organization of the image, it is hard to justify deferring the imposition of structure until a depth map is computed.

The frail, quantitative, local character of current algorithms stands in marked contrast to the robust, qualitative, holistic nature of human perception as we experience it. One is forced to conclude that something fundamental must be missing from our current models of vision. But what?

3. The Role of Structure

Figure 10 shows an image that, for most people, suggests no three-dimensional interpretation, and portrays no familiar objects. Although the picture is devoid of meaning at these levels, our percepts of the picture are rich, deep, and – somehow – still meaningful. Although difficult to capture in words, the entities we see might be variously described as textured bands, channels, swirling patterns, puffs, spots, and so forth. What we see is *structure* – orderly, regular, or coherent patterns and relationships. The perception of structure is dynamic; as we shift our attention from the coarse sweep to fine details, we notice new and less obvious relationships. With effort, we may continue beyond the less obvious, through the tenuous and on to the fanciful. Where several alternative organizations are powerful but mutually exclusive, we experience not a trailing off to ever subtler and finer relationships, but the familiar tension of multistability, as in the Necker cube, or Marroquin's pattern (Figure 11). The perception of structure is ubiquitous; whenever we strip away the perception of familiar objects and three-dimensional shapes that routinely dominates our experience, this kind of "pure structure" nearly always seems to be lurking just below the surface.

When we perceive "pure structure" as in Figure 11, what we see is not a collection of familiar namable objects, nor anything that we can label concretely. It is not an array of surfaces in 3-space. It surely is not an undifferentiated intensity array. What then *do* we see? What does all this organization *mean?* What, if anything, is it good for? If it is useful, how do we get it, and once gotten, use it?

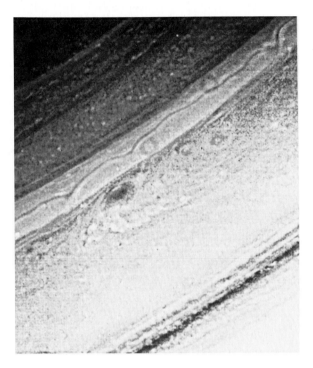

Figure 10. An image for which the absence of tri-dimensionality and familiar objects brings primitive structure to the fore.

All of these are deep questions. All have been considered, to some degree, both by psychologists (the Gestalt movement in particular) and in machine vision (in connection with segmentation, edge detection, and grouping [Rosenfeld & Kak, 1976; Ballard & Brown, 1982]), yet the basic questions have never been answered satisfactorily.

The Gestalt movement suffered because the principle of Prägnanz and the associated laws of perceptual organization were never made sufficiently concrete to offer much predictive or explanatory power. Several related reasons account for this failure; most obviously, the Gestaltists were addressing a fundamentally computational problem before computers or the associated conceptual framework were available. Moreover, although they sought unified laws to account for the phenomena of perceptual organization, their explanations were not cast in functional terms – they were not inclined to ask what perceptual organization is *for*.

In contrast, research in computational vision has been very much concerned with explanations at the functional level. However, interest gravitated away from perceptual organization precisely because it is very difficult to say what perceptual organization is for, and to characterize it in terms of well-

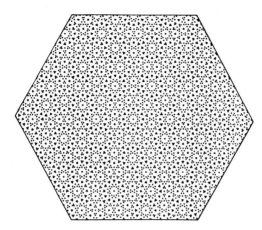

Figure 11. Marroquin's ambiguous pattern, providing evidence for the existence of active grouping processes.

defined inputs and outputs. Consider, in contrast, the problem of identifying all and only trees in an image, or of determining depth or albedo at every point. These both seem like eminently sensible things to do, and however hard it may be to accomplish them, it is clear what one will have accomplished if one succeeds – it is clear, at least intuitively, what this kind of percept says about the world (in particular, what it means for an answer to be right or wrong), and also why these are useful things to say. In consequence, any process that accomplishes these tasks may be meaningfully viewed as a solution to a problem, not only in terms of what it does but in terms of why it does what it does. This is not obviously so for perceptual organization. What, for example, does it mean to group a bunch of dots? What, if anything, is your visual system telling you about the outside world when it imposes one versus another grouping? Is there any objective sense in which some groupings are true or accurate, while others are false or inaccurate?

No answers instantly spring to mind. This doesn't mean that perceptual organization plays no functional role, only that that role is less obvious than, say, those of depth perception or recognition. A prerequisite to treating perceptual organization at a functional level is thus to discover the computational questions to which perceptual organization provides answers. This in itself appears to be a very difficult problem. It isn't really surprising, then, that having been seduced by the notion of functional explanation and by the appeal of well-defined problems, many of us felt uneasy with perceptual organization, preferring the warm secure feeling induced by problems – e.g. quantitative shape recovery – for which a precise statement of the objective came nearly for free. Nevertheless, the phenomena of organization pervade human perception, and should therefore not be dismissed lightly. That the problem is difficult to frame at the functional level makes it no less important or interesting. In avoiding the problem, we have been, as the story goes, looking for our keys where the light is.

It is universally recognized that some description in terms of primitive features must bridge the gulf between the raw image and high-level knowledge and goals. All of the structural entities we identified above – edges, coherent regions, groupings, flow patterns, parallelism, symmetry, etc.– have been recognized as candidates for this level of description. However, our ability to discover and exploit structure at this level has been hampered by the lack of a unified account of its nature and meaning. Our primary objective is to understand the contribution of primitive structure to the process of perceptual inference. Rather than proposing new or different features, we will suggest a more unified account of what these diverse structural elements are, why they're important, and above all what they *mean,* when viewed as hypotheses or assertions about the world.

We first ask, what is structure? Must we settle for a haphazard list of special cases to describe the elements of human perceivers' structural descriptions, or is there some more unified account of their content? We will suggest that an important unifying principle is spatiotemporal regularity: a number of salient structural features are simply identities over space and time. More broadly, perceived structure appears, as a generalization of these special cases, to capture a very restricted and literal kind of similarity, such as rough parallelism. We call this kind of approximate relationship "fuzzy identity." Although it is difficult to formalize fuzzy identity in a principled way, the notion is captured reasonably well by the various "least-distortion" or "smooth-as-possible" criteria that have been invoked over the years in many forms and many contexts.

The perception of fuzzy identity appears to pervade perception, and smooth-as-possible solutions have often been invoked in machine vision. Why, then are these relationships useful things to discover, and what have we learned about the world when we discover them? A number of answers to this question have been proposed, but none is convincing, and none implies a central role for structure.

We suggest an alternative account, in which the perception of structure contributes centrally to interpretation at every level: regular relationships are so unlikely to arise by chance that, when such relationships can be consistently postulated, they almost certainly reflect some underlying causal relationship, and therefore *should* be postulated. We conjecture that as a least-distortion solution approaches strict identity, the likelihood that the relation is non-accidental increases. The minimization of change is therefore a primary basis for discovering causal relationships at a primitive level.

Because perceived structure captures underlying causal relationships, we view the primitive structural description as a source of semantic precursors – an alphabet soup of structural relationships that deserve or demand semantic interpretation. This view is born out by the observation that primitive structure tends to survive, largely intact, to the highest level of interpretation; the things we perceive naively *do* turn out to be meaningful.

Although our account is preliminary and conjectural, it has direct implications for perception. At the level of obtaining structural descriptions, we are able to reformulate some problems of current interest and suggest alternative solutions. We can also suggest a more unified account, both conceptually and computationally, of a number of apparently unrelated problems in early and intermediate vision. At higher levels, the availability of primitive semantic precursors makes possible a more qualitative style of computation that could surmount many of the difficulties associated with current quantitative approaches.

3.1. What Is Structure?

Webster's dictionary defines structure as "figuratively, the interrelation of parts as dominated by the general character of the whole." On this basis, we regard the word as one bearing appropriate connotations, but whose strict definition is still up for grabs. We have no intention to define structure formally here, because we are using the term to refer to a class of perceptual phenomena, not to set up a math problem. We can, however, delimit that class of phenomena quite clearly, and identify some of its distinguishing properties. Our objective, beginning with this characterization, is to understand the phenomena: what they are, what they're for, and how they work. Until our understanding is essentially complete, any formal definition we pull from thin air will almost certainly be *wrong*. A formal definition of perceived structure that succeeds in capturing the essence of the phenomena would be more nearly a solution to the problem we pose than an initial statement of the problem.

By structure, we refer to relationships over time and space that people perceive or impose on image data more or less spontaneously. By *primitive* structure we refer in particular to those relationships we are able to perceive even when we can't interpret them in terms of familiar objects, high level knowledge of physical processes, etc. Frequently it appears possible to dissociate the perception of primitive structure from the perception of three-dimensional shape, albedo, illumination etc., although this needn't be so; we will discuss the relationship between structure, surfaces, and three-dimensional space in a later section.

Thus instances of primitive structure include perceived edges and regions (with the internal organization we impose on the shapes of edges and regions), smooth or coherent variation (of intensity, texture density, velocity, disparity, etc.), parallelism, symmetry, repetion, flow, rigid or viscous motion, smooth continuations, dot groupings, and so forth.[2]

---

[2]It is useful to view the imposition of this kind of structure as *decomposing* – or in some very primitive sense, explaining – the image data in a particular way. When we perceive a smooth curve, we are replacing or augmenting our representation of the individual points on the curve with an explicit

3.1.1. Fuzzy Identity

Significantly, all of these structural entities can be described in terms of a very literal sort of spatiotemporal regularity or coherence: the relationships we notice are characterized by shapes, patterns, or configurations that replicate or continue with little or no change over an interval of space or time. Strict parallelism, for example, entails the replication of curves with translation but with no change in shape. Rigid motion (in the image plane) is the temporal analogue to parallelism. A patch of perceived smooth variation in intensity, texture, etc. displays coherent change in a local measure as a function of position. An edge is a curve separating two internally coherent regions, or a curve along which an intensity profile or other property changes coherently (and thus, in this sense, edges reflect a kind of parallelism). Linearity is identity of tangent direction from point to point along an edge. And so forth.

This common thread of spatiotemporal coherence or regularity naturally suggests that it ought to be possible to treat these superficially distinct structural relationships as manifestations of something more general and more unified. Moreover, such a unified characterization, if we had one, would provide a parsimonious account of perceived structure.

Of the regular relationships comprising perceived structure, the easiest to describe are special cases: parallelism, rigidity, collinearity, strict homogeneity of brightness or texture. These are nothing more than identities. They are distinguished from each other only by the elements over which the identity holds and by the dimension over which the elements are arrayed. Thus, parallelism is identity of curve shape over translation in space; image-plane rigidity is identity of shape over time; collinearity is identity of tangent over position on a curve; constant brightness is identity of brightness over position in the image; and so forth. A fair amount of work has addressed the discovery and use of these identities for image interpretation, largely treating each identity as an isolated special case (e.g., Kender, 1979; Stevens, 1981; Ullman, 1979).

More difficult to capture is the generalization of these strict identities into approximate or "fuzzy" identities: to the human perceiver, each of the strict identities is a limiting case of something far more flexible. By whatever means

---

representation of a connected entity with a particular shape. When we notice parallelism or symmetry among a set of curves, we are replacing or augmenting our representation of the individual curves with a description of the common replicated shape and the manner of its replication. To a first approximation, our decompositions must be consistent with the data, in the same sense that an intrinsic image interpretation must be consistent. Within that constraint, there are infinitely many ways to carve up the data, but the human perceiver has some particular systematic way of doing so. It bears emphasizing that, by whatever means we arrive at these structural descriptions, we are, in imposing them, inventing, discovering, or hypothesizing new entities that, while consistent with the image data, are not in any direct or necessary sense *in* the data. To the extent that we attach any meaning to the constituents of our structural description, it may be viewed as an interpretation rather than a mere description of the data.

the human perceiver orders relationships, e.g., from "very roughly" to "almost" to "exactly" parallel, the ordering appears regular and consistent from person to person. The ordering also appears, at least to a first approximation, consistent from one to another mode of regularity: rough parallelism is quite a lot like viscous flow, and in turn resembles the "rough smoothness" of a curve with irregular fine structure.

That people have *some* systematic basis for perceiving "fuzzy identity" is vividly illustrated by Figure 12. No two curves are strictly parallel, by virtue of their irregular fine structure and the gross distortion from each profile to the next. Yet the perceived relation from profile to profile is immediate and compelling. We use the fact of the relation to infer a surface, we use the gross distortion from profile to profile to interpolate its shape, and we use the fine-grained irregularity to infer its rough texture.

An appealing account of fuzzy identity, and one with a long and colorful history (see Attneave [1982] for a bibliography of the psychological literature), is the notion that of all relationships that may be consistently postulated, we perceive the one that in some sense entails the least change or distortion from entity to entity. For any reasonable measure of change or distortion, the strict identities – true parallelism, etc.– are limiting cases for which the least-distortion solution entails no distortion at all. Although some treatments, particularly those cast in terms of economy or simplicity, have been vague, the smooth-as-possible notion may be, and has been, expressed in many concrete quantitative forms, all of which seem *prima facie* plausible. Subjective contours, monocular contours perceived in three dimensions, and interpolated stereo surfaces have all been modeled as variatonal minimizations of one or another curvature measure (Barrow & Tenenbaum, 1981a; Ullman, 1976; Grimson, 1981; Brady & Grimson, 1981; Brady & Horn, 1981). Similar minimizations have been used in stereo correspondence (Marr & Poggio, 1976), and in the computation of optical flow (Horn & Schunck, 1981; Hildreth, 1982).[3] We will follow Attneave (1982) and refer good-naturedly to this kind of minimization as soap-bubble solutions, even though the minimal surface (which is what literal soap bubbles compute) does not appear to be relevant to perception. When we refer to a fuzzy identity, we will refer to the relationship obtained by performing such a minimization constrained by a set of structural elements (with the further implication that the optimal solution is a near-identity).

To capture the notion of least-change or smooth-as-possible solutions, it is necessary not only to define a measure for distortion or smoothness, but to characterize the "constraints imposed by the data." Ideally, these constraints

---

[3]Taken at face value, the notion that we organize the data to minimize some distortion measure implies some grand global variational optimization, leading to the best of all possible descriptions. This view is surely over-simplified. Although variational minimizations in direct numerical form are familiar, and established techniques are available to compute them, there may be more attractive alternatives.

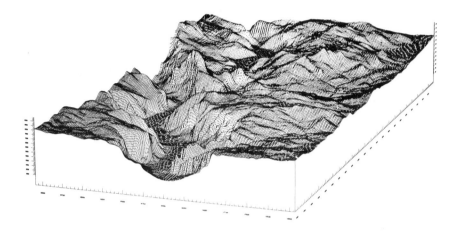

Figure 12. A profile plot of a complicated surface.

need amount to nothing more than consistency: the description must be able to regenerate the image data, in the same sense that a hypothesized scene model or intrinsic image decomposition must be able to regenerate the data when run through an image synthesis procedure. Two conditions qualify this idealization, though. First, strict consistency may apply only at a coarse scale, or only in some statistical sense. For example, when we describe a patch of the image in terms of a smooth gradient in texture density, the description must consistently account for the variation in density at some scale of description, but not for every constituent dot or texture element. Similarly, the regular profile-to-profile relation in Figure 12 obtains only at a sufficiently coarse scale. We distinguish the coarse-scale regularity from the fine-grained irregular "microstructure." Thus we view scale of description as an issue that arises primarily in establishing the degree to which a postulated regular relationship accounts for the data.

Second, the nature of the constraints changes if a model of the image-forming process is available. Viewed as a projection of a point in space, a dot in the image is constrained to lie on a view line, but may lie anywhere on that line. Thus, in the case, e.g., of rigid or approximately rigid motion of a collection of dots, a least-distortion solution might be obtained directly in the image, or directly in three-dimensional space using a model of projection. In the latter case, the position of each dot along its respective view line becomes an additional free parameter over which optimization is performed. The solution obtained in this way directly in three dimensions will in general be different from the two-dimensional one, and there appear to be cases in which regularity that is not evident directly in the image is revealed in three dimensions. Thus, although we will be discussing the search for regularity primarily as a task performed directly in the image domain, this needn't be so, and in some cases clearly is not so. Although conducting the search in three dimensions alters the space of possible explanations, the criteria by which an explanation is judged may be no different in three dimensions than in two. This relation between perceived regularity and perceived tri-dimensionality will be discussed at greater length in a subsequent section.

To sum up so far, perceived structure captures spatiotemporal regularity or coherence, in the sense of fuzzy identity over space and time. Characterizing the strict identities – parallelism, rigidity, collinearity, etc.– is straightforward, but the generalization to more flexible approximate relationships is less so, because many quantitative instantiations are possible, and there is no compelling reason to prefer any particular one. In many contexts, however, fuzzy identities appear to be reasonably well captured by selecting interpretations that by some measure minimize distortion or maximize smoothness, within constraints imposed by the data. Considerable bodies of work have addressed both the discovery and utilization of the strict identities, treated as isolated special cases, and the computation of "soap-bubble" (smooth-as-possible) solutions by variational means.[4]

3.2. Why Soap Bubbles?

Although soap-bubble solutions seem to pervade perception, and have been widely employed in machine vision, it is not immediately clear *why* they are worth computing or how they fit in to the process of perceptual inference. Are they simply a default in the absence of "real" constraints (as in filling in an incomplete surface) or do they play a more powerful functional role?

We know that matter has some tendency toward coherence over space and time, that things do hang together. Surely this tendency has something to do with the empirical significance of perceived regularity. But beyond this gross connection, is there any compelling reason to select the smoothest or most regular consistent interpretations?

The coherence of matter has been invoked to justify smooth-as-possible solutions in several quite different ways. Köhler's field theory (see, e.g., Köhler, 1929) explained the tendency to segregate the visual field into smooth

---

[4] In offering this generalization, we intend to identify as a common principle the search for regularity, and the tendency to perceive as much regularity as the data allow. We are not suggesting that perceiving motion, stereo disparity, parallelism, etc. are identical problems, and certainly not that they are subserved by common mechanisms in biological vision.

entities by appeal to electrical force fields in the brain that behave, by direct physical analogy, like soap bubbles. Regardless of its merits as a physiological theory, this is not an explanation at the functional level, and so need not concern us here.

Another line of argument appeals to Occam's Razor: all else being equal, the most economical explanation should be chosen (e.g., Attneave, 1982). Since the discovery of regularity reduces redundancy, the most regular description is the most economical, and therefore the best. This line of argument suffers from the well-known dependence of the size of a description on the form in which it is represented. Hence, the argument is quite empty until the descriptive terms are specified, and the economy argument, by itself, offers no help specifying them.

More recently, the argument has been made, with varying degrees of explicitness, that, due to the coherence of matter, smoother or more regular shapes or structures are more likely to be observed then less smooth ones, and that therefore the smoothest or most regular interpretation should be chosen, within the additional constraints. The argument was made quite explicitly for the three-dimensional interpretation of contours in (Barrow & Tenenbaum, 1981a). Similar arguments have been made in the context of visual motion (Horn & Schunck, 1981; Hildreth, 1982).[5] Even if this argument is true, it cannot be very powerful, because it appeals to the prior likelihoods of more and less smooth surfaces. How much more likely *a priori* is any one curve, surface, or flow field than any other, judging only by some integral smoothness measure? Such a measure surely has *some* distribution over the sum total of curves, surfaces, and motions we will ever observe. However, the world is a diverse place, and we *do* encounter jagged, irregular, non-rigid, and chaotic structures, as well as straight lines, planes and spheres. Consequently, that mythical distribution must surely be quite broad. Thus, by itself, it cannot provide a very compelling basis for choosing any one interpretation over another. This kind of argument is perhaps more plausible in its negative form, i.e. that without evidence to the contrary, one might as well choose the smoothest interpretation. This "why not?" approach relegates the search for regularity to the role of filling in where "real" constraints fail.

Still another line of argument was advanced by Grimson (1981) in support of a variational technique for surface interpolation from sparse stereo information. In essence, his principle that "no news is good news" states that the absence of local evidence (in this case, zero-crossings) is itself evidence that the surface is smooth. He argues that the surface minimizing quadratic variation is the most consistent with this principle. The argument is cast explicitly in terms

---

[5]These "smoothness assumptions" should not be confused with continuity assumptions. Just assuming continuity, in the case of motion or 3-D contour shape, still allows an infinity of consistent solutions. In general only one will be smoothest by some integral measure.

of the Laplacian operator, depends on details of the irradiance equation, and makes a number of assumptions (e.g. of roughly constant albedo and simple illumination). It is therefore quite specialized and frankly seems a bit strained. For example, no argument of this kind can ever constrain interpolation across occlusions, where the irradiance equation does not depend on the occluded surface at all; yet we do perceive complete surfaces across occlusions (Marr, 1982, p. 287).

Thus, we have arguments in terms of force fields in the brain, which, credible or not, are not germane; arguments for economy of description, that in their most general form are vague; arguments in terms of prior likelihood of more and less smooth surfaces, which are weak; and an argument relating surface shape, through the irradiance equation, to the probability of zero-crossings in the image's Laplacian, which is too specialized. All of these postulate or justify some version of a smooth-as-possible optimization, and none is particularly convincing. Perhaps the strongest conclusion we can draw from these arguments is that the coherence of matter lends *some* grain of empirical validity, however small, to soap-bubble solutions: all else being equal, smooth surfaces may be more likely than wildly convoluted ones, but in any absolute sense, only slightly so.

3.3. Non-accidentalness and Inferential Leverage

Although past arguments supporting soap-bubble solutions are weak, we should not give up on the idea too quickly: whether we understand them or not, such computations still pervade perception. Even if the coherence of matter lends these computations only weak empirical support, might there not be some way to leverage this small grain of truth into a powerful basis for doing inference?

Just such a source of leverage is provided by a very different kind of argument than those reviewed above. This "non-accidentalness" argument has been applied with considerable success to several of the special-case identities. In its bare form, the argument says that what *looks* parallel or rigid really *is* parallel or rigid, less because parallelism is likely in any absolute sense, than because the spurious appearance of parallelism is extremely *un*likely to arise among causally unrelated curves. We will see that the strong point of the non-accidentalness argument is its ability to leverage a very weak prior empirical constraint into a very strong conditional inference. Its weak point, in the form it has been used, is its narrowness, applying only to special-case properties such as strict rigidity and parallelism. The prospect of "generalizing" the argument by compiling an enormous list of such singular properties is not appetizing.

Perhaps the canonical example of the non-accidentalness argument is Ullman's treatment of the recovery of three-dimensional shape from rigid motion. Ullman begins with a collection of isolated dots in an image, which are the projections of dots in space. The dots are tracked over a discrete series of views, like the frames of a movie. Analyzing the relation between the dots' projected motions and their interpretation as points on a moving rigid body in space, Ullman shows that, given enough views of enough dots, the solution for rigid motion is overdetermined. That is, under ordinary conditions there can be at most a single consistent rigid interpretation, but in general no consistent interpretation will exist. Therefore, given an arbitrary collection of (non-rigid) moving dots, the probability that a consistent rigid solution will exist is effectively zero (exactly zero assuming perfect data.) Consequently, *given* that a consistent rigid interpretation exists, the dots almost certainly lie on a true rigid structure. The rigid solution then supplies the structure's motion in space, and the three-dimensional spatial relations among the dots. If no rigid solution exists, then nothing can be concluded about the dots' three-dimensional motions, beyond the fact that they cannot be rigidly connected.

A somewhat similar argument was made by Stevens (1981) for parallelism. Stevens argued that parallelism occurs sufficiently often in the three-dimensional world, and the likelihood that non-parallel space curves will appear parallel in projection is sufficiently remote, that parallelism observed in the image strongly implies parallelism in the scene, rather than an accident of projection. Related arguments have been made by Witkin (1982) to classify edges by intensity correlation across the edge; and by Binford (1981) to explain line-junction constraints in terms of general position assumptions.

How do these arguments differ from the simple claim that parallelism or rigidity is likely to be observed because matter is coherent? Apart from any claim that actual rigid structures are likely to be observed, it is possible on formal grounds to argue that, because the rigid solution is overdetermined, rigid solutions will exist only for particular combinations of moving dots, and not in general for arbitrary ones. Therefore, the likelihood that a rigid solution will exist for dots whose motions are actually independent is vanishingly small (and in fact non-existent if measurement error is neglected). This is true regardless of the frequency with which real rigid solutions are observed, and so does not depend on assumptions about the coherence of matter. Now, the relative likelihood that the dots are actually rigidly connected, *given* the observation that a rigid solution exists, is the ratio of the prior likelihood of rigid connection to the prior likelihood that a spurious rigid solution will exist. Since the latter quantity is near zero, the inference of true rigidity may be a near certainty even if the prior likelihood of rigidity is quite small. Thus, a very strong conclusion is drawn, making only very weak assumptions about spatiotemporal coherence, by conditioning the conclusion on a consistency test performed on the data. This is what we mean when we say that non-accidentalness arguments provide inferential leverage.

The non-accidentalness argument, in its most general form, is very general indeed. It is, in fact, no more than a utilization of conditional probability as a source of inferential leverage. Suppose, for example, that we had an "aardvark detector" that almost never fired in the absence of an aardvark, but always fired when an aardvark was actually present. Now presumably, the prior probability at any moment of encountering an aardvark is extremely small. However, if our aardvark detector is well designed, then the probability that it will fire in the absence of an aardvark (i.e. a false alarm) should be enormously smaller. Therefore, the odds that we have really observed an aardvark, *given* that our detector has fired, will be enormous, even though aardvarks, in an absolute sense, are rare. This kind of argument works as long as the probability of a false alarm is much smaller than the probability that the condition to be detected will actually occur. Obviously, if aardvarks were *never* observed, the argument would not work; however low the false alarm rate, a false alarm would always be more likely than a hit.

These examples illustrate both the good and bad side of the non-accidentalness argument. What is good, as we have already seen, is the leveraging of weak *a priori* assumptions into strong conditional conclusions. What is bad is the argument's open-endedness. By itself, the argument is just a formal shell, only as good as the empirical assumptions to which it is applied. Thus, we can use the argument identically to discover rigid motion or to detect aardvarks. In fact, the list of properties to which the argument *might* be applied is plainly open-ended, and thus it can never lead to a final or absolute notion of "accidentalness" or "randomness": any pattern, however random or meaningless it appears, may at the next moment become meaningful if we obtain some new model or piece of knowledge. This crucial point has sometimes been missed. For example, Lowe and Binford (1982) sought a general and complete criterion for evaluating randomness. Although they chose a particular list of "non-random" properties to look for, they viewed the choice as a pragmatic matter of reducing combinatorics. In fact, in choosing that particular list they adopted a particular definition of non-randomness.

3.4. Non-accidentalness and Fuzzy Identity

Thus, the non-accidentalness argument provides a general tool for exerting inferential leverage, and may be applied to almost any property that bears a grain of empirical validity. However, the argument is worthless without *some* valid properties to apply it to. If it is applied only to specialized properties, it will be applicable only to special situations. To use it as a basis for doing primitive perceptual inference in a general and unified way, we must do better than enumerating a host of special-case properties and relationships; we have to start with a general and unified expression derived from primitive properties of the world. We believe that fuzzy identity provides just such an expression.

We have seen that the existence of a consistent strictly rigid or strictly parallel solution is compelling evidence of true, non-accidental rigidity or parallelism. Intuitively, it would seem by extension that a spurious "almost rigid" or "almost parallel" interpretation is nearly as unlikely, with the confidence of the conclusion decreasing as the least-change solution departs further from strict identity. Thus, intuitively, there ought to be some generalization of special-case non-accidentalness arguments, like Ullman's, to the more flexible and general fuzzy identities.

What does this intuition imply? Consider rough parallelism as a simple case. Given any two curves, we may compute a least-distortion transformation of one curve into the other, by variational minimization on some measure, e.g., of total strain. Let δ be the integral measure that has been minimized, and let its minimized value be denoted by δ_{min}. In deciding whether a relation is accidental, we are choosing one of two hypotheses: H_1, that the curves are causally related in a manner reflected by the least-change transformation, or H_0, that the curves are causally independent – or at least have no simple relationship at the primitive structural level. Our intuition is that the relative likelihood of H_1 increases, as δ_{min} decreases.[6]

What condition would have to hold for this intuition to be valid? Over the universe of all pairs of curves, there are two density functions, $F_1(\delta_{min})$ = p.d.f.($\delta_{min} \mid H_1$) and $F_0(\delta_{min})$ = p.d.f.($\delta_{min} \mid H_0$) (where p.d.f. denotes probability density function). The relative likelihood of H_1 given δ_{min} is proportional to the ratio $F_1(\delta_{min})/F_0(\delta_{min})$. In these terms, our intuition is that this likelihood ratio increases as δ_{min} decreases. More specifically, in arguing earlier that the coherence of matter, by itself, provides only a weak basis for expecting smoothness, parallelism, etc., we were arguing that the distribution $F_1(\delta_{min})$ is broad. In that case, if our intuition is valid, it is primarily because $F_0(\delta_{min})$ becomes very small for small δ_{min}. Since $F_0(\delta_{min})$ is the likelihood of δ_{min} for unrelated curves, the behavior of $F_0(\delta_{min})$ for small δ_{min} would provide the leverage typical of non-accidentalness arguments.

Unfortunately, it is difficult to imagine quantifying $F_0(\delta_{min})$ and $F_1(\delta_{min})$ in a principled way. We could derive *some* distributions by adopting a simple model for the processes generating the image, but since no simple model stands much chance of capturing the diversity and unpredictability of the natural world, this would be little more than an empty formal exercise. This limitation of parametric statistics, requiring a "closed world" assumption in which all the relevant prior distributions are fully known, is by no means specifi : to the current problem, and has motivated the extensive work in distribution-free methods. Perhaps some method of this kind exists, or will exist, that will support a satisfactory formal treatment of our problem, but we know of none that does. We will therefore be content with the conjecture that, for some measure

---

[6]In actuality, we would have to normalize the least-change solution by some measure of the complexity of the curves; a close relation between two very simple curves (e.g., in the extreme, straight lines) is intuitively less "surprising" than an equally close relation between complex curves. We will avoid this issue for now by assuming that δ_{min} has been so normalized.

of change, smoothness, or distortion, the likelihood that a least-change relationship reflects an underlying causal relationship increases as the amount of change or distortion decreases. Tentatively accepting this conjecture, we will move on to consider its implications for the role of structure in perception.

3.5. Semantic Precursors

Our initial motivation for this investigation was the desire to understand the role of structure for perceptual inference: what, if anything, does a primitive structural description say about the world? We now have an answer for that question, because the meaning of structure follows directly from the non-accidentalness argument: when we strongly perceive a structural relationship, we are implicitly asserting that there is a corresponding causal relationship. Whatever the relationship means, it means *something*.[7]

For example, having discovered a compelling parallelism relationship, we may be free to interpret it in many ways – a curved surface, a waving flag, etc.– but we can never dismiss it. As we proceed from the level of primitive structure to a high-level semantic interpretation, the bare assertion of non-accidental parallelism may evolve into a more specific assertion of surface-hood, and the detailed shape description of the parallel curves may be reinterpreted as the shape of a surface. But to whatever the primitive relationship is eventually attributed, it is almost certain to survive in *some* form to the highest levels of interpretation.

In effect, perceived structural relationships are "semantic precursors," deserving and demanding explanation by subsequent interpretation. In this role, the primitive structural description provides constraints on subsequent interpretation – in the form of facts that ought to be explained – and also provides a bootstrap by which higher-level interpretations may be obtained, because the structural relationships resemble the underlying causal ones. Thus, a swirl may turn out to be a storm; parallelism may denote a curved surface or an oriented marking process; a coherent gradient of intensity or texture density may denote shading or projective distortion induced by surface curvature.

We might therefore think of structural descriptions as providing an "alphabet soup" of descriptive chunks that are almost certain to have *some* fairly direct semantic interpretation. As interpretation proceeds, the primitive chunks are assimilated more or less intact into the full interpretation.

3.6. Naive and Informed Perception

We have now presented, in preliminary form, an account of perceptual organization as a primitive inference mechanism that provides "semantic precursors" for subsequent interpretation. If this is true, we ought to observe a

---

[7]Once this assertion is attached, structural descriptions have truth values: if the perceived relation in fact turns out to be spurious, the structural description will be wrong.

close relation between naively perceived structure and semantically informed percepts. In fact, evidence for this close relation is ubiquitous. We find it significant, for instance, that being told that Figure 10 is actually an image of the surface of the planet Jupiter barely alters one's initial perceptions. Even a planetary physicist sees the same bands, swirls, and so forth, that the naive observer spontaneously perceives. These are indeed the very entities the expert is often interested in naming and explaining; frequently the names bear a direct relation to the perceived structure, as in "The Great Red Spot"!

A number of experiments have been performed confirming the close agreement between the perceptions of expert and naive observers. Rosenfeld, for example, showed that secretaries were about as competent as meteorologists in outlining clouds in weather photographs. He thus concluded that task performance was based on "non-purposive" low-level operations, rather than specific domain knowledge (Rosenfeld, 1964). More recently, Greene et al. (1982) studied how subjects with no formal knowledge of speech taught themselves to interpret spectrograms. Given a training set of spectrograms, each labeled with an English word, the subjects invented their own discrimination criteria, which were based, not surprisingly, on the perceptually prominent structures they saw in the data. These structures bore a striking resemblance to formant tracks, release bursts, frications, and other features commonly used by trained phoneticians.

Parallel observations apply to the relations between structure and three-dimensional interpretation. In Figure 13, sets of closely-spaced, parallel curves impart a strong sense of coherence to major regions of this image; such structure is seen as far too regular to have arisen by accident. Conversely, abrupt, spatially extended discontinuities in the parallel structure are seen as equally significant indications of incoherence.

Figure 13. A complex natural image illustrating the relation between primitive structure and tri-dimensionality.

Although devoid of semantic meaning, these continuities and discontinuities admit readily to interpretation in terms of surfaces and surface boundaries. The parallel curves are seen as surface contours whose local direction of curvature provides strong clues to which way the surface is locally bending (Stevens, 1981); their discontinuities are seen as ridges along which the surfaces meet. This level of perception is unchanged by knowledge that the surfaces are, in fact, geologic structures carved out by physical processes acting at a variety of scales. Not surprisingly, since surface structure is what geologists see, it is what they attempt to explain.

These examples, and many others like them, demonstrate that naively perceived structural elements are frequently incorporated, nearly intact, into our informed explanations. The primitive structural relationships must, in these cases, therefore be isomorphic or analogous to semantically interesting causal relationships. Where such analogy is present, interpreting primitive structure can amount to little more than labeling it – e.g. a swirling blob becomes a storm, the blob's outline becomes the storm's perimeter, and the internal swirling structure directly becomes a description of the turbulent flow. Where the relation between structure and interpretation is less direct, interpreting primitive structure may entail augmenting or rearranging the primitive elements, as by cut-and-paste operations. For example, in the course of imposing alternative figure-ground organizations, we may join and sever fragments according to the perceived occlusion relationships. Ambiguous figures are particularly interesting, because the same primitive structure is typically incorporated into each interpretation – with different labels, and perhaps having undergone different transformations (see, for example, Figure 14). The Rorschach figures provide singularly good examples of the assimilation of the same structural elements into many possible interpretations.

That naively perceived structure should ever capture semantically important relationships so directly is striking. Obviously the relation between primitive structure and semantic interpretations is typically more fragmentary, and sometimes quite remote. In fact, some meaningful phenomena simply are not manifest in terms of primitive spatiotemporal regularity. Thus, discovering primitive regular structure does not trivialize the task of interpretation; rather, it often makes interpretation possible by providing fragmentary "islands of meaning." How isomorphisms between these islands and our stored models are discovered, and how the whole is knitted together, will remain difficult problems, even once the process of primitive perceptual organization is understood.

4. Implications for Computational Vision

Although our account of structure has been preliminary and conjectural, its implications for computational vision are considerable. In the remainder of the paper, we will explore these implications, both at the level of discovering structure and of interpreting it. Our discussion, though speculative, will address

Figure 14. An ambiguous line drawing, illustrating the survival of primitives across alternative semantic interpretations.

substantive issues. At the level of discovering structure, we will be able to reformulate a broad range of superficially unrelated problems from early and intermediate vision in a unified way. At the level of interpretation, we will argue that the richness and meaningfulness of structural descriptions permits interpretation to proceed more directly and at a higher level than has heretofore been possible.

Viewing the discovery of structure as a primitive form of inference evokes a very different picture of perception than that implied by current hierarchic models. The hallmark of structure-oriented perception, from the most primitive levels, is the ongoing search for regularity and coherence, and the attempt to explain the observed structure in terms of models at many levels. The goal at early levels is to decompose the image into primitive descriptive chunks whose regularity or coherence deserve or demand explanation. Subsequent interpretation in terms of more specialized models – of image formation, and of particular physical processes and objects – is driven by these "semantic precursors," and acts to label, transform, elaborate, and assimilate them. The persistence of primitive structure to high levels suggests a central representation built up by invoking these four operations, rather than a strict hierarchy of representational levels.

For historical reasons, our discussion will be organized around the levels of the (by now) traditional hierarchic model: early, intermediate, and high-level vision. However, the boundaries separating these levels do not correspond very well to the distinction between the discovery and interpretation of structure. For example, the discovery of primitive structure may proceed almost indifferently in two dimensions or three, blurring the distinction between early and intermediate vision. In consequence, some issues of three-dimensional perception are more naturally treated together with problems that belong traditionally to early vision.

4.1. Early Vision[8]

At a primitive level, the goal of perception is to capture basic spatial regularity in the data, as manifest in terms of coherent regions, edges, symmetries, repetitions, smooth gradients, flow patterns, groupings, etc. The outcome at this level is a decomposition of the image into discrete chunks each of which reflects some underlying cause or process, and therefore must be explained.

That early vision should be concerned with regularity and coherence is, of course, hardly a new idea; essentially all of the early description techniques that have been proposed and investigated draw on these concepts in one way or another, but in a haphazard way. What has been missing is any sort of unified account of the nature and meaning of these concepts. Lacking this foundation, much of what has been attempted has been ad hoc and ineffectual, barely scratching the surface of what human perceptual organization is capable of. While we are far from having such a unified account, we do feel a good start has been made. In particular, the principles enunciated in Section 3, even though still informal, provide immediate insights into limitations of past approaches. More significantly, they may permit a broad range of superficially unrelated problems in early and intermediate vision to be stated and solved in a much more unified way.

The failure of early attempts to partition an image more or less directly into namable objects underscores the need for a clear understanding of goals. As was pointed out earlier, this task turned out to be prohibitively difficult, because it sought to bridge in a single leap the enormous gulf separating primitive image properties and high-level semantic descriptions. Moreover, the task has been criticized as ill-defined, because what constitutes an ''object'' depends on the context of the moment (Marr, 1982).

The goal of direct image segmentation gradually gave way to goals that appeared more reasonable for early vision – the description of low-level image features such as edges, lines, and uniformly textured regions. Despite a massive amount of work in this area, far too much to enumerate, the tangible successes (in the form of working computer models) were few and far between. There have also been attempts to develop computational models of specific

---

[8]We use the term ''early'' to connote a process that operates more or less directly on the raw data, and that relies on sufficiently general properties of the world (e.g. spatial coherence) to be applicable without knowing anything specific about where the data came from. Perhaps ''primitive'' would be a more precise characterization, since ''early'' carries the unwanted implication of a fixed sequence of processing stages or representations.

human organizational phenomena, such as the perception of dot clusters (Zahn, 1971; Zucker, 1982), Glass patterns (Stevens, 1977), and textures composed of oriented line segments (Beck, Prazdny, & Rosenfeld, in this volume). Many of the more recent modeling efforts were motivated by Marr's concept of a "full primal sketch."

Lacking unifying principles, the work on image description degenerated into a large number of ad hoc techniques, each concerned with a narrow and specialized manifestation of structure. There was little motivation for why particular features were thought to be of interest, in terms of what they said about the world, what their role was in a vision system, why they were any better than other features, and so forth.[9] Similarly, the criteria used to identify and evaluate structural features (e.g., homogeneity, collinearity) were weakly motivated, mainly by vague appeals to statistics or psychophysics. There was, in particular, no principled basis for deciding how likely it was that a given organization truly reflected some causal relationship in the scene. The actual modeling was often at a superficial level e.g., focusing on the shapes of subjective contours rather than the structural relationships that gave rise to them (Ullman, 1976; Brady & Grimson, 1981). Moreover, since models dealt with particular, narrow classes of structure, there were few, if any, attempts to exploit the rich redundancy of organizational cues that are typically available (e.g, edge finding programs did not exploit the coherence within adjacent regions, and region finding programs did not exploit the breakdown of coherence across edges.)

In short, early vision has been (and largely remains) in a very unsatisfactory state. Can we not do better than an open-ended repertoire of frail and narrow ad hoc techniques?

4.1.1. Toward a Unified Account of Perceived Structure

As previously observed, the common thread underlying many forms of primitive structure perception appears to be the discovery of non-accidental fuzzy identities. If this is really so, it could provide a basis for placing the seemingly diverse phenomena of early vision – edges, curves, texture regions, dot clusters, flow patterns, subjective contours, and so forth – into a unified and principled framework. We do not yet have such a framework, but we do have what could be the beginnings of one.

---

[9]Marr may have had some principled goals in mind for early vision when he initially proposed the primal sketch. However, he never stated clearly what semantics should be attached to the primal sketch's descriptive elements and why. In any event, Marr's original conception of the primal sketch was deferred in favor of his subsequent work on edge detection (Marr & Hildreth, 1979) and the 2-1/2 D sketch (Marr, 1978). Describing an image by the zero-crossings of its convolution with a difference of Gaussians may have physiological and mathematical motivation, but it is not clear what, if anything, such zero-crossings say about the world.

A necessary part of a framework for early vision is a common language for describing all the various manifestations of structure in terms of fuzzy identities. In addition to a descriptive language, we will also need a criterion for evaluating the applicability (or "goodness") of particular descriptions to particular images, and a means of discovering descriptions that are "best" by that criterion. The fuzzy identity principle contributes directly to description and evaluation, but only indirectly to discovery.[10]

Describing structure. All the common manifestations of primitive structure (coherent regions, symmetries, repetitions, smooth gradients, flow patterns, groupings, and so forth) can be characterized as a shape, pattern, or configuration that replicates or continues with little or no change over an interval of space or time. What distinguishes these cases is the nature of the elements involved (e.g. dots, intensity profiles, curves, patches,) and the manner in which those elements are laid out – over space, time, between stereo pairs, etc.

To make this concrete, we have to be able to express, for any particular phenomenon, (a) what pattern is continuing or replicating (e.g. a texture descriptor, an intensity profile, a three-dimension volume, etc.), (b) the domain of its replication or continuation (e.g. the trajectory of a moving object through space-time, or the simultaneous replication of parallel contours over space, etc.), (c) the change or distortion of the pattern over its domain.[11] The domain of replication might be thought of as providing a context with respect to which the local deformation of each replication can be economically described.

Although we have not attempted a systematic classification of organizational phenomena within this framework, we have looked at enough examples to convince ourselves of the framework's descriptive power and generality. A powerful example, discussed earlier, is the case of parallel curves, such as the profiles in Figure 12 or the surface contours in Figure 15. Here, the elements are the curves and the domain of replication is parallel translation. (It does not matter whether the replication occurs over space or time.) The description can be applied recursively. One level down, the curves themselves can be described as point or line elements replicating over a smooth curvilinear trajectory. This

---

[10]The discovery process is inherently combinatorial and, therefore, computationally intensive. As Lowe and Binford point out, it would be prohibitive to examine all possible groupings in an image. The human visual system appears to detect only certain classes of patterns, that can be accommodated using local operators at a hierarchy of scales (Julesz, 1981; Lowe & Binford, 1982).

[11]We can think metaphorically of the base pattern as a paint brush, the domain as a description of a brush stroke, and the distortion as a description of changes in the brush along the path of the stroke (e.g. if we twist or bend the bristles). By this metaphor, each descriptive chunk denotes the stroke of a smoothly changing brush. The literal paintbrush is of course only a metaphor; the "brush" might, e.g., be a description of a three-dimensional structure that is smoothly flowing or flexing over time, and the "brushstroke" may be a description of its rotation and translation. The base pattern, domain of replication, and distortion represent a decomposition of a pattern into (roughly) independent causal components, as discussed in Section 3.1.

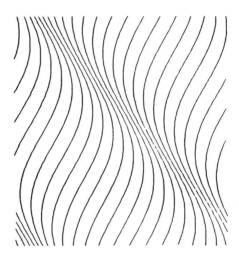

Figure 15. Parallelism can evoke the strong impression of contours lying on a three-dimensional surface.

level of organization is made explicit in Figures 16a and 16b, which show that the trajectory need not be continuous. Moreover, as shown in Figure 16c, the element that is being replicated is independent of the trajectory. In Figure 16d, the basic elements are once again dots, but the domain of replication is different. There are two plausible descriptions: organize the dots first into radial lines and then replicate these lines rotationally; alternatively, organize the dots along concentric arcs and then replicate these elements radially.

Other regular patterns can be described just as easily as the above examples, in terms of their elements, a domain and distortion. Natural patterns seem

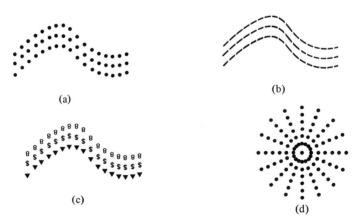

Figure 16. Examples of the grouping of discrete elements: (a) dotted curves, (b) dashed curves, (c) curves composed of arbitrary abstract elements, (d) a radial dot pattern.

only slightly trickier. Consider some examples. For a roughly parallel linear pattern (e.g. a river, or wood grain) the basic element could be a cross-section of the pattern, and the domain could describe the trajectory over which the cross-section sweeps (e.g., the path of a river, winding its way through a plain). A roughly circular pattern (e.g., the growth rings on a tree stump) could be similarly described in terms of a radial cross-section and a rotational domain, or a circular cross-section and a dilational domain.

Evaluating structural descriptions. The basis for evaluating the applicability of a structural description to an image (or for choosing among alternative descriptions) is the least-distortion/non-accidentalness argument developed in Section 3.1. In essence, the argument states that for some appropriate measure of distortion, the likelihood that two structural elements are causally related increases as the amount of distortion decreases. Given alternative descriptions in the above language, one would therefore prefer the one for which a measure of total distortion was minimized.

We have not argued for a specific distortion measure, beyond implying that some form of smooth-as-possible or soap-bubble computation seemed appropriate. To say exactly what measure should be minimized, the argument will need to be developed further into a formal theory. We can, however, be somewhat more specific about the way a semantically motivated evaluation measure should behave.

Clearly, for any reasonable measure, confidence that two entities are causally related should increase with their degree of similarity, approaching certainty in the limiting case of a strict identity and degrading gracefully for near-misses. Moreover, confidence should also reflect the likelihood that a relationship is non-accidental. This implies that confidence should increase with the complexity of the entities involved in a relation: given identical parallel curves, the more complex the curves, the more sure one can be that they are causally related. Confidence should also increase with the range of scale over which an identity relationship holds: it should be possible to discover relationships that hold only over some limited range of scale (e.g., at coarse scales); if the same relationship is also observed at other scales, so much the better. In short, an evaluation measure should, at the least, take into account similarity, complexity, and scale.

How important are these considerations in practice? Numerous ad hoc evaluation measures have been developed for comparing structural elements, many specifically in the context of curve matching (e.g., correlation, variational minimization of distortion metrics, qualitative comparison of distinguished feature points, etc.). These measures have focused almost exclusively on similarity, taking little account of complexity and scale. Predictably, their performance in the various tasks to which they were applied was erratic; there simply was no basis for associating evaluation scores with meaningfulness.

By contrast, an evaluation measure having the desired behavior was recently applied, with considerable success, to the task of describing a one-dimensional signal (e.g., an intensity profile). This problem is not unlike describing a (one-dimensional) image and has, in the past, proved to be equally intractable. At each point on the signal, imagine a vector containing values for the first n derivatives taken at various scales, and perhaps the values of other differential operators. These vectors can now be treated just like the profiles in Figure 12 and used to carve up the signal into meaningful segments.

In Figure 17, the upper waveform in each pair is an original signal and the lower waveform is an abstraction of the important large-scale events in that signal. These events represent intervals over which first and second derivatives are continuous both along the signal and over a contiguous range of scales.[12] Several things are noteworthy about these results. First, they appear to be sig-

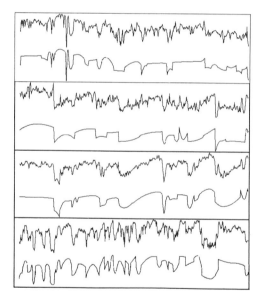

Figure 17. Examples of a signal-description technique applied to natural waveforms. Below each waveform is shown the idealized description.

---

[12]Traditional mathematical definitions of continuity and discontinuity (i.e., in the limit as interval size shrinks to zero) are not applicable to images, both because images are discrete, and because they depict physical events at multiple scales. A surface, for instance, that appears smooth at a coarse scale may appear discontinuous at a finer scale, then smooth again at a still finer scale. Witkin has introduced the notion of "natural scales" at which continuities and discontinuities observed in the data correspond veridically to some physical structure in the scene. He has observed that natural scales occur where the description (i.e., locations of discontinuities) is stable over a range of scales (Witkin, 1983). Such stability can, in fact, be viewed as just another instance of a structural relation (i.e., replication over scale) that would be unlikely to arise by accident.

nificantly better than those obtained with previous ad hoc waveform segmentation algorithms – i.e., they really seem to capture the perceptually salient events; in particular, unlike conventional linear filtering approaches, high frequency detail has been suppressed without appreciably blurring the desired structure. Second, the decomposition shown is not unique; alternative decompositions can be obtained by evaluating continuity at different ranges of scale (Figure 18). In this way, a discrete space of alternative descriptions, each meaningful at some "natural scale," can be generated and examined under control of higher-level processes. For instance, in the case of a broken contour, reasonable descriptions can be obtained at both coarse and fine scales. A full report on this signal description technique will be forthcoming shortly (Witkin, 1983).

A unified view of early vision. Many traditional problems of early vision (and many that are traditionally associated with intermediate vision as well) can be recast in terms of the descriptive framework presented above – i.e., discovering relations among elements replicating over space and time. Expressed in these terms, a number of superficially unrelated problems appear reducible to a common form. Indeed, to a first approximation, all the essential problems of early and intermediate vision appear to be manifest in one canonical inference problem: discovering a non-accidental, fuzzy identity relation between a pair of roughly parallel curves.

Take edge and region finding, for example. Assume, for a moment, that the curves in Figure 12 represent profiles of intensity. Without knowing any-

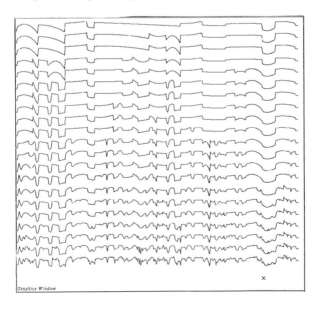

Figure 18. A waveform described at several scales.

thing about the details of the imaging transformation that produced those pro-files, the presence of similar events at various scales and at corresponding loca-tions on adjacent profiles provides strong *prima facie* evidence that such pro-files are causally related and should be grouped. Conversely, corresponding sections of adjacent profiles that are highly uncorrelated are probably unrelated. This profile-to-profile coherence can be directly exploited by both region finders (which seek coherence within an area) and edge finders (which can seek either coherence of a discontinuity from profile to profile along an edge, or the break-down in coherence across profiles parallel to and straddling the edge). Not only are edge and region finding unified within this framework; it becomes hard to conceive of them as independent problems.[13]

By ascribing different meanings to the curves in Figure 12 the underlying unity of a number of other problems in early and intermediate vision becomes obvious. If, for example, the profiles happened to represent values of depth (as might be obtained doing stereo matching on epipolar lines) then identifying corresponding points on adjacent profiles would provide the basis for surface interpolation.[14] Alternatively, the same matching process could be used to parti-tion the range data into segments with smoothly varying cross-sections, each suitable for description as a generalized cylinder (Agin & Binford, 1976). If, on the other hand, the curves represented surface contours (as they do in Figure 15), matching would then provide the basis for recovering three-dimensional shape (Stevens, 1981). This line of generalization appears to extend quite far, since curve matching, in one guise or another, plays an important role in so

---

[13]A number of researchers have successfully exploited the structure inherent in intensity profiles for basic early vision functions such as edge, line and region extraction. The Binford/Horn line finder, for example, was probably the most successful program for segmenting blocks-world scenes (Horn, 1971). It detected discontinuities within each profile, and matched similar discontinuities on adjacent profiles. However, it relied heavily on assumptions peculiar to blocks-world scenes (e.g., only straight edges, producing step and gradient discontinuities of intensity). Ehrich and Schroeder (1981) developed a similar profile-matching approach to segmentation of natural scenes based on a much richer hierarchical description of profiles in terms of peaks and valleys. Quam (1978) applied a form of numerical correlation to profile matching for the purpose of tracking linear features such as roads. Witkin (1982) used the presence or absence of correlated structure across a putative edge to verify its presence and infer things about the physical nature of the corresponding boundary in the scene. As a group, these programs achieved superior performance by exploiting the detailed struc-ture inherent in image data. (Most computer vision programs, by contrast, cannot cope with massive amounts of image detail and simply throw it away using thresholding, edge operators, or some other context-independent form of preprocessing.) Nevertheless, these techniques leave much to be desired, in terms of both absolute performance, and their ad hoc approaches to describing and match-ing intensity profiles.
[14]Although interpolation is usually not thought of as a structure matching problem, when reformulat-ed in terms of fuzzy identity and non-accidentalness, it fits naturally into our framework. (Picture a surface as having been swept out by moving one boundary curve into another, deforming the curve so as to minimize some measure of the total distortion incurred.) This reformulation is contrasted with conventional approaches to surface interpolation (e.g., Grimson, 1981) in Section 4.2.1.

many tasks, e.g., stereo correspondence (for eliminating local ambiguities [Baker, 1982]), figure-ground organization (for recognizing parallel and symmetric bounding contours), edge classification (Witkin, 1982), and even shape recognition (e.g., recognizing facial profiles [Harmon, 1977] or geographic coastlines [Davis, 1979]). Moreover, because of the close relation between parallelism and rigidity (both are identities, one over space, the other over time), the generalization extends directly to such problems as the discovery of coherent motion and the recovery of depth from rigid body motion.

Understanding relations among problems and common underlying principles of their solutions is important because it provides a clear indication of where research leverage lies. Of course, if all we succeed in doing is reducing a dozen intractable problems to one intractable problem (i.e., curve matching) we have not accomplished very much. The promising results on signal description, a problem which can itself be formulated in terms of profile matching, suggests that our conceptual framework can contribute more than just evocative examples.

4.1.2. Recapitulation

Early vision has been characterized as a compromise between what is *possible* to extract from an image and what is *desirable* for meaningful interpretation (Marr, 1982). The discovery of primitive spatio-temporal regularity appears to serve this role well; it is a level of description that is reasonable to recover directly from an image – richer than edges and regions, more primitive than surfaces and objects. It is also the right level of description on which to base subsequent interpretation; the resulting entities correspond to causal events and processes in the world; although more primitive than objects, they are just as meaningful.

We have suggested a framework for describing and evaluating primitive structure, based on the least-distortion/non-accidentalness arguments developed earlier. Primitive structure was characterized as a fuzzy identity relationship among elements replicating smoothly over space and time. The presence of such an identity relation, conversely, provides compelling evidence that the relationship is meaningful and worthy of explanation. We reformulated some problems of early and intermediate vision within this framework, and in the process, discovered that a number of superficially unrelated problems could be reduced to a common form: the discovery of fuzzy identity relationships. All of the examples of early (and intermediate) description given in this section are instances of fuzzy identities, differing only in the nature of the elements over which regularity is sought (e.g., two-dimensional curves, intensity profiles, dots), and the dimension over which the elements are arrayed (i.e., two-dimensional space, three-dimensional space, or time).[15] Although we are far from having a formal theory, much less algorithms based on such a theory, the examples support our contention that early vision can be unified and put on a sound footing.

4.2. Intermediate Vision

We have previously argued that the recovery of local, quantitative surface characteristics should, at best, be viewed as a peripheral goal of intermediate vision. Our account of structure suggests, as more significant goals: discovering new three-dimensional structural relations that are not obvious from their image projections; and accounting qualitatively for observed structural relations (two- and three-dimensional) as projections of three-dimensional surfaces. These alternative goals appear more feasible, more useful to higher levels, and more consistent with human vision. We shall now discuss each of them in turn.

4.2.1. Discovering Structure in Three Dimensions

Not all regular relations will be obvious in their two-dimensional image projections; some three-dimensional relations are revealed only by applying knowledge of the image-forming process. There are many examples demonstrating that humans do recover such structure. A particularly well-known one is our ability to attribute what, in two dimensions, appear to be randomly moving dots, to the projection of a rigidly moving body. Another is our ability to infer a smooth surface from sparse features in a random-dot stereogram. Beyond the fact that these structural relations are inherently three-dimensional, the principles for discovering such structure are essentially the same as for the two-dimensional case.

The direct analogy of discovering two-dimensional structure in two-dimensional data is discovering three-dimensional structure in three-dimensional data (e.g. the volumetric image produced by a CAT scanner); just as points and curves can replicate to produce two-dimensional structures, areas can replicate to produce solid structures (i.e., generalized cylinders).[16] The more relevant intermediate case – that of discovering three-dimensional structure given a two-dimensional *projection* – is less straightforward, because the discovery of structure is intermixed with the recovery of the third dimension. It is possible, how-

---

[15]It is tempting to conclude, on the basis of our examples, that a good deal of the superficial diversity in perceptual problems may reflect diversity in the data rather than the perceiver. This conclusion is reminiscent of Simon's well-known explanation for the complex behavior of ants (Simon, 1969). It is, of course, a speculation and at best no more than an idealization: We recognize that each problem has its own peculiarities, and by no means intend to suggest that all the problems are identical either conceptually or at the level of mechanism.

[16]In this case, the trajectory of the cross-section would be three-dimensional, and the shape and orientation of the cross-section could change as it moved along the trajectory.

ever, using a model of the image forming process, to construct the family of three-dimensional structures consistent with a two-dimensional projection. Searching for regularity in this family is then conceptually no different than doing so directly in two- or three-dimensional data.

In computational vision, problems involving three-dimensional structure are usually not thought of in terms of discovering regularity. Instead, they are generally cast in terms of quantitative surface recovery. Putting structure to the fore places many of the classical recovery techniques in a different light; the recovery of depth and other quantitative properties becomes subsidiary to the discovery of new inherently three-dimensional structure.

Ullman's formulation of the shape-from-motion problem (Ullman, 1979) can be readily recast as a technique for discovering three-dimensional regularity. In his original formulation, rigidity, projection, and non-accidentalness were used to solve quantitatively for the depths of dots painted on a rigidly moving surface. Using exactly the same computation, however, we can turn the problem around, stating as the primary objective discovering that a collection of dots are causally connected (in a three-dimensional rigid structure). We would then be using rigidity and non-accidentalness as we did in two dimensions, as a means for evaluating the meaningfulness of an organization: i.e., if a rigid (or roughly rigid) organization can be recovered, it can be assumed to represent something significant. Because we happened to employ a model of projection in the discovery process, we still recover three-dimensionality, but more nearly as a by-product of the evaluation process.

A number of other computations, whose original goal was the recovery of local quantitative three-dimensional properties, can similarly be recast in terms of discovering structural relations. In particular, the reformulation applies across the board to any computation that employs a "smooth-as-possible" solution. Such computations have been used extensively in intermediate vision both for resolving the indeterminacy brought about by the imaging process and for filling in from boundary conditions in the absence of other constraints. In these roles, the computation has typically been invoked to determine a best solution subject to constraints (e.g., the best surface across sparse stereo data). In our reformulation, the same computation would be used for evaluating the likelihood of a fuzzy identity relation, e.g., between an image curve and its possible three-dimensional projections, or directly between some three-dimensional surface contours. As such, its primary role would no longer be in determining *how* to interpolate, map, etc., but *whether* to do so.

Replacing a goal of *re*covering a quantitative property with that of *disc*overing a structural relationship has implications both for how the computation should be performed and how the results should be interpreted. In the case of smooth-as-possible computations, for example, it may require that a markedly different distortion measure be optimized (e.g., one that takes into account considerations that impact non-accidentalness, such as the complexity of a relation and the range of scales over which it holds). Even when actual computations remain the same, this reformulation of goals is far from academic: discovering regularity in three dimensions is a fundamentally different task than recovering depth maps. Which goal to adopt depends, of course, on whether depth arrays or structural relationships are seen as more important. In Section 2.2, it was argued that a numeric array, by itself, has limited utility – that even if you had one, it would still be necessary to explicate its structure. On the other hand, discovering a structural relation is important, even without depth as a byproduct, as it allows the component entities to be treated as a unit for almost any purpose thereafter. Such structure is essential for qualitative surface description and object recognition; it can even prove helpful in recovering a depth map.

Using 2-D structure in the discovery of 3-D structure. We have shown, through examples, that the human visual system is able to discover both two-dimensional structure and three-dimensional structure in image data. When three-dimensional structure is present, most of the time it will also be directly manifest in the image as some form of two-dimensional regularity. This two-dimensional regularity may be easier to discover, and would appear to be quite helpful in constraining the search for a three-dimensional relationship.

Two-dimensional structure could, for example, provide powerful and reliable spatial constraints for use in conventional quantitative recovery methods (e.g., shape from contour and shape from shading). As discussed in Section 2.2, such techniques require constraining assumptions to tie together scene parameters at different spatial locations. This role traditionally has been filled by local continuity assumptions or artificial domain restrictions. Such methods also rely heavily on local discontinuities (i.e., edge fragments) to supply boundary conditions and to delimit interpolation. Both requirements would seem better served by finding coherent entities in the image that span large spatial extents, and that correspond to meaningful structures in the scene. At the least, such entities can provide continuity assumptions that are sensitive to the content of the image, so that spatial integration does not cross perceived discontinuities. They can also prove helpful in inferring the physical nature of discontinuities; for example, the continuation of a regular texture across a boundary can be used to infer the presence of transparency or shadow. We will have considerably more to say about such inferences shortly in connection with qualitative interpretation.

Another possibility is that by treating structurally-related image features as entities, the combinatorics involved in finding a unifying three-dimensional relationship can be significantly reduced. For example, once a set of dots has been associated with a common motion, the possibility that that motion is attributable to a rigid body transformation is readily established. One simply solves a set of overconstrained equations for the relevant parameters (i.e, trajectory of translation, axis of rotation, etc.) (Ballard & Kimball, 1982); the existence of a consistent solution confirms the hypothesis. Similarly, once it has been established that a pair of parallel image curves are causally related, one can hypothesize that those curves are attributable to parallel cross-sections of a smooth ruled surface, and solve directly for the surface orientation along lines connecting corresponding curve features (Stevens, 1981). In a somewhat different vein, matching coherent image fragments rather than isolated points should reduce significantly the correspondence ambiguities in recovering depth maps from stereo or motion data. In all of these cases, the key thing is to discover what things are related; the actual numerical computation of depth or orientation values then becomes straightforward – and perhaps superfluous.

4.2.2. Qualitative Surface Description

The availability of rich and meaningful primitive structure (in two or three dimensions) makes possible a qualitative and holistic style of interpretation that departs markedly from that of local quantitative surface recovery techniques. In simple cases, where the structure observed in an image admits a fairly direct three-dimensional interpretation, interpretation can amount to little more than labeling the primitive structure – e.g., attributing the regularity of parallel wavy lines to an undulating surface and at a finer scale, a localized distortion in some of the lines to a bump or dent in that surface. The interpretation can be further elaborated, for example by using the rough shape of the lines to make qualitative statements about the shape of the underlying surface (e.g., "gently curving") and the direction of their replication to say which way it is bending. Simple qualitative statements of this type could well be more useful at higher levels of interpretation than traditional quantitative descriptions.

While qualitative interpretation may well be rooted in an underlying quantitative description (indeed, we have argued that such a description could be a byproduct of discovering structure), there is no reason why this must be so. A structural relation with reasonable spatial extent will probably have relatively few plausible interpretations, perhaps just a few discrete alternatives (e.g., parallel curves might be attributed to surface markings, surface contours, extremal boundaries, or a few other possibilities). The range of possible interpretations for fine-scale structure in the context of a global interpretation would appear to be similarly constrained. Moreover, as we shall demonstrate shortly, qualitative image features such as the direction of a brightness gradient or the shape of a highlight appear to act like switches, selecting among the possible interpretations. Qualitative surface interpretation thus has many of the characteristics of a classic labeling process (Rosenfeld et al., 1976; Tenenbaum & Barrow, 1977).

Accounting for image structure in terms of surface properties may reveal additional relationships that were not previously noticed. Such relationships can most often be built up out of the original elements of structure by performing simple cut-and-paste transformations on them. Thus, one would paste together surface fragments that adjoin along a shadow or that protrude (with smooth continuation) from behind an occluding body. Similarly, one would relate two individually coherent moving dot clusters belonging to different surfaces of the same rigid body. Since richly structured entities can combine in only a limited number of plausible ways, this transformation process, like interpretation, has the flavor of exploring a discrete space of alternatives. The possibility of transforming a structural description to conform with an interpretation means that the original description need not be perfect, only good enough to get started.

In summary, the availability of rich and meaningful primitive structure suggests the possibility of a qualitative style of interpretation involving labeling, elaboration, and transformation. The hallmark of this style is the use of qualitative photometric and geometric features to select among a discrete set of alternative interpretations.

Qualitative photometry and geometry. A number of qualitative photometric and geometric image features have been suggested (or actually used) for interpreting structure. These include junction geometry (Barrow & Tenenbaum, 1981a; Binford, 1981), the direction of shading and texture gradients relative to boundary contours (Barrow & Tenenbaum, 1981a), the correlation of intensity profiles across edges (Witkin, 1982), and the shapes of highlights (Stevens, 1979; Barrow & Tenenbaum, 1981a).

Junction geometry provides a powerful non-photometric basis for interpreting edges in the image as extremal, occluding, or intersecting surface boundaries. Knowing the type of an edge, it is usually possible to determine the nature of the surface(s) it bounds. For example, a junction formed by two straight lines implies a planer surface, and a junction formed by three straight lines implies a trihedral vertex formed by the intersection of three planes. Also, a straight line tangent to a elliptic arc (Figure 19) corresponds to a vertex where an extremal edge meets an intersecting edge and an occluding edge. This occurs, most commonly, where the curved side of a cylinder meets the planar visible end.

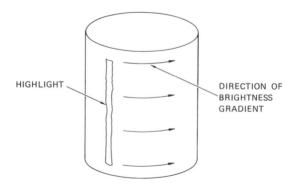

Figure 19. Qualitative cues to surface shape and boundary type.

The direction of brightness or texture gradient on a surface frequently corresponds approximately to the direction of maximum curvature of the surface. At extremal boundaries, the gradient normal to the boundary is thus likely to be very high, while at occluding contours the gradient is likely to be high along the boundary in the direction of maximum curvature (see Figure 19). The experiment on artificial shading gradients, discussed in Section 2.2, demonstrates that such qualitative photometric features play an important role in human surface perception. Of particular relevance is the fact that people were confused when the gradient direction was inconsistent with other qualitative cues (Figure 9f).

The coherence of intensity profiles taken along both sides of an edge has proven effective in distinguishing shadow and occlusion edges. Along the edge of a shadow, the profiles will be highly correlated but differ in absolute intensity, while along an occlusion edge they will be uncorrelated (Witkin, 1982). Coherent profiles (i.e., highly correlated with similar intensities) are indicative of no edge. This technique is thus closely related to techniques discussed in Section 4.1 for partitioning images by matching nearby intensity profiles.

The presence or absence of a few isolated bits of internal structure, interpretable as highlights, can cause an entire region to be perceived as a glossy, curved surface or a flat, matte one (see Figure 20, from Beck [1972]). A further clue to the nature of specular surfaces comes from the shapes of the

Figure 20. Beck's demonstration of local cues to surface properties: the addition of a few localized highlights changes the perception of an entire surface from matte (right) to glossy (left).

highlights. Highlights are actually images of the light source. With a point source, a doubly curved surface (e.g. a sphere) yields a point highlight, a singly-curved surface (e.g. a cylinder) yields a linear highlight aligned with the direction of zero curvature (Figure 19), and a planar surface is unlikely to yield a highlight at all (Stevens, 1979).

Beck's experiment raises the possibility that humans rely on qualitative features to switch among global interpretations. It suggests, as did our own artificial shading experiment, that the primary role of photometric cues in human vision may be in qualitative determination of boundary and surface types, rather than in quantitative determination of shape.

(a)

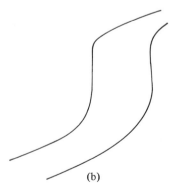

(b)

Figure 21. (a) Road scene with snow-covered road; (b) primitive structure corresponding to the road in (a).

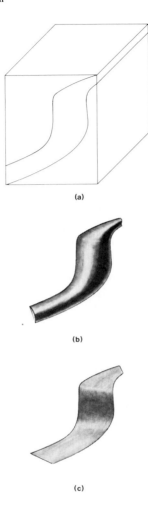

Figure 22. The same primitive structure may be interpreted in many ways, depending on context: (a) planar surface, (b) cylinder, (c) ruled surface.

Qualitative (and semi-quantitative) photometry can be used to elaborate and refine interpretations, as well as to select them. For instance, inflections in brightness are likely to correspond to inflections in surface curvature. At arbitrary points on a surface, intensity gradients and their derivatives in two orthogonal directions can provide clues to curvature, indicating whether the surface is locally planar, cylindrical, elliptic (e.g., spherical), or hyperbolic (inflexive) (Turner, 1974; Pentland, 1982b; Smith, 1982). They can indicate the presence of a bump or dent on an otherwise smooth surface, as well as the relative height of the anomaly.

Note that structural descriptions can also be very helpful in recovering more quantitative surface characterizations. For example: as mentioned previously, surface orientations can be interpolated on cylindrical sections from known values along extremal boundaries; structural features such as parallelism (Stevens, 1981) and skew symmetry (Kanade & Kender, 1980) have been found to strongly constrain local orientation on planar and ruled surfaces; matching labeled structure fragments would simplify the correspondence problem in stereo ranging; and so forth. We still contend, however, that most of the time such quantitative descriptions will be unnecessary, indeed less useful, than more qualitative descriptions.

Examples of qualitative interpretation. We shall now illustrate, with some evocative examples, how the qualitative interpretation of structure might proceed. Suppose that some primitive low-level grouping process has succeeded in extracting as a structural entity the coherent S-shaped region in Figure 21a (which we recognize as a snow-covered road). Out of context, this region might be represented by the structure sketched in Figure 21b.

Given such an image structure, a primary concern is the nature of the physical scene structure it corresponds to. Is it a shadow or a surface? If a surface, is it two- or three-dimensional, e.g., shaped like a flat S-shaped cutout, an undulating ruled surface, or a generalized cylinder? The appropriate interpretation can be selected from among these and other possible alternatives, by examining qualitative photometric and geometric features associated with the structure as well as the context in which it is embedded.

If the S-shaped structure were embedded in a context that was clearly planar, it would itself appear to be a flat surface marking lying on that plane. (This is literally the case for curves printed on a piece of paper; see also Figure 22.) If, on the other hand, there were intensity or texture gradients running across the contours and, perhaps, an elongated highlight aligned with them (Figure 22b), then the structure would likely be a cylindrical section. Finally, if intensity and texture vary most rapidly along the bounding contours, perhaps with a highlight oriented cross-wise (Figure 22c), then the structure would likely be an undulating ruled surface (a reasonable interpretation in the context of Figure 21a). Notice how easy it is (for us as humans) to flip among these interpretations, applying alternative labels to the same primitive structure.[17]

Once a fragment of image structure has been interpreted as a particular type of surface structure, simple models of projection can then be used to make qualitative statements about what the surface is doing in three dimensions.

---

[17]Unlike the traditional hierarchic model, surfaces and volumes are no longer distinguished as levels of representation; they are merely alternative three-dimensional interpretations for structural relations discovered in an image. A pair of parallel lines, for example, might be interpreted either as the extremal edges of a cylinder or as the contours of a surface patch. The related distinction between viewer-centered and object-centered descriptions, emphasized by Marr and Nishihara (1977), also does not really arise here. Unlike a range map, the structural features that we are concerned with (e.g., continuity, parallelism, symmetry axes, etc.) already provide a viewpoint-independent description.

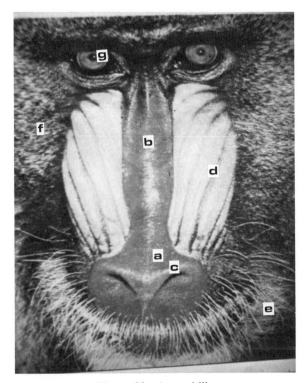

Figure 23. A mandrill.

Assume, for example, that the structure depicted in Figure 21b is to be interpreted as a ruled surface (Figure 22c). It then follows that the surface is bending faster along contour A than along contour B, and slowest of all along contour C. (Contour A is a local maximum of curvature and contour B is an inflection point. The parallel axes imply zero local curvature along contour C.) Conversely, had we assumed that the structure was a cylindrical section a la Figure 22b, then the surface would be bending faster along contour C than along contour B, and still bending fast along contour A, though in a different direction.

As a second example, we will use Figure 23 to illustrate how the qualitative interpretation of structure might proceed using actual qualitative image features. We again assume that primitive, low-level grouping processes have succeeded in pulling out interesting structures that exhibit coherence at a coarse scale. One might expect such entities to include the large dark central region (A), the white elongated and somewhat fragmented regions within it (B,C), the surrounding white fluted region (D), the ensemble of thin, parallel, curving white regions (E), as well as other prominent regions such as the large textured background region (F) and the two circular regions (G).

If we now label region B as a highlight, its elongated shape gives strong evidence that the surrounding part of region A is locally cylindrical, bending across the highlight, with zero curvature along it. Not much more can (or need be) said about the local shape of region A. Similarly, if region C is interpreted as a highlight, then region A is locally cylindrical in that vicinity, again bending across the highlight. Switching attention to region D, if the dark curving lines are perceived as shadows, than the interleaved white regions are most naturally interpreted as raised (i.e., fluted) ridges casting the shadows. The parallel curving structure of the elements labeled E conveys a strong impression of tri-dimensionality, as if they were surface contours on a generalized cylinder whose principal axis curves horizontally across the image. Finally, the texture gradients in region F convey a qualitative impression of shape; in each local area, the surface is seen to curve most rapidly in the direction of the gradient.

4.2.3. Recapitulation

We began this section by asking: What is the role of structure in intermediate vision? Perhaps a better question might have been: What is the role of intermediate vision? The plausibility arguments and demonstrations presented lead us to conclude that intermediate vision is not fundamentally concerned with recovering numeric arrays of surface characteristics, in the sense of Marr's 2-1/2 D sketch or Barrow and Tenenbaum's intrinsic images. Instead, a more appropriate goal may be the discovery and interpretation of image structure based on photometric and geometric models of imaging, where both the interpretations and the models can be highly qualitative. By way of illustration, we described one way such an interpretation process might work, involving three aspects: labelling, elaboration, and transformation.

Our reformulation of intermediate vision in terms of structure still shares, with intrinsic images, the basic goal of decomposing the image into underlying constituents attributable to the generating process. However, intrinsic images decompose the image numerically, point by point, whereas the structural decomposition is expressed in terms of entities of significant spatial extent. This reformulation has considerable appeal from a computational standpoint. First, it involves far fewer degrees of freedom, both with respect to the number of things that must be interpreted (global structures vs. local surface elements) and the space of possible interpretations (a small, discrete set of high-level alternatives vs. a continuum of numerical values). For example, a region of smoothly varying brightness can be treated as an entity, attributable, perhaps, to the curvature of a cylindrical surface or an illumination gradient on a plane. Second, it imposes spatial constraints that should prove more powerful and more reliable than local continuity assumptions. Finally, the process is driven by qualitative photometric and structural features, the kind likely to be useful in real scenes, and likely to be preserved by a wide variety of imaging processes. All of these advantages depend on the availability of a structural description.

4.3. High-level Vision

In complete analogy with the goals set forth for intermediate vision, we see two primary goals for high-level vision: discovering new structural relations that are not obvious without specialized (i.e., semantic) models, and using such models to attribute meaning to previously discovered structure.

Two well-known examples of semantically imposed structure are: Gregory's overexposed image of a spotted dog on a spotted background (reproduced in Figure 24); and Johansson's demonstration in which dots apparently moving at random are suddenly perceived as lights on the joints of a dancing couple (Johansson, 1975).[18] In both examples, the gulf between primitive and high-level organization is too wide to bridge unless the semantic concepts are known a priori. Situations such as these are difficult to perceive (often requiring long scrutiny or coaching). Fortunately, they do not arise very often in natural scenes.

The interpretation of previously observed structure can proceed in essentially the same way as it did at intermediate levels – through labelling, elaboration, and transformation. Where a structural description has successfully captured a meaningful concept, interpretation can again involve mainly hanging a semantic label on the structure (e.g., snake or magic carpet for Figures 21b and 21c respectively). However, as we move to higher levels, the models get more specialized, and the match between naive and informed structure is less likely to be perfect. Hence, in addition to labeling, the explanation process is also likely to involve the imposition of new relations among the naively perceived structural elements to bring the description into correspondence with stored models. A relation might be imposed, for example, between regions that can be interpreted as parts of the same object (e.g., the seat and back of a chair), or alternatively, between regions that can be interpreted as visible fragments of a partially occluded object. These imposed relations might serve as global (object-centered) coordinate frames (as illustrated in Figure 25). This style of interpretation extends beyond single objects to aggregations of objects (e.g., a room full of chairs).

Note that nothing has been assumed regarding prior three-dimensional interpretation. It seems quite reasonable to invoke semantic models directly to account for two-dimensional structure (as in, e.g., interpreting a black blob on a desk as a telephone, to cite a well-known example from the computer vision literature [Garvey, 1976], or more generally, when interpreting cartoon caricatures – e.g., of magic carpets). In addition to labeling the structure, the model might also help elaborate its three-dimensional form. The concept "magic car-

---

[18]While mechanisms for discovering non-primitive structure are beyond the scope of this paper, it is intriguing that semantic model matching can itself be viewed as a search for non-accidental relations (recall the aardvark example in Section 3.3). We are led to wonder whether this unity is more than conceptual.

Figure 24. Gregory's Dalmatian dog: an illustration of the gap that sometimes separates primitive structure from semantic interpretations.

pet,'' for example, carries with it rich associations of shape (e.g., a magic carpet is a developable surface) that can be transferred to any image structure bearing that label.

The commonalities we have observed at early, intermediate and higher levels of interpretation make these traditional distinctions seem quite arbitrary. What emerges is a much more unified view of visual perception as a process of discovering and interpreting structure.

4.4. Speculations on Representation and Control

The tendency for primitive structure to survive in high-level interpretations, and the indications that the operations of discovery and interpretation may be uniform at all levels, together argue for a central representation built up by transformation and elaboration, rather than a fixed hierarchic sequence of representations.[19] This kind of central representation in turn implies a system organization in which models of many types are applied to the central representation, each attempting to label, elaborate, or transform it in accordance with their knowledge (Figure 26). Models can involve everything from active group-

---

[19]Although we reject the notion of a strict representational hierarchy, there is still a logical ordering of levels in terms of generality, from primitive grouping and regularity, through image-formation models, to highly specialized object models. Moreover, discovering structure at a primitive level may often be a prerequisite to high-level explanation.

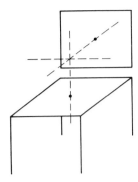

Figure 25. Global coordinate frame for a chair.

ing processes (both local and global) that operate primarily on raw imaeg structure, to the models of imaging and semantic objects discussed in this paper, to
very high-level schemata concerned with the type of scene (e.g., "kitchen
scene," "landscape") or scenario (e.g., "couple dancing") being depicted. As
suggested in the figure, models need not be limited strictly to visual perception,
but can extend to other sensory modalities including auditory (e.g., "a ringing
black blob"), tactile (e.g., a hard surface), heat, and odor perception.

The notion of a central representation of structure where various perceptual modalities come together has strong introspective appeal. Closing an eye
may cause things to appear slightly flatter (due to loss of stereopsis), but basically the world looks pretty much the same. Similarly, degrading sensory data
by suppressing color, texture, or motion (or sound, heat, and odor) will make
the world seem progressively less rich, but will generally have limited impact
on one's perception of three-dimensional structure and semantic content. Even
adding or deleting semantics (e.g., by telling a person what he is looking at in
an unfamiliar scene) will usually not have a first-order effect on perceived scene
structure. Each of these examples contributes to the subjective impression that

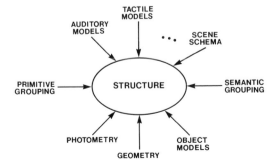

Figure 26. A system organized around a central representation.

some basic representation of image structure is being elaborated by whatever sensory data and models are available. When both eyes are closed, there is no longer any image structure to hang interpretations on, and the world appears instantly to vanish.

A system organized around a central representation of structure could, in principle, overcome some of the computational problems traditionally associated with the sequential, data-driven, hierarchical organizations of current vision theories. Potential advantages of a centrally organized system include responsiveness, reliability, and graceful degradation, all vital to survival. With a central representation, a large, rapidly moving blob with yellow and brown stripes could immediately invoke the semantic label "tiger" (and an appropriate avoidance action) withtout having to wait for a detailed surface reconstruction. Such a system also would not be derailed by a degraded image or missing knowledge source that made interpretation impossible at some level. Thus, although Figure 27 is too degraded to recover surface structure directly (from shading, texture, contour, etc.), there is still enough information to invoke a face model, and then fill in surface detail on semantic grounds. (In the absence of knowledge about faces, all one would see is a set of high-contrast blobs.) In general any knowledge source (i.e., model) that has something to contribute can do so productively, affording very flexible and opportunistic control.[20]

The organization depicted in Figure 26 is somewhat reminiscent of the well-known Hearsay speech understanding system (Erman et al., 1980), with the discrete space of structural descriptions serving the role of a global blackboard. Such an organizaton, in which all knowledge sources (from grouping to scene models) contribute to a common representatin, facilitates knowledge integration. First, since shading, texture, and other manifestations of image structure derive from the same scene structure, it is reasonable to expect that various structure building processes will yield consistent, if not identical, results. Moreover, one would expect that qualitative interpretations of shape, derived on the basis of global structure (e.g., brightness and texture gradients over a region), will be much more consistent, and thus easier to integrate, than the noisy local values of surface characteristics now computed.

5. Summary and Conclusions

We began with a critical look at the hierarchical models of vision proposed by Marr at MIT and by Barrow and Tenenbaum at SRI, and found some fundamental weaknesses that derive from their lack of concern with structure. A keystone of these models was an intermediate description of local surface

---

[20]Leaping to conclusions can of course lead to erroneous interpretations (and inappropriate actions) that might have to be taken back on closer scrutiny: e.g., that wasn't your friend you just waved to. Such errors, though embarrassing, are seldom fatal. Since opportunism can improve responsiveness in critical tasks (like avoiding tigers), it would appear to be a good design compromise.

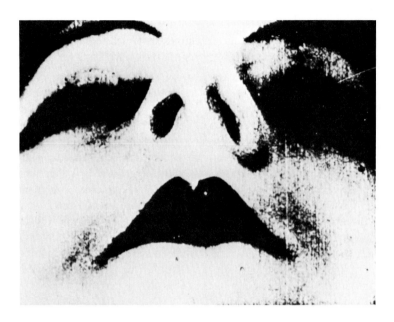

Figure 27. A degraded image, demonstrating that semantic interpretation can proceed even where surface perception is impossible.

characteristics in the form of numeric arrays (known as intrinsic images or the 2-1/2 D sketch). It had been believed that such a description was the proper goal of early vision, that it could be recovered using arrays of local processes, and that it would greatly faacilitate higher level description into surfaces, volumes and objects. We questioned these beliefs, because we suspect that in complex scenes, surface characteristics cannot be readily obtained by local means, and that even if they could be, the resulting intrinsic images are not that much more directly useful, for many tasks, than the original image.

We next explored the nature of primitive structure and its role in vision. We observed that people's ability to perceive structure in images exists apart from both the perception of tri-dimensionality and the recognition of familiar objects; we can impose organization on data (noticing flow fields, regularity repetition, etc.) even when we have no idea what is being organized. Such naively perceived structure is consistent across observers. Moreover, the naive observer often sees essentially the same things an expert does, the difference between naive and informed perception often amounting to little more than labeling the perceptual primitives. That naively perceived structure survives more or less intact once a semantic context is established implies that the visual system has some basis for guessing what is important without knowing why. We argued that the discovery of structure was a primitive form of inference, the basis for which lies in the relation between structural and causal unity: the appearance of spatiotemporal coherence or regularity is so unlikely to arise by the chance interaction of independent entities that such regular structure, when observed, almost certainly denotes some significant underlying process or event.

Finally, we considered in some detail the implications of structure for computational theories of vision. Viewing the discovery of structure as a primitive form of inference evokes a very different picture of perception that that implied by current hierarchic models. The hallmark of structure-oriented perception, from the most primitive levels, is the ongoing search for regularity and coherence, and the attempt to explain the observed structure in terms of models at many levels. The goal at early levels is to decompose the image into primitive descriptive chunks whose regularity or coherence deserve or demand explanation. Subsequent interpretation in terms of more specialized models – of image formation, and of particular physical processes and objects – is driven by these "semantic precursors," and acts to label, transform, elaborate, and assimilate them. Perceived structure provides a richer and more meaningful initial description than, say, local surface characteristics, permitting interpretaton to proceed more directly and at a higher level than in current models.

The discovery of primitive structure cuts across the traditional boundaries of early and intermediate vision, because the search for regularity proceeds in essentially the same way whether conducted in two dimensions or three. Indeed, thinking about the search for regularity as a primitive form of inference has allowed us to reformulate some problems of early and intermediate vision, and in the process, discover that a number of superficially unrelated problems might be reducible to a common form. To a first approximation, all the essential problems at these levels appear to be manifest in one canonical inference problem: discovering a non-accidental, fuzzy identity relation between a pair of roughly parallel curves. Progress on this problem would impact edge and region finding, edge classification, stereo and motion correspondence, interpolation, subjective contours, surface description, shape matching, and a host of other problems. Fundamentally, all these problems involve the discovery of fuzzy identities. What distinguishes them is primarily the nature of the elements over which regularity is sought (e.g., two-dimensional curves, depth profiles, dots), the way those elements are arrayed over space and time, and the constraints on the solution (e.g., the projective constraint); the basis for inference and the nature of the inferences drawn are the same in every case. If this promise of unification is fulfilled, it will provide not only conceptual and theoretical unity, but more pragmatically it will allow us to focus our research activities more efficiently; a great deal of effort has been dissipated developing ad-hoc solutions to narrow problems. If we can identify truly fundamental problems (or at least *more* fundamental problems), then whatever success we have in solving them will by definition have general and enduring applicability.

The interpretation of structure also crosses traditional boundaries, in this case between intermediate and high level vision. For example, having discovered a compelling parallelism relationship, we may be able to interpret it in many ways and at many levels – as surface markings, a curved surface, a waving flag, even just noisy data – but we can never dismiss it. In the easiest cases, where a primitve structural description directly captures what is important, an explanation may entail little more than hanging label(s) on it (e.g., an undulating surface, a flag). Where the match between naive and informed structure is less perfect, imposing an explanation may also entail performing simple holistic transformations, such as cut-and-paste, to bring the description into correspondence with stored models. Almost always, however, the elements of primitive structure will be assimilated, nearly intact, into all levels of interpretation. This persistence of primitive structure to the highest levels suggests a central representation built up by a process of elaboration and transformation, rather than a strict hierarchy of representational levels.

In conclusion, perceptual organization plays a central role in human vision. We feel strongly that it deserves a more central role in machine vision as well. We have made a start toward a principled account of the basis for discovering and interpreting structure. Turning our ideas into a formal theory of perceptual inference will not be easy, and it is almost inevitable that in the process, many of our speculations will be proved wrong. However, it is important to persevere because we believe that perception will never be understood deeply without understanding it at this basic level.

Acknowledgment

We thank our colleagues Harry Barrow, Richard Duda, and Peter Hart for their insightful comments.

References

Agin, G. J., & Binford, T. O. Computer description of curved objects. *IEEE Transactions on Computers,* 1976, **25**, 439-449.

Attneave, F. Prägnanz and soap-bubble systems: A theoretical exploration. In J. Beck (Ed.), *Organization and Representation in Perception.* Hillsdale, NJ: Erlbaum, 1982.

Baker, H. A system for automated stereo mapping. Proceedings, DARPA Image Understanding Workshop, 1982, 215-222.

Ballard, D. H., & Brown, C. M. *Computer Vision.* Englewood Cliffs, NJ: Prentice-Hall, 1982.

Ballard, D. H., & Kimball, O. A. Rigid body motion from depth and optic flow. University of Rochester, Computer Science Department, TR-70, 1982.

Barrow, H. G., & Tenenbaum, J. M. Recovering intrinsic scene characteristics from images. In A. R. Hanson & E. M. Riseman (Eds.), *Computer Vision Systems*. New York: Academic Press, 1978.

Barrow, H. G., & Tenenbaum, J. M. Computational vision. *Proceedings of the IEEE,* 1981, **69**, 572-595. (a)

Barrow, H. G., & Tenenbaum, J. M. Interpreting line drawings as three-dimensional surfaces. *Artificial Intelligence,* 1981, **17**, 75-116. (b)

Beck, J. *Surface Color Perception.* Ithaca, NY: Cornell University Press, 1972.

Binford, T. O. Inferring surfaces from images. *Artificial Intelligence,* 1981, **17**, 205-244.

Brady, M., & Grimson, W. E. L. The perception of subjective contours. MIT AI Memo 666, 1981.

Brady, M., & Horn, B. K. P. Rotationally symmetric operators for surface interpolation. MIT AI Memo 654, 1981.

Brooks, R. A. Symbolic reasoning among 3-D models and 2-D images. *Artificial Intelligence,* 1981, **17**, 285-348.

Davis, L. S. Shape matching using relaxation techniques. *IEEE Transactions on Pattern Analysis and Machine Intelligence,* 1979, **1**, 60-72.

Duda, R. O., Nitzan, D., & Barrett, P. Use of range and reflectance data to find planar surface regions. *IEEE Transactions on Pattern Analysis and Machine Intelligence,* 1979, **1**, 259-271.

Ehrich, R. W., & Schroeder, F. H. Contextual boundary formation by one-dimensional edge detection and scan line matching. *Computer Graphics and Image Processing,* 1981, **16**, 116-149.

Erman, L. D., Hayes-Roth, F., Lesser, V. R., & Reddy, D. R. The Hearsay II speech understanding system: Integrating knowledge to resolve uncertainty. *Computing Surveys,* 1980, **12**, 213-253.

Fischler, M. A., & Barrett, P. An iconic transform for sketch completion and shape abstraction. *Computer Graphics and Image Processing,* 1980, **13**, 334-360.

Garvey, T. D. Perceptual strategies for purposive vision. SRI International, Artificial Intelligence Center, Technical Note 117, 1976.

Gennery, D. B. A stereo vision system for an autonomous vehicle. Proceedings, 5th International Joint Conference on Artificial Intelligence, 1977, 576-582.

Greene, B. G., Pisoni, D. B., & Carrell, T. D. Learning to identify visual displays of speech: A first report. *Journal of the Acoustical Society of America,* Supplement 1, 1982, **71**, S96.

Grimson, W. E. L. *From Images to Surfaces.* Cambridge, MA: MIT Press, 1981.

Harmon, L. D. Automatic recognition of human face profiles. *Computer Graphics and Image Processing,* 1977, **6**, 135-156.

Hildreth, E. C. The integration of motion information along contours. Proceedings, IEEE Workshop on Computer Vision: Representation and Control, 1982, 83-91.

Horn, B. K. P. The Binford-Horn LINE-FINDER. MIT AI Vision Flash 16, 1971.

Horn, B. K. P. Obtaining shape from shading information. In P. H. Winston (Ed.), *The Psychology of Computer Vision.* New York: McGraw-Hill, 1975.

Horn, B. K. P., & Schunck, B. G. Determining optic flow. *Artificial Intelligence,* 1981, **17**, 185-203.

Johansson, G. Visual motion perception. *Scientific American,* 1975, **232** (6), 76-88.

Julesz, B. Textons, the elements of texture perception, and their interactions. *Nature,* 1981, **290**, 91-97.

Kanade, T., & Kender, J. R. Mapping image properties into shape constraints: Skewed symmetry, affine transformable patterns, and the shape-from-texture paradigm. Proceedings, National Conference on Artificial Intelligence, 1980, 4-6.

Kender, J. R., Shape from texture: A computational paradigm. Proceedings, DARPA Image Understanding Workshop, April 1979, 134-138.

Köhler, W. *Gestalt Psychology.* New York: Liveright, 1929.

Lowe, D. G., & Binford, T. O. Segregation and aggregation: An approach to figure-ground phenomena. Proceedings, DARPA Image Understanding Workshop, 1982, 168-178.

Marr, D. Representing visual information. In A. R. Hanson & E. M. Riseman (Eds.), *Computer Vision Systems.* New York: Academic Press, 1978.

Marr, D. *Vision.* San Francisco, CA: Freeman, 1982.

Marr, D., & Hildreth, E. C. Theory of edge detection. *Proceedings of the Royal Society, London,* 1980, **B207**, 187-217.

Marr, D., & Nishihara, H. K. Representation and recognition of the spatial organization of three-dimensional shapes. *Proceedings of the Royal Society, London,* 1977, **B200**, 269-294.

Marr, D., & Poggio, T. Cooperative computation of stereo disparity. *Science,* 1976, **194**, 283-287.

Nevatia, R., & Binford, T. O. Description and recognition of curved objects. *Artificial Intelligence,* 1977, **8**, 77-98.

Pentland, A. P. Private communication, 1982. (a)

Pentland, A. P. Local computation of shape. Proceedings, National Conference on Artificial Intelligence, 1982, 22-25. (b)

Quam, L. Road tracking and anomaly detection. Proceedings, DARPA Image Understanding Workshop, May 1978, 51-55.

Rosenfeld, A. Final Report on Contract NAS5-3461, 1964.

Rosenfeld, A., Hummel, R. A., & Zucker, S. W. Scene labeling by relaxation operations. *IEEE Transactions on Systems, Man, and Cybernetics,* 1976, **6**, 420-433.

Rosenfeld, A., & Kak, A. C. *Digital Picture Processing.* New York: Academic Press, 1976.

Simon, H. A. *The Sciences of the Artificial.* Cambridge, MA: MIT Press, 1969.

Smith, G. The recovery of surface orientation from image irradiance. Proceedings, DARPA Image Understanding Workshop, 1982, 132-141.

Stevens, K. A. Computation of locally parallel structure. MIT AI Memo 392, 1977.

Stevens, K. A. Surface perception from local analysis of texture and contour. MIT AI TR-512, 1979.

Stevens, K. A. The visual interpretation of surface contours. *Artificial Intelligence,* 1981, **17**, 47-73.

Tenenbaum, J. M., & Barrow, H. G. Experiments in interpretation-guided segmentation. *Artificial Intelligence,* 1977, **8**, 241-274.

Terzopoulos, D. Multi-level reconstruction of visual surfaces. MIT AI Memo 671, 1982.

Turner, K. J. Computer perception of curved objects using a television camera. Doctoral dissertation, University of Edinburgh, 1974.

Ullman, S. Filling-in the gaps: The shape of subjective contours and a model for their generation. *Biological Cybernetics,* 1976, **25**, 1-6.

Ullman, S. *The Interpretation of Visual Motion.* Cambridge, MA: MIT Press, 1979.

Witkin, A. P. Recovering surface shape and orientation from texture. *Artificial Intelligence,* 1981, **17**, 17-45.

Witkin, A. P. Intensity-based edge classification. Proceedings, National Conference on Artificial Intelligence, 1982, 36-41.

Witkin, A. P. Scale-dependent signal description. In preparation.

Zahn, C. T. Graph-theoretical methods for detecting and describing Gestalt clusters. *IEEE Transactions on Computers,* 1971, **20**, 68-86.

Zucker, S. W. Toward a theory of orientation selection and grouping: Evidence for Type I and Type II processes. McGill University, Computer Vision and Graphics Laboratory, TR-82-6, 1982.

Zucker, S. W., Rosenfeld, A., & Davis, L. S. General purpose models: Expectations about the unexpected. Proceedings, 4th International Conference on Artificial Intelligence, 1975, 716-721.

Computational and Psychophysical Experiments in Grouping: Early Orientation Selection

Steven W. Zucker

McGill University
Montreal, Canada

Abstract

Oriented entities are a fundamental construct of early vision, and we consider the oriented structures that can explicitly arise from collections of dots; i.e., dot grouping. Psychophysical demonstrations suggest that there are two types of such grouping processes, separated according to several accuracy, or specificity, requirements. We then discuss the what, why, and how questions associated with these observations: what is being constructed (abstractly, a vector field of orientations); why it is being constructed (to enable surface inferences from monocular cues); and how it could be constructed (in a way that is not inconsistent with basic neurophysiological constraints). The final result is a computational model that essentially uses lateral inhibition among orientation-selective operators to satisfy a given optimization criterion. But since this model is essentially syntactic in its operation, criteria must exist that delimit when its results could be valid. One criterion is formulated as a size/density constraint that functions conjointly with the existence of orientation structure.

1. Introduction

When do collections of dots group into lines? Into curves? How? Why? These questions were first explicitly raised by the Gestalt psychologists, who noted that "one sees a series of discontinuous dots upon a homogeneous ground not as a sum of dots, but as figures" (Wertheimer, 1923, pp. 72-73). Their answer was formulated as a series of principles of perceptual organization, such as the tendency of dots to group with their nearest neighbors. But these principles were purely descriptive; of what are they a result? In this paper we shall re-examine these questions within the context of what appear to be certain very early – almost primitive – forms of dot pattern grouping. We shall be concerned with one kind of structure into which dots can group – namely, oriented entities; as well as why they should, and how they might, form these groupings.

Dot grouping into oriented entities is a complex of tasks, and our first two results are aimed at delimiting functionally and psychophysically distinct sub-tasks. The first of these is formulated as a size/density constraint, and it separates conditions under which collections of dots appear to form smooth curves from those in which they only form piecewise straight ones. We take this difference to be substantial, because it has many other psychophysical manifestations as well as deep modeling implications. We are only concerned, in this paper, with patterns of dots that fall on the smooth, dense side of this constraint.

Our second result is that there are two fundamentally different types of processes for grouping dots into oriented entities. The first of these results in curves that are highly accurate in position and curvature, and for which very small variations cause a large apparent difference. The resultant curves are essentially one-dimensional. The second type is much more tolerant of positional errors and of variation in the dot tokens, but results in lower curvature resolution. It produces information more connected with two-dimensional processing. We refer to these processes as Type I and Type II, respectively.

The rationale for two types of grouping processes comes from the surface inferencing task. Type I processes can be viewed as capturing information about the contours that separate surfaces, while Type II processes provide information about within-surface variations. Random textures, such as grass and hair, provide examples that illustrate why Type II processes are inherently less specific than Type I processes.

There is an important theoretical connection between Type I / Type II grouping, surface inferencing, and the size/density constraint. The grouping processes can provide a basis for "first guesses" about surfaces, such as where they are discontinuous in orientation or reflectance, and how they are curving. But such first guesses are notoriously coarse; they are based on syntactic kinds of processing that may be wrong. Therefore, the mechanisms that compute them must also have built-in safeguards delimiting when they are likely to be valid. These safeguards are provided by the size/density constraint and the fact that natural physical events give rise to image structures that possess an inherent orientation. It is extremely unlikely that collections of physical events will form images with oriented structure randomly, provided the events are of significant size and density.

Our third result is a computational model for accomplishing Type I and Type II grouping. The model uses asymmetric (i.e., elongated) operators to signal putative orientation information. Its goal is the establishment of a vector field of orientations, and it also uses these operators to represent this vector field. However, since the responses of these operators are ambiguous, a strategy must be developed for their interpretation. This arises from the identification of constraints about how different-sized operators signal curvature, and leads to an optimal interpretation strategy using relaxation techniques. The result, in the end, looks very much like a lateral inhibitory network running among differently oriented operators at neighboring spatial positions.

The computational model provides a framework for formally differentiating between Type I and Type II processes based on the positional specificity of the operators. It also suggests how orientation selection could be connected to symmetry detection, to texture, and to motion.

2. On the Role of Early Grouping Processes

The need for two early types of grouping processes is suggested by the surface-inferencing task, or the problem of interpreting intensity-based cues into (constraints about) physical surfaces. This is perhaps the most basic problem that the visual system must address, and it provides the context within which orientation becomes a truly meaningful, and necessary, notion. Without it, there would be no way to characterize either the structure of surfaces or the retinal projection of salient features of surfaces.

The surface-inferencing task readily separates into two different sub-tasks on the basis of accuracy or specificity requirements. The first of these pertains to constraints about intersections between surfaces (Binford, 1981). Topologically such intersections are (unions of) one-dimensional contours. They will be most useful when they are very accurately positioned, because then they can serve as tight boundary conditions against which many different sources of surface discontinuity information can be integrated. For example, the more tightly boundary-condition contours can be specified, the more tightly they can constrain the interpolation of disparity data (Grimson, 1981; Terzopoulos, 1982), and the more accurately they will constrain other descriptions of the physical world (Barrow & Tenenbaum, 1978). Clearly, the visual system's goal should be to express them as tightly as possible, and the structure of the physical world is such that this will normally be very tight indeed.

The second sub-task is to provide information about the structure within a surface, rather than about the boundaries between surfaces. This information seems more closely aligned with two-dimensional constraints of an inherently looser nature. Consider, for example, a grassy field, in which the individual blades signal a general orientation, but which overlap and sway to the extent that individual differences are substantial. They will be roughly aligned, and of roughly the same length, but not precisely so. Stereo will also be inherently less accurate in such situations. Information about surfaces is available, however, and should be integrable into a smooth surface, at least to a first approximation. This second sub-task is closely related to similarity grouping and texture, two of the domains from which monocular surface cues derive. Our basic position is that orientation information is an essential component of these cues, and that, therefore, it should be one of the first kinds of information inferred from images. But, as the differences between these two sub-tasks suggest, inferences about orientation are not likely to be the same in why they are made, how they are made, and precisely what they signify. Some discussion of their relation to similarity grouping and texture is presented in Section 7.2.

3. Size and Density Constraints for Very Early Grouping Processes

We shall take grouping to be a generic term for the different processes, early in vision, through which local entities give rise to less local ones, or by which abstract entities are constructed out of more primitive ones. This is certainly a large class of processes. The two types that we shall consider in this paper seem to be extremely early, almost primitive, ones. In a functional sense their role is connected to the surface-inferencing task, but only to provide rough "first guesses" for the two sub-tasks discussed in Section 2. These guesses must then be integrated with other sources of information for proper interpretation.

Since we are assuming that the grouping processes under study are very early, essentially syntactic ones, it follows that they should be data-directed, and rather inflexible in their behavior. But syntactic behavior only works under well-defined conditions. Consider, for example, the frog, which jumps syntactically at distinct spots within given size and velocity ranges. When these "spots" turn out to be flies, the frog has a meal; when they are birds much farther away, the frog has just wasted energy.

The conditions under which the early grouping processes should function satisfactorily must be expressed in the language of the data, and should reflect their physical correlates. Two seem to be active. The first of these is the size of the dots, or individual entities, and the second is their density. Clearly, if the dots could vary tremendously in either of these variables, then rather elaborate machinery would be required to deal with them. But there should be enough variation for the data to be useful. Otherwise, such special-purpose machinery would hardly be advantageous. And, if the grouping processes are related to those that effect orientation selection while finding (or from) contours, then their scale variations should correspond to these as well. It is supportive, in this regard, that when dot patterns are made from line patterns, the subjective appearance is very similar. Such similarities certainly cast doubt on complex processing in the case of dots preceding the later processing common to both classes of patterns.

Size and density covary, in the sense that we shall require at least several dots (of size roughly 1 minute of visual arc) to fall within each degree of visual angle. The dots must each be large enough to affect the intensity structure in

this area without overwhelming it. (It is, of course, the case that larger structures could work if the following arguments were applied not to them in their
entirety, but rather just to local descriptions of them, such as pieces of their
contours. This is a situation that we will not address in this paper.) And
enough of them must fall within it to affect it as well. When either of these
constraints is violated, grouping can still occur, but it appears differently in a
phenomenal sense. We shall take this to indicate that different processes are
operative. Several different observations support this distinction.

3.1. Apparent Smoothness

The first effect is one of smoothness. When dots are closely spaced,
smooth curves are very apparent; see Figure 1a. But when the dots are sparse,
the curve appears piecewise constant (Figure 1b). This latter case indicates a
form of grouping that takes place over larger distances, always involves only
pairs of dots, and appears to connect them with a "virtual line" (Marr, 1982;
Stevens, 1978). Smooth virtual curves do not arise in this fashion. We are
interested in the kinds of phenomena indicated by Figure 1a. (Compare with
Figure 2, in which these effects coincide.) Additional support for this constraint
is discussed in (Zucker, Stevens, & Sander, 1982), in the context of line vs.
pair judgements within a psychophysical task, and in (Zucker, 1982) for various
subjective configurations.

4. Type I and Type II Processes

The functional identification of Type I and Type II processes is clearly
with the two sub-tasks discussed in the context of surface inferencing: Type I
processes are responsible for the highly accurate boundary conditions, while
Type II processes are responsible for the more coarse interior constraints.
(Their functional roles may, however, become more intertwined than this.) We
shall assume that the Type I and Type II processes under study satisfy the
size/density constraint, and therefore that they are first participants in the surface inferencing process.

The existence of two distinct types of grouping processes can be readily
demonstrated with dot patterns. Several of the Gestalt phenomena, such as the
grouping of regular arrangements of dots into straight lines, appear to be Type I
phenomena, while random dot Moiré patterns appear to be Type II. Each of
these will now be demonstrated in turn, following a brief rationale of the use of
dot patterns to study grouping processes.

4.1. On the Methodological Choice of Dot Patterns

Our Type I processes bear some resemblance to the kinds of grouping, or
interpretative processes, that are implicated in the task of locating intensity
change contours (commonly called "edges") in images, or at least the process

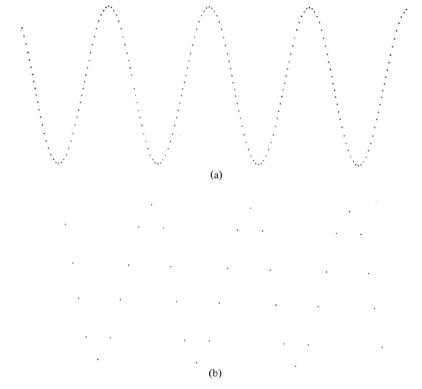

(a)

(b)

Figure 1. A sinusoid, with (a) 50 dots/cycle resolution and (b) 10 dots/cycle resolution. Note how the denser figure looks smooth, but the sparser one looks piecewise straight.

of assigning orientations to them. The reason is that most accounts of contour finding begin with the application of differential operators, like the ones modeled by Marr and Hildreth (1980), that are sensitive to these changes. Such operators respond strongly in the presence of isolated intensity changes following slowly curving contours; they respond artifactually in the neighborhood of changes following very acute contours, and they may be obscured in low contrast situations. Thus the accurate responses must be separated from the artifactual ones, leaving only segments rather than continuous contours (Marr & Hildreth, 1980, Figure 8). The segments must be interpreted into smooth, continuous contours (Zucker, 1981; Zucker et al., 1982). As these segments shrink in length and grow in density they approach, in the limit, the dot grouping task. (The relevance of the size/density constraint can now be interpreted, in a sense, as empirically indicating when this identification is acceptable as a first approximation.)

The connection with contour finding is advantageous, in that it suggests that dot grouping is an early and important phenomenon, and is in fact exercis-

Figure 2. A triangular wave with 10 dots/cycle resolution. Note how the sparse sinusoid resembles this wave, except for the apparent misplacement of the "top" points. The points of singular curvature, like endpoints, can participate in groupings separate from the curve that defines them.

ing essential machinery. This view is further supported by the existence of arguments at all three explanatory levels. But it is not advantageous, in the sense that our understanding of the contour-finding machinery is still rather vague. There is no universal agreement about how the responses of the differential operators are interpreted, or even about whether their full response, or just the "zero-crossings" in their responses, are used. We have chosen to work with dot patterns precisely because they offer us an opportunity to sidestep this problem. Our explicit assumption is that dots are so simple that they will be functionally equivalently described no matter how these earlier issues are resolved. This statement, in fact, can serve as a definition of what we mean by "dots" – those intensity events that are essentially invariant to the earlier, differential processing. The details in our theory may change, but its thrust and basic structure will remain invariant.

4.2. Type I Grouping

The first series of demonstrations is close in spirit to those in Wertheimer's (1923) original paper. Figure 1 shows how a series of dots, arranged linearly, organize into lines. Several points are noteworthy. First, the spatial arrangement of dots has given rise to a much more abstract percept, that of a line (or an arrangement of lines), and this line has an explicit orientation associated with it. (It has other abstract structure as well, such as endpoints and length.) Wertheimer explained this demonstration with the "principle" that dots tend to group according to spatial proximity. Whereas this is often the case, we shall show that this principle is neither necessary nor fundamental; rather, when it holds, it can be shown to be derivable from more fundamental considerations. Thus it becomes a descriptive, rather than an explanatory, "principle."

The sharp quality of the grouping in this example can be revealed by a few related examples. The positional differences between the dots in the two examples (Figure 1 and Figure 2) are very small, but they lead to a rather large apparent difference in the final appearance. From this we conclude that high positional accuracy is important; small changes in position may cause a large change in the resultant percept.

The above argument that the visual system is highly sensitive to the position of dots, and that it can use them to extract lines, is corroborated wonderfully by three-dot alignment tasks. These are tasks in which two dots are fixed in position, and a third can be moved until it aligns perfectly with them. Human performance for this task has been accurately assessed, with the result that alignment capability is typically in the range of a few seconds of visual angle (Beck & Schwartz, 1978). This hyperacuity implies that the representation underlying the positioning of the dots is finer, perhaps by an order of magnitude. Furthermore, since this task requires precisely the capabilities required for orientation selection in grouping, we conclude that high positional accuracy is available. As we discussed in Section 2, such positional accuracy is important for constraining descriptions of the physical world.

Given this high positional accuracy, it becomes possible for the grouping process to capture high degrees of curvature, as is illustrated by the sinusoid.

Our third observation about this kind of grouping is that it is highly sensitive to the element, or primitive, comprising it. This was also demonstrated by Wertheimer, who introduced another descriptive law – form similarity – to account for it.

These demonstrations exercise what we take to be the defining characteristics of Type I grouping:

> (4.2.1) high positional specificity
> (4.2.2) high orientational change specificity
> (4.2.3) high primitive specificity
> (4.2.4) one-dimensionality of percept
> (4.2.5) presence of endpoints

This last point, the presence of endpoints, is illustrated by the corners of the triangular wave, which could readily enter into additional groupings.

4.3. Type II Grouping

We shall now present examples of grouping phenomena that do not exhibit the above characteristics. In general, they will be much less specific. To start, we will use a technique discovered by Glass (1969) to generate what he referred to as random dot Moiré patterns (or RDMPs). A RDMP can be made by producing a random field of dots, making a copy of it, transforming the copy, and then displaying the transformed original and the copy superimposed.

If the transformation is a rigid rotation, then the result is a pattern such as the one shown in Figure 3. Note the appearance of circularity together with the absence of any well-formed circular contours. Many other transformations, such as translations or expansions, work as well.

The most striking difference between RDMPs and the examples of Wertheimer grouping shown earlier is their random and incomplete character; they give an impression rather that the crisp, clear effect that would be present with individual contours. There is a much looser requirement on the underlying spatial structure, and the result is two-dimensional, rather than one-dimensional. Dots can be moved around, e.g., by making the pattern with a different underlying random distribution, and the overall effect does not change.

A consequence of this relaxed spatial structure is that certain aspects of orientation resolution, in particular changes in orientation (or curvature), should be reduced. This follows because the exact correspondence between dots must be known to represent rapidly changing orientations, as was the case above. But now such linkings are ambiguous, and the appropriate "corresponding" pairs of dots; i.e., the dots responsible for the apparent orientation, are often not nearest neighbors. Demonstrations of this coarser curvature selectivity are shown in Figures 4 and 5, in which the RDMP transformations are sinusoidal and triangular, respectively.

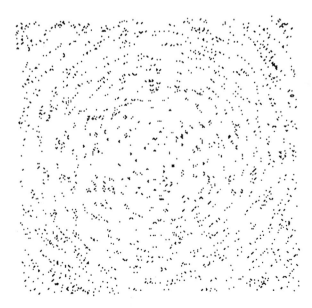

Figure 3. A circular random dot Moiré pattern. It was made by taking a pattern of 2000 random dots, and superimposing it on a copy of the pattern that was first rotated by about 1 degree. Note the apparent circularity, but the absence of complete, or aligned, circular contours. Note also how the random pairings of dots give rise to orientations that are distinct from the circular flow.

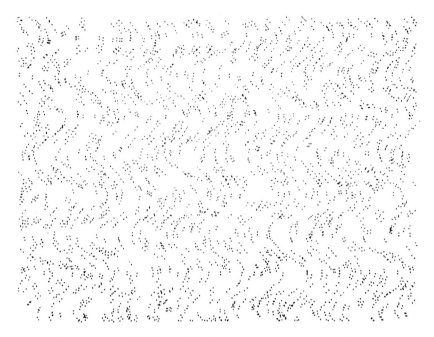

Figure 4. An RDMP in which the transformation is a sinusoid.

A second, important difference with respect to curvature changes is the lack of apparent endpoints in the triangular-wave RDMP. This suggests much more spatial smoothing for this type of grouping.

The third definitive characteristic of the Wertheimer groupings, which is also relaxed for RDMPs, is variation in the underlying primitive, or element, of which the pattern is comprised. Again, this characteristic can be relaxed within limits; see Figure 6. We take this not to be an important definitive characteristic, however, because of our presumption that the pattern consists of "dots," or entities that are invariant to earlier processing. As the primitive changes, this assumption becomes less accurate. Also, as the primitive changes, there will be more variation in spatial specificity, which is the important difference (for our present purposes).

In summary, then, there appear to be two types of grouping processes involved in the above examples, and they can be separated according to the observational criteria listed at the end of Section 4.2. For Type II grouping they become:

> (4.3.1) low positional specificity
> (4.3.2) low orientational change specificity
> (4.3.3) low primitive specificity
> (4.3.4) two-dimensionality of percept
> (4.3.5) absence of endpoints

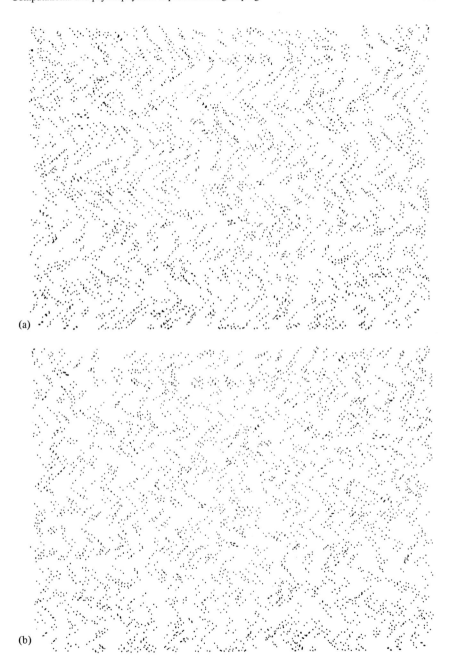

(a)

(b)

Figure 5. An RDMP composed of triangular waves. Note the apparent absence of endpoints, or points of singular curvature, in these patterns. Such points would be very apparent if the triangular wave were not created in the Type II fashion. (a) The frequency of this triangular wave is 4 cycles from the top to the bottom. (b) An increase of frequency to about 7 cycles shows the limit for perceiving this class of pattern.

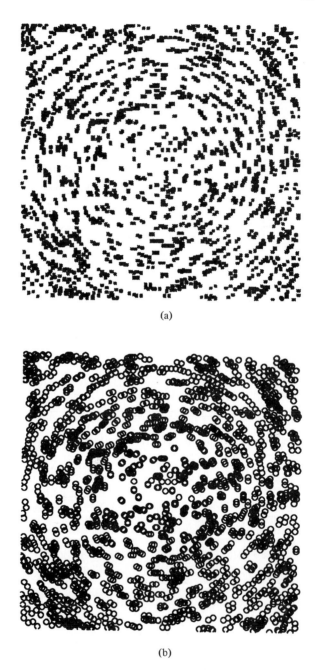

(a)

(b)

Figure 6. A series of circular RDMPs in which the "dots" comprising the original and the copy patterns have been replaced with (a) jittered squares; (b) circles; and (c) broken lines. The circular pattern is destroyed in this last example, but would still be apparent if the lines were shorter.

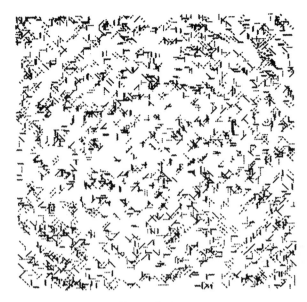

Figure 6 (c)

The term specificity has been used above because we would like to stress that the differences between Type I and Type II processes are distinct from those that would arise from differences in spatial resolution. This distinction is important because the early spatial-frequency limited channels (Campbell & Robson, 1968) provide a tempting first hypothesis, with Type I happening over the high-resolution data, and Type II happening over the low-resolution data. But this is not the case, as can readily be seen by simply blurring your eyes while looking at the displays. Furthermore, it cannot explain the similarity in orientational specificity and the difference is orientational change (i.e., curvature) specificity. Spatial averaging does not work as an explanation. (However, as we show in Section 5, it may play a role in the solution, because it does provide a mechanism for smoothing that the Type I and the Type II processes could certainly utilize.)

To summarize, then; Type I processes are high in their specificity requirements, and produce a one-dimensional result; Type II processes are (relatively) low in their specificity requirements, and produce results more intimately connected with two-dimensional processing. There is one important exception, however, which indicates the special circumstance in which the two begin to coincide – the case of perfectly smooth surface contours. This is a special form of monocular, intensity-based surface cue in which contours lie exactly on a smooth surface. As long as the contours are widely spaced, they should each produce similar results. But since Type II processes are less sensitive to positioning, when the contours are closely spaced Type II processes may simply blur them together, erasing small orientation differences, while Type I processes should still be able to resolve them.

5. A Model for Orientation Selection

Differences between Type I and Type II grouping processes, as just demonstrated, would seem to suggest that the mechanisms supporting them must differ drastically as well. This is not the case, however, as we attempt to show in this section. Our plan is to outline a generic orientation selection task, and to sketch a formal solution to it. We can then discuss computational issues related to the implementation of this solution.

The resultant model, in the end, is only a sufficient one, but it does begin to look very plausible biologically. It essentially amounts to a lateral inhibitory network running among differently oriented operators of different sizes, but with the central locations of these operators skewed slightly in position. These operators are used both to represent and to signal (i.e., compute) orientations. The orientation selection task is formally posed as an optimization problem (Hummel & Zucker, 1980), and the multiple-size operators provide curvature constraints necessary for its solution (Zucker & Parent, 1982). The position skewing is necessary for spatial continuity of solutions.

5.1. An Overview of the Model

The first step in specifying a model is a formal specification of its goals. We partition the orientation selection task into two different stages, each with its own goals, somewhat along local/global lines. The first stage is the definition of a vector field of orientations; i.e., the assignment of a direction to each point in a fixed region. (The coordinates of this region are "retinal.") This vector field is a lot like a flow pattern; the vector at each point indicates the direction in which a curve will be moving when it passes through that point. A more detailed introduction to vector fields can be found in (do Carmo, 1976).

The second stage is the determination of actual curves passing through these fields, which requires the solution of a system of differential equations. Since the latter of these stages is straightforward, we shall concentrate on obtaining the vector field.

In order to establish such a vector field, mechanisms must be provided for representing and for obtaining it. A unit line segment, centered at a well-defined spatial location, is perhaps the most common representation, but what is really required is a spatial structure that is asymmetric; the major axis can then be taken to represent an orientation. (For line segments, of course, the minor axis is negligible.) Most importantly for our purposes, such asymmetric spatial structures can also be interpreted as operators capable of signaling the presence of orientation information. Consider, for example, the result of convolving (i.e., evaluating) an eccentric operator with a high-contrast dotted line. We shall use an operator that is just an ellipsoidally-shaped region, and whose response is just a (normalized) sum of the inputs. When the operator and the line are oriented identically, it will respond most strongly; when the operator and the line are perpendicular, it will respond weakly. This is because the maximal response occurs when the maximal amount of line lies under the operator. Strong responses thus signal the orientation of straight lines.

Problems arise, however, when the lines are not perfectly contrasted with the background, when noise is present, or when they curve – situations at the heart of many Type I, and virtually all Type II, patterns. In particular, the operators give intermediate responses, neither strong nor weak. How can these be interpreted? This question can be formally recast as an optimization problem, which we call the response matching problem, in the following way. Since we know the structure of the operators, we can compute their expected response to known patterns. Then, given the response observed to arise from an unknown pattern, we assert that the unknown pattern is the one which, if present, would have given an expected response as close as possible to the observed response.

The solution to this response matching problem requires constraints as well as a mechanism for applying them. The constraints come from the different size operators. Small operators can only signal (locally) straight lines, as was implicitly shown above. But larger ones, when rotated, can actually signal curves. This is because the curve will systematically enter and leave the "receptive field" of the operator many times as the operator is rotated. Taken together, the responses of these operators, at different orientations, lead to the definition of a functional suitable for guiding the search for the underlying pattern orientation.

The search for the optimal pattern can be done cooperatively by allowing the responses to interact according to the constraints outlined above for the rotating operators. For example, a short straight segment pointing in direction N should receive positive reinforcement from large operators at that same position that are also pointing in direction N, as well as from similarly oriented small and large operators at neighboring positions (in the directions $+N$ and $-N$). It should receive negative support from all of these operators oriented in directions $N + \pi/2$ and $N - \pi/2$, since these would indicate lines running in perpendicular directions. The result is a network something like the one illustrated in Figure 7. The formal derivation of this network, or, more precisely, of a relaxation labeling algorithm for solving the response matching problem, is in Zucker (1982).

Given such a model, we can now more concisely describe the differences between Type I and Type II processes – Type I processes require very tight spatial sensitivity in the original operators, while Type II processes must be much looser. Thus Type I processes will respond only to smooth curves; as they

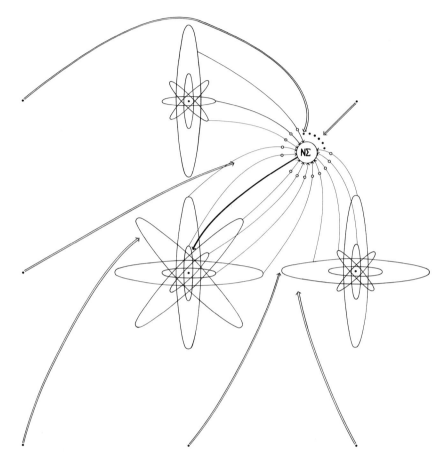

Figure 7. A sketch of a cooperative model for solving the response matching problem. The ellipsoidally-shaped receptive fields are shown (in part) at three positions, but the structure is actually repeated uniformly everywhere. Also, the scale has been spread out, so that the operators and their interconnections could be illustrated. In actuality, the operators overlap nearby spatial positions; here they are shown separately. The updating computation is shown just for the small, vertically-oriented mask, but, again, analogous computations are repeated for all operators. Each neighboring operator contributes to the updated response of this vertical one through a normalized sum, following multiplication by (possibly attenuating and sign changing) coefficients. These coefficients are determined from the response profiles of the masks evaluated over known patterns. Such computations occur in lock step, parallel fashion over all operators and positions. Or, equivalently, they could be implemented as a parallel, pipeline network. Details are in (Zucker, 1982).

become noisier, they will leave the sensitivity range to which the operators can respond. Type II processes, on the other hand, will respond to both perfect patterns and to noisy ones, but they will tend to blur the perfect ones. Note that the difference is not simply one of scale resulting from, say, a spatial averaging process; this would have other, incorrect consequences.

6. Experiments with the Model

Any implementation of the cooperative grouping algorithm must make choices about the sizes of the operators chosen and about the orientation resolution. Practical realities have limited us to considering only a coarse orientation resolution of 8 distinct orientations. We have also used operators that are ellipsoidally shaped, with purely positive (or excitatory) inputs. The ideal responses of these operators were obtained by convolving them with a binary image of two thick lines meeting at about a 30 degree angle. The size of this training image was 30×30 pixels.

A 60×60 pixel test image was obtained as a section of a much larger diagonal RDMP (see Figure 8). Note that, although the original pattern gave a strong impression of a diagonal flow to the upper right-hand portion, this sub-image is much more varied in its appearance. Spatially-adjacent pairs of dots give especially strong indications of lines, resulting in a pattern of orientations for this example that is quite varied.

The input to the model for this example was a collection of ellipsoidally-shaped operators of three different sizes: major axes of 9, 19, and 39 pixels in length. All inputs to them were excitatory, and they were modulated by weighting coefficients according to a Gaussian distribution in both directions.

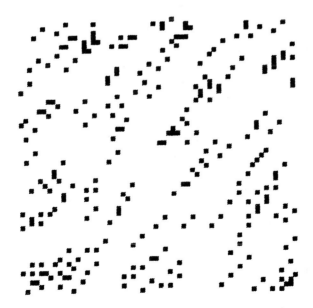

Figure 8. A 60×60 binary test image. It was obtained as a section of a much larger RDMP in which the ''apparent flow'' is toward the upper right-hand corner. Viewed from a distance, this pattern exhibits a much more varied flow, however, due to its small size and the random dot positions. The double pairs of dots are extremely strong counterexamples to the larger flow.

ITERATION: 40

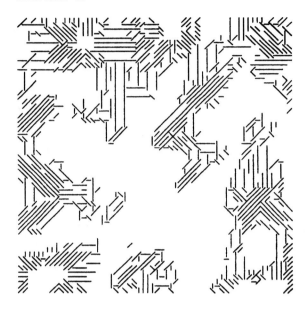

Figure 9. A preliminary result of running the orientation selection mechanism (similar to the one in Figure 7) on the image in Figure 8. The intention is to describe the field of orientations, and these are indicated as short line segments.

The result of the response matching network is shown in Figure 9. Note that it shows a dominance of orientations toward the upper right, but that it has also found the horizontal and vertical lines due to the paired pixels. Given the limited spatial extent of this example, the responses near the borders are somewhat artifactual, in that the larger operators were truncated.

This experiment with the model was highly preliminary, and was intended just to indicate that orientation selection as described above is feasible. More detailed experimentation will be described in (Zucker, 1982).

7. Consequences, Implications, and Speculations

The primacy attributed to orientation should be reflected in its use by the visual system. If it were not useful, then no argument regarding it could have enduring strength. The first, and most obvious use of orientation is in intensity-based contour identification and completion, as has already been mentioned (Section 4.1). But several other very early visual information processing tasks would seem to use it as well, as we shall now attempt to show.

7.1. Symmetry Detection

Symmetry detection is one of the most salient talents of the early visual system, and in a recent series of experiments, Barlow and Reeves (1979) studied the conditions under which symmetry was apparent in random dot patterns. In addition to confirming that humans possess a remarkable ability to detect symmetry, two additional findings seem to be of particular importance. Firstly, a statistical analysis of their results indicated that only a fraction of the dots in the patterns observed were actually involved in symmetry detection. Secondly, random perturbation of the dots within a "tolerance region" around their symmetric positions did not destroy the appearance of symmetry. That is, it showed that the comparison process across the axis of symmetry was not perfect. Their conclusion was that a "mechanism" for symmetry detection could be formulated in which the visual array was partitioned into a number of small regions, and the dots falling into these regions estimated. Symmetry detection could then be accomplished by comparing these estimates region-by-region.

We would agree with their two findings, but not with their characterization of the mechanism implied. (Barlow and Reeves also appear unconvinced that this kind of mechanism could have any neurophysiological reality). The need for the regions was both to allow for tolerances in dot position, and to deal with the combinatorics of comparison. But it is not necessary. Another model – that orientation selection is first performed, and that the resultant vector field then provides a basis for symmetry detection – seems much more plausible to us. The orientation vector field resulting from a mechanism like the one that we are proposing for Type II processes could account for the Barlow/Reeves findings in a natural way. Moreover, the vector field should contain a continuous contour of angular discontinuity, which could then provide a natural axis of symmetry. Barlow and Reeves (1979, Figure 8) also showed, in support of this claim, that the portion of the dot pattern on either side of the axis contained the information most relevant to symmetry detection. In summary, while this is probably not the only mechanism involved in symmetry detection ("clusters" of dots are also involved), it does provide a natural explanation of the current psychophysics.

7.2. Texture and Surfaces

The connection between texture and surfaces was clearly established by Gibson (1950) in a series of demonstrations of texture gradients (e.g., 1950, Figures 32 and 33), or patterns in which regular arrays of entities are projected onto the frontal plane. But it soon became clear that more than just a gradient is necessary (Attneave, 1966; Stevens, 1979). This can be shown with dot patterns; when the dots are positioned randomly, but increasing in density, then no surface percept is apparent. But when the dots form oriented structures, either (roughly) vertical or horizontal (with perspective), then the surface percept is obvious. Our conclusion is that gradients are important for surface percepts, but they presuppose an orientation structure (at least for this class of patterns). Without it the gradient loses its force. Incidentally, Gibson's examples, cited above, used dots large enough to have ellipsoidal (i.e., oriented!) projections.

Working from a different perspective, Beck (1982) concluded, in a recent review of the texture and similarity grouping literature, that "the most important of the properties associated with shape for textural segmentation is slope (p. 289)." Thus orientation fields of the kind suggested in this paper seem strongly implicated in texture description and similarity grouping processes. Without orientation (or other such cues) the connection with surfaces is missing. Moreover, the grouping process studied in this paper provides precisely the kind of "linking" rules that theories of texture presuppose (Beck, 1982).

Images of natural textures, such as hair, illustrate the need for a Type II process. These images consist of "streaks," or thin, dotted contours, each of which represents a portion of a line of curvature for the surface that they define. But complete lines rarely exist, so the incomplete structures must be grouped. Spatial selectivity must be low, so that the "parallel" segments support one another. And smoothing is permissible, because the lines of curvature that are important are the low-frequency ones, i.e., the ones that depict the surface and not the random hairs standing on end.

7.3. Grouping in Space and Grouping in Time

There are strong analogies between the processes that group entities across space and those that group entities across time; that is to say, those processes responsible for motion. Anstis (1970) has shown that there is at least a rough correspondence between certain random dot Moiré patterns and motion, and in this section we would like to just list the motion analogues of our psychophysical claims. They will be dealt with in more detail in a subsequent paper.

The size/density constraint has analogues in both time and space for motion. The former is usually characterized as a distinction between short-term and long-term motion processes, while the latter again corresponds to sparse/dense arrangements. In particular, the observation that sparse arrangements of dots in space always group into piecewise linear curves corresponds to the planarity assumption typically invoked to solve the Johansson (1975) "moving light" displays. And short term/long term motion processes are distinguished by the visual angle subtended (Anstis, 1980).

The Type I / Type II distinction corresponds to rigid and "deformable" body motions, respectively. The requirements here translate into high positional and directional selectivity, and recent evidence by Lappin (1980) shows that, indeed, for the Wallach and O'Connell (1953) "kinetic motion effect" the correlation between the true, rigid-body positions and dots randomly displaced from these positions cannot be lower than 0.993 for a rigid-body percept to result. Such accuracies are clearly not attainable for waterfalls and sandstorms, the motion analogues of Type II phenomena.

8. Summary and Conclusions

Grouping is a myriad of processes that take elements of a local nature and combine, or abstract, them into elements of another, usually less local nature. In this paper we concentrated on dot patterns, and studied the process of inferring orientation information from collections of them.

Two psychophysical observations formed the basis for our study. First, there is a size/density constraint that is active in dot grouping tasks. For sparse collections of dots, the groupings are always piecewise straight, while for dense collections the groupings are much more smooth. Secondly, dot grouping processes exist with different specificity requirements. The first of these is very high, while the second is much more relaxed. While this is the essential difference between them, secondary differences in orientation change selectivity were also apparent.

Functional differences between the two types of dot grouping processes were also considered. A particularly salient one was the presence of distinct endpoints in the first, spatially sensitive case, and their absence in the second, or spatially-relaxed process. Endpoints suggest surface semantics and occlusion, and, indeed, the surface inferencing task provided a strong rationale for separating the two types of grouping processes.

Our final contribution was a model for orientation selection and grouping, which has two stages: the construction of a field of local orientations, and then the construction of curves through these fields. This model permitted a formal statement of the difference between the two types of grouping processes as the spatial selectivity of basic asymmetric operators. The entire model, in fact, looks very much like a lateral inhibitory network running among such operators, which suggests, together with the size/density constraint, that these types of orientation selection and grouping processes are extremely primitive ones. They can be viewed as providing a "first guess" about surfaces that must be integrated with other "guesses" from other information sources.

Acknowledgment

This research was supported by the Natural Sciences and Engineering Research council. Much of the theoretical work on relaxation labeling was done in collaboration with R. Hummel. The relaxation experiments were carried out in a general testbed implemented by J. Mohammed; P. Parent implemented the specific models. P. Lamoureux and M. Parker helped with the displays, and S. Davis and Y. Leclerc provided critical comments on the manuscript.

References

Anstis, S. Phi movement as a subtraction process. *Vision Research,* 1970, **10**, 1411-1430.

Anstis, S. The perception of apparent movement. *Philosophical Transactions of the Royal Society, London,* 1980, **B290**, 153-168.

Attneave, F. Inferences about visual mechanisms from monocular depth effects. *Psychonomic Science,* 1966, **4**, 133-134.

Barlow, H., & Reeves, B. The versatility and absolute efficiency of detecting mirror symmetry in random dot displays. *Vision Research,* 1979, **19**, 783-793.

Barrow, H., & Tenenbaum, J. M. Recovering intrinsic scene characteristics from images. In A. R. Hanson & E. M. Riseman (Eds.), *Computer Vision Systems.* New York: Academic Press, 1978.

Beck, J. Textural segmentation. In J. Beck (Ed.), *Organization and Representation in Perception.* Hillsdale, NJ: Erlbaum, 1982.

Beck, J., & Schwartz, T. Vernier acuity with test dot objects. *Vision Research,* 1978, **19**, 313-319.

Binford, T. Inferring surfaces from images. *Artificial Intelligence,* 1981, **17**, 205-244.

Campbell, F., & Robson, J. Applications of Fourier analysis to the visibility of gratings. *Journal of Physiology,* 1968, **197**, 551-556.

do Carmo, M. *Differential Geometry of Curves and Surfaces.* Englewood Cliffs, NJ: Prentice-Hall, 1976.

Gibson, J. *The Perception of the Physical World.* Boston, MA: Houghton-Mifflin, 1950.

Glass, L. Moiré effect from random dots. *Nature,* 1969, **243**, 578-580.

Grimson, W. E. L. *From Images to Surfaces.* Cambridge, MA: MIT Press, 1981.

Hummel, R., & Zucker, S. On the foundations of relaxation labeling processes. McGill University, Computer Vision and Graphics Laboratory, TR 80-7, 1980.

Johansson, G. Visual motion perception. *Scientific American,* 1975, **232** (6), 76-88.

Lappin, J., Donner, J., & Kottas, B. Minimal conditions for the visual detection of structure and motion in three dimensions. *Science,* 1980, **209**, 717-719.

Marr, D. *Vision.* San Francisco, CA: Freeman, 1982.

Marr, D., & Hildreth, E. Theory of edge detection. *Proceedings of the Royal Society, London,* 1980, **B207**, 187-217.

Stevens, K. Computation of locally parallel structure. *Biological Cybernetics,* 1978, **29**, 19-26.

Stevens, K. Surface perception from a local analysis of texture and contour. Doctoral dissertation, MIT, 1979.

Terzopoulos, D. Multi-level reconstruction of visual surfaces. MIT AI Memo 671, 1982.

Wallach, H., & O'Connell, D. The kinetic depth effect. *Journal of Experimental Psychology,* 1953, **45**, 205-217.

Wertheimer, M. Laws of organization in perceptual forms. *Psychologische Forschung,* 1923, **4**, 301-350. (Translation in W. D. Ellis (Ed.), *A Source Book of Gestalt Psychology.* New York: Harcourt, Brace, 1938.)

Zucker, S. Computer vision and human perception. Proceedings, 7th International Joint Conference on Artificial Intelligence, 1981, 1102-1116.

Zucker, S. Orientation selection and grouping: Evidence for Type I and Type II processes. McGill University, Computer Vision and Graphics Laboratory, TR 82-6, 1982.

Zucker, S. & Parent, P. Multiple size operators and optimal curve finding. Proceedings, 6th International Conference on Pattern Recognition, 1982, 745-747.

Zucker, S., Stevens, K., & Sander, P. Similarity, proximity, and the perceptual grouping of dots. MIT AI Memo 670, 1982.